INTRODUCTORY STATISTICS FOR BUSINESS AND ECONOMICS

IVER E. BRADLEY
University of Utah

JOHN B. SOUTH
Arkansas State University

The Dryden Press
Hinsdale, Illinois

To Jane and Judith Ann

Copyright © 1981 by The Dryden Press
A division of Holt, Rinehart and Winston, Publishers
All rights reserved
Library of Congress Catalog Card Number: 80-65789
ISBN: 0-03-053026-1
Printed in the United States of America
123 032 987654321

Acquisitions Editor   Feeny Lipscomb
Developmental Editor   Nedah Abbott
Project Editor   Jane Perkins
Art Director   Stephen Rapley
Production Manager   Peter Coveney

Text and Cover Designer   Alan Wendt
Copy Editor   Carol Gorski

PREFACE

Our goals in preparing this text were to present statistics as an interesting, important discipline and to make it easily understandable to the student. In order to create interest and motivation, we have tried to devise examples and problems that are lively and realistic. They are neither trivial nor too difficult, and they relate to actual applications. Fundamental propositions and concepts are presented in detail, lucidly, and as simply as possible. Where appropriate, they appeal to the students' intuition. We do not simplify by making incorrect statements or omitting important assumptions; the material is technically correct but with a primary emphasis on accessibility. If we have succeeded in our aim, then clearly the instructor will find the teaching job much easier.

The book is designed for a one- or two-term course in undergraduate statistics for business and economics, and it presupposes only basic algebra and advanced arithmetic. Access to a computer would be a definite plus in solving certain problems but is not required. However, the use of a hand calculator is highly recommended and, for some problems, necessary if a computer is not available.

This text offers the instructor unusual flexibility in selecting both content and level of coverage. The first 12 chapters discuss the main themes and the topic emphases covered in most business and economics statistics courses. Two additional features allow the instructor to tailor use of the book to particular goals. First, nearly all chapters have appendixes. Some cover technical developments and derivations, and others discuss additional topics related to the material covered in the chapter. Whether a topic was included in an appendix, or in the chapter body, depended on its importance to the main theme. Second, we included chapters on special topics beyond Chapter 12. They are written at a level consistent with the rest of the book.

Introductory statistics texts are frequently classified by mathematics prerequisites. Our book presupposes only basic algebra, but we do not think this is a very important point. Certainly, the derivations of many key propositions used in a basic statistics course require either calculus or postcalculus background, and we in no way wish to discount the importance of proof and derivation. However, when we consider the sheer magnitude of material to be covered—the conceptual subtleties in such topics as probability theory, applied sampling, and hypothesis testing, and the importance of applied statistics in the world of business and economics—we are inclined to make no apologies for having omitted most of the mathematically sophisticated derivations and proofs. Actually, it is our experience that most students of business and economics benefit more from time spent on realistic examples with intuitive appeal than they do from time spent on mathematical development, even when calculus is required in their program. Several good introductory mathematical statistics books are available to the student who may wish to supplement the material in this text. The instructor may find it worthwhile to hold several special sessions on the mathematical developments for those few interested students.

For the instructor, we have prepared a comprehensive *Solutions Manual* and a complete *Test Bank*.

ACKNOWLEDGMENTS

We are indebted to many of our colleagues for their incisive reviews and suggestions. Many of these suggestions were used in the several rewrites of our textbook. We extend our thanks to James A. Xander, University of Central Florida; Timothy J. Killeen, University of Connecticut; George F. Rhodes, Jr., Colorado State University; William D. Warde, Oklahoma State University; John T. Sennetti, Texas Tech University; Justin Stolen, University of Nebraska at Omaha; and Henry F. Ander, Arizona State University.

We would also like to thank Wallace R. Blischke, statistical consultant of Van Nuys, California, for his review and suggestions; Boyd L. Fjeldsted, University of Utah, for his review of Chapter 12; and Russell C. Richards, statistician, Salt Lake City, Utah, for his review of Chapter 13 and assistance with the computer-related aspects of the textbook.

We want to extend our sincere appreciation to Charles Ford of Arkansas State University and Henry Crouch of Pittsburg State University (Kansas) for their encouragement and support of our textbook project.

Typing a statistics manuscript is the ultimate test of a typist's patience and skill; we would like to thank Teresa Hood, Carol Castelli, Renae Rigler, and Fran Lingle for their yeoman work at the typewriter. We would also like to thank Rebecca Priester for her assistance with the index.

Finally, we would like to recognize The Dryden Press staff who worked so expertly with us and express our appreciation for the very professional manner with which all aspects of editing and publishing our book were handled.

We want to thank Feeny Lipscomb, acquisitions editor, Nedah Abbott, developmental editor, Jane Perkins, project editor, Alan Wendt, designer, and Peter Coveney, production manager. We are also most grateful to our copy editor, Carol Gorski, and our proofreaders, Jean Babrick and Robert Beran.

Iver E. Bradley
University of Utah

John B. South
Arkansas State University

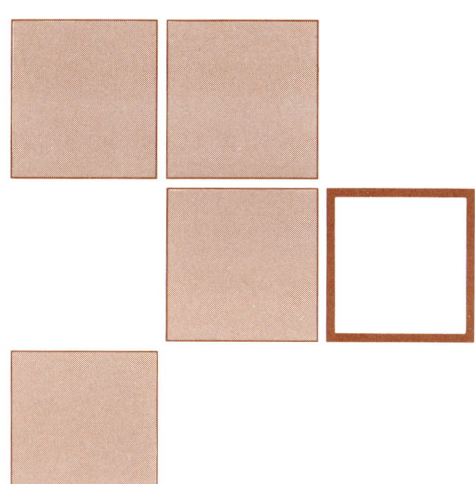

CONTENTS

CHAPTER 1
INTRODUCTION   1

1.1 Statistics and Modern Business   2
1.2 The Population and the Sample   3
1.3 Errors   4
1.4 Measurement, Accuracy, and Rounding   7
1.5 Your Career and Statistics   8

CHAPTER 2
DESCRIPTIVE STATISTICS   9

2.1 Measures of Centrality   10
2.2 Measures of Dispersion or Variability   16
2.3 Percentiles, Quartiles, and Deciles   21
2.4 Graphical Presentation of Data   22
2.5 Relative Frequencies, Pie Charts, and Population Trees   30
2.6 Charting Time Series   34
2.7 Calculating Descriptive Measures Using Grouped Data   38
2A Appendix: The Summation Symbol (Σ)   45
2B Appendix: Some Algebraic Proofs   46
2C Appendix: Change of Scale for a Frequency Distribution, or ''Coding''   48
    Problems   51

CHAPTER 3
PROBABILITY   **61**

3.1 Some Basic Terminology   63
3.2 The Classical Definition of Probability   66
3.3 The Relative Frequency and the Subjective Definitions of Probability   70
3.4 The Formal Mathematics of Probability Theory   73
3.5 Solving Probability Problems   82
3.6 More Problem Solving   87
3.7 Bayes' Theorem   91
3A Appendix: Combination Mathematics and the Classical Definition of Probability   94
3B Appendix: Probability Trees   95
3C Appendix: Extension of Bayes' Theorem   98
3D Appendix: Proof of Theorems   99
Problems   100

CHAPTER 4
RANDOM VARIABLES AND
THE BINOMIAL DISTRIBUTION   **109**

4.1 Discrete Random Variables   110
4.2 Expected Value and Variance of a Random Variable   112
4.3 The Binomial Distribution   117
4.4 Using the Binomial Table   124
4.5 Continuous Random Variables   126
4A Appendix: Expected Value and Variance Theorems with Selected Proofs   130
4B Appendix: Mathematical Formulation for the Continuous Random Variable Case   133
4C Appendix: The Hypergeometric Probability Distribution   134
4D Appendix: The Poisson Distribution   136
4E Appendix: The Binomial Distribution and the Binomial Theorem   138
Problems   139

CHAPTER 5
THE NORMAL DISTRIBUTION   **149**

5.1 General Properties of the Normal Distribution   150
5.2 The Standard Normal Distribution   152

5.3 Normal Random Variables: The General Case — 157
5A Appendix: The Normal Approximation to the Binomial Distribution — 162
Problems — 165

CHAPTER 6
SAMPLING THEORY — 171

6.1 The Distribution of the Sample Mean — 173
6.2 The Central Limit Theorem — 182
6.3 The Student-t Distribution — 184
6.4 The Distribution of the Sample Proportion — 186
6.5 Sampling from a Finite Population — 190
6A Appendix: Selected Proofs and Additional Theorems — 193
Problems — 196

CHAPTER 7
STATISTICAL ESTIMATION — 199

7.1 Point Estimation — 200
7.2 Interval Estimation for μ: The Large-Sample Case — 202
7.3 Interval Estimation for μ: The Small-Sample Case — 207
7.4 Estimating the Population Proportion — 211
7.5 Sample Size Determination — 215
7A Appendix: The Derivation of a Confidence Interval — 219
7B Appendix: Finite Population Interval Estimation and Sample Size — 221
Problems — 225

CHAPTER 8
HYPOTHESIS TESTING — 231

8.1 Two Types of Error and Classical Hypothesis Testing — 232
8.2 Hypothesis Tests Concerning the Mean—Large Sample Size — 234
8.3 Hypothesis Tests Concerning the Mean—Small Sample Size — 243
8.4 Hypothesis Tests Concerning the Population Proportion — 246
8.5 Additional Comments — 248

| 8A | Appendix: Type II Error in Hypothesis Testing | 249 |
| | Problems | 253 |

CHAPTER 9
STATISTICAL INFERENCE: TWO POPULATIONS — **261**

9.1	Testing Hypotheses about Two Population Means—Large Sample Size	262
9.2	Testing Hypotheses about Two Population Means—Small Sample Size	268
9.3	Testing Hypotheses about Two Population Proportions	272
9.4	Estimating the Difference between Two Population Parameters	276
9A	Appendix: The Paired Difference Test for Equality of Two Population Means	283
9B	Appendix: More on Hypothesis Tests Concerning Two Population Means	287
9C	Appendix: Pooled Variance	289
	Problems	290

CHAPTER 10
REGRESSION ANALYSIS — **299**

10.1	Linear Regression	301
10.2	Linear Regression—A More Comprehensive Example	304
10.3	The Classical Linear Regression Model	308
10.4	Inference from Regression	309
10.5	The Coefficient of Correlation and the Coefficient of Determination	317
10.6	An Alternative Set of Computational Formulas (Optional)	319
10.7	A Linear Regression by Computer	323
10.8	Multiple Linear Regression	327
10A	Appendix: The Analytical Geometry of a Straight Line	333
10B	Appendix: Derivation of Least Squares Estimators	339
10C	Appendix: Notes and Comments on Nonlinear Regression Models	341
	Problems	341
	Printout 1	349

Printout 2   351
Printout 3   353

CHAPTER 11
TIME SERIES   355

11.1 The Components of a Time Series   357
11.2 Trend Analysis (The Linear Model)   359
11.3 Moving Averages, Cycles, and Irregulars   361
11.4 Measuring Seasonal Effects   365
11A Appendix: Nonlinear Trend Functions   374
11B Appendix: A Multivariable Model That Includes Trend and Seasonal Patterns   380
11C Appendix: Forecasting   381
Problems   384

CHAPTER 12
INDEX NUMBERS   391

12.1 Interpretation of Index Numbers   394
12.2 Construction of Index Numbers: Price Relatives, Laspeyres Price Index, Paasche Price Index, and Fisher's Ideal Index   395
12.3 Quantity Indexes: Quantity Relatives, Laspeyres and Paasche Quantity Indexes   401
12.4 Laspeyres versus Paasche and Other Comments   403
Problems   407

CHAPTER 13
ANALYSIS OF VARIANCE AND STATISTICAL INFERENCE RELATED TO VARIANCE   411

13.1 The Chi-Square Distribution and F-Distribution   412
13.2 One-Way Analysis of Variance   416
13.3 Two-Way Analysis of Variance   423
13.4 Statistical Inference on the Pairwise Differences   426
13.5 Analysis of Variance on the Regression Problem   427
Problems   429

CHAPTER 14
NONPARAMETRIC STATISTICS — 433

- 14.1 A General Chi-Square Test Statistic — 434
- 14.2 Goodness-of-Fit Test—The Multinomial Population — 436
- 14.3 Contingency Table Tests—Tests of Independence — 440
- 14.4 Goodness-of-Fit Test—The Normal Distribution — 445
- 14.5 Runs Tests for Randomness — 448
- 14.6 The Wilcoxon Test — 452
- 14.7 The Kruskal-Wallis Test — 457
- 14.8 Spearman's Rank-Correlation Coefficient — 460
- 14A Appendix: Goodness-of-Fit Test—The Poisson Distribution — 463
- Problems — 467

CHAPTER 15
DECISION THEORY — 481

- 15.1 The Expected Monetary Value Criterion — 485
- 15.2 The Expected Value of Perfect Information — 488
- 15.3 The Expected Opportunity Loss Criterion — 489
- 15.4 Decision Criteria when Probabilities Are Not Known — 492
- 15.5 Decision Trees — 494
- 15.6 Insurance, Gambling, and Utility Theory — 500
- 15A Appendix: Deriving a Utility Function — 506
- Problems — 509

CHAPTER 16
SAMPLING — 515

- 16.1 Stratified Random Sampling — 516
- 16.2 Estimation of Proportions with Stratified Sampling — 524
- 16.3 Sample Size Estimation with Stratified Sampling — 526
- 16.4 Implementing Randomness in Sampling — 529
- 16.5 Systematic Sampling — 532
- 16.6 Cluster Sampling — 533
- Problems — 537

CHAPTER 17
STATISTICAL QUALITY CONTROL — 541

17.1 $\bar{x}$-Control Charts — 543
17.2 $\bar{x}$-Charts and Hypothesis Testing — 547
17.3 Acceptance Sampling — 552
17.4 Operating Characteristic Curves — 557
17A Appendix: Acceptance Sampling and the Poisson Distribution — 558
17B Appendix: Notation — 560
Problems — 560

TABLES — 567

Table A	Binomial Probabilities	568
Table B	Standard Normal Curve	572
Table C	Student-t Distribution	573
Table D	Chi-Square Distribution	574
Table E-1	F-Distribution	576
Table E-2	F-Distribution	578
Table F	Poisson Probabilities	580

CHAPTER 1

INTRODUCTION

In this chapter we discuss the scope of statistics, explore the importance of this tool in the world of business, and introduce some of the problems with which the statistician is concerned.

Specific objectives are—

to present some notion of the broad spectrum of statistics (Section 1).

to illustrate the typical sequence of activities in the statistical process as carried out in today's business world (Section 2).

to present the fundamental notion of a population versus a sample from the population (Section 3).

to discuss the kinds of errors with which statisticians are concerned (Sections 4 and 5).

Statistical methods and procedures are important in almost every facet of American life today. The decision-making processes of modern business and industry are replete with statistical applications. For example, the quality of much of our industrial output is controlled by statistical methods; many major marketing decisions are based upon statistical surveys; and a number of standard auditing and accounting procedures are statistical in nature.

Anyone who has watched election returns come in on the major TV networks has probably been impressed by the accuracy of the networks' predictions of final results, frequently based upon a very small percentage of the returns. Many governmental decisions, particularly those related to budgeting, are a function of statistically determined revenue forecasts. Television shows live and die on the basis of TV ratings—ratings determined by a surprisingly small sample.

Biologists and agriculturists used statistical procedures very early. Remember nineteenth-century botanist Gregor Mendel and his early work in genetics, as he counted the green and yellow peas to "prove" his theory of dominance? Agricultural scientists were innovators in statistical procedures designed to test the effects of various levels of fertilizer, water application, etc., on the growth of crops.

The examples above by no means represent an exhaustive list of the applications of statistical methods; but before going on to other topics, we should mention that the most modern and widely accepted theory of matter and energy in our physical world is fundamentally statistical and probabilistic in concept.

A second dimension to the statistical spectrum is that related to the level of sophistication in the field. Working at the low end of the conceptual scale are those many people calling themselves statisticians, and frequently classified as such within their industry, who work exclusively in the area of graphical methods and descriptive statistics. At the other end of the applied statistics spectrum are the innovators who are developing and using new methods and models frequently requiring high-level and quite modern mathematical concepts. In order to appreciate and understand some of these new developments, considerable specialization could be required. The everyday world of applied statistics includes descriptive statistics along with the use of a wide variety of quite standard procedures and models. If you cover most of the material in this textbook, you will have at least a brush with, if not the mastery of, many of the well-known standard procedures.

1.1 Statistics and Modern Business

The science of measurement from incomplete data and the probabilistic inferences drawn from this implied impreciseness form the core of statistical applications in today's world of business. To illustrate this point, we consider the several steps in the typical statistical process.

1. Data are gathered; they are the individual measurements taken from the elements of a sample from a larger population or universe. The data may be measured numbers (e.g., completion times for a given task); they may be integers or whole numbers (e.g., the number of persons in a household); or they may simply be tallies into a set of categories (e.g., type of transportation used to and from work). Proper collection of these data is extremely important if not conceptually difficult.

2. These data are then organized and summarized. This process may include making certain kinds of graphical presentations. Sample summary measures, such as the average, the range, or the percentage distribution in the case of categorical data, are calculated. The whole of Chapter 2 and parts of Chapter 12 are devoted to this area, known as *descriptive statistics*. The fact that the data come from a sample of a larger population is, of course, the source of the impreciseness mentioned above. For example, we would not expect the sample average balance from 100 charge accounts to be a precise measure of the average balance of a universe of 200,000 accounts.

3. Inferences are made from the descriptive statistics, and these inferences become the basis for decisions. The estimated magnitude of the impreciseness, the *statistical error*, is conveyed by the statistician to the decision maker in the form of probabilistic statements. The inferential aspect of the statistical process includes a wide variety of tools, methods, and ideas. In Chapters 3, 4, and 5 we are concerned with the tools, the basic theories of probability, and the related statistical models; in Chapter 6, the theoretical foundations for inferential statistics are presented, and the major portion of the remaining material is directed to the elements of inference.

1.2 The Population and the Sample

The population or universe is the totality of all elements being studied. In spite of the straightforward simplicity of this statement, it is important to give careful consideration to the definition and delineation of the population. Consider the following example. A large department store wants to study its delinquent charge accounts. There are approximately 7,000 delinquent accounts on the books, and some 1,000 accounts move in and out of this ledger each month. Delinquency time ranges from 1 month to 12 months; if an account is more than 12 months delinquent, it is turned over to a collection agency. The company desires an estimate of: (1) the mean dollar balance of all delinquent accounts (the estimated mean balance multiplied by the total number of delinquent accounts will provide an estimate of total dollar balance in all delinquent accounts); and (2) the mean delinquency time in months. In regard to the first problem, it seems quite clear that the 7,000 currently delinquent accounts form the population to be studied. This is correct. To some it seems just as clear that this same population will yield the correct answer to the second problem. Not so! A moment's reflection will show that an account in the ledger for 10 months will have the same time weight as 10 accounts that remain in the delinquency ledger for just 1 month. Moreover, delinquency time in the current ledger is incomplete. The population to be studied for the second problem must be the accounts as they move from the delinquent ledger.

A sample is a proper subset of the population; i.e., it is a collection of elements that number something less than the population size. (A 100 percent sample is called a census.) The primary reason for sampling is the cost-related aspects of data collection and analysis. In addition, it may not be possible to take a census. In the above charge account example, accounts leaving the delinquent ledger would be viewed by the statistician as a process, similar to that of tossing a coin; a process can go on ad infinitum.

In order to make inferences from the sample to the population, we must have a *random* sample. The term *random* means that the sample trials are independent of each other. For a finite population, this means that all possible samples have the same chance of being selected; or at each step in the selection all elements in the population are equally likely to be chosen.

Descriptive measures, such as the average and the range, from the collected sample data are called *sample statistics*; the corresponding population measures are *population parameters*. The sample statistic is a variable; it is a function of the sample selected. A population's parameter, albeit more often than not unknown, is a constant. For example, if several samples were taken from the 7,000 delinquent charge accounts, the sample average balance would probably be different for each sample; without a census, the mean balance of the current ledger of 7,000 accounts would be an unknown constant.

 ## 1.3 Errors

Statisticians give consideration to several kinds and sources of error related to the general problem of sampling from a population. An overview and an awareness of these kinds of errors should help you appreciate the subject of statistics as you progress through the textbook.

Introduction

Response Error

A survey of human subjects is always fraught with response error. It isn't necessarily that people don't tell the truth; perhaps, more often than not, they just don't know the exact answer. For example, consider the question, "How much money did your family spend on groceries last week?" Most can give a reasonably good estimate, but only a few would know the exact amount. Consider a second question, "How old are you?" With a modest mental effort, all of us could give this answer to the number of years, months, and days. To some this is a very sensitive question, and they might trim a few years. An interesting observation related to this point was brought out in a comparison of the 1960 census and the 1970 census. The number of very old people in 1970 (90 years of age and older) was too large compared with the 1960 data. It would seem that many old people understated their ages in 1960 or that they overstated their ages in 1970. Consider a third question, "Do you favor the elimination of sales tax on food items?" This requires a yes, no, or don't know answer. It could well be that you haven't even thought about the matter; but, if asked this question by a member of a survey team, in order to avoid a show of ignorance, you might well give a firm yes or no. If the question were to appear on the ballot in the next election, you might have changed your mind after having given careful consideration and study to the issue.

If the response error is consistently in one direction or the other, it is said to be a *response bias*. For example, there seems to be a tendency among adults to understate their ages. If this were the case, sample mean ages would tend to cluster around a number somewhat smaller than the true mean age in the population. The difference between the true mean and the mean of all sample means would be the bias.

Nonresponse Error

If, for any number of possible reasons, response is not forthcoming from all elements in the sample, a nonresponse error or bias is highly probable and must be considered. Mail and telephone surveys are particularly vulnerable to this problem. Response to mail questionnaires might be as low as 5 percent and will rarely exceed 35 percent in the absence of some sort of leverage. The assumption that the nonresponse group is similar to the response group certainly simplifies the problem, but more often than not this assumption will lead to significant error and bias in the statistical conclusions. To illustrate the point, consider the case of the survey team commissioned to study the impact of tourism in a certain state. Length of stay, expenditures, and activities are illustrative of items of interest to the state planning board. The survey team stopped out-of-state automobiles as they entered the state and asked the people to complete a survey instrument in the form of a daily diary. Some 25 percent of the sample returned the questionnaire, and only 80 percent of the questionnaires received were usable. With 20 percent response, the consulting group reported their findings as if they were dealing with a random sample of the total population of tourists. Even the statistically uneducated person should have been nervous about this sort of reporting. Further study revealed that, the longer people stayed, the less inclined they were to complete the survey form, and younger persons failed to respond in the same proportion as older persons. The major objective of the study was to estimate total tourist expenditures in the state; the bias created by the nonresponse error resulted in an estimate that was later shown to be only a fraction of actual tourist expenditures.

Related to the problem of nonresponse error is another real-world problem: the nonavailability of certain components of the population. For example, in voter polls it is very expensive if not almost impossible to include the overseas military. Fortunately this is a very small component of our eligible voting group, and bias from such a small group can usually be ignored. As a second example, telephone books are frequently used as a population basis in a given area. Several groups are not represented; low-income people frequently do not have phones, and many others choose to have unlisted numbers. The statistician can only speculate as to the bias that might be created by these nonlisted groups, and perhaps a better answer would be to address statistical conclusions to that percentage of the population covered by the phone book.

Statistical Error

Perhaps the major emphasis throughout this text and the statistics course you are about to begin is the estimation and measurement of statistical error. The term error in this context may be misleading to some who think it implies a mistake. Actually, we will see that statistical error is unavoidable, and the concept is fundamental to the science of statistical sampling. In order wholly to understand this notion, you must stay with us for a term or two; but we will attempt to illustrate the idea. Consider a large population, 1 million or so families, where the 1979 mean family income was $21,500. A random sample of 300 families is taken, and the sample mean family income is calculated. Assuming other errors to be negligible, would we expect the sample mean income to be exactly $21,500? And if not, how close would we expect the sample mean to be to our population mean? Also, if the sample size were 30 instead of 300, would our expectations change with respect to the sample mean?

As we proceed with our statistics course, we will find some interesting answers to the above questions; and, in a general way, they will support what most of you, with your limited experience, intuitively feel. Sample means of any size tend to cluster about the population mean, and this clustering becomes tighter with larger sample sizes. We will develop a very specific measure, the *standard error of the mean*, for this spread. This is an example of statistical error. The standard error of the mean, like many other statistical errors, is inversely proportional to the square root of the sample size. We cannot guarantee that the 300-unit sample mean will be closer to the true mean than the 30-unit sample mean; but we will wager in favor of the larger sample, and, moreover, we will have the tools to be precise as to the nature of the wager.

1.4 Measurement, Accuracy, and Rounding

The accuracy of sample measurements of such quantitative dimensions as time, distance, weight, velocity, etc., is constrained by the quality of the measuring instrument. If we assume that there is no human error in the use of the instrument, then the accuracy of the measurement is in terms of the number of significant digits. For example, if task completion time were measured to the nearest .1 second, a measurement of 123.1 seconds would really be 123.1 ± .05 seconds, and we would say there are 4 significant digits. If sacks of feed were weighed to the nearest pound, a measurement of 94 pounds would be 94 ± .5 pounds. (There is one exception to this tradition, and that is the measurement of age. Unless we stress that a person should state his or her age to the nearest year, the statement "I am 25," will mean 25.0 to 26.0 years.)

The rules of operations with measured numbers are quite simple if we are working strictly in the area of multiplication (and/or division) or strictly in the area of addition (and/or subtraction). These are as follows:

1. Relative or percentage error in the measurement is the important consideration for multiplication and division—i.e., the minimum number of significant digits in any one of the numbers dictates the accuracy of the result. For example, consider

    $$\frac{(24.26)\,(0.185)}{9.8}$$

    where all 3 numbers are measured numbers. There are only 2 significant digits in the denominator number; therefore the result should contain only two significant digits. Using a hand calculator you will get 0.4579693; the answer should be expressed as .46.

2. The absolute value of the rounding is the significant consideration in the addition-subtraction rule. For example, how should

the sum, 183.4 + 91.17, be expressed? We should round 91.17 to 91.2 because the maximum rounding error of ±.05 in the first number, 183.4, swamps the rounding error, ± .005, in the second number. In addition and subtraction, the result should be rounded to the decimal place of the number of least absolute accuracy.

As you proceed through your course in statistics, you will find many important calculations that combine addition-subtraction with multiplication-division. In order to maintain the accuracy of the original data in all calculations that combine these operations, we recommend that the traditional rules given above be ignored and that as many digits as possible be carried through the complete series of calculations. The final result should, of course, be rounded to agree with the accuracy of the original data. If a computer is being used, this problem is handled automatically. The printout will also contain more significant digits than are usually necessary, and these final results can be rounded to fit the problem. If a hand calculator is being used, rounding too soon can lead to incorrect and even spurious results. For example, the calculation of one of the statistical measures, the coefficient of correlation, combines all of the basic arithmetic operations. A correlation coefficient in excess of unity is meaningless, and from the definition this is a mathematical impossibility. Early rounding that appears to be entirely consistent with the two rules given above may lead to correlation coefficients that exceed unity when the correct value is something slightly less.

1.5 Your Career and Statistics

Assuming that your primary career interest is in business and economics, the chances of your becoming a professional statistician are slim indeed. It is, however, almost certain that you will have frequent contact with statistical reports; and the ability to understand, criticize, and evaluate the results could be extremely important to your career. Many of you will need to actually use some of the standard statistical procedures and techniques in papers and reports you are assigned to prepare. You can rest assured that the use of quantitative methods in business has not yet peaked but is still growing. Statistics is a major part of the general area of quantitative methods, and you who learn and understand the basic concepts of your first statistics course will certainly reap some rewards in your career ahead.

CHAPTER 2

DESCRIPTIVE STATISTICS

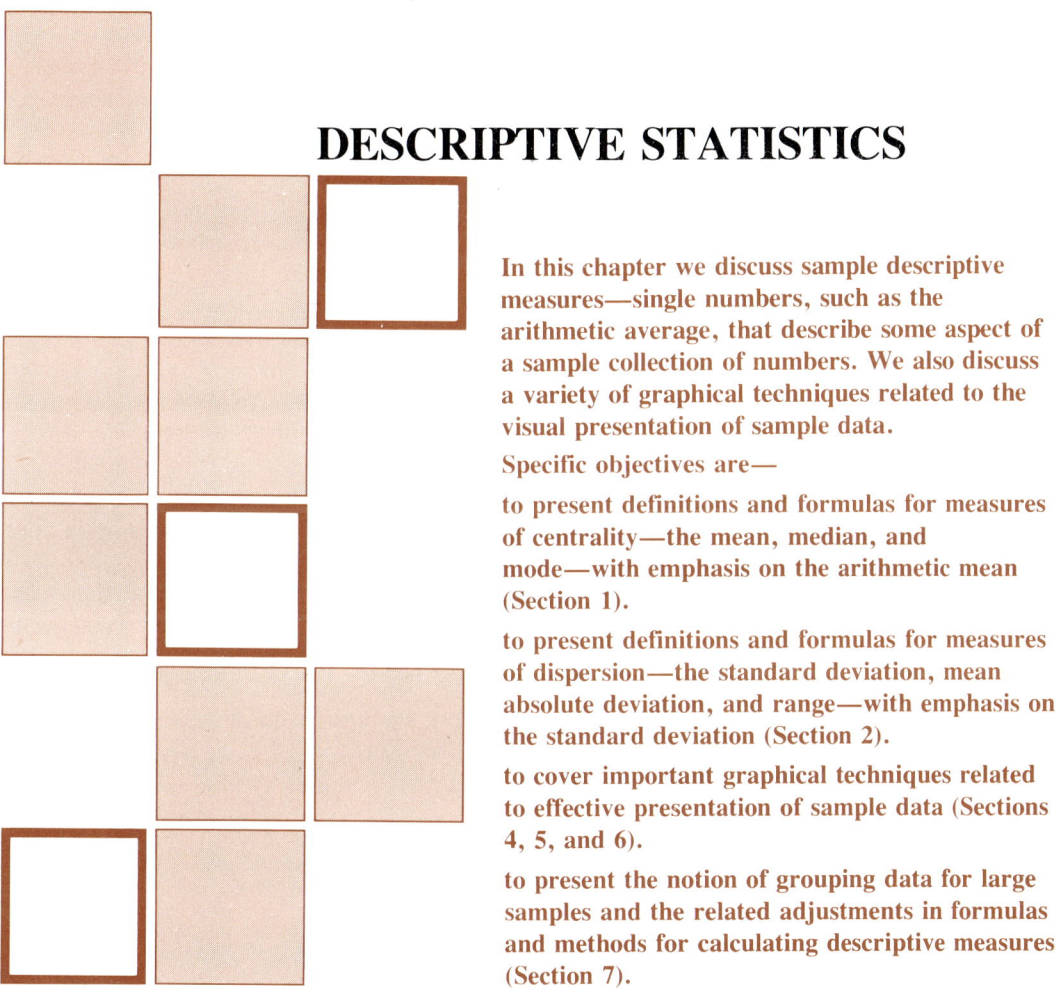

In this chapter we discuss sample descriptive measures—single numbers, such as the arithmetic average, that describe some aspect of a sample collection of numbers. We also discuss a variety of graphical techniques related to the visual presentation of sample data.

Specific objectives are—

to present definitions and formulas for measures of centrality—the mean, median, and mode—with emphasis on the arithmetic mean (Section 1).

to present definitions and formulas for measures of dispersion—the standard deviation, mean absolute deviation, and range—with emphasis on the standard deviation (Section 2).

to cover important graphical techniques related to effective presentation of sample data (Sections 4, 5, and 6).

to present the notion of grouping data for large samples and the related adjustments in formulas and methods for calculating descriptive measures (Section 7).

A descriptive statistic is an ordinary number abstracted from a collection of numbers—the data—that summarizes or describes some characteristic of the data. As discussed in Chapter 1, the data collected and analyzed are typically a sample from a larger population or universe. Descriptive measures defined and analyzed in this chapter will be the sample *statistics;* analogous population measures are known as *parameters*.

A sample statistic may be an end unto itself; it is frequently important to describe in a simple and efficient manner some characteristic of a data set. Finding a sample statistic is also the first step leading to the more sophisticated statistical analyses we begin in Chapters 6 and 7.

This chapter also includes a discussion of graphical techniques. As a supplement to the descriptive statistics, a graph can be an invaluable summary tool.

2.1 Measures of Centrality

Perhaps the most important descriptive measure in all statistical operations and certainly the one most frequently used is the *arithmetic average* or *arithmetic mean*.[1] *Mean* and *average* are synonyms, and throughout the text the use of either of these terms will refer to the arithmetic mean or average.

Most of you are already familiar with the concept of an average; we will formalize this notion along with those relating to other descriptive measures. In so doing it becomes necessary to introduce and use a certain amount of symbolism and notation with which you may not be familiar. It is our hope that you will approach new mathematical notations as you would a new language. Learn to translate the algebra to English; the student who fails to separate notation from the notion may well lose track of concepts he already understands.

To illustrate our point that you probably understand the idea of an average, consider the question, If 4 tests were given during the term in a particular class, how would you find your average test score? (Answer: Add the scores and divide by 4.) In a more formal way, we define the sample arithmetic mean as the algebraic sum of the numbers in the sample divided by the sample size. Let $x_1, x_2, \ldots x_n$ be the collection of n numbers and let $\bar{x}$ (read "x-bar") be our symbol for the average value of x.

---

1. The word *arithmetic* implies that there are other kinds of averages. This is the case, but they are not important in the everyday use of statistics. See problems 36 through 39 at the end of the chapter for the definitions of the geometric and harmonic means along with illustrations.

Definition of Sample Mean

The definition above may be written

$$\bar{x} = \frac{\sum_{i=1}^{n} x_i}{n}.$$

The capital sigma, Σ, indicates that numbers should be summed;

$$\sum_{i=1}^{n} x_i$$

is read as "the summation of x-sub-i, i going from 1 to n," and means $(x_1 + x_2 + \cdots + x_n)$ or, simply, sum the n numbers in the sample. The range of summation, indicated by $i = 1$ to n, may be deleted when it is clear that we are summing all n values of the sample; in lieu of writing

$$\sum_{i=1}^{n} x_i,$$

we could write Σx_i. For a detailed discussion of the summation symbol, see Appendix 2A at the end of this chapter.

The following examples should clarify any problems you might have understanding and calculating the mean values of a set of numbers.

Example 2.1.1

In Mathematics 1AX, Jean received scores of 92, 81, and 95 on her 3 midterm tests. What was her test average going into the final?

The 3 scores total to 268; this divided by 3 is 89.3, Jean's test average. In our formal notation $x_1 = 92$, $x_2 = 81$, and $x_3 = 95$; and

$$\bar{x} = \frac{\sum_{i=1}^{n} x_i}{3} = \frac{x_1 + x_2 + x_3}{3} = \frac{92 + 81 + 95}{3} = 89.3.$$

Example 2.1.2

Professor Smith drives to school each morning. Being a statistics teacher, he measured the driving time for 6 different mornings. The results were 11.5, 15.0, 12.2, 11.0, 17.8, and 11.9 minutes. What was the professor's mean driving time for the 6 mornings?

The sample mean is readily calculated using the formula given above:

$$\bar{x} = \frac{\sum_{i=1}^{6} x_i}{6} = \frac{11.5 + 15.0 + 12.2 + 11.0 + 17.8 + 11.9}{6} = 13.2.$$

Measures of Centrality

The professor's average driving time for the 6 days was 13.2 minutes.
The sample mean of the following problem has an interesting interpretation.

Example 2.1.3

A highway department employee sitting at the side of Highway 66 records "1" when a domestically manufactured vehicle passes and "0" for each foreign-made vehicle. The first 25 values recorded were as follows: 1, 1, 1, 0, 1, 0, 0, 1, 1, 1, 1, 0, 1, 1, 0, 1, 0, 0, 1, 1, 1, 1, 1, 0, 0. Find the average of this sample and interpret the result.

The sample mean for the sample of 25 values is

$$\bar{x} = \frac{\sum_{i=1}^{25} x_i}{25} = \frac{1+1+1+0+1+0+0+1+\cdots \text{etc.}}{25} = \frac{16}{25} = .64.$$

The sample mean is the proportion of ones in the sample of 25, i.e., there are 16 ones and 9 zeroes in the sample. For this problem the sample mean is also the proportion of domestically manufactured vehicles in the sample.

If the sample numbers are not of equal importance, we need a variant of the ordinary mean, the *weighted mean*. For example, if the math professor in Example 2.1.1 told Jean that the third test would count double the first test and the second test was worth 50 percent more than the first, what would her test average be? Recall that her test scores were 92, 81, and 95, in that order. Her average would be calculated as follows:

$$\bar{x}_w = \frac{(1)(92) + (1.5)(81) + 2(95)}{4.5} = \frac{92 + 121.5 + 190}{4.5} = 89.7.$$

Note that the divisor is not the number of test scores but the sum of the weights.

Definition of Weighted Mean

The formal definition of a weighted mean is

$$\bar{x}_w = \frac{\sum_{i=1}^{n} w_i x_i}{\sum_{i=1}^{n} w_i}$$

where

$\bar{x}_w$ = the weighted sample mean
x_i = the ith sample value
w_i = the weight assigned to the ith sample value
n = the sample size.

This formula is not difficult to apply if you distinguish carefully between the sample values and the weights to be assigned to the sample values.

Example 2.1.4   A delivery van driver buys gas at 3 different stations during the day. She buys 6 gallons at the first station for 75 cents per gallon. Next she pays 84 cents per gallon for 12 gallons of gas. The third price paid was 76 cents per gallon for 5 gallons of gas. What was the average price per gallon paid by the driver?

We have 3 different prices for gasoline, 75, 84, and 76 cents per gallon. It would not be correct to simply average the 3 prices because different amounts were purchased at each of the stations. The prices need to be weighted according to the amounts purchased. For this problem,

$x_1 = .75 \quad w_1 = 6$
$x_2 = .84 \quad w_2 = 12$
$x_3 = .76 \quad w_3 = 5$

Using the formula for the weighted mean, $\bar{x}_w$,

$$\bar{x}_w = \frac{\sum_{i=1}^{n} w_i x_i}{\sum_{i=1}^{n} w_i} = \frac{w_1 x_1 + w_2 x_2 + w_3 x_3}{w_1 + w_2 + w_3}$$

$$\frac{6(.75) + 12(.84) + 5(.76)}{6 + 12 + 5} = \frac{18.38}{23} = .799,$$

we obtain the average price per gallon paid by the van driver, 79.9 cents.

Example 2.1.5   If you received 3 different grades last semester, an A, a B, and a C, can we conclude that your average grade was B; i.e., if, on a 4-point scale, A = 4.0, B = 3.0, and C = 2.0, can we conclude your semester grade point average was 3.0? If all courses were of equal credit hours, your GPA would be 3.0. What would your GPA be if the A were received in a 5-credit-hour class, the B in a 3-hour class, and the C in a 2-hour class?

Most of you would be quite perturbed if this were your performance and a GPA of 3.0 were reported to your parents. The GPA is a classic example of the weighted mean. You would want to weight the point values with the credit hours:

Measures of Centrality

$$x_1 = 4.0 \quad w_1 = 5$$
$$x_2 = 3.0 \quad w_2 = 3$$
$$x_3 = 2.0 \quad w_3 = 2.$$

$$\bar{x}_w = \frac{\Sigma w_i x_i}{\Sigma w_i} = \frac{5(4) + 3(3) + 2(2)}{5 + 3 + 2} = \frac{33}{10} = 3.3.$$

As you expected, the true GPA exceeds the unweighted and incorrect value of 3.0.

Other frequently used measures of centrality are the median and the mode; the median is the more important.

Definition of Median

The *median*, MD, is the middle number in the sample. In a more technical way, the median is defined as any value such that half of the numbers in the sample are greater than or equal to this value and the other half are less than or equal to it.

Consider the example:

Example 2.1.6

The following are outside diameter measurements in centimeters from 10 metal disks randomly selected from the output of a stamping machine: 3.81, 4.07, 3.98, 4.15, 3.76, 4.02, 3.88, 3.91, 4.05, and 3.67. Find the median.

In order to find the median, efficiency is usually served if we array the numbers from smallest to largest or vice versa. The 10 values rearranged in ascending order are: 3.67, 3.76, 3.81, 3.88, 3.91, 3.98, 4.02, 4.05, 4.07, and 4.15.

Any number between the fifth and sixth numbers, 3.91 and 3.98, will satisfy the definition given above and is therefore a median.

The idea that the sample median is not unique seems a bit unsatisfactory. If the sample size is an even number, it is traditional to define *the* median as the average of the two middle numbers. In this example, the median,

$$MD = \frac{3.91 + 3.98}{2} = 3.945.$$

If the sample size is an odd number, there will be a clear-cut middle number in the array; the sample median will be unique. Assume the 4.15 measurement in our example were never taken; the middle measurement of the remaining nine diameters is 3.91 centimeters; this is the median.

We should note that the mean value of the 10 numbers is 3.93 centimeters (verify this). The median value, 3.945 centimeters, is very

Descriptive Statistics

close to the mean; in this example both describe the centrality of the data set at about the same position.

As we proceed through the material in the text we will find that the statistician's primary measure of centrality is the mean. From the standpoint of a purely descriptive measure, the median is more representative of the "typical" member of the data set than the mean if there are extreme values in the sample. This point is illustrated by the next example.

Example 2.1.7

In an interview with the local newspaper editor, a businessman in a small town claimed that the average salary paid by his firm was $20,000 per year. The editor was duly impressed but, knowing something about the five employees, he decided to investigate further. From an undisclosable source the editor found the five individuals' annual salaries to be $4,000, $4,000, $5,000, $7,000, and $80,000. (The $80,000 salary went to the manager, who was also the owner's son-in-law.) Was the average really $20,000? Find the median.

The median salary was $5,000, the middle number in the array. The total wage bill, $\Sigma x_i = 4,000 + 4,000 + 5,000 + 7,000 + 80,000$, is $100,000; this divided by 5 gives the average salary of $20,000.

This example was contrived to emphasize the point that different measures of centrality may well locate this centrality at quite different positions. Perhaps some judgment and experience are required for a reasonable interpretation and presentation of results like this. The median salary best represents the welfare level of the "typical" person employed by the firm. The mean value of $20,000 multiplied by 5 gives the total wage bill of $100,000; this money is paid to workers in the community, and this fact cannot be ignored in evaluating the community welfare. (Note that the median multiplied by the sample size yields a number with no real meaning; i.e., $5 \times \$5,000$ is far short of the aggregate salary bill.) Knowledge of both the mean and median can be useful; see the discussion in Section 3 of this chapter on asymmetry and skewness.

Definition of Mode

The mode is defined as the sample value of greatest frequency—the fashionable or the most popular measurement.

The mode is not a very useful statistic if the sample size is small, because most of the values in a small sample will be different from each other. The mode is meaningful when sample sizes are large and the data are grouped into several categories. See the discussion of frequency distributions and related graphical methods later on in the chapter. The mode is also the most likely value; i.e., if one were to select randomly

Measures of Centrality

an item from a sample or population, the value most likely to be selected would be the mode.

2.2 Measures of Dispersion or Variability

Faced with the problem of describing a collection of numbers with only two measures, we would choose a measure of centrality and a measure of dispersion. The statistician's primary measure of dispersion is the standard deviation, which will be defined and discussed in some detail in this section. A very simple measure of dispersion and one with which most of you are familiar is the range.

Definition of Range

There are various ways to define the *range*, Rg.[2] It is simply defined as the difference between the largest number in the set and the smallest number; or, let x be the variable, then

$$Rg = x_{\max} - x_{\min}.$$

Example 2.2.1

As an example, consider the 6 measurements of Professor Smith's driving times in Example 2.1.2. The maximum time was 17.8 minutes and the minimum time was 11.0 minutes.

The range, $Rg = 17.8 - 11.0 = 7.8$ minutes. The mean driving time of 13.2 minutes along with the range of 7.8 minutes conveys a better description of the sample data than does the mean alone.

The range is used frequently as the measure of dispersion in quality control applications. It has the advantage of being easy to calculate and, for small sample sizes, the sample range can be used as an estimate of the population variability with roughly the same confidence as more complicated sample measurements.

The standard deviation, s, as well as the mean absolute deviation, MAD, a third measure of dispersion, are both defined in terms of the deviations of the data points about the mean, $x_i - \bar{x}$. Some of these deviations will be positive and some negative; in fact, the algebraic sum of the deviation is always 0; i.e., $\Sigma(x_i - \bar{x}) = 0$. See the technical appendix of this chapter for the proof.

---

2. There is good rationale for a slightly more sophisticated version: $Rg = x_{\max} - x_{\min} + 1/10^r$ where r is the accuracy in number of decimal places in the measurements. For example, if the tallest member of the Trail Blazers team is 6 feet 10 inches tall and the shortest is 6 feet 1 inch, and we know that these measures are rounded to the nearest inch, we might be inclined to report the range as 6 feet 10 inches − 6 feet 1 inch + $1/10^r$, or 10 inches as opposed to 9 inches. As accuracy of measurement increases, the difference resulting from the 2 definitions becomes smaller. If the reported heights of the 2 players above were 6 feet 10.3 inches and 6 feet 0.9 inches, the range would be either 9.4 inches or 9.5 inches according to the definition preferred.

Definition of Sample Standard Deviation

Squaring each of the deviations results in a set of non-negative numbers;[3] this is fundamental to the definition of the sample standard deviation:

$$s = \sqrt{\frac{\Sigma(x_i - \bar{x})^2}{n - 1}}.$$

Note the following:

1. $\Sigma(x_1 - \bar{x})^2 = (x_1 - \bar{x})^2 + (x_2 - \bar{x})^2 + \ldots + (x_n - \bar{x})^2$ and, inasmuch as each of the members is non-negative, the sum will usually be positive; it is zero in the uninteresting case where all $x_i = \bar{x}$.
2. If the denominator were n instead of $n - 1$, the number under the radical would be an "average squared deviation";[4] even so, the number under the radical is approximately the average of the squared deviations.
3. The square root operation brings the units of s to the same dimension as those of the original variable. For example, if x were in dollars, s^2 would be in square dollars and s would be in dollars.

The standard deviation is the square root of the approximate average squared deviation. This is quite obviously a measure of dispersion, but one that doesn't readily relate to your experience. At this point, most students feel quite comfortable with the mean, median, mode, and range, but they see the standard deviation formula as an unpleasant arithmetic exercise. We can only assure you that this is a very important statistical measure.

The discussions and related problems on sampling and statistical inference in Chapters 6, 7, and 8 will convince you, we think, of this importance.

The square of the standard deviation is called the *sample variance*. Technically speaking, the variance is not a basic descriptive measure, but you should become familiar with the term *variance*. Separate and apart from the use of the sample standard deviation, the variance is the crucial statistic for several important areas of statistical analysis.

Definition of Variance

The definition of variance is very simply the square of the standard deviation:

$$s^2 = \frac{\Sigma(x_i - \bar{x})^2}{n - 1}.$$

---

3. Actually all the squared deviations will be positive if none of the x_i equals the sample mean; zeroes result only for $x_i = \bar{x}$.

4. The definitional choice of $n - 1$ in the denominator has to do with the estimation properties of the sample statistic and will be discussed in Chapters 6 and 7.

In order to illustrate the calculation of variance and standard deviation, consider the following problem:

Example 2.2.2

Find the sample standard deviation and variance of Professor Smith's driving times in Example 2.1.2. The observed times were 11.5, 15.0, 12.2, 11.0, 17.8, and 11.9.

The sample mean was calculated in Example 2.1.2 to be $\bar{x} = 13.2$. The sample variance can be calculated after the sample mean has been calculated:

$$s^2 = \frac{\sum_{i=1}^{n}(x_i - \bar{x})^2}{n-1} = \frac{\sum_{i=1}^{6}(x_i - \bar{x})^2}{6-1}$$

$$= \frac{(11.5 - 13.2)^2 + \cdots + \cdots + \cdots + \cdots + (11.9 - 13.2)^2}{5}$$

$$= \frac{2.89 + 3.24 + 1.0 + 4.84 + 21.16 + 1.69}{5} = \frac{34.82}{5} = 6.96$$

$$s = \sqrt{6.96} = 2.64 \text{ minutes.}$$

Alternate Formula for Standard Deviation

An alternate formula for calculating the variance and/or the standard deviation is provided by the identity, $\Sigma(x_i - \bar{x})^2 = \Sigma x_i^2 - n\bar{x}^2$. (See Appendix 2B for the proof of this identity.) This formula is usually computationally more efficient than the definitional formula.

The computational formula for the sample standard deviation would be

$$s = \sqrt{\frac{\Sigma x_i^2 - n\bar{x}^2}{n-1}}$$

where

s = sample standard deviation
$\bar{x}$ = sample mean
n = sample size
x_i = the value of the ith sample observation.

(It is important to note that:

$$\sum_{i=1}^{n} x_i^2$$

calls for squaring each of the sample values before summing; i.e.,

$$\sum_{i=1}^{n} x_i^2 = x_1^2 + x_2^2 + \cdots + x_n^2.$$

Also, $n\bar{x}^2$ calls for the sample size times the square of the mean value.) Consider the following example:

Example 2.2.3

Refer to Example 2.1.6 where the following outside diameter measurements in centimeters were obtained from 10 metal disks: 3.81, 4.07, 3.98, 3.76, 4.15, 4.02, 3.88, 3.91, 4.05, and 3.67 centimeters. Find the sample mean and the sample standard deviation.

We first calculate the Σx_i and the Σx_i^2:

$\Sigma x_i = x_1 + x_2 + \cdots + x_n = 3.81 + 4.07 + 3.98 + 3.76 + 4.15 + 4.02 + 3.88 + 3.91 + 4.05 + 3.67 = 39.30;$

and

$$\sum_{i=1}^{n} x_i^2 = x_1^2 + x_2^2 + \cdots + x_n^2 =$$

$(3.81)^2 + (4.07)^2 + (3.98)^2 + (3.76)^2 + (4.15)^2 + (4.02)^2 + (3.88)^2 + (3.91)^2 + (4.05)^2 + (3.67)^2 = 154.6558.$

The sample mean, $\bar{x}$, and sample standard deviation, s, are calculated using these computed summations:

$$\bar{x} = \frac{\Sigma x_i}{n} = \frac{39.30}{10} = 3.93 \text{ centimeters}$$

$$s = \sqrt{\frac{154.6558 - 10(3.93)^2}{10 - 1}} = \sqrt{\frac{0.2068}{9}} = \sqrt{0.02298}$$

$s = 0.15$ centimeters.

The sample mean is 3.93 centimeters and the sample standard deviation is .15 centimeters.

The computational formula is identically equal to the definitional formula; any difference between the results if both were used to calculate s would be due to rounding errors and/or errors in arithmetic. In order to illustrate this general proposition, we present the calculations from the definitional formula in tabular form in Table 2.1. As expected, the same value, $s = .15$ centimeters, was obtained. Note also that

I. $\Sigma(x_i - \bar{x}) = 0$, and
II. $\Sigma(x_i - \bar{x})^2 = 0.2068 = \Sigma x_i^2 - n\bar{x}^2.$

Measures of Dispersion or Variability

Table 2.1 Calculation Format for Standard Deviation

	x_i	$x_i - \bar{x}$	$(x_i - \bar{x})^2$
	3.81	−.12	.0144
	4.07	.14	.0196
	3.98	.05	.0025
	3.76	−.17	.0289
	4.15	.22	.0484
	4.02	.09	.0081
	3.88	−.05	.0025
	3.91	−.02	.0004
	4.05	.12	.0144
	3.67	−.26	.0676
Totals	39.30	0	.2068

$$\bar{x} = \frac{\Sigma x_i}{n} = \frac{39.30}{10} = 3.93$$

$$s = \sqrt{\frac{\Sigma(x_i - \bar{x})^2}{n-1}} = \sqrt{\frac{.2068}{9}} = \sqrt{.02298} = .15 \text{ centimeters}$$

You may recall from previous courses that the absolute value of a real number (indicated by parallel vertical lines about the number, e.g., $|y|$) is simply the magnitude of the number; the absolute value of a positive number is the number itself and the absolute value of a negative number is the positive of that number. For example, $|6| = 6$ and $|-4| = 4$.

Definition of Mean Absolute Deviation

This leads to another idea for a measure of dispersion called the *mean absolute deviation, MAD*:

$$MAD = \frac{\Sigma |x_i - \bar{x}|}{n}.$$

As with the standard deviation, a set of non-negative numbers is being summed; the resulting sum is always positive with the exception of the special case where all $x_i = \bar{x}$. The mean absolute deviation is the average magnitude by which the sample values differ from the sample mean.

Example 2.2.4

Find the mean absolute deviation of the outside diameter measurements in the previous example: 3.81, 4.07, 3.98, 3.76, 4.15, 4.02, 3.88, 3.91, 4.05, and 3.67 centimeters. Compare the *MAD* with the standard deviation.

See Table 2.2 for the solution to this problem. The mean absolute deviation is 0.124 centimeters, which is less than the standard deviation of 0.15 centimeters. The standard deviation is typically larger than the mean absolute deviation. As a measure of dispersion the mean absolute deviation has an intuitive appeal over the standard deviation; i.e., most of us are quite comfortable with the notion of average. As we will see later, the standard deviation is the natural measure in terms of our statistical models. On the surface, the standard deviation with the squares and square roots may seem to you more complex than the *MAD* with the absolute value operation. This is not necessarily the case. There are mathematically unmanageable aspects related to the algebraic manipulations of absolute values.

Table 2.2 Calculation Format for Mean Absolute Deviation

	x_i	$x_i - \bar{x}$	$|x_i - \bar{x}|$
	3.81	−.12	.12
	4.07	.14	.14
	3.98	.05	.05
	3.76	−.17	.17
	4.15	.22	.22
	4.02	.09	.09
	3.88	−.05	.05
	3.91	−.02	.02
	4.05	.12	.12
	3.67	−.26	.26
Totals	39.30	0	1.24

$$\bar{x} = \frac{39.30}{10} = 3.93 \text{ centimeters}$$

$$MAD = \frac{\sum_{i=1}^{n} |x_i - \bar{x}|}{n} = \frac{1.24}{10} = .124 \text{ centimeters}$$

2.3 Percentiles, Quartiles, and Deciles

The concept and use of percentiles are not common in business and economics; at the same time, the notion is simple enough, and percentiles are frequently used measures in the fields of education and psychology. The median is the fiftieth percentile. If we let P_k be the kth percentile, then

P_k = the kth percentile, a number such that k percent of the values are less than or equal to that number.

Quartiles, Q, and *deciles*, D, refer to specific percentiles; e.g., the median is the fiftieth percentile which is the second quartile, Q_2, and the fifth decile, D_5.

Q_k = the kth quartile. Q_1 is P_{25}, Q_2 is P_{50}, and Q_3 is P_{75}
D_k = the kth decile. D_1 is P_{10}, D_2 is P_{20}, etc.

As we know, the median is a measure of centrality; other percentiles are classified as measures of location but, of course, do not measure central tendency. The first and third quartiles are also used for another measure of dispersion, the *semi-interquartile range*, which is defined as follows:

Semi-interquartile range = $\dfrac{Q_3 - Q_1}{2}$.

Note that $Q_3 - Q_1$ is a range that includes the middle 50 percent of the numbers. Procedures for estimating percentiles from raw data are similar to those for the median. These estimates[5] are possible for small samples but have more meaning in reference to large samples and populations. For this reason we postpone examples and illustrations to Section 6 where the discussion centers on calculation of descriptive measures from large samples tallied in a frequency distribution.

2.4 Graphical Presentation of Data

In conjunction with and supplementary to the summary descriptive measures discussed in the previous two sections, it frequently is necessary for the statistician to prepare a graphical presentation of his or her results. Surely a wise man must have once said that a good picture is worth a thousand numbers. Unlike pure mathematicians and theoretical physicists, whose discourse is only within their professional group, applied statisticians must, more often than not, communicate the results of their professional work to the lay persons of the community. An important vehicle of this communication is a well-planned graphical presentation. Spend a half hour sometime thumbing through news magazines, newspapers, and annual reports from large corporations. You will find a number of graphical or pictorial presentations, visual summaries of a variety of statistical data. Figures 2.1, 2.2, 2.3, and 2.4 are some examples of typical graphical presentations from the printed media.

---

5. The ith number in an ascending array estimates the $(i/n + 1)^{th}$ percentile.

Figure 2.1 Family Incomes of Common Shareholders of Utah Power and Light Company

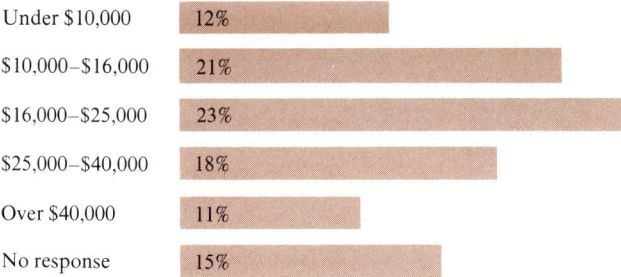

Under $10,000 — 12%
$10,000–$16,000 — 21%
$16,000–$25,000 — 23%
$25,000–$40,000 — 18%
Over $40,000 — 11%
No response — 15%

SOURCE: Utah Power and Light Company, *Who Owns Utah Power?* (Salt Lake City, 1977).

Figure 2.2 Occupations of Individuals Owning Stock in Utah Power and Light Company

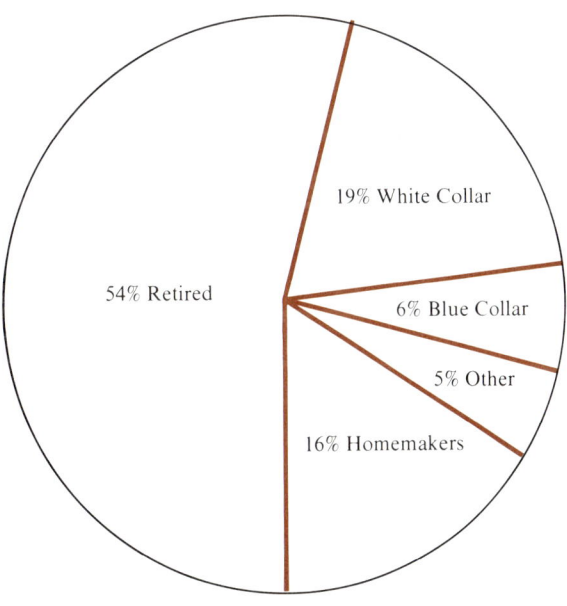

54% Retired
19% White Collar
6% Blue Collar
5% Other
16% Homemakers

SOURCE: Utah Power and Light Company, *Who Owns Utah Power?* (Salt Lake City, 1977).

Graphical Presentation of Data

Figure 2.3 The Nation's Transportation in 1979
Percentage of Total Passenger-Miles (Hypothetical Nation)

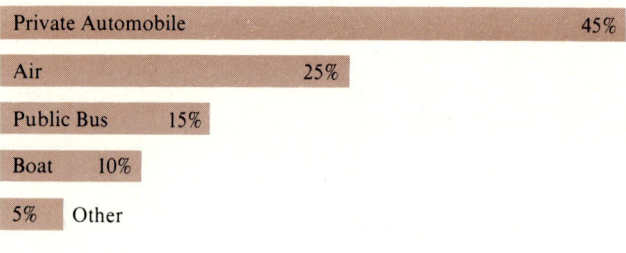

In this section we will discuss bar charts, line charts, and frequency polygons as methods for presenting data. The first step in preparing a large group of numbers for visual presentation is to define groups or classes and determine the number of sample values that fall in each group or class. A collection of numbers that has been tallied into groups or classes is called a *frequency distribution*. A frequency distribution and the corresponding graph are not only important in the communication mentioned above but they may also be a first step in the analysis of the sample results. Frequency distributions are referred to as either a *natural* frequency distribution or an *artificial* frequency distribution depending upon how the groups or classes are determined.

There is a large class of problems where the grouping is natural. For example, if the variable takes on only whole number values and the range of possible values is quite small, the responses can simply be

Figure 2.4 Home Building
New Housing Starts (Hypothetical Country)

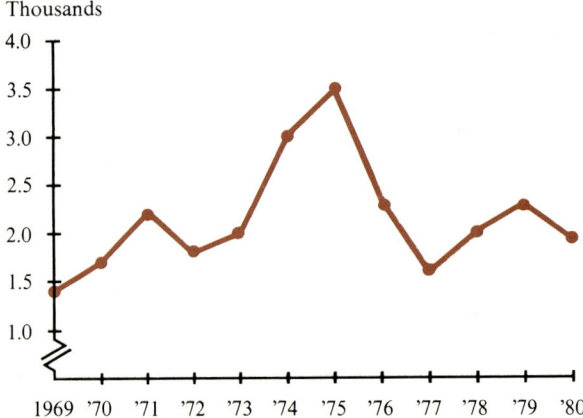

Descriptive Statistics

tallied to the natural values of the variable. Consider the following example using hypothetical data:

Example 2.4.1 In Boise, Idaho, 50 households were sampled to study family size and the related impact on public school enrollments. There were 8 families with no children, 19 with 1 child, 12 with 2 children, 7 with 3 children, 2 with 4 children, 1 with 6, and 1 with 11.

These data would be naturally tallied as indicated in Figure 2.5. A line chart has been chosen for the presentation of this natural frequency distribution in Figure 2.6. Later in this section we will justify the choice of a line chart as preferable to the possible use of a bar chart or a frequency polygon for a visual description of this particular type of data. Careful study of the tabular form of the frequency distribution in Figure 2.5 will convey much of the same information to most of us as that conveyed by the line chart. With considerably less effort, however, we get the same and perhaps more information from the line chart: e.g., the most popular number of children is one; there are a number of families with no children; very few families have more than four children; etc. Remember, this is only a sample from a much larger population.

In preparing a chart or graph you will want to keep the following points in mind: (1) there should be a title; (2) both axes should be clearly labeled and scaled; and (3) the source of data should be given. In addition, the esthetic appeal of the graph is improved if the paper allocated to the graph is more or less fully utilized; this is handled quite simply by proper scaling of the axes.

Figure 2.5 Number of Children in Family (Tally)

Number of Children x	Tally	Frequency f
0	⁄⁄⁄⁄ ⁄⁄⁄	8
1	⁄⁄⁄⁄ ⁄⁄⁄⁄ ⁄⁄⁄⁄ ⁄⁄⁄⁄	19
2	⁄⁄⁄⁄ ⁄⁄⁄⁄ ⁄⁄	12
3	⁄⁄⁄⁄ ⁄⁄	7
4	⁄⁄	2
5		0
6	⁄	1
7		0
8		0
9		0
10		0
11	⁄	1
		$n = 50$

Figure 2.6 Number of Children in Family (Line Graph)

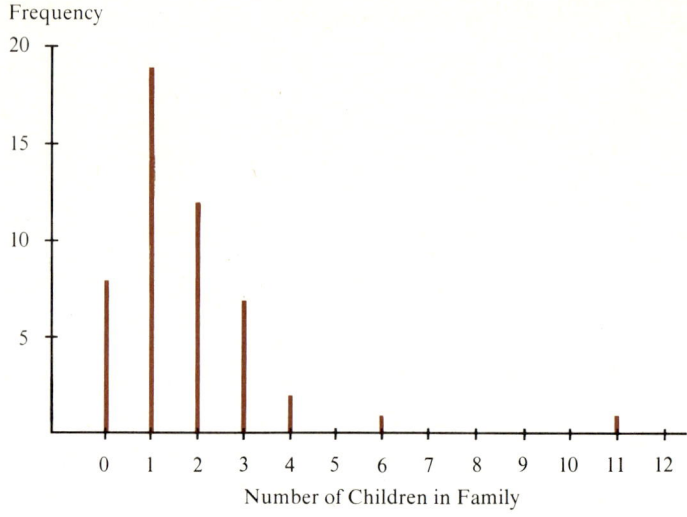

Consider this second example which illustrates through hypothetical data what we mean by an artificial frequency distribution.

Example 2.4.2

From the several hospitals in a large city, a random sample of 44 nurses reported annual income in thousands of dollars as follows:

10.5, 8.4, 8.3, 9.1, 9.0, 9.5, 7.7, 12.0, 12.3, 9.0, 11.0, 13.0, 8.2, 10.2, 8.0, 9.5, 9.0, 8.1, 8.1, 10.6, 10.1, 11.3, 9.5, 9.4, 10.6, 9.2, 9.5, 9.8, 7.7, 11.1, 8.3, 8.5, 8.6, 10.0, 9.3, 9.7, 12.0, 9.6, 7.8, 7.5, 10.2, 8.6, 8.8, 9.5.

There is no natural grouping of the annual incomes in this sample. It is our problem to decide on how many intervals or classes to use and what length each of these should be. Assuming all interval lengths to be the same size, it is clear that the product of the number of intervals and the length of each interval must be at least as big as the range in order to include all numbers. Let us use k for the number of intervals and c for the interval length. Once the number of classes, k, has been determined, we know approximately what the class size, c, should be. The relationship between sample size, n, and the number of intervals, k, is not precise.[6] See Table 2.3 for our recommendations for values of k for a given sample size.

---

6. One rule that has been suggested for determining the number of classes is known as Sturgess's Formula: $k = 1 + 3.3 \log n$.

Descriptive Statistics

Table 2.3 k as Determined by n

k Number of Classes	n Number of Items in the Sample
6-8	30-50
7-10	50-100
9-12	100-200
11-15	200-400
14-20	over 400

Once the numbers are tallied into a frequency distribution, the individual identity of each number has been lost. This procedure potentially introduces error; it is desirable to keep this error to a minimum. Several suggestions will help in this respect:

1. The products of the numbers of intervals and the class size, kc, must cover the range, but the excess should be kept to a minimum.
2. No sample value or number should fall at the endpoint of an interval. This is accomplished by making the endpoints of the intervals one decimal place beyond those of the data.
3. The total interval, kc, should be a continuum; i.e., the upper endpoint of one interval is the lower endpoint of the next.

The maximum sample value in the above example is 13.0. The minimum value in the sample is 7.5. Thus, the sample range is $13.0-7.5 + 0.1 = 5.6$.[7] The number of items in the sample, n, is 44. From Table 2.3 we observe that the preferable number of classes, k, is 6, 7, or 8. If we select $k = 7$ and $c = .8$ we obtain a total interval of $7(.8) = 5.6$, which precisely covers the range. The frequency table for the sample of 44 annual salaries is shown in Figure 2.7.

The frequency distribution creates an image of the annual wage rate of nurses in the city. A bar chart or histogram enhances this picture (see Figure 2.8). The histogram is constructed so that it has the following properties: (1) there are no gaps from bar to bar and the boundaries of the bars are at the endpoints of the income intervals; and (2) the height of each bar is proportional to the frequency in each interval. The frequencies need not be written in each bar, as has been done in Figure 2.8, because the height of the bar can be read from the vertical scale. In Figure 2.9 a frequency polygon chart has been constructed from the bar chart. The frequency in each category is marked at the midpoint of the respective interval. The frequency marks are then connected in sequence by straight lines.

---

7. In order to cover all numbers in a frequency distribution, one must use the second definition of range in Section 2, $rg = x_{max} - x_m + 1/10^r$ where r is the number of decimal places in the data.

Graphical Presentation of Data

Figure 2.7 Frequency Distribution—Nurses' Income Data

Income Categories	Interval Endpoints	Tally	Frequency
7.5–8.2	7.45–8.25	|||| |||	8
8.3–9.0	8.25–9.05	|||| ||||	10
9.1–9.8	9.05–9.85	|||| |||| ||	12
9.9–10.6	9.85–10.65	|||| ||	7
10.7–11.4	10.65–11.45	|||	3
11.5–12.2	11.45–12.25	||	2
12.3–13.0	12.25–13.05	||	2

Inspection of each of the above charts records an immediate mental image of the distribution. The centrality and spread of nurses' incomes as well as the relative popularity of the various income levels are, to a degree, perceived at a glance. Which is a better presentation of the data? Would a line chart be preferred to either of these two? A bar chart is clearly preferable to a frequency polygon when dealing with an artificial frequency distribution. Because the bases of each bar are of equal length and the heights are frequency measures, the areas of the bars are proportional to the frequencies. In Figure 2.10 the bar and the frequency polygon charts are plotted together. Note how the polygon underrepresents the true frequencies in the first, last, and the modal income intervals.

There are adjustments to the frequency polygon or the histogram that lead to a graph that may be preferable to either; they center on the notion that a smooth curve may be the best representation of the distri-

Figure 2.8 Nurses' Salaries (Bar Chart or Histogram)

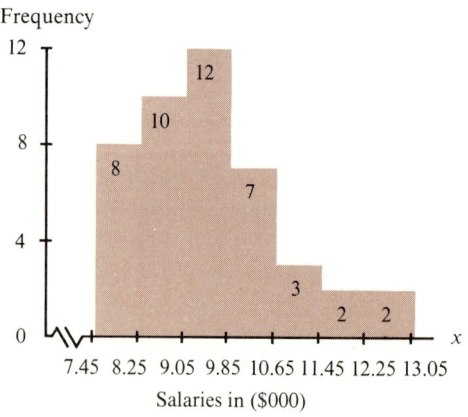

Descriptive Statistics

Figure 2.9 Nurses' Salaries (Frequency Polygon Chart)

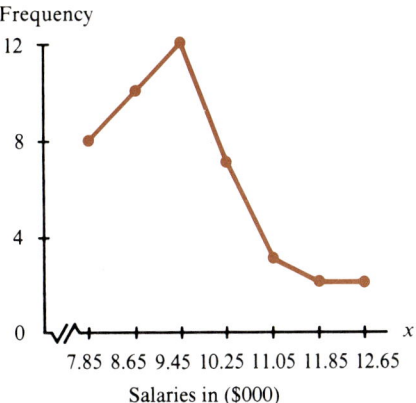

bution. This curve would be fitted so that the area under the curve over a given interval is equal to the area of the bar chart rectangle for that interval. Sophisticated techniques, not to be covered here, exist for this sort of curve fitting. As an example, see Figure 2.10, where we have simply superimposed a judgment fit on the bar chart and the frequency polygon.

The line chart (see Figure 2.6) is obviously preferable for the natural frequency distribution. Both the histogram and the frequency

Figure 2.10 Histogram and Frequency Polygon Comparison with Smooth Curve Frequency of Nurses' Salaries

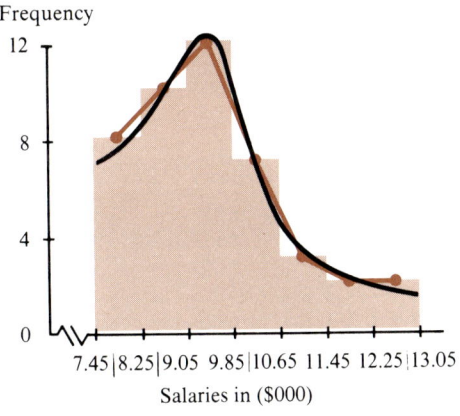

Graphical Presentation of Data

polygon would imply that scores other than integers could exist. The reverse argument ranks the bar chart over the line chart for artificial distributions, i.e., the line chart would imply that only integer values are possible. Variables such as nurses' salaries can be measured to five or six significant digits; from the standpoint of practical graphical presentation, a nurse's salary might exist at any point along the continuum.

 ## 2.5 Relative Frequencies, Pie Charts, and Population Trees

Once a frequency table has been calculated for a set of data, the relative frequency distribution is obtained by calculating the proportion of sample values that falls in each of the respective classes. If in Example 2.4.2 (Nurses' Salaries) each of the frequencies is divided by 44, the sample size, we get the proportion of cases in each salary group. The result of this operation is the relative frequency distribution presented in Table 2.4.

If the relative frequency distribution were graphed as a histogram, the shape would be identical to that of the frequency histogram of the actual frequencies. The only difference would be the scale markings on the vertical axis. The sum of the relative frequencies should, of course, be unity. Rounding errors can easily create, as in our example, an error in the last digit. Without serious risk, we could arbitrarily adjust the largest relative frequency or frequencies in order for the summation to be 1.000. In the above example, we might increase the third relative frequency from .273 to .274.

Figure 2.11, which is the same as Figure 2.3, is reproduced here as an example of a relative frequency bar chart with the relative frequencies shown as percentages rather than decimal fractions as they are in Table 2.4. In Figure 2.11 we have a relative frequency bar chart for

Table 2.4 Relative Frequency Distribution

Income Categories	Frequency	Relative Frequency
7.5–8.2	8	0.182
8.3–9.0	10	0.227
9.1–9.8	12	0.273
9.9–10.6	7	0.159
10.7–11.4	3	0.068
11.5–12.2	2	0.045
12.3–13.0	2	0.045
Totals	44	0.999

Figure 2.11 The Nation's Transportation in 1979
Percentage of Total Passenger-Miles (Hypothetical Nation)

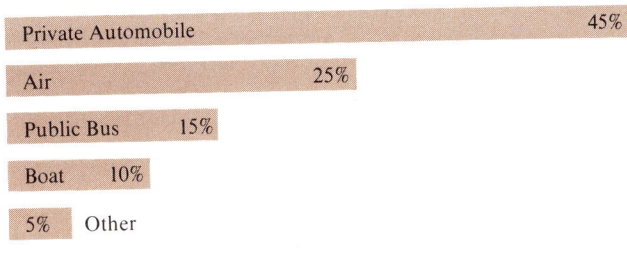

what is called a *categorical* or *attribute* variable rather than a variable that is counted or measured. (See the discussion in Chapter 1 on this point.) The modes of transportation are listed in five nonoverlapping exhaustive categories with the percentages and the length of the bars in terms of passenger-miles. Earlier it was stated that there should be no gaps in a bar chart presentation; we were, however, discussing frequency distributions. The gaps in Figure 2.11 seem effective and are perhaps even necessary when charting categorical data.

A percentage distribution of categorical data is also effectively presented in what is called a *pie chart*. For example, see Figure 2.12, a reproduction of Figure 2.2. The percentage of stockholders in the several groups is simply shown as a proportionate cut of the pie. A protractor is needed to construct a pie chart. Recall that there are 360 degrees in a circle; each slice of the pie is cut to equal its corresponding percentage of 360 degrees.

As a final comment, notice the unequal income intervals on the income axis in Figure 2.13, which is a reproduction of Figure 2.1. The income axis is not truly scaled, and, in one sense, this becomes a categorical presentation rather than a true frequency distribution. In the earlier discussion of the frequency distributions, the problem of unequal intervals on the variable axis (or *x*-axis) was not considered. What would be the "honest" or "fair" way to construct a bar chart in the case of unequal interval lengths? To illustrate what should be done, let us arbitrarily modify the data in Figure 2.13. Assume total responses of 85 with no nonresponses. The percentages in the various income classes will be directly translated to number of responses as given in Table 2.5. Also, change the one endpoint from $16,000 to $15,000.

The smallest common divisor of the income classes is $5,000, that from $10,000 up to $15,000. The frequency axis will be scaled to fit the frequency of 21 in this interval. Let's assume that incomes under $10,000 may be as low as 0. There are now two intervals of $10,000 on either side of the $5,000 interval, one interval of $15,000, and one open-ended interval. Assume the midpoint of the open-ended interval

Figure 2.12 Occupations of Individuals Owning Stock in Utah Power and Light Company

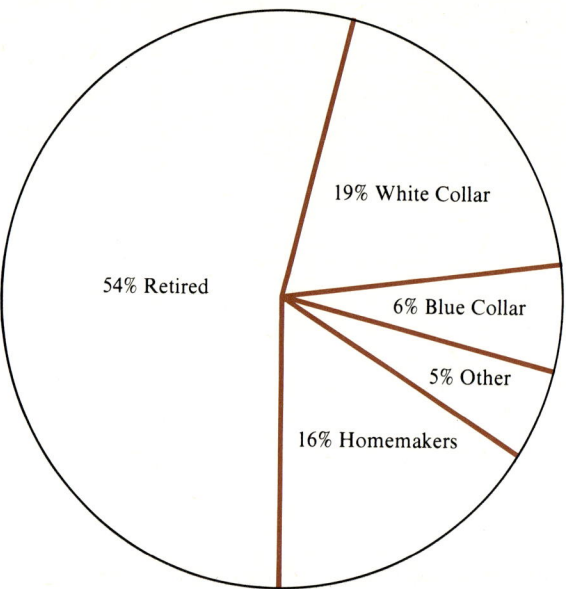

SOURCE: Utah Power and Light Company, *Who Owns Utah Power?* (Salt Lake City, 1977).

Figure 2.13 Family Incomes of Common Shareholders of Utah Power and Light Company

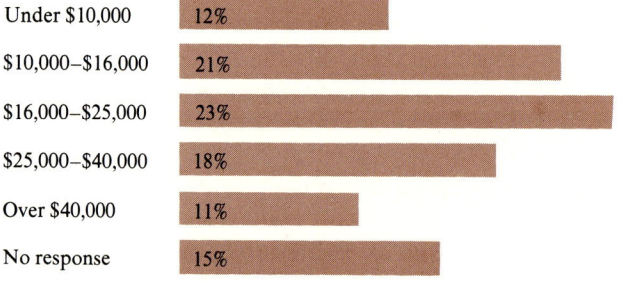

SOURCE: Utah Power and Light Company, *Who Owns Utah Power?* (Salt Lake City, 1977).

Descriptive Statistics

Table 2.5 Adjusted Income Groupings of Utah Power and Light Company Stockholders

Income Class	Frequency f
under $10,000	12
10,000–15,000	21
15,000–25,000	23
25,000–40,000	18
40,000 and over	11
	$n = 85$

is $50,000, and that this is a $20,000 interval. The frequencies in the $10,000 intervals must be scaled at ½ value, those in the $15,000 interval at ⅓ value, and those in the $20,000 class at ¼ value in order that the area of the bars fairly represent the frequencies; i.e., instead of the interval from 0 to $10,000 having a frequency of 12, there would be 2 intervals, one from 0 to $5,000 and the other from $5,000 to $10,000, with a frequency of 6 each. See the bar chart in Figure 2.14 for the implementation of what we have discussed. As with Figure 2.10 we have also sketched in a smooth judgment fit curve. The income picture in Figure 2.14 is quite different from that perceived in Figure 2.13. Admittedly there were adjustments and assumptions in the translation

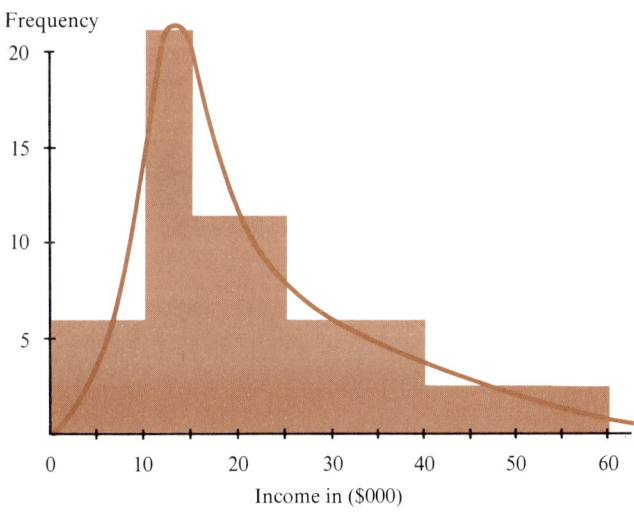

Figure 2.14 Family Incomes of Common Shareholders of Utah Power and Light Company

Source: Utah Power and Light Company, *Who Owns Utah Power?* (Salt Lake City, 1977).

Relative Frequencies, Pie Charts, and Population Trees

Figure 2.15 Number of Students by Class and by Sex, Fall Term, 1978

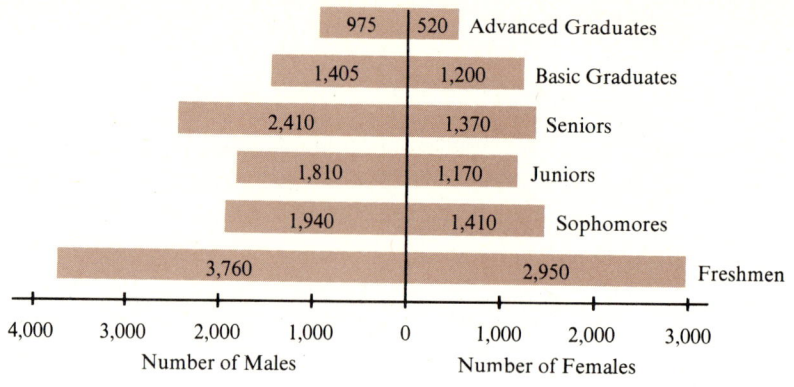

from 2.13 to 2.14, but with the exception of ignoring the 15 percent nonresponse, they led to relatively minor changes.

Another interesting and effective variant of the bar chart is the *population tree*. This scheme is used to present the age-specific and sex-specific frequencies of populations. This kind of chart is presented in Figure 2.15 with hypothetical data from Snake River University given in Table 2.6. The data are sex-specific, but, in lieu of age-specific data, we have used class-specific data in order to illustrate the flexibility of this type of pyramid chart.

Table 2.6 Enrollments by Class and by Sex, Snake River University, Fall Term, 1978

Class	Male	Female	Total
Freshmen	3760	2950	6710
Sophomore	1940	1410	3350
Junior	1810	1170	2980
Senior	2410	1370	3780
Basic Graduates	1405	1200	2605
Advanced Graduates	975	520	1495
Totals	12,300	8620	20,920

 ## 2.6 Charting Time Series

Time series are considered in some detail in Chapter 11. Inasmuch as many of the graphic presentations in the media are of time series, it seems appropriate to address the charting aspects at this point. The

Descriptive Statistics

measurement of some variable taken at intervals over time is called a *time series*. In Figure 2.4 we have a time series chart, a picture of the change in an economic variable over time. This graph could provide excellent support to a written article, but one should be critically aware of the fact that media charts and graphs are frequently legalistic; i.e., they are presented to support an argument. Notice in the chart that the vertical axis is "broken"—it is not scaled continuously from 0. The effect is, of course, to exaggerate the relative variability of the series. This is not to say that we should never break an axis in order to be scientific in our presentations. If our main interest were in the absolute change comparisons in the variable through time, we might decide to break the axis. We should, however, be very much aware of this as we evaluate presentations by other people and also as we make a decision as to how to present our own data.

To further emphasize this point, we call your attention to Figures 2.16 and 2.17, which are two charts of the same hypothetical data, the

Figure 2.16 Full-time Equivalent Enrollments, Snake River University

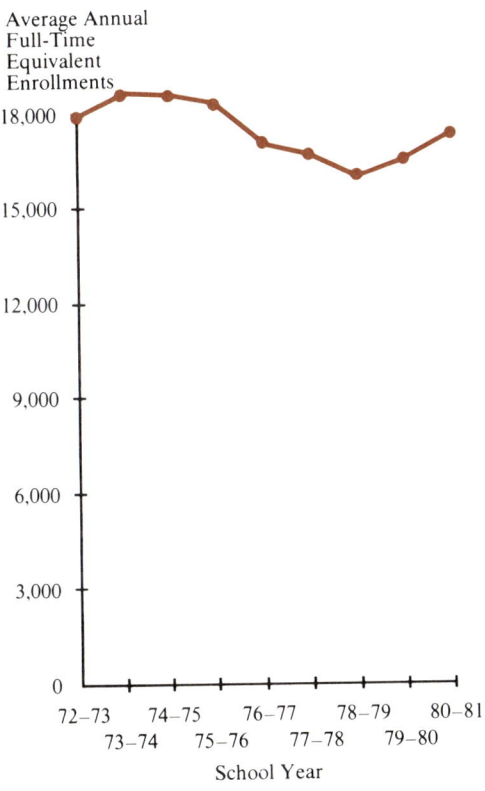

Figure 2.17 Full-time Equivalent Enrollments, Snake River University

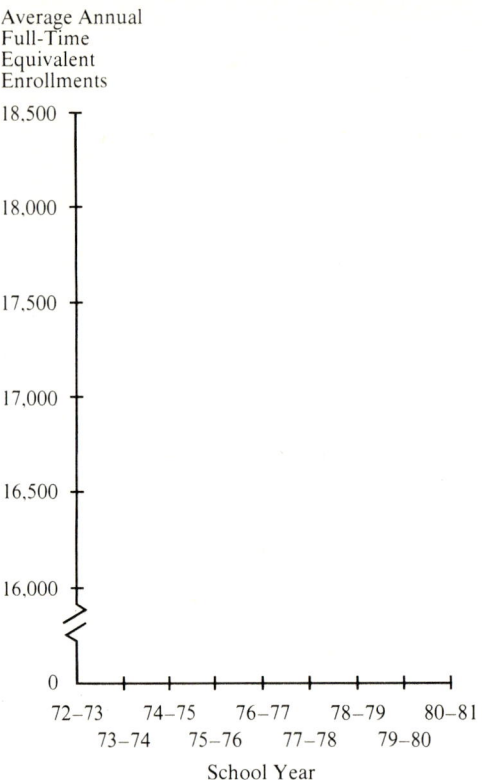

annual average full-time equivalent (FTE) enrollments at Snake River University from academic year 1972–73 through 1980–81. Table 2.7 presents the same data in tabular form. Figure 2.16 shows a rather mild wave working its way through time. This is an "honest" presentation of the percentage or relative change in enrollments. Without careful analysis, one might conclude from Figure 2.17 that this enrollment wave is very violent. Figure 2.17, however, shows a better picture of the absolute enrollment changes. Which is better? We take no position on this point. It depends on what one wants to show. In 2.17 we are being scientifically accurate by showing very clearly the break in the vertical axis.

A formal measure of skewness is not needed for our work in this textbook, but it is helpful to have an understanding of the term. A distribution is symmetrical about the mean, if for every x-value above the mean there is a mirror image point below the mean. A distribution

Table 2.7 Average Annual Full-Time Equivalent Enrollments, Snake River University

Academic Year	FTE
1972–73	17,930
1973–74	18,550
1974–75	18,470
1975–76	18,250
1976–77	17,310
1977–78	17,050
1978–79	16,480
1979–80	16,710
1980–81	17,420

that is asymmetrical is generally referred to in terms such as *skewed, positively skewed, skewed to the right, negatively skewed,* or *skewed to the left*. You should have a general idea of what is meant by these terms. For example, the distribution of nurses' income is positively skewed: it tails off to the right (see Figure 2.10). Figure 2.18 has simple

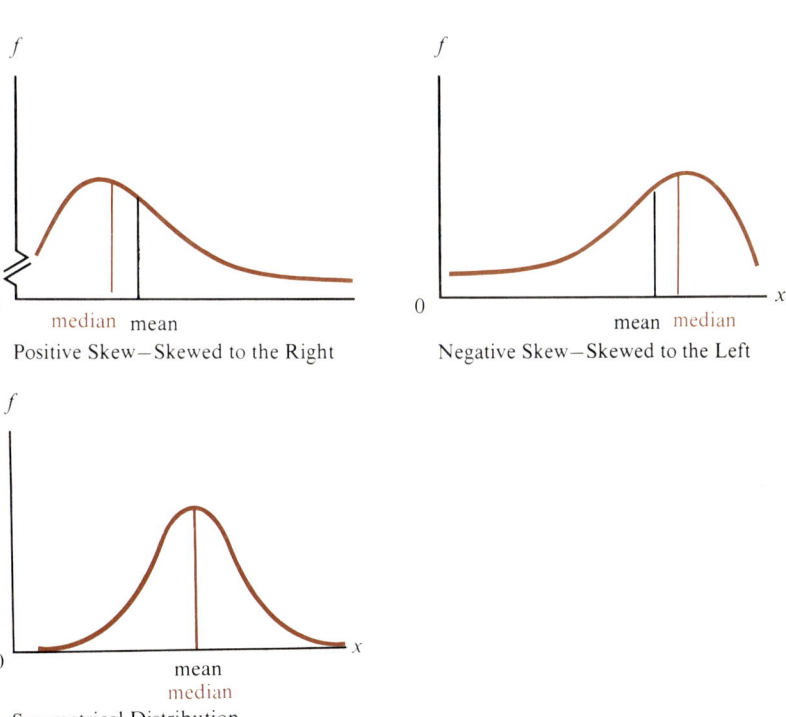

Figure 2.18 Types of Distribution

Charting Time Series

sketches of positive skewness and of negative skewness, along with a symmetrical distribution. Note that the extreme values for a positively skewed distribution are relatively large, and thus they cause the mean value to be larger than the median. On the other hand, the extreme values of a negatively skewed distribution are relatively small, causing the mean value to be less than the median value.

2.7 Calculating Descriptive Measures Using Grouped Data

In the age of computers and sophisticated pocket-size calculators, the notion of grouping data in order to increase the efficiency of calculations is considered by some to be of questionable value. This would have to be considered an optional topic. There are, however, situations where the analysis from the grouped data might be preferable to that from the raw data. We have already seen the value of a frequency distribution in the visual analysis and presentation of the data. The mode, the median, and other percentiles of a population are perhaps more easily estimated from the grouped sample data than from the raw data. It is not always convenient and sometimes not possible to use a computer; in this situation, grouping data for large samples significantly increases the efficiency with which means, standard deviations, etc. are calculated. (See Appendix 2C for the most efficient method and related formulas.) Several examples are discussed to develop the ideas and necessary adjustments to the formulas.

Example 2.7.1

Refer to Example 2.4.1 where 50 households were sampled with the following results: There were 8 households with no children, 19 with 1 child, 12 with 2 children, 7 with 3 children, 2 with 4 children, 1 with 6 children, and 1 with 11 children. Let x be the number of children per household. Find the mean and standard deviation of x. Let f_i represent the frequency in the ith class. In lieu of adding the 50 numbers, multiplication may be used to shortcut the addition problem; i.e.,

$$\sum_{j=1}^{50} x_j = \sum_{i=0}^{11} x_i f_i$$

where

x_j = the jth sample observation
x_i = the ith class number of children
f_i = the number of sample observations in the ith class.

The tabular organization of this calculation is presented in Table 2.8. The mean is a special case of the weighted mean discussed earlier:

$$\bar{x} = \frac{\sum_{i=0}^{11} x_i f_i}{n} = \frac{89}{50} = 1.78.$$

Note that

$$\sum_{i=1}^{11} f_i = 50,$$

the sample size, and $\Sigma x_i f_i = 89$, the total number of children in the sample.

For frequency distributions, the use of the computational form of the standard deviation formula is easier and more efficient than the use of the definitional form. With frequencies, the formula is modified as

$$s = \sqrt{\frac{\sum_{i=0}^{11} x_i^2 f_i - n\bar{x}^2}{n-1}}.$$

Each of the x_i^2 is also weighted by its frequency. In organizing the problem, it is not necessary to calculate an x^2 column, if we observe that $x(xf) = x^2 f$. Calculating the standard deviation, we get

$$s = \sqrt{\frac{319 - 50(1.78)^2}{49}} = 1.81.$$

Table 2.8 Calculations for Mean and Standard Deviation of Natural Frequency Distribution

x	f	xf	$x^2 f$
0	8	0	0
1	19	19	19
2	12	24	48
3	7	21	63
4	2	8	32
5	0	0	0
6	1	6	36
7	0	0	0
8	0	0	0
9	0	0	0
10	0	0	0
11	1	11	121
Totals	50	89	319

Example 2.7.2 For a second and a more comprehensive example, consider again the data on nurses' salaries presented in Example 2.4.2. Calculate the median, several other percentiles, the mode, the mean, and the standard deviation.

In order to calculate the mean and standard deviation for the artificial frequency distribution, we make the assumption that the scores within each interval average to the midpoint of the interval. For any specific interval, there could well be a significant error between the actual average of the scores within the interval and the midpoint; we plan and hope for compensating errors. Typically, frequencies increase to the mode and then decrease. It seems reasonable to think that midpoints would tend to underestimate the interval averages below the mode, and this would tend to be offset by overestimates above the mode. With this assumption, and letting x_i be the midpoint of the ith interval, the problem, the calculations, and the formulas are identical to those given above for the natural frequency distribution; we have assumed, in essence, that all scores in each interval are exactly at the midpoint of the interval.

Formulas for Mean and Standard Deviation of Group Data

The formulas for calculating the mean and standard deviation from artificially grouped data are shown below. (Note the use of the symbol $\cong$ and its implication for the measures calculated from an artificial frequency distribution. This symbol means "is approximately equal to"; as already mentioned, potential error is introduced by grouping.)

$$\bar{x} \cong \frac{\sum_{i=1}^{k} x_i f_i}{n}$$

and

$$s \cong \sqrt{\frac{\sum_{i=1}^{k} x_i^2 f_i - n\bar{x}^2}{n-1}}$$

where

x_i = the midpoint of the ith interval
f_i = the frequency or number of observations in the ith interval
k = the number of intervals
n = the total number of observations in the sample.

In Table 2.9 we have set up the format and done the calculations for the

Table 2.9 Frequency Distribution Calculations

Midpoint x	f	xf	$x^2 f$
7.85	8	62.80	492.98
8.65	10	86.50	748.23
9.45	12	113.40	1071.63
10.25	7	71.75	735.44
11.05	3	33.15	366.31
11.85	2	23.70	280.85
12.65	2	25.30	320.04
Totals	44	416.60	4015.48

a. $\bar{x} = \dfrac{\sum_{i=1}^{n} x_i f_i}{n} = \dfrac{416.60}{44} = \$9.47$ thousands.

b. $s_x = \sqrt{\dfrac{\sum_{i=1}^{n} x_i^2 f_i - n\bar{x}^2}{n-1}} = \sqrt{\dfrac{4015.48 - 44(9.47)^2}{43}} = \$1.28$ thousands.

nurses' salaries problem. (If a calculator or a computer is not available, significant time saving with an increased chance of accuracy can be accomplished by introducing a new variable; see Appendix 2C at the end of this chapter for a discussion of this idea.)

In order to estimate the median and other percentiles, the frequency distribution from Figure 2.7 is reproduced in Table 2.10 with an additional column, the cumulative frequency, F. The cumulation is from lowest to highest income categories and is called a *less than* cumulative frequency. There are F_i incomes less than the upper endpoint of the ith interval. For example, the cumulative frequency for the interval 9.05–9.85 is 30; therefore, we know there are 30 incomes in the sample that are less than 9.85 thousand dollars (\$9,850). The cumula-

Table 2.10 Cumulative Frequency Calculations

Interval Endpoints	Frequency f	Cumulative Frequency F
7.45–8.25	8	8
8.25–9.05	10	18
9.05–9.85	12	30
9.85–10.65	7	37
10.65–11.45	3	40
11.45–12.25	2	42
12.25–13.05	2	44
	$n = 44$	

tive frequency is useful for both the algebraic and graphical estimates of the median and/or any other desired percentile.

From the ungrouped 44 incomes, the median would be calculated as the average of the twenty-second and twenty-third score in an array. The average value of the twenty-second and twenty-third of the sequentially ordered sample values is 9.45, or $9,450. The calculation of the sample median from the frequency distribution requires a slightly different approach. The median is obtained from a frequency distribution by estimating the income where the cumulative frequency is 22, exactly 50 percent of 44. The twenty-second score is located in the interval 9.05–9.85. This is evident from a simple inspection of the cumulative frequency, F_i. Using the bar chart assumption, i.e., incomes are uniformly distributed throughout the interval, we simply interpolate into the critical interval. There are 18 incomes less than 9.05. Of the 12 scores in the third interval, we need 4 more to get to 22; 4/12 or 1/3 of the interval length, .8, is added to 9.05.

Median $= 9.05 + \frac{1}{3}(.8) = 9.32$.

The general formula would be:

Median $= L_i + \dfrac{.5n - F_{i-1}}{f_i} c$

where

$L_i =$ the lower endpoint of the critical interval
$F_{i-1} =$ the cumulative frequency up to the critical interval
$f_i =$ the frequency in the critical interval
$c =$ the interval length.

We recommend that the student try to understand the notion of interpolation, rather than using the formula for the median (the fiftieth percentile). If you understand the idea, then it is possible, by the same method, to estimate any desired percentile. For example, to find the seventy-fifth percentile (or the third quartile, Q_3) we estimate the income at the cumulative thirty-third score (3/4 × 44 = 33). The thirty-third score is in the fourth interval; it is 3/7 of the way into the fourth interval. The seventy-fifth percentile or third quartile is calculated as follows:

$Q_3 = 9.85 + \dfrac{3}{7}(.8) = 10.19$.

It is estimated that 75 percent of the nurses make less than $10,190 per year and, of course, 25 percent have annual salaries in excess of this amount.

A graphical method that is very useful if a number of percentiles are needed—in a hurry and not too accurately—is a graph of the cumulative function. This is called the *ogive,* or for the F_i, in Figure 2.19, the "less-than" ogive.

The F_i are plotted at the upper endpoints of the ith interval and are connected by straight lines. There are no scores below the lowest endpoint, 7.45, and the ogive starts with a zero at that point. The cumulative graph is a nondecreasing function of x, starting at zero and climbing to n (44 in our example) which is plotted at the upper endpoint of the highest interval.

The median may be located graphically by starting at 22 on the F-axis, going horizontally to the ogive, then reading down to the median on the x-axis. This is conceptually the same as the algebraic method; any difference in the result would be due to the problem of reading the graph accurately. Any percentile can be located in a similar manner. We have located the median and Q_3 on the ogive chart in Figure 2.19.

The mode is traditionally estimated as the midpoint of the interval of greatest frequency—the modal interval. In the example there are 12 salaries between 9.05 and 9.85, and the mode would therefore be 9.45. The most popular salary is estimated to be $9,450. A modification of this simple rule would normally give a better estimate of the population mode. It is a matter of interpolating within the interval of greatest frequency. The interpolation is proportional to the differences between the modal frequency and the two adjacent frequencies. In our example these differences are: (1) $12 - 10 = 2$; and (2) $12 - 7 = 5$. The modal

Figure 2.19 Less-than Ogive

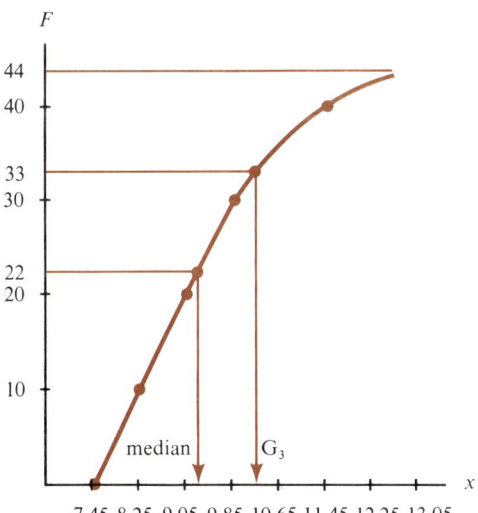

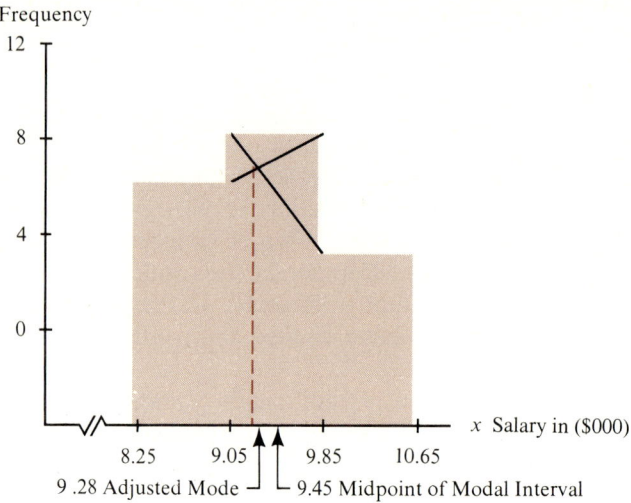

Figure 2.20 Graphical Adjustment for Mode Nurses' Salaries

interval is divided by the ratio of 2 to 5, and the adjusted mode would equal 9.05 + 2/7(0.80) which equals 9.28. A clever construction, illustrated in Figure 2.20, yields the same interpolation. Any difference between the graphical result and the algebraic answer would be due to the quality of the graph paper and the skill of the draftsman.

We have considered only the case where the frequency distribution has equal-length intervals. Minor problems arise if intervals are of unequal length and/or the top interval is open-ended. Most of the distributions of family income are presented with unequal intervals and an open-ended upper interval. For both the mean and the standard deviation, the midpoint of any closed interval is used and assumed to be the average of all numbers in the interval. One must know (or have an estimate of) the mean value of all scores in the open-ended interval. There is no added problem in finding the median or most percentiles. However, if a percentile is high enough to fall in the open-end interval, then one would need to use some judgment.

2A Appendix: The Summation Symbol (Σ)

The notation

$$\sum_{i=1}^{n} x_i$$

simply means to sum or add the x's from x_1 through and including x_n, and this is read as the "summation of x-sub-i, i going from 1 to n." The letter i is the index of summation, the identification mark. If, in the problem under discussion, it is clear that all of the x's are being summed, as it usually is when calculating the mean, the index is often deleted and we write

Σx.

The complete notation is convenient and useful in several ways:

1. Sums can be truncated; for example

   $$\sum_{i=3}^{5} x_i \text{ calls for } x_3 + x_4 + x_5.$$

2. The index itself can call for an operation; e.g.,

   $$\sum_{i=1}^{5} i \text{ calls for } 1 + 2 + 3 + 4 + 5 = 15.$$

3. And as another example,

   $$\sum_{i=1}^{3} x_i^i = x_1 + x_2^2 + x_3^3.$$

The laws or rules of operation with the summation symbol follow directly from the several laws learned in algebra and arithmetic courses:

1. $\sum_{i=1}^{n} c x_i = c \sum_{i=1}^{n} x_i$ where c is an arbitrary constant.

2. $\sum_{i=1}^{n} (x_i \pm y_i) = \sum_{i=1}^{n} x_i \pm \sum_{i=1}^{n} y_i.$

3. $\sum_{i=1}^{n} c = nc.$

The first rule follows from the distributive law of algebra:

$$\sum_{i=1}^{n} cx_i = cx_1 + cx_2 + \cdots + cx_n = c(x_1 + x_2 + \cdots + x_n) = c\sum_{i=1}^{n} x_i.$$

The second rule follows from the associative law of algebra:

$$\sum_{i=1}^{n} (x_i \pm y_i) = (x_1 \pm y_1) + (x_2 \pm y_2) + \cdots + (x_n \pm y_n)$$
$$= (x_1 + x_2 + \cdots + x_n) \pm (y_1 + y_2 + \cdots + y_n)$$
$$= \sum_{i=1}^{n} x_i \pm \sum_{i=1}^{n} y_i.$$

The third law is proved by observing that for each value of i the constant term c is the same; thus:

$$\sum_{i=1}^{n} c = c + c + \cdots + c = nc.$$

The following observations may be of some help in the understanding of this notation:

1. The $\left(\sum_{i=1}^{n} x_i\right)^2$

    means to sum the x's and square the sum; whereas,

    $$\sum_{i=1}^{n} x_i^2$$

    calls for squaring each x_i and summing the squares. These are generally not equal.

2. $\sum_{i=1}^{n} x_i \sum_{i=1}^{n} y_i,$

    by the same token, is different from and in general does not equal

    $$\sum_{i=1}^{n} x_i y_i.$$

2B Appendix: Some Algebraic Proofs

1. Prove that

$$\sum_{i=1}^{n} (x - \bar{x}) = 0.$$

Descriptive Statistics

$$\sum_{i=1}^{n}(x_i - \bar{x}) = \sum_{i=1}^{n} x_i - \sum_{i=1}^{n} \bar{x}. \quad \text{(Rule II, Appendix 2A.)}$$

$$\sum_{i=1}^{n} x_i - \sum_{i=1}^{n} \bar{x} = \sum_{i=1}^{n} x_i - n\bar{x}. \quad \text{(Rule III, Appendix 2A.)}$$

$$\sum_{i=1}^{n} x_i = n\bar{x}. \quad \text{(From the definition of } \bar{x} \text{ we have } \bar{x} = \Sigma x_i / n \text{; solve algebraically for } \sum_{i=1}^{n} x_i.\text{)}$$

$$\sum_{i=1}^{n} x_i - n\bar{x} = n\bar{x} - n\bar{x} = 0.$$

2. Prove that

$$\sum_{i=1}^{n}(x_i - \bar{x})^2 = \sum_{i=1}^{n} x_i^2 - \frac{\left(\sum_{i=1}^{n} x_i\right)^2}{n} = \sum_{i=1}^{n} x_i^2 - n\bar{x}^2.$$

$$\sum_{i=1}^{n}(x_i - \bar{x})^2 = \Sigma(x_i^2 - 2x_i\bar{x} + \bar{x}^2).$$

(Square the expression within the summation symbol.)

$$\sum_{i=1}^{n}(x_i^2 - 2x_i\bar{x} + x^2) = \sum_{i=1}^{n} x_i^2 - \sum_{i=1}^{n} 2x_i\bar{x} + \sum_{i=1}^{n} \bar{x}^2$$

(Rule II.)

$$\sum_{i=1}^{n} x_i^2 - \sum_{i=1}^{n} 2x_i\bar{x} + \sum_{i=1}^{n} \bar{x}^2 = \sum_{i=1}^{n} x_i^2 - 2\bar{x}\sum_{i=1}^{n} x_i + n\bar{x}^2.$$

(Rule I and Rule III.)

$$\sum_{i=1}^{n} x_i = n\bar{x}. \quad \text{From the definition of } \bar{x}.$$

$$\sum_{i=1}^{n} x_i^2 - 2\bar{x}\sum_{i=1}^{n} x_i + n\bar{x}^2 = \Sigma x_i^2 - 2\bar{x}(n\bar{x}) + n\bar{x}^2 = \Sigma x_i^2 - n\bar{x}^2.$$

Thus, we have

$$\sum_{i=1}^{n}(x_i - \bar{x})^2 = \sum_{i=1}^{n} x_i^2 - n\bar{x}^2.$$

Substituting the definition of $\bar{x}$,

$$n\bar{x}^2 = n\left(\frac{\sum_{i=1}^{n} x_i}{n}\right)^2 = \frac{\left(\sum_{i=1}^{n} x_i\right)^2}{n}.$$

Therefore,

$$\sum_{i=1}^{n}(x_i - \bar{x})^2 = \sum_{i=1}^{n} x_i^2 - n\bar{x}^2 = \sum_{i=1}^{n} x_i^2 - \frac{\left(\sum_{i=1}^{n} x_i\right)^2}{n}.$$

The proof of the latter proposition provides us with the computational formula for the sample variance, s^2. The sample variance is defined as

$$s^2 = \frac{\sum_{i=1}^{n}(x_i - \bar{x})^2}{n-1}.$$

Substituting the expression for

$$\sum_{i=1}^{n}(x_i - \bar{x})^2$$

obtained in proof 2 gives us the computational formula for the sample variance:

$$s^2 = \frac{\sum_{i=1}^{n} x_i^2 - n\bar{x}^2}{n-1}.$$

2C Appendix: Change of Scale for a Frequency Distribution, or "Coding"

The introduction of a convenience variable when calculating the mean and standard deviation from a frequency distribution can be a great time saver, particularly if a hand calculator is not available. Also, the chances of making an arithmetic mistake are considerably reduced because the products and sums will be relatively small integers. This procedure is often called "coding," but it amounts to a simple and arbitrary change of origin and scale on the x-axis—a *linear transformation*. We will consider only the case of equal-length intervals.

Let v be the new scale variable. Set $v = 0$ corresponding to some midpoint of one of the intervals. The placement of this new origin does not require good judgment. It is an arbitrary choice; however, if the new origin is placed at the midpoint closest to the mean, the products and sums will be smaller than those for any other choice. From this new origin, let v take on positive integral values, 1, 2, 3, etc., to correspond to increasing midpoints of the original x-variable, and -1, -2, etc., to correspond with decreasing midpoints. The equation that describes this transformation will always be of the form

$$x = x_o + cv$$

where

x_o = the midpoint corresponding to $v = 0$, the origin of the scale variable, v
c = the length of the interval
v = the change-of-scale variable.

For this type of transformation, it can be shown that

$$\bar{x} = x_o + c\bar{v}$$

and

$$s = cs_v$$

where

$\bar{x}$ = the sample mean
$\bar{v}$ = the mean of the scale variable, v
s = the sample standard deviation
s_v = the standard deviation of the scale variable, v.

The mean and standard deviation of the scale variable, v, are calculated from the following formulas:

$$\bar{v} = \frac{\sum_{i=1}^{k} v_i f_i}{n}$$

$$s_v = \sqrt{\frac{\sum_{i=1}^{k} v_i^2 f_i - n\bar{v}^2}{n - 1}}$$

where

v_i = the value of the scale variable for the ith interval
f_i = the frequency or number of observations in the ith interval
k = the number of intervals
n = the total number of observations in the sample.

The nurses' salary example, Example 2.7.2, is used to illustrate the "coding" technique. The first two left-hand columns of Table 2.10 are reproduced from Table 2.9. The third interval was arbitrarily selected as the origin interval or the interval where $v = 0$ because it has the largest frequency. Let us stress again that this is an arbitrary criterion for selecting the origin interval and that the calculated values for $\bar{x}$ and s will not change with a different origin interval. The scale variable, v, is given in Table 2.10 along with the calculation of

$$\sum_{i=1}^{k} v_i f_i$$

and

$$\sum_{i=1}^{k} v_i^2 f_i.$$

Using those values in the above formulas, $\bar{v}$ and s_v can be calculated:

$$\bar{v} = \frac{\sum_{i=1}^{k} v_i f_i}{n} = \frac{1}{44} = .023$$

$$s_v = \sqrt{\frac{\sum_{i=1}^{k} v_i^2 f_i - n\bar{v}^2}{n-1}} = \sqrt{\frac{111 - 44(.023)^2}{43}} = 1.606.$$

Table 2.10 Frequency Distribution Calculations Using Transformed Variable

Interval Endpoints	Interval Midpoints	Scale Variable v_i	Frequency f_i	$v_i f_i$	$v_i^2 f_i$
7.45–8.25	7.85	−2	8	−16	32
8.25–9.05	8.65	−1	10	−10	10
9.05–9.85	9.45	0	12	0	0
9.85–10.65	10.25	1	7	7	7
10.65–11.45	11.05	2	3	6	12
11.45–12.25	11.85	3	2	6	18
12.25–13.05	12.65	4	2	8	32
		Totals	44	1	111

The calculated values of $\bar{v}$ and s_v have no real world meaning and therefore should not be reported as statistical measures or results. They are intermediate values from which the sample mean, $\bar{x}$, and the sample standard deviation, s, can be calculated using the linear transformation formulas given above:

Descriptive Statistics

$$\bar{x} = x_o + c\bar{v} = 9.45 + .8\bar{v} = 9.45 + .8(.023) = \$9.47 \text{ thousands}$$
$$s = cs_v = .8s_v = .8(1.606) = \$1.28 \text{ thousands.}$$

These are the same values for the sample mean and sample standard deviation that were obtained in Table 2.8. This is what we expected. It can be shown that no error is introduced by the coding procedure. You should compare the formulas used in Table 2.8 with the formulas for $\bar{v}$ and s_v used here. They are essentially the same with v_i substituted for x_i.

Problems

1. a. Find the arithmetic mean and the standard deviation of the following set of numbers: $(-1, 3, 0, 0, 2, 5, 3, 1, 0, \text{ and } -3)$.
   b. Compute the mean and variance of the first five positive integers.

2. Find the mean, median, range, and standard deviation of the following set of numbers: $(8, -4, 0, 7, 3, -5, 12)$.

3. Mr. Peters, the owner-manager of Peters Fish & Chips, had recently received complaints from some of his customers regarding how long they had to wait for service at the drive-in window. Picking the busiest time of day, Mr. Peters recorded the following times in minutes, from arrival to service completion, for 10 customers: 3.2, 15.1, 10.9, 8.3, 8.8, 2.5, 9.4, 12.7, 18.4, and 6.6. Calculate the mean waiting time and the standard deviation of waiting time.

4. The following data are from a random sample of twelve revolving accounts at the local department store for May, 1979, billings. (All rounded to the nearest dollar.)

Account	T-Ending May Balance	X-Amount Paid During Month	Y-Amount Charged During Month
1	318	60	14
2	0	39	0
3	73	10	24
4	165	100	120
5	158	0	93
6	492	57	269
7	210	25	46
8	54	10	0
9	119	20	0
10	283	14	194
11	177	150	88
12	256	50	37

Find:

a. (i). $\sum_{i=1}^{12} T_i$; (ii). $\sum_{i=1}^{12} X_i$; (iii). $\sum_{i=1}^{12} Y_i$;

b. $\overline{T}$, $\overline{X}$, and $\overline{Y}$;

c. the standard deviation of each of the three variables.

5. Of the 45-member traveling squad of the Hong Kong Flyers, the 25 linemen average 260 pounds, and the 20 backfield and wide receivers average 190 pounds. What is:
   a. the average weight of the whole squad;
   b. the total weight of the squad?

6. In a certain class, a teacher grades 30 percent on weekly test average, 30 percent on daily work and 40 percent on the final exam. Joe had 83 percent, 62 percent and 94 percent respectively on three items. What was his average percent?

7. Fifty motors were randomly selected, and a pull test of the endshields was performed on each motor to test the strength of the epoxy holding the endshield. The strength was measured in pounds of pressure required to separate the endshield from the motor casing. The following values were obtained from the 50 motors:

521	479	512	607	436
543	484	585	507	485
493	560	389	505	661
510	519	653	522	493
401	534	500	489	419
534	551	481	677	648
567	465	642	558	531
429	573	452	398	526
488	410	528	577	473
521	488	598	509	537

a. Construct a frequency distribution for these data.
b. Prepare a histogram from this frequency distribution.
c. Calculate the mean and standard deviation using the frequency distribution.

8. Consider the following data (in thousands of dollars) on total revenues in the several departments of Scrug's Drugs for the month of November, 1979: Drugs, $381; Toys, $196; Hardware, $78; Notions and Sundries, $239. Construct a pie chart showing the percentage distribution of revenues in the four departments.

9. Consider the following frequency distribution resulting from a study of the U.C. Bowling League. The variable, x, is time in

minutes to completion of a 3-game series between two 4-person teams.

Midpoint x	Frequency f
111.5	2
113.0	12
114.5	16
116.0	8
117.5	6
119.0	3
120.5	2
122.0	1

Find:
a. the sample mean of x.
b. the sample variance of x.
c. the sample median value of x.
Estimate:
d. the range of x.
e. the mode of x.

10. a. Construct a bar chart for the data on bowling times in the previous problem. Find the adjusted mode of x using the construction method.
b. Plot a less-than ogive. Locate Q_1, Q_2, and Q_3.

11. Consider the following 2-year time series on the Consumer Price Index, Base Year 1967.[8]

	J	F	M	A	M	J	J	A	S	O	N	D
1977	175.3	177.1	178.2	179.6	180.6	181.8	182.6	183.3	184.0	184.5	185.4	186.1
1978	186.9	188.3	189.8	191.5	193.2	195.1	196.7	197.7	199.1	200.7	201.8	202.9

a. Plot this series on arithmetic graph paper. (Break the CPI axis to exaggerate the change.)
b. Calculate the 12 annual percentage changes; e.g., January, 1977, to January, 1978, is

$$\left[\frac{186.9}{175.3} - 1.000\right] 100 = 6.6\%$$

c. Plot this time series on arithmetic grid.

8. Bureau of Labor Statistics, *Consumer Price Index for All Urban Consumers* (Washington, D.C.: U.S. Government Printing House, 1979), m.p.

12. Construct a frequency table for the following scores:

```
20  24  37  46  52
21  25  37  47  53
21  27  37  47  54
22  30  38  48  56
23  31  40  49  56
23  32  42  51  56
24  32  42  51  59
24  33  44  52  60
```

   a. Calculate the sample mean, median, and standard deviation.
   b. Do these same calculations with the ungrouped data and compare the results.

13. Consider the following frequency distribution of age in months of a sample of undergraduate college students:

Midpoint x	Frequency f
216	4
231	50
246	22
261	15
276	10
291	4
306	3
321	0
336	1
351	1

   a. Calculate the arithmetic mean and standard deviation of x.
   b. Calculate the median. The twentieth percentile.
   c. Estimate the range.

14. Five coins were tossed 18 times with the following results, where x is the number of heads;

Midpoint x	Frequency f
0	1
1	3
2	6
3	4
4	0
5	4

   a. Compute the average number of heads, $\bar{x}$.
   b. Compute the standard deviation of x.
   c. Find the median.

Descriptive Statistics

15. Consider the following frequency distribution:

Midpoint x	Frequency f
117.5	3
123.5	5
129.5	12
135.5	10
141.5	28
147.5	14
153.5	8
	n = 80

    a. Calculate the mean.
    b. Calculate the median.
    c. Calculate the standard deviation.

16. a. Try to obtain the data from your school on fall quarter headcount by class and by sex. Plot a pyramid chart of these data.
    b. If you are unable to obtain data from your own school, following are hypothetical data from a small four-year college:

	Male	Female	Total
Freshmen	329	400	729
Sophomores	177	171	348
Juniors	150	111	261
Seniors	142	83	225
Total	798	765	1563

17. Some 40 students were asked to rate Professor K on his overall effectiveness as a statistics teacher. The ranking scale was a set of integers from 1 to 7 inclusive, 1 for "lousy," 7 for "fantastic"; numbers in between were described accordingly—e.g., 4 for "it's a toss-up between him and the textbook." The results were:

x (scale number)	1	2	3	4	5	6	7
f (frequency)	0	3	8	7	15	5	2.

    a. Find the mean, median, and mode of x.
    b. Find the standard deviation of x.

18. Sketch a distribution where:
    a. the mean and the mode are the same;
    b. the mean exceeds the median.

19. A farmer divides his herd of n cows among his 4 sons. The eldest son gets ½ the herd, the second son gets ¼, the third son ⅕, and the fourth son the balance, which is 7 cows. How large is n?

20. The arithmetic mean (average) of a set of 50 numbers is 38. If 2 of

the numbers, namely 45 and 55, are discarded, what will be the mean of the remaining set of numbers?

21. On an exam of n questions, a student answers correctly 15 of the first 20. Of the remaining questions, he answers ⅓ correctly. Altogether he gets ½ the questions correct. How large is n?

22. An attempt was made to determine the duck population around Sutters Swamp during the time of the year when there is practically no duck migration in or out of the local area. Researchers caught, tagged, and turned loose 100 ducks. A couple of days later, another random sample of 100 was taken. Of these, 12 had tags from the previous tagging. Estimate the population of ducks around Sutters Swamp.

23. Find the mean and variance of the first 7 positive odd integers, i.e., 1, 3, 5, etc.

24. Eleven numbers averaged 16. One of the numbers, 36, was considered to be extreme and was discarded. What is the average of the remaining 10 numbers?

25. The sum of a set of 16 numbers is 48. The sum of the squares of this set of numbers is 160. Find:
   a. the arithmetic mean;
   b. the variance;
   c. the standard deviation.

26. The mean of a set of 15 numbers is 2.80. The sum of the squares of this set of numbers is 180. Find:
   a. the sum of the set of numbers;
   b. the standard deviation of the set of numbers.

27. The sum of a set of 25 numbers is 450, and the sum of their squares is 8,250. What are the mean and standard deviation of the set of numbers?

28. a. The average of a set of 12 numbers is 8.50. What is the sum of the twelve numbers?
   b. The sum of the squares of a set of numbers (6 of them) is 55. The average of the set is 1.50.
      (i). What is the variance?
      (ii). What is the standard deviation?

29. Construct a line chart for the data in:
   a. Problem 14.
   b. Problem 17.

■ The remaining problems require material from the appendices.

30. The following hypothetical frequency distribution summarizes the results of 120 tests on operating time in hours, x, of the SP-197 transistor:

Midpoint x	Frequency f
1600	4
1675	12
1750	27
1825	30
1900	18
1975	8
2050	15
2125	3
2200	2
2275	1

    a. Compute the arithmetic mean and standard deviation of operating time, x. (Use a transformed variable.)
    b. Find the median, 90th percentile, and range.

31. If (4, 3, 0, −2, 1, −4) are values of x_i respectively for $1 \le i \le 6$, find:

    a. $\sum_{i=1}^{6} x_i$ b. $\sum_{i=2}^{4} x_i$ c. $\sum_{i=1}^{4} x_i$ d. $\left(\sum_{i=2}^{6} x_i + 1\right)^2$

32. If x_i takes on these respective values (1, −2, 0, 3, −1) and (1, −1, 1, 0, 2) are the respective values of y_i, find:

    a. $\sum_{i=2}^{4} (x_i - y_i);$ b. $\sum_{i=1}^{4} (y_i - 1);$

    c. $\sum_{i=1}^{2} x_i^2;$ d. $\left(\sum_{i=1}^{2} y_i\right)^2;$

    e. $\sum_{i=1}^{4} c$ if $c = 10$.

33. Below is a frequency distribution of the psychological scores of 130 students:

Midpoint x	Frequency f
36	3
48	5
60	8
72	22
84	26
96	30
108	21
120	8
132	4
144	2
156	1

Given also: For $v = 0$ corresponding to $x = 96$, the $\Sigma vf = -67$ and the $\Sigma v^2 f = 487$.
  a. Find the mean and the standard deviation.
  b. Plot the more-than and less-than ogive on the same chart and estimate the median from your charts.
  c. Estimate Q_1 and Q_3 from your chart.

34. $x_1 = 1$, $x_2 = 3$, $x_3 = 5$, etc., such that x_i is the ith odd positive integer. Find each of the following:

  a. $\sum_{i=1}^{5} x_i$; b. $\sum_{i=3}^{5} x_i^2$; c. $\sum_{i=3}^{5} (x_i = i)$; d. $\sum_{i=2}^{6} x_i - 10$.

35. Given the following data:

x_i	y_i
9	225
16	36
25	25
9	64
144	31
64	4
81	1

Evaluate:

  a. $\sum_{i=1}^{7} (\sqrt{x_i + y_i})$; b. $\sum_{i=1}^{7} (\sqrt{x_i y_i})$;

  c. $\sum_{i=1}^{7} \sqrt{x_i}$; d. $\sum_{i=1}^{7} \sqrt{y_i}$.

36. The geometric mean *(GM)* is the nth root of the product of the n numbers; i.e.,

$$GM = \sqrt[n]{x_1 \cdot x_2 \cdots x_n}.$$

Inspection and consideration of the formula shows clearly that all of the values under the radical should be positive; i.e., the *GM* is quite useless otherwise. If any one of the numbers is 0, the *GM* = 0. If all numbers are positive except for 1 negative, then the nth root of the negative number is either negative or imaginary according as n is odd or even. These are just two examples of the trouble we face with the *GM* if any of the measurements are 0 or negative. For you who understand logarithms, it is of interest to note that the log *GM* is the arithmetic mean of the logarithms of the *x*-values.

For the past 3 years, salaries at a certain firm have increased by 8.3 percent, 6.1 percent, and 12.3 percent. What is the average rate of increase over the past 3 years? Hint: Find *GM* of 1.083, 1.061, and 1.123.

37. The geometric mean may be a useful descriptive measure when one is working with a time series that exhibits exponential or geometric growth or decay. The series of dollar amounts in the compound interest problem is an example of geometric growth; if there is continuous compounding, we have exponential growth.

    At the end of 10 years a $1,000 deposit at compound interest showed a balance of $2,250. How much was in the account at the end of 5 years, the midpoint of the term of the investment?

38. The harmonic mean *(HM)* is the reciprocal of the average reciprocal; i.e.

$$HM = \frac{1}{\frac{\frac{1}{x_1} = \frac{1}{x_2} + \cdots + \frac{1}{x_n}}{n}}$$

$$= \frac{n}{\Sigma \frac{1}{x}}.$$

    A. J. drove 30 miles per hour for 1 hour and then 90 miles per hour for 1 hour. What was his average speed? We will define average speed as total distance divided by total time. So, it is a simple matter to see that A. J. drove 120 miles in 2 hours for an average speed of 60 miles per hour. Note that this is the arithmetic mean of the 2 rates.

    A. J. drove 30 miles per hour for 1 mile and 90 miles per hour for 1 mile. What was his average speed? The average speed is not quite so obvious as in the preceding example, and, if your intuition tells you 60 miles per hour, you are wrong, but not alone. Try the harmonic mean, then check by dividing total distance by time.

39. Find $\bar{x}$, the *GM*, and the *HM* of 2 and 8; of 9, 81, and 729.

CHAPTER 3

PROBABILITY

In this chapter we look at the several different but not inconsistent philosophical bases for developing probability theory. Considerable emphasis is given to the several basic definitions and rules governing probability theory, and simple examples are used to illustrate these definitions and rules.

Specific objectives are—

to present some of the basic terms and notions to be used throughout (Introduction and Section 1).

to present the classical definition of probability (Section 2).

to present the empiricist's and the subjectivist's views of probability (Section 3).

to present the rules and definitions that govern theory and applications (Section 4).

to show how the preceding theoretical development is used in problem solving (Sections 5 and 6).

to present the notion of revising probabilities with partial information through the use of Bayes' Theorem (Section 7).

Probability theory forms the basis for statistical inference, which is a major topic in a basic statistics course. It also stands by itself as a topic of interest for the decision maker in industry and government. We are constantly faced with choices for which the outcome is uncertain. Probability theory can be a valuable tool for selecting the best alternative under such circumstances. In this chapter we will present a thorough discussion of the basics of probability theory. Extended applications of probability theory as a tool in decision making will be covered in Chapter 15, "Decision Theory."

The manager is faced with many situations where a knowledge of some elementary probability concepts can contribute to better decisions. For example, consider a small manufacturing firm that has developed a new product. The firm incurred large debts in developing the product. In order to stay in business so that they can reap the benefits of their new discovery, the owners of the manufacturing firm have determined they must sell 3,000 units of their new product by the end of this year. They are considering two alternatives: either sell exclusive rights to a large retailing chain or sell the rights to three smaller chains in separate independent regions of the same geographic area covered by the large chain. They feel that there is a 40 percent chance the large chain will sell 3,000 units and that each of the three small chains has a 60 percent chance of selling 1,000 units. Without a knowledge of probability theory, a person might be inclined to choose the small chains because their individual probabilities of 60 percent are considerably higher than the 40 percent probability for the large chain. However, if the manufacturing firm is to keep from going bankrupt, all three of the small chains must sell 1,000 units. The probability of this occurring is not much more than half the probability that the big chain will sell the necessary 3,000 units (21.6 percent chance[1] versus 40 percent). If the assumed probabilities are correct, then the better alternative is to go with the large firm.

To become proficient at solving probability problems, it is necessary to acquire a good understanding of the basic terminology and theorems or rules of probability theory. For most beginning students, this intuitive feeling or understanding is best attained by initially referring to simple activities involving uncertainty with which the student is familiar and which can readily be visualized. Thus, the definitions and the basic concepts are developed in the first part of this chapter using examples of the games-of-chance type. The latter part of the chapter discusses the solution of more general types of probability problems.

---

1. This probability was found by multiplying .6 by itself 3 times, i.e., $(.6)^3 = .216$. The reason for doing this is discussed in Section 3.4.

3.1 Some Basic Terminology

Many students consider probability theory to be one of the more difficult topics covered in a basic statistics course. It is not really as difficult as it first appears. If you take the time and effort to learn the basic terminology, you will find that you will be better able to follow the discussion in this chapter. After studying this material and working through a few problems, you will acquire a basic mastery of the topic and, in addition, have the satisfaction of being capable of solving some fairly challenging probability problems. Not only is probability theory a useful topic, it can also be enjoyable.

Before we can consider some formal definitions of probability, it is necessary to define some terms that will be used throughout the chapter. The definitions of five basic terms are presented below and then discussed in detail.

Definitions

An *experiment* is a process by which a measurement or observation is obtained.

An *outcome* is a measurement or observation obtained when an experiment is performed.

A *probability space* is a collection of mutually exclusive and exhaustive outcomes of an experiment.

An *event* is any collection of one or more outcomes of the probability space.

The notation $P(\ )$ will be used for the probability that the event in parentheses will occur when the experiment is performed.

The most basic element of any probability problem is the experiment. If the experiment of interest has more than one possible outcome, then we have uncertainty. The intent of probability theory is to provide a methodology for assigning probabilities to the outcomes of an experiment. There are a variety of types of probability experiments that are encountered every day in government and business: put a new product on the market; measure the life of a light bulb; run for political office; measure the inside diameter of a motor casing; etc. Unnecessary confusion can occur when dealing with probability problems if one fails to clearly define the experiment of interest. "Flip a coin once" and "flip a coin twice" are two different experiments. It seems obvious but is still worth noting.

The outcome of an experiment is a measurement or an observation resulting from the experiment. The length of life of a light bulb can be measured in hours, minutes, seconds, or fractions of a second. The experiment in this case would be to burn a light bulb until it fails. If the experiment is to flip a coin once, we don't take a measurement, we make an observation. We observe whether the heads side or the tails side of the coin shows. Two outcomes of an experiment are mutually exclusive if both can *not* occur simultaneously when the experiment is performed.

The requirement that a probability space have only mutually exclusive outcomes means that the outcomes making up the probability space have to be so defined that only one outcome will occur when the experiment is performed. The collection of outcomes is exhaustive if at least one outcome occurs when the experiment is performed once. These concepts are illustrated in the solution to the following example.

Example 3.1.1 A marble is randomly selected from a cup containing 3 marbles, each of which is a different color. The colors are dark red, dark brown, and light yellow. Define a probability space for this experiment.

A probability space for this experiment can be defined as follows:

$S = \{R, B, Y\}$

where

S = the probability space
R = the marble selected is red
B = the marble selected is brown
Y = the marble selected is yellow.

We observe that this collection of outcomes is exhaustive: if the experiment is performed, i.e., select a marble from the cup, then 1 of the 3 outcomes in the probability space must occur. This collection of outcomes is also mutually exclusive in that no 2 outcomes can occur simultaneously when the experiment is performed once.

The definition of a probability space is not necessarily unique for an experiment. An alternative definition for this experiment would be as follows:

$S = \{D, L\}$

where

S = the probability space
D = the marble selected is dark colored
L = the marble selected is light colored.

The 2 outcomes in this probability space are mutually exclusive and exhaustive, i.e., any possible outcome of the experiment is classified by 1 of these 2 outcomes. If we are interested only in whether a dark or a light marble is selected, then this probability space may be satisfactory. If we want to distinguish between the 2 dark colored marbles, as would be the case if we were to use the classical definition of probability provided in the next section, then the first definition of the probability space given for this example is the one we would want to employ.

The definition of an event given at the beginning of this section can be discussed using the following example.

Example 3.1.2   A die is rolled once, and the number of spots showing on the top side is observed. Define a probability space for this experiment and each of the following events: an even number of spots occurs; an odd number of spots occurs; 5 spots occur; and either 1 or 2 spots occur.

The die is a cube with each of the 6 sides identified by a different number of spots ranging in number from 1 to 6. If we identify each possible outcome of the experiment, roll a die once, as the number of spots on the top side of the die, then the most elemental probability space, S, is defined as follows:

$S = \{1, 2, 3, 4, 5, 6\}$.

One of the indicated outcomes must occur when the die is rolled once, and no 2 can occur simultaneously.

Let

A be the event that an even number of spots will occur;
B the event that an odd number of spots will occur;
C the event that 5 spots occur; and
D the event that either 1 or 2 spots occur.

These events can be defined in terms of the probability space outcomes as follows:

$A = \{2, 4, 6\}$
$B = \{1, 3, 5\}$
$C = \{5\}$
$D = \{1, 2\}$.

Each of these 4 events satisfies the definition of an event given above—"any collection of 1 or more outcomes of the probability space." The event C contains only 1 outcome from the probability space. The probability space itself, S, is also an event.

It is important to understand that an event occurs when any one of the collection of outcomes defining the event occurs. For example, the event "$A = \{2, 4, 6\}$" occurs if 2 spots show when the die is rolled. The event A also occurs if 4 spots show or if 6 spots show. Beginning students sometimes erroneously assume that it is necessary for all 3 outcomes to occur before event A can occur. This would, of course, be impossible since the experiment consists of rolling the die once. An event is a collection of outcomes from a probability space, which means no 2 outcomes can occur simultaneously when the experiment is performed once. For this particular example, *A* represents the event that an even number of spots comes up. This event occurs if the outcome is an even number of spots. Also observe that if the outcome of the experiment should be 2 spots then both event *A* and event *D* occur even though the die is rolled only once.

Throughout the remainder of this chapter, much of the discussion will be centered on probability values that are assigned to outcomes and events. The last definition in the list given at the beginning of this section provides us with a notation for referring to the probability of an event; thus, $P(A)$ is read "the probability the event *A* will occur when the experiment is performed." For the event *A* as defined above, this would be the probability an even number of spots occurs when a die is rolled once. The event *C* is defined as $C = \{5\}$; therefore, $P(C)$ would be the probability that five spots show when we roll the die.

3.2 The Classical Definition of Probability

The definition of probability first used by the seventeenth-century mathematicians and probabilists[2] is referred to as the Classical Definition of Probability. Early work in probability was motivated by the nobility's interest in games of chance. The following definition was used to assign probabilities to events arising from games of chance:

Classical Definition

If an experiment has *N* equally likely outcomes and *K* of these outcomes satisfy the event *A*, then

$$P(A) = \frac{K}{N}.$$

Note that this definition provides us with a method for assigning probabilities to events. The classical definition is somewhat limited in its use. Nevertheless, there are applications to some types of problems encountered in business and economics, and the axioms of the more sophisticated axiomatic definition of probability are motivated by the Classical Definition of Probability. The following examples will help you to understand the assignment of probabilities based upon the classical definition.

---

2. Blaise Pascal (1623–62) and Pierre Fermat (1601–65) were both great seventeenth-century French mathematicians; together they created probability theory. Pascal also made significant contributions in physics, and Fermat's contributions in number theory were outstanding. Both worked on significant problems in analytic geometry, and their work set the stage for the invention of calculus by Newton and Leibniz. Pascal invented the first adding machine in 1642, the same year that Galileo died and Newton was born.

Example 3.2.1   A balanced coin is flipped once. What is the probability that heads will show?

In order to use the classical definition, it is necessary to specify or assume that the coin is balanced, i.e., it is just as likely that heads will appear as tails. The experiment has two equally likely outcomes, one of which satisfies the event "heads occurs." Thus, for this experiment we have $N = 2$ and $K = 1$. The probability that heads will occur, $P(H)$, can be calculated using the classical definition of probability:

$$P(H) = \frac{K}{N} = \frac{1}{2}.$$

Most people would say, without seriously considering the problem of balance, that the probability of heads occurring in this experiment is "fifty-fifty" or 1/2; they are using, probably unknowingly, the Classical Definition of Probability.

Example 3.2.2   It is known that 4 of the motors in a lot of 25 motors are defective. If a motor is selected at random from the lot, what is the probability that it will be a good motor?

If a motor is randomly selected, then there are 25 equally likely outcomes to this experiment; i.e., each of the motors can be selected as an outcome. Twenty-one of these outcomes satisfy the event G, a good motor is selected. The probability that a good motor will be selected, $P(G)$, can be calculated using the Classical Definition of Probability:

$$P(G) = \frac{21}{25}.$$

Example 3.2.3   In example 3.1.2, a die is rolled once, and the number of spots showing on the top side is observed. The following events are defined:

$A = \{2, 4, 6\}$
$B = \{1, 3, 5\}$
$C = \{5\}$
$D = \{1, 2\}$.

Find the respective probabilities for each of these events.

The sample space for this experiment, $S = \{1, 2, 3, 4, 5, 6\}$, consists of 6 equally likely outcomes. (We are assuming the die is balanced.) Three of these outcomes satisfy the event A, "an even number of spots occurs," and $P(A)$ is calculated as follows:

$$P(A) = \frac{3}{6} = \frac{1}{2}.$$

We repeat the comment made earlier—if one of the outcomes in the collection of outcomes defining the event occurs, then the event has occurred. It is *not* necessary (or possible) for all the outcomes defining the event to occur. Thus, if the die is rolled and the side with 2 spots occurs, we say that the event A has occurred. The probabilities of the events B, C, and D are given below.

$B = \{1, 3, 5\}$ $P(B) = \frac{3}{6} = \frac{1}{2}$

$C = \{5\}$ $P(C) = \frac{1}{6}$

$D = \{1, 2\}$ $P(D) = \frac{2}{6} = \frac{1}{3}.$

The probability that an odd number of spots will occur is 1/2, the probability that 5 spots will occur is 1/6, and the probability that either 1 spot or 2 spots will occur is 1/3.

When using the Classical Definition of Probability, one has to be careful with the delineation of the elemental, equally likely outcomes in the probability space.

Example 3.2.4

The experiment is to flip 2 balanced coins. What is the probability both coins will come up heads?

The following probability space,

$S = \{$both heads, both tails, one of each$\}$,

is a valid one for this experiment. One and only 1 of the outcomes will occur when 2 coins are tossed. If you were to assume that these outcomes are equally likely, you would conclude that the probability of both coins coming up heads is 1/3. You would be wrong in your conclusion, but you would be in good company. The famous eighteenth-century mathematician d'Alembert was said to have made the same mistake.

A probability space with equally likely events can be defined for this experiment as follows:

$S = \{HH, TT, HT, TH\}.$

Here we are keeping track of the 2 individual coins (let 1 be a nickel, the other a dime) and noting all possible outcomes. Two heads can only occur 1 way—the nickel can show heads and the dime can show heads. Similarly, 2 tails can only occur 1 way—the nickel can show tails and the dime can show tails. One of each can occur 2 ways—dime-heads and nickel-tails, or dime-tails and nickel-heads. These 4 outcomes are equally likely, and we can use the Classical Definition of Probability to obtain the following probabilities:

$$P(2 \text{ heads}) = \frac{1}{4}$$

$$P(2 \text{ tails}) = \frac{1}{4}$$

$$P(1 \text{ of each}) = \frac{2}{4} = \frac{1}{2}.$$

A similar type of situation is apparent when we consider the probabilities arising from throwing a pair of dice. We can list as our probability space all possible sums of spots showing on the pair of dice:

$S = \{2, 3, 4, 5, 6, 7, 8, 9, 10, 11, 12\}$.

One and only 1 of these outcomes will occur when a pair of dice is thrown. These are not, however, equally likely outcomes. To obtain a probability space with equally likely outcomes, it is necessary to consider all possible physical combinations of the 2 dice. There is only 1 way that the first outcome, a sum of 2, can occur—1 spot on the red die and 1 on the green die. The spots summing to 3 can occur in 2 ways—the red shows 1 and the green 2, or the red shows 2 and the green 1. The sum of the spots totals 3 in both cases, but they are 2 different physical outcomes.

Example 3.2.5

A pair of dice is thrown once. What is the probability both dice will show 1 spot? What is the probability that the sum of the spots showing will equal 7? What is the probability the spots showing will sum to 9?

Table 3.1 lists all possible paired combinations that can occur when a pair of dice is thrown once. There are 36 equally likely outcomes. There is only 1 way that the 2 dice will have 1 spot each, thus,

$$P(2 \text{ spots}) = \frac{1}{36}.$$

There are 6 possible combinations that result in a total of 7: (1, 6), (2, 5), (3, 4), (4, 3), (5, 2), and (6, 1). Thus,

$$P(7 \text{ spots}) = \frac{6}{36} = \frac{1}{6}.$$

Similarly, we can identify 4 combinations in Table 3.1 that sum up to 9 spots:

$$P(9 \text{ spots}) = \frac{4}{36} = \frac{1}{9}.$$

The Classical Definition of Probability

Once the probability space in Table 3.1 with equally likely outcomes has been identified, the probabilities of any event can readily be calculated.

Table 3.1 Outcomes from Throwing a Pair of Dice

First Die	Second Die					
	1	2	3	4	5	6
1	1, 1	1, 2	1, 3	1, 4	1, 5	1, 6
2	2, 1	2, 2	2, 3	2, 4	2, 5	2, 6
3	3, 1	3, 2	3, 3	3, 4	3, 5	3, 6
4	4, 1	4, 2	4, 3	4, 4	4, 5	4, 6
5	5, 1	5, 2	5, 3	5, 4	5, 5	5, 6
6	6, 1	6, 2	6, 3	6, 4	6, 5	6, 6

Unfortunately, the Classical Definition of Probability is limited in its usefulness. It applies only to a certain class of experiments—those experiments where a probability space with a finite number of equally likely outcomes can be identified. The classical definition doesn't help answer such questions as: What is the probability the next part produced will be defective? What is the probability it will rain tomorrow? What is the probability your tire will blow out the next time you drive your car? Attempting to find these probabilities using the classical definition will convince you that it is necessary to have other definitions of probability.

There is a philosophical problem with the classical definition; in a sense it is circular. The term *equally likely* is an undefined term and its meaning is essentially the same as *equally probable,* i.e., the outcomes are equally likely if each has the same probability of occurring. We are using an undefined synonym of probability to define probability. As we have seen, however, the classical definition often works well in practice. Most people will agree that, unless a coin or a die is deliberately loaded for bias, the assumption of equally likely outcomes is almost automatic, and these people are willing to wager accordingly. The axioms of the axiomatic definition of probability become less abstract in meaning if one compares them with the Classical Definition of Probability.

3.3 The Relative Frequency and the Subjective Definitions of Probability

The Relative Frequency Definition of Probability is based on empirical observation. We perform the experiment a large number of times and observe the frequency of an outcome or an event. The proportion of times that the event A occurs, or in other words, the relative frequency of the occurrence of A, becomes an estimate of the probability of A. If, in an inspection of several hundred parts, we observed that 2 percent of the parts produced by a machine were defective, then we would estimate that the probability that the next part produced by the machine will be a defective part is .02. This concept is summarized in the following definition.

Relative Frequency Definition   If an experiment is repeated N times (N is large), and the event A occurs K times, then

$$P(A) \cong \frac{K}{N}.$$

The "approximately equal to" symbol, $\cong$, is used because the above probability is generally not exact. In order to get the exact probability, the experiment must be repeated ad infinitum. Using the limit notation of calculus, we would write this exact probability as

$$P(A) = \lim_{N \to \infty} \frac{K}{N}.$$

The latter is the exact relative frequency definition. The expression

$$\lim_{N \to \infty} \frac{K}{N}$$

is read "the limiting ratio of K/N as N goes to infinity." In other words, repeat the experiment an infinite number of times and the probability of A will be the proportion of times the event A occurs. The reader should note that N and K have different meanings in this definition than they did in the classical definition. This definition has appeal because in theory it puts probability theory on a solid empirical foundation; however, in practice it is never possible to get an exact probability this way because it is impossible to repeat an experiment an infinite number of times. For practical applications, the "approximately equal to" definition is employed. The experiment cannot be repeated an infinite number of times, but if it is repeated a large number of times, then a usable approximation of the true probability can be determined.

Example 3.3.1   An unbalanced coin is flipped 10,000 times. Heads is observed on 3,586 of those flips. What is the probability of obtaining a heads if the coin is flipped once?

The classical definition cannot be used to find the probability that an unbalanced coin will be heads because the outcomes "heads" and "tails" are not equally likely events if the coin is unbalanced. The relative frequency definition can be used to obtain an approximation of the probability of heads. The experiment "flip a coin once" has been repeated 10,000 times, and the event heads has occurred 3,586 times. The probability of heads when the coin is flipped once can be calculated from the formula for the Relative Frequency Definition of Probability:

$$P(\text{heads}) = \frac{K}{N} = \frac{3{,}586}{10{,}000} = .3586.$$

Probabilities that are sometimes referred to as *historical probabilities* are of the relative frequency type. For example, it is observed that 5 percent of a particular brand of television set sold during the past year by a large store came back for repairs before the warranty period ended. A salesman in the store selling a customer this brand of TV might say to himself (never to the customer), "The probability this set will come in for repairs before the end of the warranty period is .05." The experiment is to sell a customer a TV set. The event of interest is whether the TV set comes in for repairs before the warranty period ends. The probability was obtained by repeating the experiment a large number of times; i.e., a large number of TV sets were sold to individual customers during the last year, and the proportion of customers bringing in their TV set for repairs was observed to be .05 (possibly 100 repaired sets out of 2,000 sold). The probabilities given for living past retirement age, getting cancer, having a car accident, etc., are determined using the relative frequency definition.

The Relative Frequency Definition of Probability is of particular value in conceptualizing what is meant by the probability of an event. Even though we can't physically perform an experiment an infinite number of times, we can conceive of doing it. If you were asked the question, What is meant by saying that the probability is 1/2 that heads will show when a balanced coin is flipped? your answer would probably be something to the effect that if the coin were flipped a large number of times, heads would show approximately 1/2 the time; and if the coin were flipped an infinite number of times, then heads would occur exactly 1/2 the time. Also, as the number of tosses increased, you might expect, on the average, the proportion of heads to become closer to 1/2.

Subjective Definition

$P(A) = $ What you "feel" or "think" about the chance of event A occurring.

The Subjective Definition of Probability is also referred to as the Bayesian definition because of Bayes' Theorem (see Appendix 3C) and its extensive use in the relatively new field of statistical decision theory.

There is a problem in that different people will assign different probabilities to the same event. If a dime is being tossed, users of the classical definition would assign the probability of 1/2 to each of the 2 possible outcomes. Given certain experiences, a person might "feel" that the probability of heads is different from 1/2, and we might find several people who would all assign different probabilities to the occurrence of heads. If, however, probability theory is to be useful and successful in business decision making, the assigned probabilities must realistically relate to the real-world problem. It would seem that a useful assignment would be made by an "expert." Inasmuch as this word implies experience and knowledge, the useful subjective assigned probability is similar to, though not so formal as, that made by the relative frequency definition. If the probabilities derived using the Subjective Definition of Probability are to have practical value, they will be understood in terms of the relative frequency approach to probability.

In the use of the Subjective Definition of Probability, one is not completely free in the assignment of probabilities to events. There is a restriction that the probabilities assigned to the events be "rational." As always, what is meant by rational is arbitrarily defined. In this case, probabilities assigned to events are said to be rational if they satisfy the Axiomatic Definition of Probability discussed in the next section. If your statistics professor were to say that the probability of rain tomorrow is 1.5 or −.5 you would probably accuse the professor of being irrational. After making the appropriate notation next to your name on the class records, he or she might ask what caused you to make such an outburst. You might comment that probabilities are never greater than 1 or less than 0. You would be hard put to find a prohibition of negative probabilities that is inherent in the Subjective Definition of Probability. The best you can say is that rational probabilities must satisfy the Axiomatic Definition of Probability.

3.4 The Formal Mathematics of Probability Theory

In this section we want to develop the rules that are used for finding probabilities of various combinations of events given that the probabilities for the individual events have been determined. You should read and study this section carefully, paying particular attention to the definitions, but don't get "bogged down" here. In the next section (3.5), we will work a variety of types of problems using the definitions and theorems developed in this section. As you work through the examples in Section 3.5, the concepts covered in this present section will become less abstract and more understandable.

The Axiomatic Definition of Probability is introduced in this section because the rules for solving complex probability problems are derived from this definition. The classical and relative frequency definitions both generate probabilities that satisfy the axiomatic definition; and, as indicated above, the user of the Subjective Definition of Probability is required to select probabilities that satisfy the axiomatic definition of probability if he or she is to get reasonable and rational results. The axiomatic definition provides the common ground for discussing probability theory regardless of which of the other three definitions we employ to assign probability values to events.

Before the axiomatic definition can be formally stated, it is necessary to define some terms, ideas, and the associated notation borrowed from the mathematics of set theory. These will be used throughout the chapter and will prove useful in solving probability problems. Five terms are defined below and discussed.

Definitions

$\overline{A}$, read "the complement of A," is the collection or set of all outcomes in the probability space S that are not in A.

ϕ is the null or empty collection or set of outcomes, i.e., it is the impossible event.

The event AB occurs if and only if both event A and event B occur when the experiment is performed. This is read as "A and B" or simply "AB"; in set theory it is the "intersection" of A and B.

The event $A \cup B$ occurs if either event A or event B or both events occur when the experiment is performed. This is read as "A or B"; it is the "union" of A and B.

The events A and B are *mutually exclusive* events if it is impossible for both A and B to occur when the experiment is performed once, i.e., $AB = \phi$, the null or empty set.

The following example will be helpful for the discussion of these definitions.

Example 3.4.1

A die is rolled once. Use the following events to illustrate the terms defined above:

$S = \{1, 2, 3, 4, 5, 6\}$
$A = \{1, 2, 3\}$
$B = \{2, 4, 6\}$
$C = \{1, 3, 5\}$
$D = \{5, 6\}$.

The complement of A — $\overline{A}$ — is also referred to as "not A." It is defined as the event that occurs when A does *not* occur. The probability space for this experiment is already known to us from previous examples:

$S = \{1, 2, 3, 4, 5, 6\}$.

The complement of A consists of all the outcomes in S that are not in A:

$\overline{A} = \{4, 5, 6\}$.

Similarly,

$\overline{B} = \{1, 3, 5\} = C$.

Verbally, what this latter expression is conveying is simply that, when a die is rolled once, either an even number of spots, the event B, comes up or an odd number of spots, the event C, comes up. The events B and C are complements of each other.

Figure 3.1, part a, illustrates the concept of the complement of an event using what is known as a Venn Diagram.[3] The rectangle represents the probability space. The circle labeled A represents the collection of outcomes in S that define the event A. The complement of A would then be all the outcomes in S outside of the event A. The complement of A is indicated by the shaded area.

Figure 3.1 Venn Diagrams

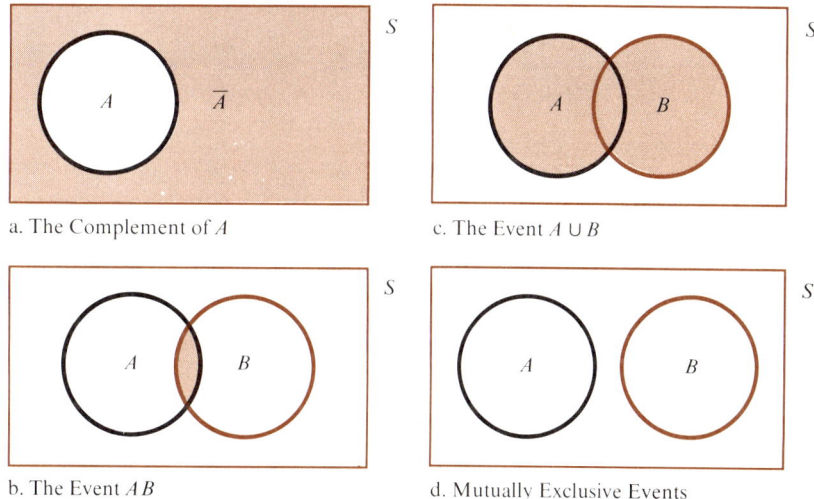

a. The Complement of A

b. The Event AB

c. The Event $A \cup B$

d. Mutually Exclusive Events

---

3. Mathematicians will probably refer to these diagrams as Euler Diagrams. Leonard Euler (1707–83), a famous Swiss mathematician, was said to have been the first to utilize them. John Venn, an English logician working around the turn of this century, revived their use and apparently added the notion of hatching. Almost universally today we call them Venn Diagrams.

The Formal Mathematics of Probability Theory

It is necessary for the mathematical development of probability theory to define the impossible event, ϕ. Like zero in the real number system, ϕ is important. The probability assigned to ϕ will in fact be 0.

Using set theory notation, the symbol $A \cap B$ is the "intersection" of events A and B. Our notation, AB, means the same thing, and we will read it as "A and B." The probability of A and B, $P(AB)$, is the "joint probability" of events A and B. For practical applications it is important that you understand that the symbol AB represents an event that could occur when the experiment is performed, and that by definition the event AB occurs only if *both* event A and event B occur when the experiment is performed once. In the die-throwing example above, the events A and B were defined as follows:

$A = \{1, 2, 3\}$
$B = \{2, 4, 6\}$.

Both event A and event B will occur if 2 spots show when the die is rolled once, and this is the only way both A and B can occur simultaneously. Thus, we would write,

$AB = \{2\}$.

The event AB is illustrated by the shaded area in Figure 3.1, part b. The event AB is the collection of outcomes in S that are common to both A and B.

In set theory the symbol $A \cup B$ is read as "A union B"; for our work in probability, we say "A or B." The "or" in this statement is always the "inclusive or," i.e., A or B or both. The event $A \cup B$ occurs if either one or both of the events occur.

The event $A \cup B$ is the collection of all outcomes in A or in B or in both and is illustrated by the Venn Diagram in part c of Figure 3.1. In the example above, we had

$A = \{1, 2, 3\}$
$B = \{2, 4, 6\}$.

For the event $A \cup B$ we get

$A \cup B = \{1, 2, 3, 4, 6\}$.

If one of the five outcomes—1, 2, 3, 4, or 6 spots—occurs, then the event $A \cup B$ has also occurred. Note that the event $A \cup B$ is much more likely to occur than the event AB. The latter requires the simultaneous occurrence of the two events, while $A \cup B$ only requires one of the two events to occur.

The Venn Diagram in part d of Figure 3.1 illustrates two mutually exclusive events. The events A and B in the diagram have no outcomes common to both events; thus, it is impossible for both events to occur simultaneously. This means that, if we know one of two mutually exclusive events has occurred, we can be sure the other event has *not* occurred. In the above die-throwing example, the following events were defined:

$A = \{1, 2, 3\}$
$B = \{2, 4, 6\}$
$C = \{1, 3, 5\}$
$D = \{5, 6\}$.

The events B and C are mutually exclusive; i.e., if the die is rolled once, it is impossible for the number of spots to be both an even number and an odd number. Also, observe that if the event B, an even number of spots, has occurred, we know for sure that event C, an odd number of spots, has *not* occurred. This is an example where the impossible event notation, ϕ, can be used:

$BC = \phi$.

The events A and D are also mutually exclusive. This is not the case for any other pair of the above events; i.e., all other pairs have at least one outcome in common, and therefore it is possible for them to occur simultaneously.

The three definitions of probability discussed up to this point—classical, relative frequency, and subjective—all satisfy the axioms of the Axiomatic Definition of Probability.[4] Therefore, the rules and laws of probability derived from the Axiomatic Definition of Probability are also applicable to the probabilities obtained from the other three definitions.

Axioms

P is a probability measure defined on the probability space S if the following axioms are satisfied for all events in the probability space S:

1. $P(A) \geq 0$.
2. $P(S) = 1$.
3. If A and B are mutually exclusive events in S, then $P(A \cup B) = P(A) + P(B)$.

(A and B are events defined on the sample space S.)

Theorems

The following theorems can be proved from the axioms. (The proofs are in Appendix 3D at the end of this chapter.)

1. $P(\phi) = 0$.
2. $P(\bar{A}) = 1 - P(A)$.
3. $P(A) \leq 1$.

---

[4] An axiomatic or deductive system is characterized by a set of undefined terms, a set of axioms (assumptions), and a set of definitions. From these, theorems are developed. Theorems can be proved based upon the original assumptions or axioms.

The Formal Mathematics of Probability Theory

It is helpful to consider what the above axioms and theorems require for probabilities to be valid. Axiom 1 and Theorem 3 state that probabilities must be between 0 and 1, inclusive. Negative probabilities and probabilities greater than 1 are not allowed. Axiom 2 states that the probability of the certain event is 1. (S must occur because by definition the outcomes in S are exhaustive, which means one of them must occur when the experiment is performed.) Theorem 1 states that the probability assigned to the impossible event or the empty set is 0. Theorem 2 gives us the probability for the complement of A when the probability of A is known; note that this also implies that $P(A) + P(\overline{A}) = 1$. Axiom 3 combined with Axiom 2 assures that the probabilities assigned to any combination of events will not exceed 1. It can also be deduced from these axioms[5] that the sum of the probabilities assigned to all outcomes in the probability space, S, will equal 1.

After we discuss the definitions and probability theorems presented below, we will be prepared to work basic probability problems.

Definitions

$P(A|B)$ is read "the probability the event A will occur given that the event B has occurred or will occur."

The conditional probability of A *given* B *is defined as*

$$P(A|B) = \frac{P(AB)}{P(B)}.$$

The events A and B are *independent events* if $P(A|B) = P(A)$. This also implies that $P(B|A) = P(B)$.

Theorems

4. $P(AB) = P(A)P(B|A) = P(B)P(A|B)$.
5. If A and B are independent events, then $P(AB) = P(A)P(B)$, and conversely.
6. $P(A \cup B) = P(A) + P(B) - P(AB)$.

It should be observed that the symbols $P(A)$ and $P(A|B)$ are both referring to the probability that the event A will occur. $P(A)$ is the probability assigned to the event A without any knowledge about any other events occurring when the experiment is performed. On the other hand, $P(A|B)$ is the probability we assign to the event A if we know that the event B has occurred or will occur. The definition for the conditional probability of A, given that B has occurred,

$$P(A|B) = \frac{P(AB)}{P(B)},$$

is simply treated as a definition and nothing more in the axiomatic development of probability theory.

---

5. Strictly speaking, Axiom 3 has to be extended to include the union of any number of mutually exclusive events in S.

It can be shown to appeal to our intuition if we think in terms of the Classical Definition of Probability. Treat the probability space in the Venn Diagram in Fig. 3.2 as if it were made up of equally likely outcomes evenly spread over the area of the rectangle. In this case, the ratio of the area of A to the area of S would be the probability of A. If it is known that the event B has occurred, then we know that none of the outcomes outside of B has occurred. The event B is a probability subspace. The outcomes in B that satisfy the event A are those outcomes in both A and B—i.e., the outcomes in the event AB. The probability that A will occur given that B has or will occur is the ratio of the area of AB to the area of B, or

$$\frac{P(AB)}{P(B)}.$$

The probability of event A occurring may not be affected by whether or not the event B occurs. If this is the case, we say that A and B are *independent* events. In other words, if $P(A|B) = P(A)$, then A and B are independent. If $P(A|B) \neq P(A)$, then the events A and B are said to be *dependent* events. The idea of dependent and independent events cannot be clearly illustrated with a Venn Diagram. For practical applications of probability theory, it is sometimes apparent that two events are independent, and other times it is a matter of empirical observation. If two people are working at two different locations on the same problem and they don't communicate with each other, it is reasonable to assume that the probability of one individual's solving the problem is independent of whether or not the other individual solves the problem. If two coins are tossed, the probability of heads on the second coin is independent of what occurred with the first coin. Whether or not the probability that Joe gets to school on time is dependent on his alarm's going off in the morning has to be empirically observed; it would seem that these would be dependent events for most people, but it could well be the case that Joe's arrival at school on time is not in the least affected by whether or not his alarm goes off.

Figure 3.2   Venn Diagram

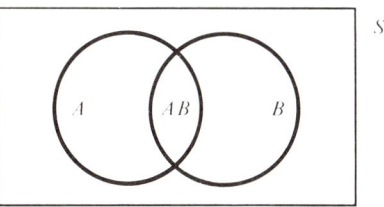

The following example indicates the mathematical nature of the definition of independent events.

Example 3.4.2 A single die is thrown once. Are the events $A = \{2\}$ and $B = \{2, 4, 6\}$ independent? Are the events $C = \{1, 2\}$ and B independent?

To determine the independence of A and B, it is necessary to refer to the definition of independence and compare $P(A|B)$ with $P(A)$. From the Classical Definition of Probability, we have

$$P(A) = \frac{1}{6}$$

$$P(B) = \frac{3}{6} = \frac{1}{2}$$

$$P(AB) = \frac{1}{6}.$$

The last probability is determined by observing that $AB = \{2\}$. The $P(A|B)$ can be calculated from the definition of $P(A|B)$, namely,

$$P(A|B) = \frac{P(AB)}{P(B)} = \frac{\frac{1}{6}}{\frac{1}{2}} = \frac{1}{3}.$$

The $P(A|B)$ does not equal the $P(A)$; therefore, A and B are not independent events. It can be shown, however, that B and C are independent events. The event $BC = \{2\}$; therefore,

$$P(BC) = \frac{1}{6}.$$

Also,

$$P(C) = \frac{2}{6} = \frac{1}{3}.$$

From the definition of conditional probability, we have

$$P(C|B) = \frac{P(BC)}{P(B)} = \frac{\frac{1}{6}}{\frac{1}{2}} = \frac{1}{3}.$$

We conclude that B and C are independent events because

$$P(C|B) = P(C) = \frac{1}{3}.$$

Theorem 4, $P(AB) = P(A)P(B|A) = P(B)P(A|B)$, is readily obtained from the definition of a conditional probability. The $P(AB)$ is solved for algebraically from the definition of $P(A|B)$:

$$P(A|B) = \frac{P(AB)}{P(B)}.$$

Therefore, solving algebraically for $P(AB)$,

$$P(AB) = P(B)P(A|B).$$

Similarly, for $P(B|A)$:

$$P(B|A) = \frac{P(AB)}{P(A)}.$$

Therefore,

$$P(AB) = P(A)P(B|A).$$

If the events A and B are independent, then by definition $P(A|B) = P(A)$ and $P(B|A) = P(B)$. Substituting these values in Theorem 4 gives us Theorem 5,

$$P(AB) = P(A)P(B) \text{ if } A \text{ and } B \text{ are independent events.}$$

The proof of Theorem 6 requires some theorems from set theory. However, a glance at the Venn Diagram for A ∪ B, shown in Figure 3.3, will convince you that this theorem is intuitively correct. The $P(A)$ includes the $P(AB)$, and $P(B)$ also includes $P(AB)$. Thus, the sum $P(A) + P(B)$ includes $P(AB)$ twice. It is therefore necessary to subtract $P(AB)$ from the sum to get $P(A \cup B)$:

$$P(A \cup B) = P(A) + P(B) - P(AB).$$

If A and B are mutually exclusive events, $P(AB) = 0$, and the above formula reduces to Axiom 3.

Before working probability problems, it can be helpful to distin-

Figure 3.3 Venn Diagram

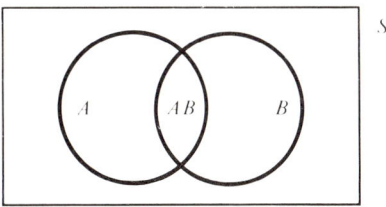

guish clearly between mutually exclusive events and independent events. Beginning students tend to confuse the two notions.

If 2 events with non-0 probabilities are mutually exclusive, it is impossible for them also to be independent. If 2 events are mutually exclusive, then, by definition, they cannot both occur when the experiment is performed once. Thus, $P(AB) = 0$ and

$$P(A|B) = \frac{P(AB)}{P(B)} = 0,$$

which means $P(A|B) \neq P(A)$. This latter condition means that the definition for independence is not satisfied, and A and B must be dependent events. The probability of A occurring is 0 if a mutually exclusive event B has already occurred; therefore, the $P(A)$ is very much dependent on whether or not B has occurred.

3.5 Solving Probability Problems

Five basic rules for working probability problems have been selected from the axioms and theorems discussed above and they are listed below. All of the definitions of probability discussed in this chapter satisfy the Axiomatic Definition of Probability. This permits the use of these rules for solving probability problems regardless of the definition used to arrive at the initial probabilities. Probability problems can be both challenging and fun to work. They can also be confusing at times. An approach to applied probability problems that follows essentially the steps listed here can be helpful in avoiding some of the confusion that might arise.

1. Initially, ignore the numerical probability values that are given.
2. Define the events of interest.
3. State the probability question in terms of the defined events using probability event notation.
4. Apply the appropriate probability rule from the Summary of Basic Probability Formulas below.
5. Substitute the appropriate numerical probability values.

Summary of Basic Probability Formulas

Basic probability formulas are as follows:

1. $P(A) + P(\overline{A}) = 1$.
2. $P(AB) = P(A)P(B|A) = P(B)P(A|B)$.
3. $P(A \cup B) = P(A) + P(B) - P(AB)$.
4. If A and B are *independent* events, then $P(AB) = P(A)P(B)$.
5. If A and B are *mutually exclusive,* then $P(A \cup B) = P(A) + P(B)$.

Once you become adept at working probability problems, you may not always want to follow the five steps; but, whenever confusion starts to arise, you will find it helpful to return to this basic approach.

Example 3.5.1

Based on past experience, a small contractor feels that there is a .6 probability that his bid on a federal government contract will be accepted and a .9 probability that his bid submitted to the local school board will result in a contract. The acceptance or rejection of either bid is independent of the outcome of the other bid. What is the probability that both bids will result in contracts for the contractor?

As indicated in the steps given above, rather than trying to decide immediately what should be done with the given probability values, confusion can be avoided by a systematic approach. First define the events of interest:

A = The bid with the federal government is accepted.
B = The bid with the local school board is accepted.

Using the probability event notation, the question of interest can be stated as

$P(AB) = ?$

Note that $P(AB)$ is the probability that both A and B will occur. Referring to the Summary of Basic Probability Formulas, it is apparent that either Rule 2 or Rule 4 would apply here because both rules deal with the intersection of two events. Rule 4 is a simpler rule to use and is applicable for this problem because the likelihood of a bid's being accepted is independent of what happens to the other bid. We would therefore apply Rule 4 in the Summary of Basic Probability Formulas.

$P(AB) = P(A)P(B)$.

From the statement of the problem, the probabilities of the defined events are obtained:

$P(A) = .6$
$P(B) = .9$.

The answer to the question, What is the probability both bids will be accepted? is then calculated to be

$P(AB) = P(A)P(B) = (.6)(.9) = .54$.

Example 3.5.2

Fifty percent of the patrons entering Disneyland go on the Matterhorn ride and 70 percent go on the Haunted Mansion ride. It has also been observed that 40 percent go on both rides. What is the probability that a person entering Disneyland selected at random will go on at least 1 of the 2 rides?

In this problem, three different hypothetical probability values are given. In order to determine whether they should be multiplied together or added or subtracted, a systematic approach is recommended. The events of interest can be defined as:

M = Patron goes on the Matterhorn ride.
H = Patron goes on the Haunted Mansion ride.

The event "the patron goes on at least one of the rides" occurs if the patron goes on the Matterhorn ride or the Haunted Mansion or both. This is the event defined by the union notation, $A \cup B$, i.e., $A \cup B$ occurs if A or B or both occur. Thus, using probability event notation, the question asked can be stated as

$P(M \cup H) = ?$

Rules 3 and 5 in the Summary of Basic Probability Formulas deal with finding the probability of the union of two events. Rule 3 will have to be used for this problem because events M and H are not mutually exclusive. We know this is the case because it is possible for a person to go on both rides. From Rule 3 we have

$P(M \cup H) = P(M) + P(H) - P(MH).$

From the information given in the problem, the probabilities for the defined events can be obtained:

$P(M) = .5$
$P(H) = .7$
$P(MH) = .4.$

The probability that a person selected at random will go on at least 1 of the 2 rides can now be calculated using these probabilities:

$P(M \cup H) = P(M) + P(H) - P(MH) = .5 + .7 - .4 = .8.$

In the last example, percents were used as probabilities. This was appropriate given the experiment described by the problem. The experiment is randomly to select a patron of Disneyland and observe his or her behavior with regard to the 2 rides mentioned. It has been observed that 50 percent of the Disneyland customers go on the Matterhorn ride. In essence, the experiment—a person entering Disneyland—has been repeated a large number of times, and 50 percent of the times, the result of the experiment has been that the patron goes on the Matterhorn ride. Thus the proportion of times this event occurs is .5, and using the Relative Frequency Definition of Probability results in $P(M) = .5$. A percent can also become a probability using the Classical Definition of Probability. If the experiment is to randomly select 1 adult from a community where it is known that 40 percent of the adults are for fluoridating the drinking water, then the probability the adult so selected will be for fluoridating the drinking water is .4.

Example 3.5.3 In Example 3.5.1, the probability that the federal government will accept the submitted bid is .6 and the probability that the local school board will accept the contractor's bid is .9. What is the probability that neither of the 2 bids submitted by the contractor will be accepted?

In the solution to Example 3.5.1, we let A be the event the bid with the federal government is accepted and B the event the bid with the local school board is accepted. If we use the events A and B as defined for Example 3.5.1, then

$\bar{A}$ = The bid with the federal government is not accepted
$\bar{B}$ = The bid with the local school board is not accepted.

The question asked can be stated as

$P(\bar{A}\bar{B}) = ?$

Because the outcomes on the bids are independent of each other, we use Rule 4 from the Summary of Basic Probability Formulas:[6]

$P(\bar{A}\bar{B}) = P(\bar{A})P(\bar{B})$.

To get $P(\bar{A})$ and $P(\bar{B})$, Rule 1 in the summary of probability formulas can be used:

$P(\bar{A}) = 1 - P(A) = 1 - .6 = .4$
$P(\bar{B}) = 1 - P(B) = 1 - .9 = .1$.

The probability that neither the federal government bid nor the local school board bid will be accepted is

$P(\bar{A}\bar{B}) = P(\bar{A})P(\bar{B}) = (.4)(.1) = .04$.

Example 3.5.4 Refer again to Example 3.5.1. What is the probability the contractor will have at least 1 of the 2 bids accepted?

This problem can be done in 2 ways. First, the procedure can be similar to the solution to Example 3.5.2—i.e., the event that at least 1 of the 2 bids is accepted is $A \cup B$, where

A = The bid with the federal government is accepted
B = The bid with the local school board is accepted.

Applying Rule 3 from the Summary of Basic Probability Formulas, as we did for Example 3.5.2, we get

$P(A \cup B) = P(A) + P(B) - P(AB)$.

Because A and B are independent,

$P(AB) = P(A)P(B)$

---

6. Note that if A and B are independent, then $\bar{A}$ and $\bar{B}$ are independent as are A and $\bar{B}$ and $\bar{A}$ and B.

Solving Probability Problems

(see solution to Example 3.5.1), and the probability that the contractor will have at least 1 of the bids accepted is

$$P(A \cup B) = P(A) + P(B) - P(A)P(B)$$
$$= .6 + .9 - (.6)(.9) = .6 + .9 - .54 = .96.$$

Another approach to this problem is to note that, if the contractor does not have at least 1 of the bids accepted, then it must be the case that neither of them is accepted; i.e., these 2 events are complements of each other. Applying Rule 1 in the Summary of Basic Probability Formulas, we have

$P(\text{at least 1 bid is accepted}) + P(\text{neither bid is accepted}) = 1.$

Solving for the probability that at least 1 bid will be accepted, we get

$P(\text{at least 1 bid is accepted}) = 1 - P(\text{neither bid is accepted}).$

In the previous example, the probability that neither bid would be accepted was calculated to be .04. Using this result, the desired probability can now be computed:

$P(\text{at least 1 bid is accepted}) = 1 - .04 = .96.$

Note that .96 agrees with the answer found by using the straightforward approach.

Extensions of Rules 4 and 5

Sometimes problems arise where the events of interest consist of a combination of a number of independent events or a number of mutually exclusive events. The extensions of Rules 4 and 5 in the Summary of Basic Probability Formulas can be used for solving these types of problems.

4′ If $A_1, A_2, \ldots, A_K$ are K independent events, then $P(A_1 A_2 \ldots A_K) = P(A_1)P(A_2) \ldots P(A_K).$

5′ If $A_1, A_2, \ldots A_K$ are K mutually exclusive events, then $P(A_1 \cup A_2 \cup \ldots \cup A_K) = P(A_1) + P(A_2) + \ldots + P(A_K).$

Example 3.5.5

A balanced coin is tossed 5 times. What is the probability that heads will occur all 5 times?

We can arbitrarily define the 5 events of interest as follows:

$H_i = i^{\text{th}}$ toss comes up heads.

The probability that all 5 tosses will come up heads can be stated as follows:

$P(H_1 H_2 H_3 H_4 H_5) = ?$

The outcome of each toss is independent of the outcomes on the other 4 tosses; therefore, Rule 4' in the Extension of Rules 4 and 5 can be employed:

$$P(H_1H_2H_3H_4H_5) = P(H_1)P(H_2)P(H_3)P(H_4)P(H_5)$$
$$= \left(\frac{1}{2}\right)\left(\frac{1}{2}\right)\left(\frac{1}{2}\right)\left(\frac{1}{2}\right)\left(\frac{1}{2}\right) = \left(\frac{1}{2}\right)^5 = \frac{1}{32}.$$

The probability that all 5 tosses will be heads is 1/32. It should be carefully noted what is meant by independent events. If a balanced coin is tossed 4 times and heads comes up 4 times, one often hears the not-too-uncommon fallacy, "The law of averages says there is a high probability the next one will be heads." This is completely wrong. The likelihood of heads on the fifth toss is independent of what occurred for the previous 4 tosses. The probability of heads on the fifth toss remains the same, 1/2. Before the 5 tosses are made, the probability of all heads is 1/32. If it is known that the first 4 tosses are heads, then the probability that 5 heads will have occurred after the fifth toss is made is 1/2.

Example 3.5.6   A balanced coin is tossed 5 times. What is the probability tails will occur on at least 1 of these tosses?

Here is a good example of the value of Rule 1 in the Summary of Basic Probability Formulas. To answer the question asked in a straightforward manner becomes a long, complex task. The event "at least 1 heads" occurs if exactly 1 heads occurs, or exactly 2 heads occur, or exactly 3 heads, etc. The solution to this problem is best found by noting that the complement of at least 1 tails is no tails or, equivalently, all heads. We can use the result from Example 3.5.5 above to obtain the solution to this problem.

$P(\text{at least 1 tails}) + P(\text{no tails}) = 1.$

Therefore,

$$P(\text{at least 1 tails}) = 1 - P(H_1H_2H_3H_4H_5) = 1 - \frac{1}{32} = \frac{31}{32}.$$

The probability is 31/32 that 1 or more of the 5 tosses will be tails.

3.6 More Problem Solving

The material covered up to this point gives the reader a good understanding of basic probability theory. The examples in this section demonstrate how a combination of the rules in the basic probability formulas and their extensions can be used to solve more complex probability problems. The next section, on Bayes' Theorem, discusses a method for solving a different type of probability problem from those discussed in this section. Bayes' Theorem has played an important role in the development of decision theory and Bayesian statistical theory.

Example 3.6.1

The manager of an exclusive department store is concerned that the shipment of 10 electric toasters just received meet the high standards his customers demand (the price is 50 percent higher than at a discount store). He can't open each of the individual boxes because his customers will not accept appliances in opened boxes. He randomly selects 2 toasters for inspection which he will also use as display models. If 6 of the 10 toasters satisfy the manager's rigid standards, what is the probability that the 2 toasters he selects for inspection will both be satisfactory?

S_1 = Satisfactory toaster is selected on first draw
S_2 = Satisfactory toaster is selected on second draw
D_1 = Defective toaster is selected on first draw
D_2 = Defective toaster is selected on second draw.

Then the question can be stated:

$P(S_1 S_2) = ?$

Because sampling is done without replacement, the probabilities for the second toaster are dependent on whether a satisfactory or defective toaster was drawn first. We must use Rule 2 in the basic probability formulas to find this probability:

$P(S_1 S_2) = P(S_1) P(S_2 | S_1).$

The Classical Definition of Probability can be used to find that

$$P(S_1) = \frac{6}{10}$$

$$P(S_2 | S_1) = \frac{5}{9}.$$

Remember that the first toaster drawn is not replaced, which leaves 9 in the shipment. The second toaster is randomly selected from the remaining 9 toasters. If the first toaster drawn was satisfactory, then there are 5 satisfactory toasters left. Thus, the probability that the second toaster drawn will be satisfactory, given that the first toaster drawn was satisfactory, is 5/9. The solution to the problem is

$$P(S_1 S_2) = P(S_1) P(S_2 | S_1) = \left(\frac{6}{10}\right)\left(\frac{5}{9}\right) = \left(\frac{30}{90}\right) = \frac{1}{3}.$$

The probability is 1/3 that the 2 toasters selected will both be satisfactory.

Example 3.6.2 Refer to the previous example (3.6.1). What is the probability that 1 of the toasters drawn will be satisfactory and the other will be defective?

Two different-quality toasters can occur in 2 ways using the events defined in Example 3.6.1. The first toaster can be satisfactory and the second defective, or the first toaster can be defective and the second satisfactory. Both of these events, $S_1 D_2$ and $D_1 S_2$, result in different-quality toasters; thus, for the event "the toasters are different," we write $\{(S_1 D_2) \cup (D_1 S_2)\}$. The question asked in this problem becomes

$$P\{(S_1 D_2) \cup (D_1 S_2)\} = ?$$

Because the events $S_1 D_2$ and $D_1 S_2$ are mutually exclusive—i.e., they cannot both occur simultaneously—we can apply Rule 5 in the Summary of Basic Probability Formulas to eliminate the brackets:

$$P\{(S_1 D_2) \cup (D_1 S_2)\} = P(S_1 D_2) + P(D_1 S_2).$$

Now Rule 2 can be used to find $P(S_1 D_2)$ and $P(D_1 S_2)$:

$$P(S_1 D_2) + P(D_1 S_2) = P(S_1)P(D_2|S_1) + P(D_1)P(S_2|D_1).$$

The required probabilities can be found using the classical definition as was done for Example 3.6.1:

$$P(S_1) = \frac{6}{10}$$

$$P(D_2|S_1) = \frac{4}{9}$$

$$P(D_1) = \frac{4}{10}$$

$$P(S_2|D_1) = \frac{6}{9}.$$

The probability that the 2 toasters drawn will be of different quality is

$$P(S_1)P(D_2|S_1) + P(D_1)P(S_2|D_1) = \left(\frac{6}{10}\right)\left(\frac{4}{9}\right) + \left(\frac{4}{10}\right)\left(\frac{6}{9}\right) = \frac{48}{90} = \frac{8}{15}.$$

Example 3.6.3 A large corporation is going to introduce 3 new products next year. Based on their experience with similar types of products, the sales managers responsible for the respective products have provided the following probabilities: the probability that product A will be successful is .8; the probability that product B will be successful is .6; and the probability that product C will be successful is .5. The probability of a product's being successful is independent of how the other products perform. What is the probability that exactly 2 of the 3 new products to be introduced will be successful?

More Problem Solving

Let

A = Product A is successful
B = Product B is successful
C = Product C is successful.

There are 3 ways that exactly 2 of the 3 products can be successful. These are: products A and B are successful and product C is not; or products A and C are successful and product B is not; or products B and C are successful and A is not successful. These 3 events can be written $AB\bar{C}$, $A\bar{B}C$, and $\bar{A}BC$, respectively. The question asked can be stated:

$$P\{(AB\bar{C}) \cup (A\bar{B}C) \cup (\bar{A}BC)\} = ?$$

Note that it isn't sufficient to write only AB to indicate that products A and B were the 2 successful ones. The experiment states that we are interested in observing 3 products, and the probability question asks about exactly 2 products being successful. Thus, we have to account for the third product. This is done by writing $AB\bar{C}$, which is read "The 3 events, 'product A is successful,' 'product B is successful,' and 'product C is *not* successful,' all occur when the experiment ['introduce three new products'] is performed." The 3 ways exactly 2 products can be successful are mutually exclusive; so the extended rule, Rule 5' from the Extensions of Rules 4 and 5 can be used to remove the brackets:

$$P\{(AB\bar{C}) \cup (A\bar{B}C) \cup (\bar{A}BC)\} = P(AB\bar{C}) + P(A\bar{B}C) + P(\bar{A}BC).$$

Using Rule 4' (the events are assumed to be independent in the problem) results in

$$P(AB\bar{C}) + P(A\bar{B}C) + P(\bar{A}BC)$$
$$= P(A)P(B)P(\bar{C}) + P(A)P(\bar{B})P(C) + P(\bar{A})P(B)P(C).$$

$P(A)$, $P(B)$, and $P(C)$ are given in the problem. $P(\bar{A})$, $P(\bar{B})$, and $P(\bar{C})$ can be calculated using Rule 1 in the Summary of Basic Probability Formulas; i.e., $P(A) + P(\bar{A}) = 1$. Using these probabilities, the probability that exactly 2 of the 3 products will be successful can be calculated:

$$P(A)P(B)P(\bar{C}) + P(A)P(\bar{B})P(C) + P(\bar{A})P(B)P(C)$$
$$= (.8)(.6)(.5) + (.8)(.4)(.5) + (.2)(.6)(.5)$$
$$= .24 + .16 + .06 = .46.$$

It is important to note that .46 is the probability that exactly 2 of the 3 products will be successful. It is *not* the probability at least 2 of the 3 products will be successful. To find this latter probability, you need to find the probability that all 3 products will be successful and add it to the probability that exactly 2 will be successful.

 ## 3.7 Bayes' Theorem

Before looking at Bayes' Theorem, it is helpful to consider the following problem.

Example 3.7.1 If your friend Joe has his house painted, there is an .8 probability he will be able to sell it by the end of the month. If he doesn't have the house painted, then there is only a .4 chance he will sell the house by the end of the month. The probability that Joe will have his house painted is .7. What is the probability that Joe will sell his house by the end of the month?

This type of problem is best approached by using the following observation:

If 1 of the 2 mutually exclusive events, B and $\bar{B}$, must occur if A is to occur, then

$A = (AB) \cup (A\bar{B})$.

If the event A occurs, it will occur in 1 of 2 ways: either A will occur when event B occurs, or it will occur when the event $\bar{B}$ occurs. This expression is applicable for this type of problem. Clearly, if Joe is going to sell his house by the end of the month, he is either going to have his house painted and sell it or he isn't going to have it painted and sell it. To take advantage of the above observation, we define events as follows:

A = Joe sells his house by the end of the month
B = Joe has his house painted
$\bar{B}$ = Joe does not have his house painted.

You should convince yourself that, with the events as defined above, it is the case that $A = (AB) \cup (A\bar{B})$ for this problem. The question asked in this problem is

$P(A) = ?$

The observation made above can be used to good advantage to answer this question. Since $A = (AB) \cup (A\bar{B})$,

$P(A) = P\{(AB) \cup (A\bar{B})\}$.

By definition of B and $\bar{B}$, the events AB and $A\bar{B}$ are mutually exclusive. Therefore, Rule 5 in the basic probability formulas can be used to remove the brackets:

$P\{(AB) \cup (A\bar{B})\} = P(AB) + P(A\bar{B})$.

The 2 events A and B, as well as the 2 events A and $\bar{B}$, are clearly dependent (the probability that A, "Joe sells his house," will occur depends very much on whether or not B, "the house is painted," occurs); so Rule 2 from the basic probability formulas is used to find $P(AB)$ and $P(A\bar{B})$:

$$P(AB) + P(A\bar{B}) = P(B)P(A|B) + P(\bar{B})P(A|\bar{B}).$$

For $P(AB)$, we could have written the alternative formula $P(AB) = P(A)P(B|A)$, but the probabilities, $P(A)$ and $P(B|A)$, are not available in this problem [the question asks for $P(A)$] so the formula $P(AB) = P(B)P(A|B)$ is used. Thus, for $P(A)$ we end up with the following formula:

$$P(A) = P(B)P(A|B) + P(\bar{B})P(A|\bar{B}).$$

The required probabilities are given in the problem:

$P(B) = .7$
$P(A|B) = .8$
$P(A|\bar{B}) = .4$.

The probability that Joe sells his house by the end of the month can be calculated using these probabilities and observing that $P(\bar{B}) = 1 - P(B) = 1 - .7 = .3$ since B and $\bar{B}$ are complements of each other.

$$P(A) = (.7)(.8) + (.3)(.4) = .56 + .12 = .68.$$

The solution to the following problem leads to the derivation of Bayes' Theorem.

Example 3.7.2   Refer to Example 3.7.1. The end of the month has arrived, and you learn that Joe has sold his house. You do not know whether or not he had it painted first. What is the probability that Joe had his house painted?

Note carefully the difference between the questions asked in Examples 3.7.1 and 3.7.2. In the first example, the experiment hasn't been performed, and the question asks for the probability of one of the final outcomes. For Example 3.7.2, the experiment has been performed, and the final outcome is known, but which of the intermediate events occurred is not known. The question that is asked concerns the probability of one of the intermediate events having occurred, given that the outcome of the experiment is known. Using the same events as defined for Example 3.7.1,

A = Joe sells his house by the end of the month
B = Joe has his house painted
$\bar{B}$ = Joe does not have his house painted,

the question can be stated as

$P(B|A) = ?$

That is, what is the probability that the event B, "Joe has his house painted," occurred, given that it is known that event A, "Joe sells his house by the end of the month," has occurred?

Referring to the definition of conditional probability given in Section 3.2, we have

$$P(B|A) = \frac{P(AB)}{P(A)}.$$

In the solution to Example 3.7.1, we derived the expression $P(A) = P(B)P(A|B) + P(\overline{B})P(A|\overline{B})$ for this experiment. Using this expression in the denominator and applying to the numerator Rule 2 in the Summary of Basic Probability Formulas, we get

$$P(B|A) = \frac{P(B)P(A|B)}{P(B)P(A|B) + P(\overline{B})P(A|\overline{B})}.$$

The latter expression is known as Bayes' Theorem, which is formally stated below.

Bayes' Theorem If one of the two mutually exclusive events, B or $\overline{B}$, must occur if the event A is to occur, then

$$P(B|A) = \frac{P(B)P(A|B)}{P(B)P(A|B) + P(\overline{B})P(A|\overline{B})}$$

and

$$P(\overline{B}|A) = \frac{P(\overline{B})P(A|\overline{B})}{P(B)P(A|B) + P(\overline{B})P(A|\overline{B})}.$$

Using the probabilities given in the problem, the probability that Joe painted his house given that he sold it by the end of the month can be found:

$$P(B|A) = \frac{P(B)P(A|B)}{P(B)P(A|B) + P(\overline{B})P(A|\overline{B})} = \frac{(.7)(.8)}{(.7)(.8) + (.3)(.4)}$$

$$= \frac{.56}{.56 + .12} = \frac{.56}{.68} = .82.$$

This should be compared with $P(B)$, the probability we assign to the event "Joe has his house painted" before the outcome of the experiment is known. The $P(B|A) = .82$ is greater than $P(B) = .7$. This is what we might expect. If we know that Joe sold his house, we are inclined to assign a higher probability to the event that he painted his house because painting his house increases the likelihood of selling it. Note that the numerator of Bayes' Theorem is the term in the denominator associated with the occurrence of event B. To find the $P(\overline{B}|A)$, the denominator remains the same, but the numerator changes to the second term in the denominator, $P(\overline{B})P(A|\overline{B})$.

3A Appendix: Combination Mathematics and the Classical Definition of Probability

The problem concerning the shipment of toasters worked as an example in this chapter can be solved using only the Classical Definition of Probability. The formula for finding the number of combinations that can be made up from a set of objects is useful in enumerating the number of equally likely outcomes in the probability space and in the events of interest.

The Combination Formula, for finding the number of combinations or subsets that can be found taking n objects r at a time, is:

$$C_r^n = \frac{n!}{r!(n-r)!}$$

where $n!$ is the symbol for "n factorial" and is defined as the product of the integers from 1 to n. For example,

$5! = 5 \cdot 4 \cdot 3 \cdot 2 \cdot 1 = 120$
$3! = 3 \cdot 2 \cdot 1 = 6$
$1! = 1$.
$0!$ is defined to be equal to 1.

The number of different pairs (unmatching of course) of socks that can be made up from 5 socks of different colors is the same as the combination of 5 things taken 2 at a time. In the Combination Formula, n is equal to 5 and r is equal to 2:

$$C_2^5 = \frac{5!}{2!(5-2)!} = \frac{5!}{2!3!} = \frac{5 \cdot 4 \cdot 3 \cdot 2 \cdot 1}{2 \cdot 1(3 \cdot 2 \cdot 1)} = \frac{5 \cdot 4}{2} = 10.$$

In Example 3.6.1 in this chapter, the shipment had 6 satisfactory toasters and 4 unsatisfactory toasters. Two toasters were selected without replacement. The question asked was, What is the probability that 2 satisfactory toasters will be selected?

To identify the number of outcomes in a probability space with equally likely outcomes, it is necessary to look at the number of ways the 10 toasters can be selected 2 at a time. This is a combination problem with 10 objects taken 2 at a time. The number of equally likely outcomes for this experiment is

$$C_2^{10} = \frac{10!}{2!8!} = \frac{10 \cdot 9}{2} = 45.$$

This is the denominator for use in the Classical Definition of Probability. For the numerator it is necessary to determine how many of these 45 equally likely outcomes satisfy the event "2 satisfactory toasters are drawn." This can be done by finding out how many combinations of 6 things taken 2 at a time are possible (there are 6 satisfactory toasters in the shipment):

$$C_2^6 = \frac{6!}{2!4!} = \frac{6 \cdot 5}{2} = 15.$$

Using the classical definition,

$$P(2 \text{ satisfactory toasters}) = \frac{15}{45} = \frac{1}{3},$$

which is the same answer we obtained in Section 3.6 using the probability rules developed in this chapter.

Example 3.6.2 referred to the previous problem with the toasters and asked the question, What is the probability that 1 of the toasters drawn will be satisfactory and the other will be defective? The number of equally likely outcomes is the same as above. The number of ways in which 1 good and 1 defective toaster can be drawn when 2 are selected is

$$C_1^6 \cdot C_1^4 = \left(\frac{6!}{1!5!}\right)\left(\frac{4!}{1!3!}\right) = 24.$$

The 2 combination formulas are multiplied together because, for each way a satisfactory toaster can be selected, there are

$$C_1^4 = 4$$

different ways a defective toaster can be selected. There are 24 different possible ways to select 1 satisfactory and 1 defective toaster. Using the Classical Definition of Probability,

$$P(1 \text{ satisfactory and 1 defective}) = \frac{24}{45} = \frac{8}{15},$$

which is the same as the probability found in Section 3.6 using the probability rules developed in this chapter.

3B Appendix: Probability Trees

The first problem, Example 3.7.1, worked in the section on Bayes' Theorem, can also be solved using probability trees. It is sometimes helpful for beginning students to see a problem of this type worked with probability trees; however, the probability event notation approach used in this chapter is the preferred technique because it can readily be adapted to problems with a large number of intermediate events. In theory the probability tree approach also can be used for a large number of intermediate events, but in practice the drawing becomes quite cumbersome.

Example 3.7.1 discussed in the Bayes' Theorem section was as follows:

If your friend Joe has his house painted, there is an .8 probability that he will be able to sell it by the end of the month. If he doesn't have the house painted, then there is only a .4 chance that he will sell the house by the end of the month. The probability that Joe will have his house painted is .7. What is the probability that Joe will sell his house by the end of the month?

The following events were defined:

A = Joe sells his house by the end of the month
B = Joe has his house painted
$\overline{B}$ = Joe does not have his house painted.

The probability tree technique is designed to identify all the possible combinations of outcomes of the events of interest. The extended branches represent conditional probabilities. This is indicated by the probabilities listed underneath the respective branches. The probability tree for this problem is shown in Figure 3.4.

Figure 3.4 Probability Tree

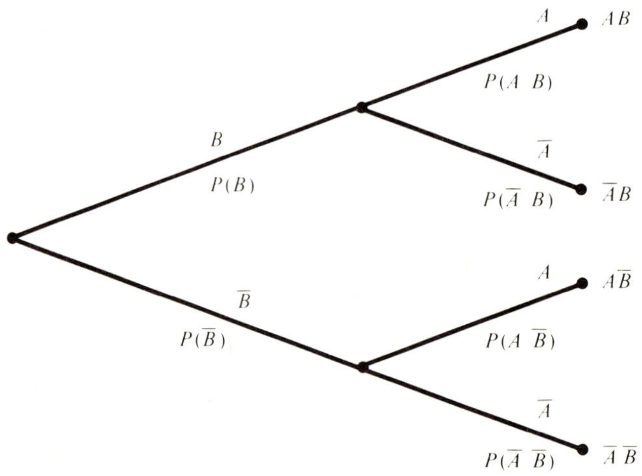

In Figure 3.5, the probability values given in the problem have been added to the probability tree, and then the probabilities at the endpoints have been calculated by multiplying the probabilities along the branches that are followed to arrive at those endpoints. Thus, to get $P(AB)$, we multiplied the values along the B and the A branches respectively, $(.7)(.8) = .56$. From Figure 3.4 it should be evident that we are applying Rule 2 in the Summary of Basic Probability Formulas—namely, $P(AB) = P(B)P(A|B)$.

To find $P(A)$, the probability that Joe will sell his house by the end of the month, the probabilities from the endpoints of branches where A has occurred are added together to get

$$P(A) = P(AB) + P(A\overline{B}) = .56 + .12 = .68.$$

This is the same result that was obtained using the probability event notation approach in this chapter. Actually, as will be revealed by carefully studying the two approaches, they are not conceptually different. The $P(A)$ was determined in Section 3.7 by observing that for this problem the event A occurs one of two ways. Therefore, we can write

$$A = (AB) \cup (A\overline{B}).$$

Then, by applying the rules in the Summary of Basic Probability Formulas, we conclude that $P(A) = P(AB) + P(A\overline{B})$, which is exactly what we came up with using the probability tree technique.

Figure 3.5 Probability Tree with Probabilities

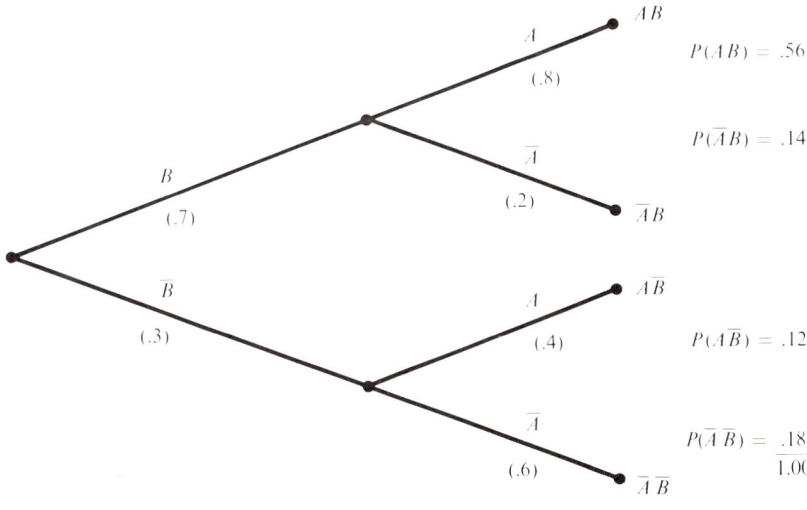

3C Appendix: Extension of Bayes' Theorem

Bayes' Theorem can be extended to handle problems with more than two intermediate events:

Bayes' Theorem

If one of the K mutually exclusive events, $B_1, B_2, \ldots, B_K$, must occur if A is to occur, then the probability of the ith intermediate event, B_i, given that A has occurred, can be found using the following formula:

$$P(B_i|A) = \frac{P(B_i)P(A|B_i)}{\sum_{j=1}^{K} P(B_j)P(A|B_j)}.$$

If the problem of interest had four intermediate events and the probability of the second intermediate event, given A has occurred, is desired, the formula becomes

$$P(B_2|A)$$
$$= \frac{P(B_2)P(A|B_2)}{P(B_1)P(A|B_1) + P(B_2)P(A|B_2) + P(B_3)P(A|B_3) + P(B_4)P(A|B_4)}.$$

Note that the numerator is the term in the denominator that has B_2 in it, the intermediate event of interest.

Example 3C.1

You know that 30 percent of the tires sold under the in-house brand name of a large retailer are made by the ARC Company. About 30 percent of the tires made by the ARC Company wear out before 30,000 miles. Fifty percent of the tires manufactured by the RRR Rubber Company wear out before 30,000 miles. They provide the retailer with 45 percent of its in-house brand tires. The Tyre Company is the supplier of the remaining 25 percent of the in-house brand tires. Only 30 percent of the Tyre Company tires last more than 30,000 miles. You purchase a set of in-house brand tires from the large retailer. They wear out with less than 30,000 miles of service. What is the probability you happened to get tires that were manufactured by the Tyre Company?

The final outcome of interest is the event—"the set of tires purchased last less than 30,000 miles." The intermediate events are that the set of tires was produced by 1 of 3 suppliers to the large retailer. Thus, to use Bayes' Theorem we define the following events:

A = The set of tires purchased wears out with less than 30,000 miles of service
B_1 = The set of tires was manufactured by ARC Company
B_2 = The set of tires was manufactured by RRR Rubber Company
B_3 = The set of tires was manufactured by Tyre Company.

The events B_1, B_2, and B_3 are mutually exclusive events, and 1 of the 3 must occur if event A is to occur—i.e., the 3 manufacturers are the only suppliers of tires to the large retailer. The question asked in this example can be stated as:

$P(B_3|A) = ?$

We know that the tires have worn out prematurely, and we are interested in the probability they came from the Tyre Company. (There are no markings on the tires themselves to indicate which of the 3 companies they came from.)

The following probabilities are given in the problem:

$P(B_1) = .30$
$P(B_2) = .45$
$P(B_3) = .25$
$P(A|B_1) = .30$
$P(A|B_2) = .50$
$P(A|B_3) = .70.$

These probabilities can be used in the extended formulation of Bayes' Theorem to find $P(B_3|A)$:

$$P(B_3|A) = \frac{P(B_3)P(A|B_3)}{P(B_1)P(A|B_1) + P(B_2)P(A|B_2) + P(B_3)P(A|B_3)}$$

$$= \frac{(.25)(.70)}{(.30)(.30) + (.45)(.50) + (.25)(.70)}$$

$$= \frac{.175}{.49} = .357 \cong .36.$$

The probability that the tires were manufactured by the Tyre Company is approximately .36. Before you determined the wearability of the tires, the probability you would assign to purchasing tires made by the Tyre Company from the large retailer is .25. But, after you have learned that they lasted less than 30,000 miles, a larger probability is assigned to the event they are Tyre Company tires because the Tyre Company tires' performance is on the average worse than the tires produced by the other 2 companies.

3D Appendix: Proof of Theorems

The first three theorems given in the text are proved using the axioms from the Axiomatic Definition of Probability. It might be helpful for the reader to refer back to the definitions of the terms at the beginning of Section 3.4 as well as the three axioms stated for the Axiomatic Definition of Probability in the same section.

Theorem 1: $P(\phi) = 0$.
Proof:

$$(S \cup \phi) = S \quad \text{Definition of } \phi$$
$$P(S \cup \phi) = P(S) = 1 \quad \text{Axiom 2}$$
$$P(S) + P(\phi) = P(S) = 1 \quad \text{Axiom 3}$$
$$P(\phi) = 0.$$

Theorem 2: $P(A) + P(\bar{A}) = 1$.
Proof:

$$(A \cup \bar{A}) = S \quad \text{Definition of } \bar{A}$$
$$P(A \cup \bar{A}) = P(S)$$
$$P(A) + P(\bar{A}) = P(S) \quad \text{Axiom 3}$$
$$P(A) + P(\bar{A}) = 1 \quad \text{Axiom 2.}$$

Theorem 3: $P(A) \leq 1$.
Proof:

$$P(A) + P(\bar{A}) = 1 \quad \text{Theorem 2}$$
$$P(A) = 1 - P(\bar{A})$$
$P(A) \leq 1$ Axiom 1; i.e., this axiom states that all probabilities are nonnegative. Subtracting a nonnegative number from 1 results in a number less than or equal to 1.

Problems

1. Given: $P(A) = 1/3$, $P(B) = 1/2$, and $P(A|B) = 1/2$.
   Find:
   a. $P(AB)$;
   b. $P(B|A)$;
   c. $P(A \cup B)$.

2. Given: (same as problem 1).
   a. Are A and B independent? Why or why not?
   b. Are A and B mutually exclusive? Why or why not?

3. There is a 3/4 probability that Joe will be home for Christmas and a 2/3 probability his brother Jim will be home for Christmas. The likelihood of one brother's being home for Christmas is independent of whether or not the other brother is home for Christmas.
   a. What is the probability that at least 1 of the brothers will be home for Christmas?
   b. What is the probability that neither will be home for Christmas?
   c. What is the probability that exactly 1 will be home for Christmas?

4. Given: $P(AB) = 0.40$; $P(A|B) = 0.80$, and $P(A) = 0.60$.
   Find:
   a. $P(B)$;
   b. $P(\bar{A})$;
   c. $P(B|A)$;
   d. $P(A \cup B)$.

5. The J & J Glass Company replaces broken auto glass. The new employee hired by the company causes the glass to crack 2 percent of the time when installing a new windshield. The manager of the company has assigned the new employee to install 2 windshields. He has decided to let the new employee install both windshields even if the employee cracks the first one, hoping that he might improve the new employee's confidence. He realizes that if the new employee cracks the first windshield, the resulting anxiety will increase the probability of breaking the second windshield to .05. What is the probability the new employee breaks both windshields?

6. A hospital has 2 generators of questionable vintage to use in case of a power failure. Each generator has sufficient capacity to supply the hospital with its power needs during a power failure. The probability that generator A will work is .6. The probability is .4 that generator B will work should it be needed. If the hospital should experience a power failure, what is the probability that it will be able to generate its own power?

7. A small-town banker has become aware of 2 investment possibilities in which he is considering investing his own money. The first investment has a .7 probability of achieving a high rate of return, and the second investment has a .6 probability of getting a high rate of return. The probabilities of success of the individual investments are independent of each other. His inclination is to put all his money in the investment with the higher probability of achieving a high rate of return; however, before he does that he would like to consider the possibility of investing an equal amount in each of the 2 investment opportunities. In order to evaluate this second alternative, he feels he needs to know the probability that at least 1 of the investments will have a high rate of return and the probability that both will have a high rate of return. Find the desired probabilities for the banker.

8. Refer to the previous problem. The banker is concerned about the probability of losing money on the 2 investments under consideration. He feels that the probability of losing money on the first investment is .05, and for the second investment it is .03. If he doesn't make a high rate of return or lose money, then he should make a slight return on his money. This latter statement is true for

each of the respective investments. If the banker chooses to invest his money equally in the two investment opportunities, what is the probability he will lose money on both investments? What is the probability he will lose money on at least 1 of the investments?

9. Consider a 1-card draw from an ordinary 52-card deck. What is the probability of:
   a. a diamond;
   b. a face card;
   c. both;
   d. either;
   e. neither?

10. An urn contains a large number of beads—30 percent red and 70 percent black. Two beads are chosen at random.
    a. What is the probability that both are black?
    b. What is the probability of 1 red and 1 black?

11. Refer to the urn in Problem 10 above and consider the following experiment: One bead is drawn; if red an honest die is tossed, but if black an honest coin is tossed. What is the probability of:
    a. red and 6;
    b. black and 6;
    c. black and heads;
    d. 6 if the bead is red;
    e. red and an odd number?

12. The head of research and development for a large corporation has an engineering problem that needs to be solved within 3 months. Her best 3-person engineering team has a .85 probability of coming up with a solution within the time limit working as a team. She is concerned, however, about the possibility that working as a team (or one might say a committee) stifles individual creativity. She knows the abilities of the individual team well enough to make probability assessments of the likelihood of each of the engineers solving the problem within 3 months if the engineer were to work alone on the problem. The respective probabilities are .75, .60, and .55. If the research manager sends the 3 engineers to 3 different laboratories to work independently of each other on the problem, what is the probability that at least 1 of the 3 engineers is going to solve the problem within 3 months?

13. The mayor of a small Texas town in the tornado belt is considering the purchase of a tornado warning siren for the town. The super-reliable, Japanese-made model will operate when needed with a probability of .995. The American-made model is reliable 98 percent of the time. The American model costs only half as much as the Japanese model. Would he be better off buying the Japanese model or two American models—i.e., what is the probability that

at least 1 of the 2 American models will operate when needed if their performance is independent of each other?

14. The quality control manager has devised a sampling plan such that there is a 30 percent chance that lots with 10 percent defective items will be accepted. Unacceptable lots are returned to the vendor. If the next 3 lots to be sampled each have 10 percent defective items and if they are sampled independently of each other, what is
   a. the probability all 3 lots will be rejected;
   b. the probability exactly 2 of the lots will be accepted;
   c. the probability that 2 or more will be rejected?

15. A stockout condition appears likely with respect to a certain diode because at current usage rates the inventory will be used up in 3 weeks. The probability that vender A will be able to make a delivery within 3 weeks is .70. The probability for vender B is .35. To improve the odds the inventory control manager stipulates that orders be placed to vendors immediately. What is the probability that within 3 weeks:
   a. both orders will be received;
   b. neither order will be received;
   c. new stock will be available?
   (Assume vendors act independently of each other.)

16. The Intelligence Unit is trying to crack the enemy code by the end of the month. They have 2 top people working independently on the code at 2 different locations. The probability that the first agent will crack the code in the allotted time is .6. The probability for the success of the second agent is .8. What is the probability the code will be cracked?

17. Thirty percent of all students at the University of Utah are freshmen. Forty percent of all students come from Salt Lake County. Fifteen percent of all students are both freshmen and from Salt Lake County. A student is selected at random. If the student is a freshman the event A occurs, and if a student is from Salt Lake County the event B occurs.
   a. Are A and B independent events? Why or why not?
   b. Find the probability that the student selected will be a freshman from Salt Lake County.
   c. If it is known that the student selected is a freshman, what is the probability he or she is also from Salt Lake County?
   d. Find the probability that the student selected will be a freshman but not from Salt Lake County.

18. Ten percent of the items in a large lot of parts are defective. The quality control technician decides to randomly select 4 items and accept the lot if all 4 selected parts are good. What is the probability he will accept the lot?

19. Two honest dice are tossed. What is the probability of:
    a. getting 2 sixes;
    b. not getting 2 sixes;
    c. getting 2 non-sixes?

20. The probability Joe passes statistics is .7 and the probability he passes algebra is .9. If these are the only 2 courses he is taking, what is the probability he will have at least 1 passing grade?

21. A marble is randomly selected from an urn containing 7 red marbles and 3 blue marbles. A marble is also selected from another urn with 2 black marbles and 3 blue marbles.
    a. What is the probability that both the marbles selected will be blue?
    b. What is the probability that both will be red?
    c. What is the probability that 1 of the marbles will be blue and the other will be red?

22. A marketing manager is trying to determine the probability that a new product he is considering will be successful. The success of the product depends on the state of the economy for the next 12 months. Based on his own assessment and that of economists, he has some confidence: he believes there is only a 20 percent chance that the economy will be so bad that it would have an appreciable adverse effect on his firm. If the economy does have an appreciable adverse effect, then the probability of a successful introduction of the new product is only .15. However, if the economy is good enough during the next 6 months so that it does not have an adverse effect on the firm, the probability is .90 that the new product will be successful. Given the marketing manager's probabilities of the events discussed above, what is the probability the new product he wants to introduce at this time will be a success?

23. The personnel manager classified the nonsalaried personnel in one of the company's plants according to occupation and residence. The plant of interest is the primary economic base of the small city in which it is located. The personnel manager obtained the following data on the plant's employees.

	Residence			
Job Title	West Suburb	South Suburb	City Center	Commutes from Another City
Machinist	15	10	5	2
Assembly Line	26	58	18	21
Maintenance	16	10	24	7

    If a nonsalaried employee is randomly selected from the plant, what is the probability that:
    a. the employee selected will be a machinist;

b. the employee selected lives in 1 of the 2 suburbs;
   c. the employee selected is either an assembly line worker or a maintenance worker;
   d. the employee selected is a maintenance worker and commutes from another city;
   e. the employee selected is a machinist or lives in the city center or both;
   f. if the employee selected is an assembly line worker, he or she will also live in the south suburb?

24. A pair of dice is to be thrown 3 times. What is the probability that the total number of spots will be between 2 and 11 inclusive on at least 1 of the 3 throws?

25. A young man hired a year ago as a production worker feels that he is being discriminated against by his firm. There were a total of 12 people hired by the firm at the same time he was hired (he is 1 of the 12). Since that time, 2 of the 12 have been promoted to supervisory positions. Both of the supervisors are women. All 5 women and 7 men hired at the same time are still working together. Before taking his complaint to the personnel manager, the male employee would like to know the probability that, if 2 people were randomly selected for supervisory positions from the 12 hired together, both would be female. What is the desired probability?

26. The president, 3 vice-presidents, and the secretary-treasurer are the only employees of the firm allowed to use the initials of their first and second names only as signatures for interoffice memos. All other employees must sign their last name. If we were to assume that each of the 26 letters of the alphabet is equally likely to be the first letter of a given name (clearly an invalid assumption), what would be the probability of at least 1 duplication of 2-initial signatures among the 5 executives of the firm? Note that there are $26 \times 26 = 676$ possible different ordered pairs of letters of the alphabet.

27. An urn contains 4 white and 3 red balls. Two balls are drawn. What is the probability that they are of the same color if:
   a. they are drawn without replacement;
   b. the first ball is replaced before the second is chosen?

28. Three new products are being introduced on the market. The products are so different that the success or failure of 1 of the products does not affect the success or failure of the other products. There is a .7 probability that the first product will be successful. The probabilities of success are .8 and .6, respectively, for the second and third products.
   a. What is the probability that exactly 2 products will be successful?

b. What is the probability at least 2 of the products will be successful?
c. What is the probability that none of the products will be successful?

29. A space system is composed of 2 stages, I and II. The probability that stage I will function properly is .9, and the probability that stage II will function properly is .95. Both stages must function if the space system is to be successful.
   a. What is the probability that the space system will be successful?
   b. If a backup for stage I is provided with a probability of .8 that it will function properly (the 2 components are independent of each other), what is the probability that the required stage I function will be performed?
   c. What is the probability that the space system will be successful if the backup system referred to in part b is provided?

30. Electronic calculators are packed in lots of 24 for shipment to small retailers. Because of time limitations, the quality control inspector only has time to inspect 2 calculators from each lot. If a lot of 24 calculators has 1 defective calculator, what is the probability that 1 of the 2 randomly selected calculators will be the defective calculator?

31. Refer to the previous problem.
   a. What is the probability that the 2 randomly selected calculators will both be good if there are 2 defective calculators in the lot of 24?
   b. What is the probability that the lot will still be passed as good (both calculators sampled are good) if 3 of the 24 calculators are bad?

32. Make a subjective probability assignment of each of the following:
   a. At least 1 of your professors will not show up for class tomorrow.
   b. A fellow student picked at random will have graduated from a high school in your state.
   c. There will be a major rainstorm or snowstorm 3 weeks from today.
   d. You will meet and chat with the president of the United States sometime in the next 5 years.
   e. A thumbtack when dropped on a hardwood floor will land point up.
   f. The next adult (not including your fellow students and your family) you will meet will have an annual income of over $50,000.
   g. In each of the above cases, formalize the basis for your assignment if you can.

33. The probability assignment in parts a and e in Problem 32 could be estimated based upon empirical data:
    a. For the next several weeks record professor absenteeism.
    b. Drop a thumbtack on a hard uniform surface 100 times or so. Record the number of times it lands point up.
    c. Compare these empirical estimates with your subjective estimates.

34. Given $P(A) = k$ and $P(B) = 3k$.
    a. If A and B are independent and $P(AB) = 0.12$, find k. What is $P(A \cup B)$?
    b. If A and B are mutually exclusive and $P(A \cup B) = 1.00$, find k.

35. Thirty-five percent of the Slobovian labor force is female. Sixty percent of the labor force is unmarried. In a random draw of one person from the labor force, and assuming that being married and sex are independent attributes, what is the probability of drawing:
    a. a married female;
    b. a person who is either married or a female;
    c. a married person, *if* the person drawn is male?

■ The remaining problems either require Bayes' Theorem (Section 3.7) or the material in the appendices or both.

36. A company has 3 stamping machines producing armature disks. Two of the machines are relatively new, and each of the new machines produces 35 percent of the total number of armature disks. The third machine is an older machine producing the remaining 30 percent of the disks. The 2 new machines are not the same with regard to quality of output; the first machine produces 1 percent defectives, and 2 percent of the disks produced by the second new machine are defective. The old machine's output is 6 percent defective. What percent of all parts produced by the 3 machines is defective?
    Hint: This problem can be worked 2 ways: (1) Find the denominator of Bayes' Theorem in Appendix 3C, i.e., $A = (AB_1) \cup (AB_2) \cup (AB_3)$ or (2) use the method developed in Appendix 3B.

37. A factory has 2 machines used in the manufacture of the same product. The numbers of pieces produced by these machines each day are 1,200 and 1,800, respectively. Suppose that the first machine is known to produce on the average 5 percent defectives, and the second produces 10 percent defectives on the average.
    a. If 1 piece is selected at random from a day's production on both machines (3,000 items) and is found to be defective, what is the probability that this piece came from the first machine?
    b. If that piece were found to be nondefective instead, what is the probability that this piece came from the first machine?

c. Are your answers complementary to each other? Why or why not?

38. A local doctor has devised a new test that shows positive with a .7 probability if the individual receiving the test is going to have a heart attack within a year. Unfortunately, there is a 10 percent chance the test will show positive when the individual is *not* going to have a heart attack within a year. The makeup of the city is such that 1 in every 600 adult males can be expected to have a heart attack within the next year. An adult male is selected at random and given the test. The test shows positive. What is the probability that the selected individual will have a heart attack within a year?

39. A marble is to be drawn from 1 of 2 urns. There is a 3/4 probability it will be drawn from the first urn and a 1/4 chance it will be drawn from the second. The first urn has 8 red and 2 white marbles. The second urn has 4 red and 6 black marbles.
    a. What is the probability that a red marble will be drawn?
    b. What is the probability that a black marble will be drawn?
    c. If a red marble has been drawn, what is the probability that it came from the second urn?
    d. What is the probability that a white marble will *not* be drawn?
    e. If a black marble is drawn, what is the probability that it came from the first urn?
    f. What is the probability that a black or white marble will be drawn?

40. If one of the weekly output of armature disks produced by the three machines in Problem 36 is selected at random and found to be defective, what is the probability that the disk was produced by the old stamping machine?

41. Ten students are eligible for the snack bar student advisory committee which will be made up of 4 students. All members will have equal responsibility on the committee.
    a. How many committees are possible if the 4 students are randomly selected?
    b. Joe and Jane are the only two Communists of the 10 eligible students. What is the probability that the committee will be 50 percent Communist?
    c. Six of the 10 students are female. What is the probability that the committee will have a majority of females?

42. Work Problem 27a using the Combination Formula given in Appendix 3A.

43. Work Problem 21 using the probability tree method discussed in Appendix 3B.

CHAPTER 4
RANDOM VARIABLES AND THE BINOMIAL DISTRIBUTION

In this chapter we give consideration to both discrete and continuous random variables, with considerable emphasis on the binomial distribution. This is the most important discrete model; the most important continuous function, the normal curve model, will be discussed in the next chapter.

Specific objectives are—

to present the definition of a discrete random variable and to discuss the related general properties and parameters (Sections 1 and 2).

to present in detail, with examples, the concepts and properties of the binomial distribution (Section 3).

to show the use of the binomial table (Table A) for finding binomial probabilities (Section 4).

to discuss the general properties of a continuous random variable (Section 5).

The concept of a random variable provides the transition from the physical probability experiments, discussed in the preceding chapter, to mathematical probability models such as the binomial model, a major topic of consideration in this chapter, and the normal distribution, discussed in the next chapter. Consideration will be given to both discrete and continuous random variables and the differences and similarities between these two classes. The concept of expectation (or expected value) provides the basis for a formal definition of the mean and variance of a random variable and a function of a random variable. Expected value theory is also basic to the understanding of statistical decision theory introduced in Chapter 15.

4.1 Discrete Random Variables

The mathematician would define a random variable as "a real-valued function defined on a probability space." A simpler but equivalent definition is:

Random Variable A random variable is a function (or rule) that assigns a number to each of the outcomes of a probability space. The value of the random variable is the assigned number.

If we let $p(x)$ be the probability function of the discrete random variable, x, then the probability that x equals some value x_i will be written as $P(x = x_i)$ or $p(x_i)$. In many cases where we are dealing with only one random variable and there is no chance for confusion, we will simply write p_i.

Properties of a Probability Function There are two important properties of this function; they follow directly from the discussion in Chapter 3 but should be formally stated:

1. $p(x_i) \geq 0$ (the probability function is nonnegative for all values of the random variable);
2. $\sum_i p(x_i) = 1$ (the sum of the set of probabilities must be unity).

These properties apply to probability functions for discrete random variables. In Section 5 of this chapter, the characteristics of continuous random variable probability functions are considered.

A discrete random variable takes on only a finite or a countable infinite number of values. (The facts that a random variable may take on infinitely many values and that the infinite sum of the corresponding probabilities is unity may be confusing to you who have not been introduced to infinite series or calculus. As a practical matter, there is a point where this infinite sum is as close to unity as we please; i.e., the sum of the remaining set of probabilities is so small that they may be effectively ignored. In many applications, the countably infinite set is reduced to a finite set, one that is adequate to satisfy the needed accuracy. As an example, consider the experiment of tossing an honest coin until heads first appears. Let x be the number of tosses; x can take on the set of positive integers from 1 to infinity and $p(x = k) = 1/2^k$. There is less than 1 chance in 1 million that the number of tosses will exceed 20. Perhaps we would be willing to ignore this very small chance if it were an applied problem, and deal with the finite set of integral values from 1 through 20.) Let us begin with a very simple example to illustrate the points we have been making.

Example 4.1.1 Many statistics teachers require their students to play the following game. A fair coin is tossed (a statistics teacher uses nothing but fair coins) and if heads occurs the student pays the teacher $10. If tails occurs, the teacher pays the student $5.

If the student's gain and loss, i.e., +5 and −10, are assigned to the possible outcomes of the experiment, "flip a coin once," then a random variable has been defined. The number −10 is assigned to the outcome "heads" and the number +5 is assigned to the outcome "tails." Let x represent the random variable and the values it might assume. Then for this random variable the question might be asked,

$P(x = 5) = ?$

This latter expression is read, "the probability that the random variable will take on the value 5 is equal to what?" An experiment is going to be performed, a rule has been established for assigning numbers to the outcomes of the experiment, and a question is asked as to what the probability is that the outcome of the experiment will be an outcome that has the number 5 assigned to it. For this random variable, the following probability statements are true:

$P(x = -10) = 1/2$
$P(x = -3.68) = 0$
$P(x = +5) = 1/2$
$P(x = \sqrt{8}) = 0.$

The values −3.68 and $\sqrt{8}$ will never occur because they are not assigned to any of the outcomes in the probability space (remember that the probability space is exhaustive as well as mutually exclusive).

It will be evident to the reader that defining a random variable makes the real number line a probability space. When this experiment is performed, one could ask, "What is the probability the number 5 or the number −10 will occur?" If a probability of 0 is assigned to all of the other values on the real number line, then a probability function, $p(x)$, has been defined for all values of x, where x is any real number.

4.2 Expected Value and Variance of a Random Variable

The expected value concept is used to define the mean and the variance of a random variable.

Expected Value

The functional notation $E(x)$ is read as "the expected value of x." For a discrete random variable, this is defined as

$$E(x) = \sum_i x_i p(x_i).$$

(Each value of the random variable is multiplied by the corresponding probability and the products are summed.)

This expectation is the mean value, μ, of the random variable. Even though this is a purely mathematical definition, we can gain some insight and understanding by thinking of the expectation as a long-run average value.

Example 4.2.1

Find the expected value of the random variable defined by the student-teacher game referred to in Example 4.1.1

There are only 2 values, -10 and 5, for which the probability function has a non-0 value. Thus the probability function for the student-teacher game can be written as

$p(x) = 1/2$ for $x = -10$
$p(x) = 1/2$ for $x = 5$
$p(x) = 0$ for all other values of x.

We note that this function satisfies the properties discussed in Section 1; i.e.,

$p(x_i) \geq 0$ for all x

and

$$\sum_i p(x_i) = p(-10) + p(5) = 1/2 + 1/2 = 1.$$

The expected value of the student-teacher random variable is calculated using the definition of expected value:

$$E(x) = \sum_i x_i p(x_i) = -10 \cdot p(-10) + 5 \cdot p(5)$$
$$= -10(1/2) + 5(1/2) = -2.50.$$

112 Random Variables and the Binomial Distribution

The expected value (for the student) of the game is a loss of $2.50. Each time the game is played the student will either lose $10.00 or win $5.00; but if the game is played a large number of times, the student will, on the average, lose $2.50 per game. If the game were played only once this is still the expected value of the game, even though the student will either end up a $10.00 loser or a $5.00 winner.

The variance, σ^2, of a random variable is defined as the expected value of the squared difference between the random variable and its expectation:

$$\sigma^2 = E[(x - \mu)^2].$$

Variance

From the definition of expectation we have the variance in summation form as

$$\sigma^2 = \sum_i (x_i - \mu)^2 \, p(x_i).$$

Standard Deviation

The standard deviation, σ, is defined as the square root of the variance:

$$\sigma = \sqrt{\sigma^2}.$$

Appendix 4A has additional discussion and related theorems on expectation. The following examples will illustrate all of the points considered in the chapter so far.

Example 4.2.2

Find the variance and the standard deviation of the random variable defined in the preceding examples (the teacher-student problem).

From the solution to Example 4.2.1, we have

$E(x) = \mu = \$-2.50.$

The variance of the random variable is calculated directly from the definition:

$$\sigma^2 = \sum_i (x_i - \mu)^2 \, p(x_i)$$

$$= [-10 - (-2.50)]^2 \cdot p(-10) + [5 - (-2.50)]^2 \cdot p(5)$$

$$= (-7.5)^2 (1/2) + (7.5)^2 (1/2) = 56.25.$$

The standard deviation is found by taking the square root of the variance:

$$\sigma = \sqrt{56.25} = \$7.50.$$

Example 4.2.3

An honest die is tossed once. Let y be the number of spots showing. Find $p(y)$, $E(y)$, and σ.

When any of the terms *honest*, *fair*, or *true* is used, we are making (or we are trying to implement) the equal-likelihood assignment to the set of possible outcomes. Thus, the probability function for the random variable, y, is

$$p(y) = \frac{1}{6} \text{ for } y = 1, 2, 3, 4, 5, 6$$

$p(y) = 0$ otherwise.

The expected value of y is calculated from the definition:

$$E(y) = \sum_i y_i p(y_i) = (1)\left(\frac{1}{6}\right) + (2)\left(\frac{1}{6}\right) + (3)\left(\frac{1}{6}\right) + (4)\left(\frac{1}{6}\right) +$$

$$(5)\left(\frac{1}{6}\right) + (6)\left(\frac{1}{6}\right) = \frac{21}{6} = 3.5.$$

The expected value, μ, is used to obtain the variance:

$$\sigma^2 = \sum_i (y_i - \mu)^2 p(y_i)$$

$$= (1 - 3.5)^2 \left(\frac{1}{6}\right) + (2 - 3.5)^2 \left(\frac{1}{6}\right) + (3 - 3.5)^2 \left(\frac{1}{6}\right) +$$

$$(4 - 3.5)^2 \left(\frac{1}{6}\right) + (5 - 3.5)^2 \left(\frac{1}{6}\right) + (6 - 3.5)^2 \left(\frac{1}{6}\right) = \frac{17.5}{6} = 2.917.$$

The standard deviation is the square root of the variance:

$$\sigma = \sqrt{\sigma^2} = \sqrt{2.917} = 1.708.$$

Example 4.2.4   An urn contains 7 white balls and 3 red balls. A 2-ball random draw is made: (a) without replacement; and (b) with replacement. Let the random variable, x, be the sample number of white balls. For each case, find the probability function, the mean, and the standard deviation.

In both cases, there are 4 possible outcomes: *WW*, *WR*, *RW*, and *RR*. *WW* maps into $x = 2$, *WR* and *RW* both map into $x = 1$, and *RR* maps into $x = 0$. The random variable takes on the discrete values, 0, 1, and 2; these will have non-0 probabilities and all other real numbers will have 0 probabilities. (Note that we could have used the number of red balls as the random variable; but, since the number of white balls plus the number of red balls is always exactly 2, knowing the distribution of x is equivalent to knowing the distribution of the number of red balls.)

First we consider the sampling-without-replacement problem. The probabilities of the 4 outcomes can be obtained using the methods covered in Chapter 3. The resulting probabilities are

$$P(WW) = \frac{7}{15}$$

$$P(WR) = \frac{7}{30}$$

$$P(RW) = \frac{7}{30}$$

$$P(RR) = \frac{1}{15}.$$

The probability the random variable x will equal 2 is 7/15. The probability it will equal 1 is $7/30 + 7/30 = 7/15$, and the probability the random variable will be 0 is 1/15. This results in the following probability function:

$$p(x) = \frac{7}{15} \text{ for } x = 2$$

$$= \frac{7}{15} \text{ for } x = 1$$

$$= \frac{1}{15} \text{ for } x = 0$$

$$= 0 \text{ otherwise.}$$

(Note that the probabilities are all nonnegative and sum to unity.)
The expected value and standard deviation can now be readily obtained:

$$E(x) = \sum_i x_i p(x_i) = (2)\left(\frac{7}{15}\right) + (1)\left(\frac{7}{15}\right) + 0\left(\frac{1}{15}\right) = \frac{7}{5} = 1.40$$

$$\sigma^2 = \sum_i (x_i - \mu)^2 p(x_i)$$

$$= (2 - 1.4)^2 \left(\frac{7}{15}\right) + (1 - 1.4)^2 \left(\frac{7}{15}\right) + (0 - 1.4)^2 \left(\frac{1}{15}\right) = .373$$

$$\sigma = \sqrt{\sigma^2} = \sqrt{.373} = .611.$$

Thus, the mean of this random variable is 1.40 and the standard deviation is .611.

Next, consider the sampling-with-replacement part of the above example. If the ball is replaced after each draw, the probability of white (or red) is the same from trial to trial. This is an example of independent trials. We have the following probabilities for the basic sample space:

$P(WW) = .49$
$P(WR) = P(RW) = .21$
$P(RR) = .09.$

The probability function can be determined from these probabilities. The probability the random variable x will equal 0 (i.e., the number of white balls will be 0) is the same as $P(RR)$. Proceeding in this manner results in the following probability function:

$$\begin{aligned} p(x) &= .09 \quad \text{for } x = 0 \\ &= .42 \quad \text{for } x = 1 \\ &= .49 \quad \text{for } x = 2 \\ &= 0 \quad \text{otherwise.} \end{aligned}$$

This probability distribution is slightly different from that obtained for the sampling-with-replacement case. However, the expected value for both cases is the same:

$$\mu = E(x) = \sum_i x_i p(x_i) = 0(.09) + 1(.42) + 2(.49) = 1.40.$$

The standard deviations are not the same:

$$\sigma^2 = \sum_i (x_i - \mu)^2 p(x_i)$$

$$= (0 - 1.4)^2 (.09) + (1 - 1.4)^2 (.42) + (2 - 1.4)^2 (.49) = .420$$

$$\sigma = \sqrt{\sigma^2} = \sqrt{.420} = .648.$$

This standard deviation, .648, is larger than σ, .611, for the sampling-with-replacement case; thus, replacement increases the dispersion of the random variable. Intuitively, we expect more dispersion when sampling with replacement because there are more possible samples. For example, if a sample of 2 marbles is being drawn without replacement from a population of 5 differently colored marbles, it is impossible to obtain a sample of 2 marbles that are the same color; but, if the sampling is done with replacement, there are 5 different ways of selecting a sample with both items selected being the same.

The first case, sampling without replacement, in the above problem is an example of the hypergeometric distribution. This distribution will be discussed in Appendix 4C. The second case, sampling with replacement, is an example of the binomial distribution; Sections 3 and 4 of this chapter are devoted to this important distribution.

Example 4.2.5 Consider the following game related to the toss of an honest die (see also example 4.2.3): If an odd number shows, the player loses $9; if an even number shows, the player wins double the number in dollars. Find the expected payoff.

Let y be the number showing on the die and w be the payoff in dollars; note that w is a random variable and y is just a device for identifying which of the six physical sides of the cube is on top. The data and the calculation of the expected value of the random variable are presented below in a tabular format.

y	w (in $)	p(w)	wp(w)
odd	−9	$\frac{1}{2}$	$-\frac{9}{2}$
2	4	$\frac{1}{6}$	$\frac{2}{3}$
4	8	$\frac{1}{6}$	$\frac{4}{3}$
6	12	$\frac{1}{6}$	2
Column sums		1.00	$-\frac{1}{2}$ = −$.50.

The monetary expectation in this game is a negative $.50. Over the long run, the player will lose an average of $.50 per game. (If the loss for an odd number had been $8 instead of $9, the expectation would have been 0. When $E(w) = 0$ for a game, the game is called a "fair" game.)

4.3 The Binomial Distribution

The binomial random variable is defined on the probability space of a binomial experiment. The binomial probability model is applicable to a class of problems that frequently occurs in business and economics. The binomial probability function provides a mathematical formula that can be used to calculate probability questions arising from a binomial experiment. In this section the binomial probability function is derived using the probability theory developed in the previous chapter; because binomial probabilities are arithmetically cumbersome to decimalize, a limited set of binomial tables is provided in Table A. The use of this table is also discussed in Section 4.4.

The reader should find the development of the probability function of interest. It is important that you should pay close attention to the characteristics of the binomial experiment so that you can recognize the type of problem for which the binomial probability model is applicable.

Binomial Experiment

The binomial experiment has the following characteristics:

1. It consists of n independent, identical trials.
2. Each trial has 2 possible outcomes: success or failure or, if you please, yes or no, favorable or unfavorable, 1 or 0, attribute A or not A, etc.

3. The probability of success, π, is the same for each trial.
4. The binomial random variable, x, is the number of successes that occur in the n trials of the binomial experiment.

The outcome of interest is arbitrarily labeled "success." Either the trial results in the attribute of interest or it doesn't. The labeling of the first outcome "success" and the other possible outcome "failure" is arbitrary. The important thing is to be consistent throughout the problem; thus, if the experiment is to flip a coin 5 times, either heads or tails may be labeled "success" for each trial, with the other outcome labeled "failure." The labeling must be the same for each trial. Note that the binomial random variable is a discrete random variable; it takes on the $n + 1$ integer values from 0 to n inclusive.

The real-world problem described by the binomial model falls into the category of what is often called *attribute sampling*. The dichotomy must be defined; i.e., the item sampled either has attribute A or it doesn't have attribute A, and it is necessary that we have independent trials. The following list should help you decide when you can use the binomial distribution as a probability model:

1. Clearly, a purely physical process, such as tossing a coin or throwing a die, must be assumed to comprise independent trials. The coin has no memory; the probability of heads is the same for each trial.
2. Many industrial processes, once set in motion, are assumed to approximate independent trials. The dichotomy usually being considered in an industrial process is defective or nondefective.
3. If a finite population is being sampled, the sampling must be done with replacement in order for the trials to be independent.
4. If the population is infinite in size or, more realistically, very large in comparison to the sample size, independent trials will be very closely approximated even if the sampling is done without replacement. As a rule of thumb, we would say that if you are doing less than 5 percent sampling there is no need to sample with replacement in order to assume independent trials.

It might be well at this point to assure ourselves that "flipping a coin 5 times" is a binomial experiment. Each time the coin is flipped is a trial. The trials are independent of each other, and the probability of success on each trial is the same, namely, 1/2, i.e., we assume it to be a "fair" coin. In this case $n = 5$. Each trial has 2 possible outcomes, heads or tails. If heads is arbitrarily labeled success and tails failure, then the possible outcomes are success and failure. The binomial random variable is the number of successes; in this case, the number of heads that occur when a coin is flipped 5 times.

There are 6 possible values that the binomial random variable defined for the above coin-flipping experiment can assume: 0, 1, 2, 3, 4, and 5. If 2 heads and 3 tails should occur when the coin is flipped 5 times, the value of the random variable would be equal to 2. If 5 tails should occur, i.e., no heads, then the value of the random variable would be 0. It is important for the student to note again that 0 successes is a possible outcome and, thus, there are $n + 1$ possible outcomes for a binomial experiment that has n trials.

Example 4.3.1 Ten percent of the parts produced by a machine are defective. The likelihood of a part being defective is independent of whether or not the previous part produced was defective. What is the probability that exactly two of the next five parts produced are defective?

This is a binomial experiment. There are 5 independent, identical trials. Each trial has 1 of 2 outcomes; the part is either defective or not defective. If we label the outcome "defective" as success, then the probability of success is the same for each trial—namely, $\pi = .10$. In terms of the random variable, a question may be asked: What is the probability the random variable will take on the value 2? i.e., What is the probability of exactly 2 successes? Another way to state this question is

$P(x = 2) = ?$,

where x is the sample number of defectives. If the probability function were known, we would simply evaluate it at $x = 2$ to find the answer to this probability question. The probability can be found using methods developed in Chapter 3, and the solution of this problem leads to the derivation of the binomial probability function.

There are 10 sequences in which exactly 2 defective parts can occur. For example, the first 2 parts may be defective and the next 3 good:

$(D_1 D_2 G_3 G_4 G_5)$.

The probability of this event's occurring is

$$P(D_1 D_2 G_3 G_4 G_5) = P(D_1) P(D_2) P(G_3) P(G_4) P(G_5)$$
$$= (.1)(.1)(.9)(.9)(.9) = (.1)^2 (.9)^3.$$

The probability of a defective part was given as .1. The complement of the event "a defective part" is the event "a good part;" so the probability of a good part is $1 - .1 = .9$. All the trials are independent; so the probability that the 5 specified outcomes will sequentially occur when the experiment is performed is the product of the simple probabilities of the respective 5 outcomes.

Another of the 10 sequences in which exactly 2 defective parts can occur is

The Binomial Distribution

$(G_1D_2G_3G_4D_5)$.

For this case, the first part is good, the second defective, the third and fourth are good, and the last part is defective. The probability of this event is

$$P(G_1D_2G_3G_4D_5) = P(G_1)\,P(D_2)\,P(G_3)\,P(G_4)\,P(D_5)$$
$$= (.9)(.1)(.9)(.9)(.1) = (.1)^2\,(.9)^3.$$

This is exactly the same probability as that obtained for the first way in which exactly 2 defectives occurred. Table 4.1 lists all 10 sequences in which exactly 2 defectives can occur along with the respective probabilities of each of these events. Each of these has a probability of $(.1)^2(.9)^3$. The 10 ways are physically different and therefore mutually exclusive events; therefore, the probability of exactly 2 defectives is the sum of the probabilities of the 10 ways 2 defectives can occur. Since the probability is the same for each of the 10 ways, we get

$$P(x = 2) = (10)\,(.1)^2\,(.9)^3 = .0729.$$

The probability that exactly 2 of the next 5 parts produced by the machine will be defective is .0729.

Combination Formula

Although a person can determine the number of ways exactly 2 defects can occur by listing all possible combinations, the result can be obtained much more easily by using the combination formula:

$$C_r^n = \frac{n!}{r!(n-r)!}$$

where

$C_r^n = $ the number of possible combinations of n things taken r at a time.

Table 4.1 The 10 Possible Ways 2 Defects Can Occur

Ways That Exactly Two Defects Can Occur	Probability of the Event's Occurring
$D_1D_2G_3G_4G_5$	$(.1)(.1)(.9)(.9)(.9) = (.1)^2(.9)^3$
$D_1G_2D_3G_4G_5$	$(.1)(.9)(.1)(.9)(.9) = (.1)^2(.9)^3$
$D_1G_2G_3D_4G_5$	$(.1)(.9)(.9)(.1)(.9) = (.1)^2(.9)^3$
$D_1G_2G_3G_4D_5$	$(.1)(.9)(.9)(.9)(.1) = (.1)^2(.9)^3$
$G_1D_2D_3G_4G_5$	$(.9)(.1)(.1)(.9)(.9) = (.1)^2(.9)^3$
$G_1D_2G_3D_4G_5$	$(.9)(.1)(.9)(.1)(.9) = (.1)^2(.9)^3$
$G_1D_2G_3G_4D_5$	$(.9)(.1)(.9)(.9)(.1) = (.1)^2(.9)^3$
$G_1G_2D_3D_4G_5$	$(.9)(.9)(.1)(.1)(.9) = (.1)^2(.9)^3$
$G_1G_2D_3G_4D_5$	$(.9)(.9)(.1)(.9)(.1) = (.1)^2(.9)^3$
$G_1G_2G_3D_4D_5$	$(.9)(.9)(.9)(.1)(.1) = (.1)^2(.9)^3$
	.0729

The symbol $n!$ means "n factorial" and is defined as $n(n - 1)(n - 2) \ldots (3)(2)(1)$, i.e., the product of the integers from 1 to n. For example, $8! = (8)(7)(6)(5)(4)(3)(2)(1) = 40{,}320$; $5! = (5)(4)(3)(2)(1) = 120$; and $1! = 1$. The 0 factorial, $0!$ is defined to be equal to 1.

In the problem above we were interested in the number of possible combinations of 5 things taken 2 at a time. Using the formula above with $n = 5$ and $r = 2$,

$$C_2^5 = \frac{5!}{2!(5-2)!} = \frac{5!}{2!\,3!} = \frac{5 \cdot 4 \cdot 3 \cdot 2 \cdot 1}{(2 \cdot 1)(3 \cdot 2 \cdot 1)} = 10.$$

Thus, the probability of exactly 2 heads when a coin is flipped 5 times can be calculated as follows:

$$P(x = 2) = C_2^5 (.1)^2 (.9)^3 = (10)(.1)^2(.9)^3 = .0729.$$

Binomial Probability Function

This latter expression suggests the general formula for the binomial probability function:

$$p(x) = C_x^n \pi^x (1 - \pi)^{n-x}$$

where

$n = $ the number of trials in the binomial experiment

$x = $ the number of successes

$\pi = $ the probability of success for each trial.

The notation $P(x|n, \pi)$ will also be used to refer to the probability of x successes when we wish to identify the 2 parameters, n and π, of the binomial distribution. This latter expression is read, "the probability of x given n and π."

For the problem just worked above, the number of trials was $n = 5$ and the probability of success on each trial was $\pi = .1$. The probability function for this particular problem is

$$p(x) = C_x^n \pi^x (1 - \pi)^{n-x} = C_x^5 (.1)^x (.9)^{5-x}$$

where

$x = $ the number of successes.

Note that this is a function of x only, once n and π have been determined. To find the probability of 2 defectives, we simply evaluate this function at $x = 2$:

$$P(x = 2) = p(2) = C_2^5 (.1)^2 (.9)^{5-2} = (10)(.1)^2(.9)^3 = .0729.$$

The Binomial Distribution

If we want to know the probability of no defects, i.e., $P(x = 0)$, we simply evaluate the probability function at $x = 0$:

$$P(x = 0) = p(0) = C_0^5 (.1)^0(.9)^{5-0} = \frac{5!}{0!(5-0)!}(.1)^0(.9)^5$$

$$= (1)(.1)^0(.9)^5 = (.9)^5 = .59049.$$

Using the binomial notational form introduced above, we could write an expression for the latter 2 probabilities as follows:

$P(x = 2|5, .1) = .0729$

and

$P(x = 0|5, .1) = .59049.$

The binomial distribution is a 2-parameter family of distributions. If the 2 parameters, n and π, are known, then the probability for any possible outcome of the binomial random variable can be found. Figure 4.1 has the probabilities calculated for all possible outcomes when $n = 5$ and $\pi = .1$, as was the case for the example we have been discussing. All possible outcomes of the random variable are considered and, as indicated, the probabilities sum to 1. The graph of the probability function is also included in Figure 4.1.

The mean and variance (or standard deviation) of the binomial random variable can be calculated using the definitions given earlier in this chapter (Section 4.2).

Mean and Variance of the Binomial Random Variable

The following formulas can be derived for the general binomial distribution from the definitions and are, of course, much easier and more efficient than calculating the mean and variance for each case. It can be shown that for a binomial distribution:

$\mu = n\pi$

and

$\sigma^2 = n\pi(1 - \pi)$

where

μ = the mean of the binomial random variable
σ^2 = the variance of the binomial random variable
n = the number of trials
π = the probability of success on each trial.

The formula for the mean agrees with our intuition. For example, if 20 percent of the apples in an orchard are spoiled and we randomly select 150 apples, we would expect on the average to obtain 30 bad apples (20 percent of 150). This represents a binomial experiment with $n = 150$

Figure 4.1 Graph of Binomial Probability Function

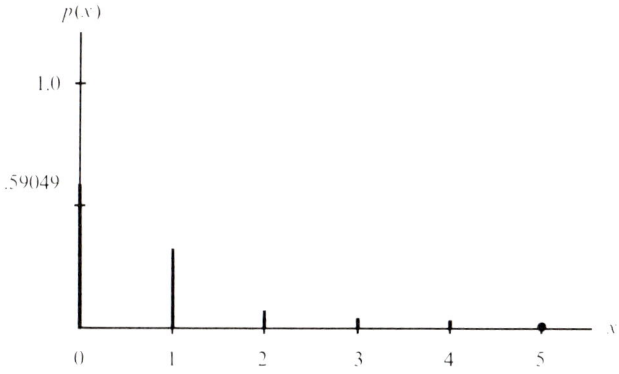

Binomial probabilities for $n = 5$ and $\pi = .1$:

$P(x = 0) = p(0) = C_0^5 (.1)^0(.9)^5 = .59049$

$P(x = 1) = p(1) = C_1^5 (.1)^1(.9)^4 = .32805$

$P(x = 2) = p(2) = C_2^5 (.1)^2(.9)^3 = .07290$

$P(x = 3) = p(3) = C_3^5 (.1)^3(.9)^2 = .00810$

$P(x = 4) = p(4) = C_4^5 (.1)^4(.9)^1 = .00045$

$P(x = 5) = p(5) = C_5^5 (.1)^5(.9)^0 = \underline{.00001}$
$ 1.00000$

and $\pi = .20$. The mean value calculated using the above is $\mu = n\pi = 150(.20) = 30$.

Example 4.3.2

Find the mean and standard deviation of the binomial random variable defined in Example 4.3.1.

For this problem we have $n = 5$ (5 parts are randomly selected) and $\pi = .10$ (10 percent of all parts produced are defective). The mean and standard deviation can be calculated using the formulas given above:

$\mu = n\pi = 5(.10) = .5$

$\sigma^2 = n\pi(1 - \pi) = 5(.10)(1 - .10) = 5(.1)(.9) = .45$

$\sigma = \sqrt{\sigma^2} = \sqrt{.45} = .67.$

The mean is .5 and the standard deviation is .67. If we take a large number of samples of 5 parts, the average number of defects per sample will be .5, although it is obvious that .5 will never occur for any single sample; we cannot get half a defective.

4.4 Using the Binomial Table

Even with an electronic calculator, calculating binomial probabilities can be tedious and time consuming. Table A has some precalculated binomial probabilities. The table is limited to binomial distributions with n equal to 5, 10, 15, 20, or 25 and π equal to 0.01, 0.05, 0.10, 0.20, ..., 0.90, 0.95, and 0.99. The probabilities given in the table are cumulative or less-than-or-equal-to probabilities. Using our binomial notation, the probabilities in the table would be $P(x \le k)$, which is equal to the sum $p(0) + p(1) + \ldots + p(k)$. For most applications the cumulative table is more efficient than a table with individual probabilities. As shown in the following examples, the individual probabilities can readily be found using the cumulative probability table.

The efficient use of the binomial tables will be clarified by a discussion of the following example. To help you gain more insight into the logic of finding individual probabilities, the probabilities associated with a closed interval on the variable and those of the greater-than-or-equal-to type of schematic are presented in Figure 4.2, which parallels the discussion. Let us reemphasize two important points:

1. Table A gives directly the less-than-or-equal-to probabilities and only those; i.e., $P(x \le k \mid n, \pi)$.

Figure 4.2 Finding Probabilities Using the Binomial Table (Table A)

a.
```
       x ≤ 2
    ┌───────┐
  0   1   2   3   4   5
  └─────┘
   x ≤ 1
```

$P(x = 2) = P(x \le 2) - P(x \le 1) = .683 - .337 = .346.$

b.
```
   x ≤ 1      2 ≤ x ≤ 4
  ┌───┐      ┌───────┐
  0   1   2   3   4   5
  └───────────────┘
          x ≤ 4
```

$P(2 \le x \le 4) = P(x \le 4) - P(x \le 1) = .990 - .337 = .653.$

c.
```
           Total probability = 1
  ┌───────────────┬───────┐
  0   1   2   3   4   5
  └───────┘       └───────┘
    x ≤ 3           x ≥ 4
```

$P(x \ge 4) = 1 - P(x \le 3) = 1 - .913 = .087.$

2. The binomial distribution is discrete; it assumes only the integral values from 0 through n.

Example 4.4.1   Consider the binomial variable with $n = 5$ and $\pi = 0.40$. Find:

a. $P(x \leq 2|5, 0.40)$; i.e., find the probability that x will be at most 2;
b. $P(x = 2|5, 0.40)$; i.e., find the probability that x will be exactly 2;
c. $P(2 \leq x \leq 4|5, 0.40)$; i.e., find the probability that x is at least 2 and no more than 4—from 2 to 4 inclusive;
d. $P(x \geq 4|5, 0.40)$; i.e., find the probability that x is at least 4.

We first locate the subtable of Table A for a sample size of 5, $n = 5$. The intersection of the column for $\pi = .4$ and the row for $k = 2$ reads 0.683. This is the answer to the first part of our problem, in decimal form with 3-place accuracy; it is the sum $P(x = 0) + P(x = 1) + P(x = 2)$.

$P(x \leq 2|5, 0.40) = 0.683$.

For the second part of the problem, refer to the schematic in Figure 4.2a as we present the logic. The event, $x \leq 1$, means that $x = 0$ or $x = 1$; the event, $x \leq 2$, as indicated above means that $x = 0, x = 1$, or $x = 2$. If the $P(x \leq 1)$ is subtracted from the $P(x \leq 2)$, we are left with the probability that x is exactly 2.

$P(x = 2) = P(x \leq 2) - P(x \leq 1) = 0.683 - 0.337 = 0.346$.

Figure 4.2b illustrates the logic used to find $P(2 \leq x \leq 4|5, 0.40)$. The interval, $2 \leq x \leq 4$, includes $x = 2, 3$, or 4; therefore, if we subtract $P(x \leq 1)$ from $P(x \leq 4)$, we are left with the probability that x will assume a value of either 2, 3, or 4.

$P(2 \leq x \leq 4) = P(x \leq 4) - P(x \leq 1) = .990 - .337 = .653$.

Lack of careful thought may lead you to subtract $P(x \leq 2)$ from $P(x \leq 4)$ for an answer to the above question. A moment's reflection will show you that this leaves only $P(x = 3) + P(x = 4)$.

For the last part of the example, see Figure 4.2c. In order to find $P(x \geq 4)$ we note that the event, $x \geq 4$, is the complement of $x \leq 3$. These two events exhaust the universe of possibilities, which means that $P(x \leq 3) + P(x \geq 4) = 1$. It follows immediately that

$P(x \geq 4) = 1 - P(x \leq 3) = 1 - .913 = .087$.

We recommend, as a first step, that probabilities involving less-than or greater-than symbols only be changed to less-than-or-equal-to or greater-than-or-equal-to probabilities. For example, change $P(x < 7|10, .60)$ to $P(x \leq 6|10, .60)$; or change $P(x > 3|15, 0.10)$ to $P(x \geq 4|15, 0.10)$.

Example 4.4.2   A state university has a student body made up of 80 percent in-state students and 20 percent out-of-state students. If 10 students are randomly selected to form a committee for studying out-of-state tuition costs, what is the probability that there will be no out-of-state students on the committee? What is the probability that the number of out-of-state students will be between 1 and 3 inclusive?

The attribute of interest for this problem is whether or not a student is an out-of-state student; thus, $\pi = 0.20$ and, because 10 students are to be selected, $n = 10$. The first probability we are asked to find can be read directly from the table:

$$P(x \leq 0) = P(x = 0) = .107.$$

There is approximately an 11 percent chance that no out-of-state students will be selected. The second probability is found by using logic similar to that presented in Figure 4.2b:

$$P(1 \leq x \leq 3) = P(x \leq 3) - P(x \leq 0) = .879 - .107 = .772.$$

There is a 77.2 percent chance that the number of out-of-state students on the committee will be either 1, 2, or 3.

4.5 Continuous Random Variables

To this point our discussion has centered on discrete random variables. The other type of random variable that is very important in statistical analysis is the continuous random variable. A continuous random variable is one that can assume any value in an interval of the real number line. An interval on the real number line has an uncountably infinite number of points which the continuous random variable can assume. An example of a continuous random variable would be the length of life of light bulbs produced by a firm. If the number assigned to the outcome is the length of time in hours, then a continuous random variable has been defined. In theory, any fraction of an hour could occur; in practice, accuracy is limited by the precision of the measuring instrument. Generally speaking, when the random variable values consist of measurements of time, distance, weight, etc., we use a continuous random variable model.

When the number of elements in a sample space are uncountably infinite, we are unable to assign non-0 probabilities to the point values of the random variable.

Properties of a Probability Density Function   The probability distribution of a continuous random variable is defined by its probability density function. The probability density function, $f(x)$, for a continuous random variable has the following properties:

1. $f(x) \geq 0$ for all x.
2. The total area under the curve defined by $f(x)$ is equal to 1.

$P(a \leq x \leq b)$

The probability that a random variable will take on a value in an interval is defined as the area under the curve over that interval, i.e.,

$P(a \leq x \leq b)$ = the area under the curve of the probability density function between the points a and b $(a \leq b)$.

The expression $P(a \leq x \leq b)$ is read, "the probability that the random variable x will assume a value between a and b, inclusively." Figure 4.3 illustrates this concept. The shaded area in Figure 4.3 represents the desired probability. Property 2 for a density function states that the total area under the curve in Figure 4.3 is 1, so the area between a and b would be less than 1.

We observe from Figure 4.3 that the $P(x = a)$ is the area under the curve over the point a. This is, of course, 0, since the area of a line is 0 by definition. Because the probability assigned to the endpoints of an interval is 0, the probability assigned to an interval is the same whether or not the endpoints are included; i.e.,

$P(a \leq x \leq b) = P(a < x < b)$.

Note that this latter statement does *not* hold in general for the discrete case; for example, the probabilities in the binomial table depend very much on whether or not the endpoints are included in the interval of interest.

The main concept to understand from this section is that the probability for a continuous random variable is equal to the area under the

Figure 4.3 A Continuous Probability Function

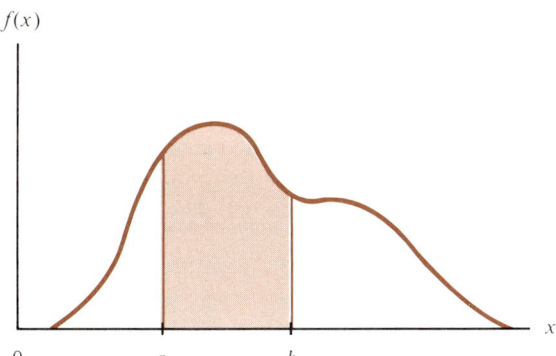

probability density function. This concept underlies the discussion of any continuous probability model and specifically the discussion of the normal distribution in the next chapter. A table of areas under the standard normal curve for selected values of the variable is available; so it won't be necessary to mathematically find the required areas. If you have some knowledge of calculus, you might be interested in Appendix 4B at the end of this chapter, which covers the more formal mathematical statements of the definitions given above.

There are a number of continuous probability models that can be used and understood by the student who has had no calculus. Most of these draw upon our knowledge of plane geometry. In order to illustrate this type of model, we have included an example using the right triangle as a model. Look for other examples in the problem set at the end of the chapter.

Example 4.5.1

Wholesale grocer Schmidt figures that the probability density function for bulk coffee beans is as follows:

$$f(x) = \frac{x}{8} \quad 0 \leq x \leq 4$$

$$= 0 \text{ otherwise.}$$

The variable x represents hundreds of pounds of beans ordered for a given day.

a. Demonstrate that the above function satisfies the necessary conditions for a probability density function.
b. What is the probability that tomorrow's demand will be less than 240 pounds?

The function given in this example is a probability density function if it satisfies the criteria listed at the beginning of this section:

1. $f(x) \geq 0$ for all x.
2. The area under the curve of $f(x)$ is equal to 1.

The function is plotted in Figure 4.4. It is positive or 0 for all values of x, i.e., the graph of the function never dips below the x-axis; therefore, the first condition is satisfied. The area under the curve is the same as the area of a triangle with base equal to 4 and height equal to 1/2; i.e., $f(4) = 4/8 = 1/2$. You will recall that the area of a triangle is equal to 1/2 the height times the base. The area of the triangle in Figure 4.4 is $(1/2)(1/2)(4) = 1$, which means that the second criterion for a probability density function is satisfied.

Because the minimum amount of bulk coffee beans that can be

Figure 4.4 Probability Density Function for Example 4.5.1

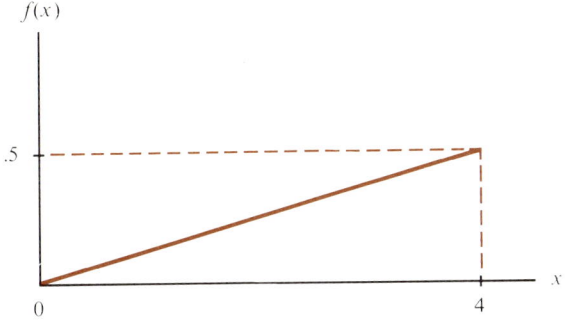

demanded for a day is 0, the probability that the demand for tomorrow will be less than 240 pounds can be stated as

$P(0 \leq x \leq 2.4)$.

Remember that the unit for x is hundreds of pounds. The probability that a continuous random variable will assume a value in an interval is equal to the area under the curve over the interval. The value of $f(x)$ at $x = 2.4$ is $(2.4)/8 = .3$. The area we need to find is the shaded area in Figure 4.5. Employing the area-of-a-triangle formula gives us $(½)(.3)(2.4) = .36$. The probability that wholesaler Schmidt will sell less than 240 pounds of bulk coffee beans tomorrow is .36.

Figure 4.5 The $P(0 \leq x \leq 2.4)$

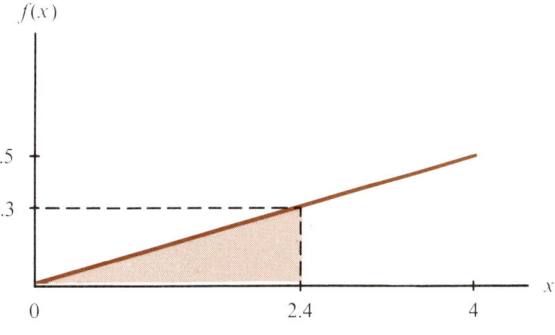

Continuous Random Variables

4A Appendix: Expected Value and Variance Theorems with Selected Proofs

The following theorems are equally applicable to both discrete and continuous random variables.

Theorem 1.   If k is a constant,
$$E(k) = k.$$
Theorem 2.   If k is a constant,
$$E(kx) = kE(x).$$
Theorem 3. $E(x \pm y) = E(x) \pm E(y).$
Theorem 4.   If x and y are *independent* random variables,
$$E(xy) = E(x)E(y).$$
Theorem 5.   If k is a constant,
$$\sigma^2_{kx} = k^2 \sigma^2_x \text{ and } \sigma_{kx} = k\sigma_x.$$
Theorem 6.   If x and y are *independent* random variables,
$$\sigma^2_{x \pm y} = \sigma^2_x + \sigma^2_y.$$

The subscripts on σ^2 indicate the variance of the random variable of interest. For example, σ^2_x is the variance of the random variable x; σ^2_{x+y} is the variance of the random variable t, where $t = x + y$; etc. If only one random variable is being discussed, then the subscript is dropped and σ^2 refers to the variance of this random variable. In Theorem 5, reference is made to the variance of a function of the random variable, kx, which is itself a random variable, and reference is also made to the variance of x itself. Thus subscripting becomes necessary to distinguish between the two variances. Theorem 6 also refers to three different variances. There is *not* a typographical error on the right-hand side of Theorem 6. It is the case that the variance of the difference of two random variables is equal to the sum of the variances of the two random variables. Theorem 6 is the theorem of most interest because it is used for some special types of applied statistics problems discussed later in this text; however, Theorems 2 through 5 are used to prove Theorem 6.

Proofs of Some of the Expected Value and Variance Theorems

Theorems 1 through 3 have similar proofs. Theorem 2 is selected as an example.

Theorem 2.   If k is a constant,
$$E(kx) = kE(x).$$

Proof:

1. From the definition of expected value,
$$E(kx) = \Sigma(kx)f(x).$$

2. $\Sigma kxf(x) = k\Sigma xf(x)$ from the properties of the summation sign. Remember, k is constant.

3. $\Sigma x f(x) = E(x)$ by definition of $E(x)$; therefore,
   $k\Sigma x f(x) = kE(x)$.

The proof of Theorem 4 requires the use of the properties of joint probability functions or random variables covered in introductory texts on mathematical statistics. The independence requirement is important. Two random variables are independent if the outcome of one random variable doesn't affect the probability distribution of the second random variable. In general, it is not the case that $E(xy) = E(x)E(y)$. Theorem 3, $E(x \pm y) = E(x) \pm E(y)$, holds for any two random variables x and y, but Theorem 4 requires that x and y be independent random variables.

The proof of Theorem 5 employs Theorem 2 and the definitions of the variance and the mean.

Theorem 5.   If k is a constant,
$$\sigma^2_{kx} = k^2\sigma^2_x.$$

Proof:

1. $\sigma^2_{kx} = E[(kx - \mu_{kx})^2]$.
2. $\mu_{kx} = E(kx) = kE(x) = k\mu_x$. This follows from the definition of the mean and Theorem 2.
3. Substituting $k\mu_x$ for μ_{kx},
   $E[(kx - \mu_{kx})^2] = E[(kx - k\mu_x)^2] = E[k^2(x - \mu_x)^2] = k^2 E[(x - \mu_x)^2]$.
   This follows from Theorem 2.
4. We note that
   $$E[(x - \mu_x)^2] = \sigma^2_x$$
   by definition of σ^2_x; so $k^2 E[(x - \mu_x)^2] = k^2 \sigma^2_x$.

Theorem 6 is really the statement of two theorems—one for the plus sign and one for the minus sign. The case for the minus sign is proved here because it is the one referred to in Chapter 9 and also because it demonstrates to the reader that the plus sign does belong on the right-hand side.

Theorem 6.   If x and y are independent random variables,
$$\sigma^2_{x \pm y} = \sigma^2_x + \sigma^2_y.$$

Proof of $\sigma^2_{x-y} = \sigma^2_x + \sigma^2_y$:

1. $\sigma^2_{x-y} = E\{[(x - y) - (\mu_{x-y})]^2\}$ by definition of variance.
2. We note from Theorem 3 and the definition of μ, $\mu_{x-y} = E(x - y) = E(x) - E(y) = \mu_x - \mu_y$. This value is substituted for μ_{x-y} in (1), the terms within the brackets are regrouped, the expression in

brackets squared, and Theorems 2 and 3 are applied to the result as follows:

$$E\{[(x-y) - \mu_{x-y}]^2\} = E\{[(x-y) - (\mu_x - \mu_y)]^2\}$$
$$= E\{[(x - \mu_x) - (y - \mu_y)]^2\}$$
$$= E[(x - \mu_x)^2 - 2(x - \mu_x)(y - \mu_y) + (y - \mu_y)^2]$$
$$= E[(x - \mu_x)^2] - 2 E[(x - \mu_x)(y - \mu_y)] + E[(y - \mu_y)^2].$$

Theorem 3 permits us to apply the expected value operator, E, to each of the individual terms. Theorem 2 was applied to the middle term to bring out the constant, 2.

3. Because x and y are independent, Theorem 4 can be used:

$$E[(x - \mu_x)(y - \mu_y)] = E(x - \mu_x) E(y - \mu_y).$$

But from Theorems 1 and 3 and the definition of μ_x we get

$$E(x - \mu_x) = E(x) - E(\mu_x) = \mu_x - \mu_x = 0.$$

Thus the middle term vanishes, and we have

$$\sigma_{x-y}^2 = E[(x - \mu_x)^2] + E[(y - \mu_y)^2].$$

Applying the definition of variance to the terms on the right-hand side of the equation gives us

$$\sigma_{x-y}^2 = \sigma_x^2 + \sigma_y^2.$$

There are two propositions,

$$\mu_{k+x} = k + \mu_x$$

and

$$\sigma_{k+x}^2 = \sigma_x^2,$$

that frequently occur as steps in proofs of applied statistical theorems. These can readily be proved from the theorems given above. See if you can determine the theorems used for each of the steps implied by the following expressions.

$$\mu_{k+x} = E(k + x) = E(k) + E(x) = k + \mu_x$$
$$\sigma_{k+x}^2 = \sigma_k^2 + \sigma_x^2 = \sigma_x^2.$$

Note that $\sigma_k^2 = 0$ because a constant has no variability; i.e., $\sigma_k^2 = E[(k - \mu)^2] = E[(k - k)^2] = E(0) = 0.$

4B Appendix: Mathematical Formulation for the Continuous Random Variable Case

The reader with a background in calculus may be interested in the mathematical formulation of this chapter's verbal discussion of a continuous random variable. A function, $f(x)$, is a probability density function of a continuous random variable if it satisfies the following two conditions:

1. $f(x) \geq 0$ for all x.

2. $\int_{-\infty}^{\infty} f(x)dx = 1$.

The first condition requires that the function never assumes a negative value; i.e., the graph of the function never dips below the x-axis. The second condition is taken from integral calculus and is equivalent to saying that the total area under the probability density curve and above the x-axis is equal to 1.

The probability of an event for a continuous random variable is defined using integral calculus:

$$P(a \leq x \leq b) = \int_a^b f(x)dx \quad (a \leq b).$$

This is the same thing that was stated verbally in the chapter: the probability that a random variable will take on a value in a specified interval is equal to the area under the probability curve over the interval. Using this definition, it can be shown that the probability that a continuous random variable takes on a point value is 0:

$P(x = a)$ can be written as $P(a \leq x \leq a)$; the latter probability is defined as

$$P(a \leq x \leq a) = \int_a^a f(x)dx = 0.$$

The mean of a continuous random variable is defined as

$$\mu = E(x) = \int_{-\infty}^{\infty} xf(x)dx.$$

The definition of the variance of a continuous random variable also employs integral calculus:

$$\sigma^2 = E[(x - \mu)^2] = \int_{-\infty}^{\infty} (x - \mu)^2 f(x)dx.$$

These definitions are analogous to those for the discrete random variable; the integral sign replaces the summation sign and $f(x)$ is used to denote a continuous probability density function in lieu of $p(x)$ for a discrete probability function.

4C Appendix: The Hypergeometric Probability Distribution

You will recall from Chapter 3 that the probability of drawing two red marbles from an urn with different colored marbles varies according to whether the marbles are being drawn with replacement or without replacement. If the marble selected is replaced before the next draw is made, then the probability of red on the second draw is the same as it was on the first draw regardless of whether or not a red marble was selected on the first draw. The selections of the marbles represent independent trials with the same probabilities. The binomial probability distribution would be applicable for this case (i.e., sampling with replacement).

On the other hand, if the first marble isn't replaced before the next draw is made, the probability of red on the second draw is not the same as that for the first draw and, in addition, it depends on whether or not the first draw was red. When sampling from a finite population without replacement, we experience dependent trials, and the binomial model is not applicable. The hypergeometric model describes this type of problem, and the desired probabilities can be calculated from the formula for $P(x)$.

For the hypergeometric distribution, the random variable x is the number of successes in n trials. As with the binomial distribution, the outcome of each trial is dichotomous—either success or failure. The parameters of the hypergeometric distribution are the population size, N, the number of successes in the population, K, and the sample size, n. The probability function, $p(x)$, is obtained by counting the number of equally likely possible outcomes and then applying the Classical Definition of Probability discussed in Chapter 3. The ratio of the number of samples with exactly x successes and therefore $n - x$ failures to the number of possible sample outcomes is the probability of x successes, or $p(x)$. There are

$$C_n^N$$

equally likely possible samples of size n selected without replacement from a population of size N.[1] There are

$$C_x^K$$

ways that x success can occur, and for each of these combinations there are

$$C_{n-x}^{N-K}$$

combinations of $n - x$ failures. The total number of samples of size n with exactly x successes and $n - x$ failures is the product of these two numbers:

$$C_x^K C_{n-x}^{N-K}.$$

Hypergeometric Probability Function

The hypergeometric distribution function is the ratio of this product to the total number of possible samples of size n:

$$p(x) = \frac{C_x^K C_{n-x}^{N-K}}{C_n^N}$$

where

x = the number of successes in the sample
n = the sample size
N = the population size
K = the number of successes in the population.

Consider the following example:

Example 4C.1

What is the probability of getting exactly 3 aces in a 5-card poker hand dealt from an ordinary 52-card deck?

The population size is 52 and the sample size is 5. There are 4 aces in the deck; so $K = 4$. The hypergeometric probability function for this problem is

$$p(x) = \frac{C_x^K C_{n-x}^{N-K}}{C_n^N} = \frac{C_x^4 C_{5-x}^{48}}{C_5^{52}}.$$

The probability of 3 successes is found by evaluating this function at $x = 3$:

[1]. The combination symbol, C_x^n, is defined in Appendix 3A.

$$p(3) = \frac{C_3^4 C_2^{48}}{C_5^{52}} = \frac{\left(\frac{4!}{3!1!}\right)\left(\frac{48!}{2!46!}\right)}{\left(\frac{52!}{5!47!}\right)} = .0017. \quad \text{(Approximately 1 chance in 600!)}$$

(If you take time to verify the decimal approximation of .0017, you will find hypergeometric calculations to be even more tedious than binomial calculations.)

The mean value or expected value of the hypergeometric variable is identical to that of the binomial:

$$\mu = E(x) = n\pi \qquad \text{(Note that } \pi = k/N\text{)}.$$

You will recall that π is the population proportion of successes. The formula for the variance of the hypergeometric random variable is

$$\sigma^2 = n\pi(1-\pi)\left(\frac{N-n}{N-1}\right).$$

For the poker hand example above, the expected number of aces, $\mu = 5(4/52) = 0.385$. (The most likely number of aces is 0 with a probability of 0.66.) Also, $\sigma^2 = .327$ and $\sigma = .572$. Note that if the sample size, n, is small compared to the population size, N, the expression $(N - n)/(N - 1)$ is close to 1 and the variance is, then, approximately equal to $n\pi(1 - \pi)$. This latter expression is the formula for the variance of a binomial distribution. This helps to justify a statement made earlier in this chapter to the effect that the binomial model may be used when sampling without replacement if the population is very large relative to the sample size.

4D Appendix: The Poisson Distribution

The Poisson Distribution is of particular interest in queuing theory or waiting-line models. The results obtained assuming a Poisson Distribution are relatively simple to apply. There are situations where the Poisson Distribution provides a fairly good approximation to the actual behavior of the random variable of interest. For this case, probabilities can be readily calculated using the Poisson probability function. Unfortunately, costly errors have been made in practice by assuming that the random variable of interest follows a Poisson Distribution when in fact it does not. The derivation of the Poisson Distribution is beyond the scope of this text, but the assumptions needed for the derivation are interesting and also helpful when trying to make a judgment about the validity of the Poisson model for a particular application. Let the variable x be the number of occurrences; i.e., $x = 0, 1, 2$, etc. If there exists a small enough unit of time such that: (1) $P(x = 1)$ is a positive number and (2) $P(x \geq 2)$ is 0 for all practical purposes, then the random variable x has a Poisson Distribution.

The Poisson random variable can assume the value 0 or any positive integer. The probability of extremely large integer values gets very small as the integer gets larger; thus, for practical applications we can assume there is essentially a finite number of outcomes. The Poisson probability function is defined as follows:

$$p(x) = \frac{\lambda^x e^{-\lambda}}{x!}$$

where

λ = the mean of the random variable x
$x = 0, 1, 2, 3, 4, \ldots$

The symbol e represents the so-called natural number encountered in calculus; it is a constant and is equal to 2.7183 to the nearest 4 decimal places. The function is only defined for $x = 0$ or a positive integer.

The classic example of the application of the Poisson Distribution has to do with calls coming into a telephone switchboard.

Example 4D.1

The average number of calls coming into a switchboard during the busiest period of the day, 9:00 to 10:00 A.M., for a small firm is 5 calls per minute. The switchboard will handle 10 calls per minute. If the number of incoming calls follows a Poisson Distribution, what is the probability that for any given minute there will be exactly 2 calls? What is the probability of 5 or less calls? What is the probability the switchboard will be jammed—i.e., receive more than 10 calls?

The Poisson random variable, x, for this problem is the number of calls coming in during a given minute. The value of the Poisson parameter, λ, is 5 for this problem: it is the average number of calls per minute. The probability of exactly 2 calls can be found from the table of the cumulative Poisson Distribution, Table F, in a manner similar to that employed for binomial probability problems, or it can be calculated directly from the Poisson probability function given above. Employing the first approach, we have

$P(x = 2) = P(x \leq 2) - P(x \leq 1) = .1247 - .0404 = .0843.$

The $P(x \leq 2)$ was found by looking under the column headed $\lambda = 5$ in Table F and along the row for $k = 2$. The probability of exactly 2 incoming calls is 0.843. We get essentially the same answer using the formula:

$$P(x = 2) = p(2) = \frac{\lambda^x e^{-\lambda}}{x!} = \frac{5^2 e^{-5}}{2!} = \frac{25 \frac{1}{(2.7183)^5}}{2 \cdot 1} = .0842.$$

The discrepancy is due to rounding in the table. If the table entries had been carried out to 5 decimal places, the answers would be the same to 4 decimal places.

Appendix to Chapter 4

The probability of 5 or less calls is read directly from Table F:

$P(x \leq 5) = .6160.$

The probability of more than 10 calls is found as follows:

$P(x > 10) = 1 - P(x \leq 10) = 1 - .9863 = .0137.$

The logic of this solution is the same as for the binomial case. The Poisson outcomes are all integer outcomes; thus $P(x > 10)$ is the same as $P(x \geq 11)$.

If the parameter n of the binomial distribution is very large and the parameter π is very small, then the Poisson Distribution with $\lambda = n\pi$ can be used to approximate binomial probabilities. For example, if x has a binomial distribution with $n = 1000$ and $\pi = .003$, the probability that x will be 5 or less can be found from the cumulative Poisson Distribution Table, Table F, using $\lambda = n\pi = 1000(.003) = 3$. Looking under the column for $\lambda = 3$, we get $P(x \leq 5) = .9161$. Because binomial tables for $n = 1000$ and $\pi = .003$ are not generally available, this method of approximation can be quite useful and does save considerable calculation. In the applied area of quality control sampling for fraction of output that is defective, the Poisson table is used almost exclusively because of the high volume of binomial tables that would be required to replace the one Poisson table.

4E Appendix: The Binomial Distribution and the Binomial Theorem

It is interesting to note how the binomial distribution is related to the general binomial expression, $(x + y)^n$, encountered in algebra courses. Let $\theta = 1 - \pi$. The binomial theorem found in college algebra texts states that

$$(\pi + \theta)^n = \sum_{x=0}^{n} C_x^n \pi^x \theta^{n-x}.$$

This is the sum of the binomial probabilities from 0 to n. This also provides a proof that the sum of the binomial probabilities is equal to 1. The sum is equal to $(\pi + \theta)^n$, which is equal to $(\pi + (1 - \pi))^n = 1^n = 1$.

Implicit in the above is the fact that n is a positive integer; it is interesting to note that, if n is not a positive integer, the general form of the expression still holds but an infinite series is generated.

The following formula is interesting and quite useful if a full set of binomial probabilities needs to be hand calculated:

$$P(x = k + 1) = \left(\frac{n - k}{k + 1}\right)\frac{\pi}{\theta} P(x = k).$$

This formula can be used to find the probabilities for the following example:

Example 4E.1 A sample of size 4 was randomly selected from a large population that has 12 percent senior citizens. Let x be the sample number of senior citizens.

Begin with $P(x = 0) = (1 - \pi)^n = (.88)^4 = .600$;

$$P(x = 1) = \left(\frac{4-0}{0+1}\right)\left(\frac{.12}{.88}\right) \quad P(x = 0) = 4\left(\frac{.12}{.88}\right)(.600) = .327$$

$$P(x = 2) = \left(\frac{4-1}{1+1}\right)\left(\frac{.12}{.88}\right) \quad P(x = 1) = \left(\frac{3}{2}\right)\left(\frac{.12}{.88}\right)(.327) = .067$$

$$P(x = 3) = \left(\frac{4-2}{2+1}\right)\left(\frac{.12}{.88}\right)(.067) = \left(\frac{2}{3}\right)\left(\frac{.12}{.88}\right)(.067) = .006$$

$$P(x = 4) = \left(\frac{4-3}{3+1}\right)\left(\frac{.12}{.88}\right)(.006) = \left(\frac{1}{4}\right)\left(\frac{.12}{.88}\right)(.006) = .0002$$

Tedious calculations may frequently be eliminated if several special cases of the binomial probabilities are kept in mind. Remember that $\theta = 1 - \pi$.

1. $P(x = 0) = \theta^n$; the probability of no successes is the probability of failure in a single trial raised to the nth power.
2. $P(x = n) = \pi^n$; the probability of all n successes is the probability of success in a single trial raised to the nth power.
3. $P(x \geq 1) = 1 - \theta^n$; at least one success is the complement of no successes.
4. $P(x \leq n - 1) = 1 - \pi^n$; not getting all successes is complementary to all successes.
5. $P(x = 1) = n\pi\theta^{n-1}$.
6. $P(x = n - 1) = n\pi^{n-1}\theta$.

Problems

1. From the binomial tables find:
   a. $P(x \geq 6 | n = 10, \pi = .70)$;
   b. $P(x \leq 1 | n = 20, \pi = .05)$;
   c. $P(x = 5 | n = 25, \pi = .20)$;
   d. $P(x = 15 | n = 15, \pi = .95)$;
   e. $P(6 \leq x \leq 10 | n = 20, \pi = .40)$.
2. 70 percent of all Peace Corps personnel are males; 20 are selected at random to go to Borneo.
   a. What is the expected number of males?

b. What is the most likely number of females and what is this probability?
c. What is the chance that all 20 are males?
d. What is the probability that exactly 3 are females?
e. What is the probability that there will be a majority of males; i.e., 11 or more?

3. From the binomial tables find:
   a. $P(x < 6 | n = 10, \pi = .70)$;
   b. $P(x = 1 | n = 20, \pi = .05)$;
   c. $P(x > 5 | n = 25, \pi = .20)$;
   d. $P(x \neq 15 | n = 15, \pi = .95)$.

4. The long-run probability of Professor Smith's being on time to class is .70. Given a month with 20 school days, what is the probability that the professor will be on time
   a. exactly 12 times;
   b. at least 16 times;
   c. no more than 10 times?
   d. What assumption did you need to make?

5. A marketing manager makes the statement that, if he were to randomly select 10 customers, 3 of those customers would prefer the deluxe model. What is wrong with this type of statement? If 30 percent of the firm's customers prefer the deluxe model, what is the probability that exactly 3 in a random sample of 10 customers will prefer the deluxe model?

6. Ten percent of the calculators prepared for shipping by AR-ELECTRO Company have minor defects. The remaining 90 percent are free of defects. Quality control department employees remove calculators with serious defects before they get to shipping, but they miss most of the calculators with minor defects. The calculators are shipped in boxes containing 25 calculators. What is the probability that 1 of AR-ELECTRO's customers, a small retailer, will receive a box with
   a. exactly 3 defective calculators;
   b. no defective calculators;
   c. no more than 5 defective calculators;
   d. at least 1, but no more than 4 defective calculators?

7. Find the mean and standard deviation of the random variable with the following probability function:

$$p(x) = .10 \quad x = -3$$
$$= .05 \quad x = -1$$
$$= .20 \quad x = 0$$
$$= .15 \quad x = 3$$
$$= .45 \quad x = 6$$

$$= .05 \quad x = 9$$
$$= 0 \quad \text{otherwise.}$$

8. The probability that Jack will be able to sell 100 of his special gingerbread houses next Christmas is .30. There is a 50 percent chance that he will sell 75 gingerbread houses, and the probability is .20 that he will sell only 50 gingerbread houses. He uses facilities he already has in his doughnut shop to make the gingerbread houses; so the only extra costs he incurs are materials costs of $3.20 per gingerbread house. He sells them for $12.50 each. Any gingerbread houses left over after Christmas are discarded. If Jack decides to make 100 gingerbread houses for the Christmas season, what is his expected profit from the gingerbread houses? (Hint: Let the random variable x be the amount of profit he makes for each of the 3 possible demand levels.) Find $E(x)$, the expected value of x.

9. The random variable x has a binomial distribution with $n = 20$ and $\pi = .4$. Find:
   a. $P(x < 10)$;
   b. $P(x \geq 3)$;
   c. $P(4 \leq x \leq 9)$;
   d. $P(x = 8)$;
   e. $P(4 < x < 9)$;
   f. $P(8 \leq x < 9)$.

10. After consulting the almanac, a farmer decided that the following table accurately represents the potential profits he might make next year.

Profit	Probability
$50,000	.02
$35,000	.12
$20,000	.50
$10,000	.25
0	.08
-$10,000	.03

    Let x be the profit the farmer makes next year.
    a. Write the probability distribution function of x.
    b. What is the expected profit for next year?
    c. Find the variance of the random variable, x.

11. Seventy percent of Joe's customers order cheesecake for dessert. The cheesecake is made using a special recipe given to him by his grandmother. Joe only has 10 servings of cheesecake left for the remainder of the day. If a customer never orders more than 1 serving of cheesecake and if his customers order independently of each other, what is

a. the probability he will not run out of cheesecake if he has 15 customers come in for the rest of the day;
b. the probability he will have 3 servings remaining if only 10 customers come in for the rest of the day?

12. A printer is preparing a pamphlet for an engineering firm describing one of their complex computerized chemical-mixing production facilities. He knows that there is a 40 percent probability that he will make at least 1 error on any given page. If the likelihood of error for each of the 5 pages in the pamphlet is independent of the other pages, find
a. the probability that there will be no errors in the pamphlet;
b. the probability that at least 3 pages will have errors;
c. the probability that all 5 pages will have errors;
d. the probability that no more than 4 pages, but at least 1 page, will have errors.

13. The internal auditor is concerned about the carelessness of the billing clerk. She feels confident that 5 percent of the statements he prepares will have arithmetic errors. The auditor feels that if she finds more than 2 statements with arithmetic errors in a random sample of 15 statements prepared by the billing clerk, she will have ample support for her claim. If the billing clerk is in fact making arithmetic errors on 5 percent of the statements he prepares, what is
a. the probability that more than 2 statements in the random sample of 15 statements will have arithmetic errors;
b. the probability that there will be no arithmetic errors in the sample?

14. The company president has asked the pension fund manager to determine the expected rate of return on the pension fund investment portfolio for the next year. The pension fund manager estimates that the rate of return will be 20 percent if the economy really booms; it will be 14 percent if there is a moderate boom; it will be 10 percent if the economy stays level with this year; it will be 8 percent if there is a slight recession, and only 5 percent if there is a steep recession in the next year. The corporation vice-president in charge of economic analysis assigns the following probabilities to the 5 possible states of the economy during the next year: the probability of a significant boom is .05; the probability of a moderate boom is .25; the probability that the economy will be the same as this year is .35; the probability of a slight recession is .30; and the probability of a steep recession is .05.
a. Let x be the rate of return on the pension fund portfolio for the possible states of the economy. Determine the probability distribution function of the random variable x.

b. What is the expected rate of return on the pension fund portfolio for next year? I.e., find $E(x)$.
c. Find the variance of the random variable in part a.

15. The random variable x has a binomial distribution with $n = 15$ and $\pi = .80$. Find:
    a. $P(x > 11)$;
    b. $P(7 \leq x \leq 13)$;
    c. $P(x < 12)$;
    d. $P(8 < x < 12)$.

16. The local grocer is advertising what appears to be an unusually good price for 20-count bags of apples. One of the grocer's regular customers asks her neighbor, who works at the Wholesale Grocery Supply Company downtown, about the quality of the apples being offered at the special price. He informs her that the shipment of apples received by the wholesaler was 20 percent culls. The culls were randomly distributed throughout the shipment. The wholesaler decided to put them in bags of 20 apples each and sell them at a good price rather than try to sort out the culls. The customer of the local grocer decides that it is still a fairly good price if the bag of twenty apples she buys has only 20 percent culls. If she happens to get a bag with 25 percent or more culls, then she feels that it would not be a good buy. On the other hand, if the bag of apples has 10 percent or less culls, then she would consider it a really good buy. If she sends her son to the local grocer to buy a randomly selected bag of 20 apples, what is
    a. the probability that he will select a bag of apples she would classify as a "really good buy";
    b. the probability that he will select a bag of apples she would classify as "not a good buy";
    c. the probability that he will select a bag of apples with exactly 20 percent culls?

17. The mayor of your town has decided to make a decision about renovating the playground equipment at the community park based on the results of a random sample of 15 of the voters in the town. If the majority of the voters in the sample favor renovation of the playground equipment, then the mayor will proceed with the project; otherwise, he'll spend the money on improving the golf course. You are certain that 60 percent of the townspeople favor the playground equipment renovation project but realize that, due to chance, it is possible that the mayor's sample will not show a majority favoring the expenditures for the playground. What is the probability that there will *not* be a majority favoring the playground project?

18. The quality control manager has decided to determine whether or

not to accept a large lot of bearings received from a vendor by taking a random sample of 25 bearings. If there is no more than 1 defective bearing in the sample, he will accept the lot. He will reject the lot if there are 2 or more defective bearings in the sample. The price paid for the bearings allows for 5 percent of the lot to be defective; so top management doesn't want to reject a lot with 5 percent or fewer defective bearings. On the other hand, the materials manager is very concerned about accepting a lot that has 10 percent or more defective bearings in it.
   a. If the lot is 5 percent defective, what is the probability that the sample result will cause the lot to be rejected?
   b. If the lot is 10 percent defective, what is the probability that the sample outcome will result in the lot's being accepted?

19. The wire-winding machines at a small electric motor manufacturing plant have a 10 percent probability of breaking down during any given day. The probability that 1 of the machines will need repairs is independent of whether or not other machines break down. The maintenance manager has hired only 1 wire-winding-machine mechanic because there are 10 machines and, thus, there should be on the average 1 machine per day needing to be repaired and serviced. The maintenance manager is considering hiring another winding-machine mechanic because it really causes problems with the production schedule when more than 1 machine breaks down and there is only 1 repairman available to service the machines.
   a. You can help the maintenance manager make a decision by providing him with the percent of days, or the probability, that more than 1 machine will need repairs.
   b. It would also be helpful to know the probability that 4 or more machines will need repairs on a given day.

20. Consider the following game: An honest coin and an honest die are tossed. If a head–6 shows, the player wins $108. If the result is tail–odd number, the player loses $40; if otherwise, the player loses $3.
   a. What is the expected value of the game?
   b. If the game is not a fair game, change the win amount to make it a fair game.

21. One can find the following game at a good many of the local pubs in Summit County. Five thousand sequentially numbered cards (0001 thru 5000) are folded with numbers on the inside, punched, and stacked in a random manner on a metal spindle. The player pulls the card; if the number ends in a 3, 6, or a 9, he must pay $.10 for the thrill; all other numbers are free, and numbers ending in 00 win $1.50. For this problem, assume independent trials. Note that this is not a bad assumption if you start with a nearly new deck,

but it may be a bad assumption as the deck depletes. Joe decides to pull 8 cards.

a. Let x be the number of $.10 cards he might get. Use the binomial formula to find

(i). $P(x = 0)$
(ii). $P(x = 4)$
(iii). $P(x \leq 2)$.

b. What is the probability that he will get at least 1 winner?

22. Four honest dice are tossed.
   a. Use the binomial formula to find the probability of 2 sixes;
   b. What is the probability of not getting 2 sixes?

■ The remaining problems require material covered in the appendices.

23. The random variable x has a mean equal to 7.8 and a standard deviation equal to 2.3. The mean and standard deviation of the random variable y are 12.6 and 4.2, respectively. Find the mean and standard deviation of the random variable w where w is defined as:
   a. $w = x + y$;
   b. $w = x - y$;
   c. $w = 3x - 4y$;
   d. $w = y - 10$;
   e. $w = 12 - 4x + y$.

24. The probability density function for the random variable x is:

$$f(x) = \frac{2x}{21} \quad 2 \leq x \leq 5$$
$$= 0 \quad \text{otherwise.}$$

   a. Verify that the above function is in fact a density function (i.e. verify that the two conditions in Appendix 4B are satisfied).
   b. Find $P(3 \leq x \leq 4)$.
   c. Find $P(x \geq 2)$.

25. An urn contains 8 red and 4 white balls. What is the probability of 2 red balls in 2 random draws
   a. with replacement;
   b. without replacement?

   Solve this problem using the appropriate probability distribution functions discussed in this chapter.

26. A lot of 100 television sets contains 5 defectives. What is the probability that a random selection of 3 will contain no defectives?

27. The random variable x has a Poisson Distribution with $\lambda = 1.7$. Find:
   a. $P(x = 2)$;
   b. $P(x > 3)$;
   c. $P(x \leq 4)$;
   d. $P(x \geq 1)$.

28. The loading dock foreman has to plan his manpower needs for unloading trucks during the day shift. He has to unload all trucks that arrived during the night up to 7:00 A.M. It takes a crew of 3 people to unload 1 truck during the regular shift. If the trucks are not unloaded during the day shift, he has to pay overtime to get all trucks unloaded before the start of the next shift. He decides to hire enough people to unload 5 trucks a day even though the average number of trucks arriving during the night has been 3.6. What is the probability that for any given day he will have to pay overtime wages; i.e., what is the probability that more than 5 trucks will need to be unloaded? Assume that the number of trucks arriving during the night follows a Poisson Distribution.

29. Refer to Problem 28. What is the probability that on any given day the loading dock foreman will have 2 crews with nothing to do (i.e., only 3 trucks need to be unloaded that day)? For any given day what is the probability that at least 1 crew will not be unloading trucks?

30. The mean number of oil tanker arrivals per week at Philadelphia from Saudi Arabia is 3. Assume that the number of arrivals per week has a Poisson Distribution.
   a. What is the probability of no arrivals; of exactly 3 arrivals; of 4 or fewer?
   b. At most, 5 tankers per week can be unloaded. What is the probability that at least 1 tanker will not be unloaded during a given week?

31. Flipco, a major television manufacturer, purchases zyks in lots of 10,000 from Slyco. Flipco samples 200 items from incoming lots; if 0 or 1 defectives are found, the 10,000 zyks are accepted; otherwise they are returned to Slyco. Slyco must go through a 100 percent inspection and return the lot of 10,000 zyks guaranteed 100 percent perfect, i.e., no defectives. What is the probability that Flipco will accept the lot of zyks if the percentage defective in the lot is
   a. 1 percent;
   b. 2 percent;
   c. 5 percent;
   d. 0.5 percent;
   e. 0.1 percent;

f. 0 percent;
   g. 100 percent.
   (Hint: Use Poisson approximation to the binomial.)
32. On a weekday evening, 2 percent of the television audience in a metropolitan area are tuned to the local educational channel, 7. In a random sample of size 60, let x be the number of television sets tuned to Channel 7. What is the probability that x is
   a. equal to 0;
   b. at least 5;
   c. no more than 2;
   d. exactly 1?
   (Hint: Use Poisson approximation to the binomial.)
33. Compare the Poisson approximation to the actual binomial probabilities for $n = 20$ and $\pi = 0.05$.
34. A free-wheeling pointer spins on an axis at the center of a circle; the circumference of the circle is scaled from 0 to 60, e.g.,

Let the variable, x, be the number designated by the pointer when it stops after a random spin. What is the nature of the density function? Find
   a. $P(x = 30)$;
   b. $P(x = \sqrt{2})$;
   c. $P(29.5 \leq x \leq 30.5)$;
   d. $P(18 \leq x \leq 30)$;
   e. μ and σ.
(Hint: Assuming we really have a free-wheeling spinner, the density function of x from $x = 0$ to $x = 60$ will be uniform or rectangular. The total area must be unity, $f(x) = 1/60$ for $0 \leq x < 60$, $f(x) = 0$ otherwise.)

CHAPTER 5

THE NORMAL DISTRIBUTION

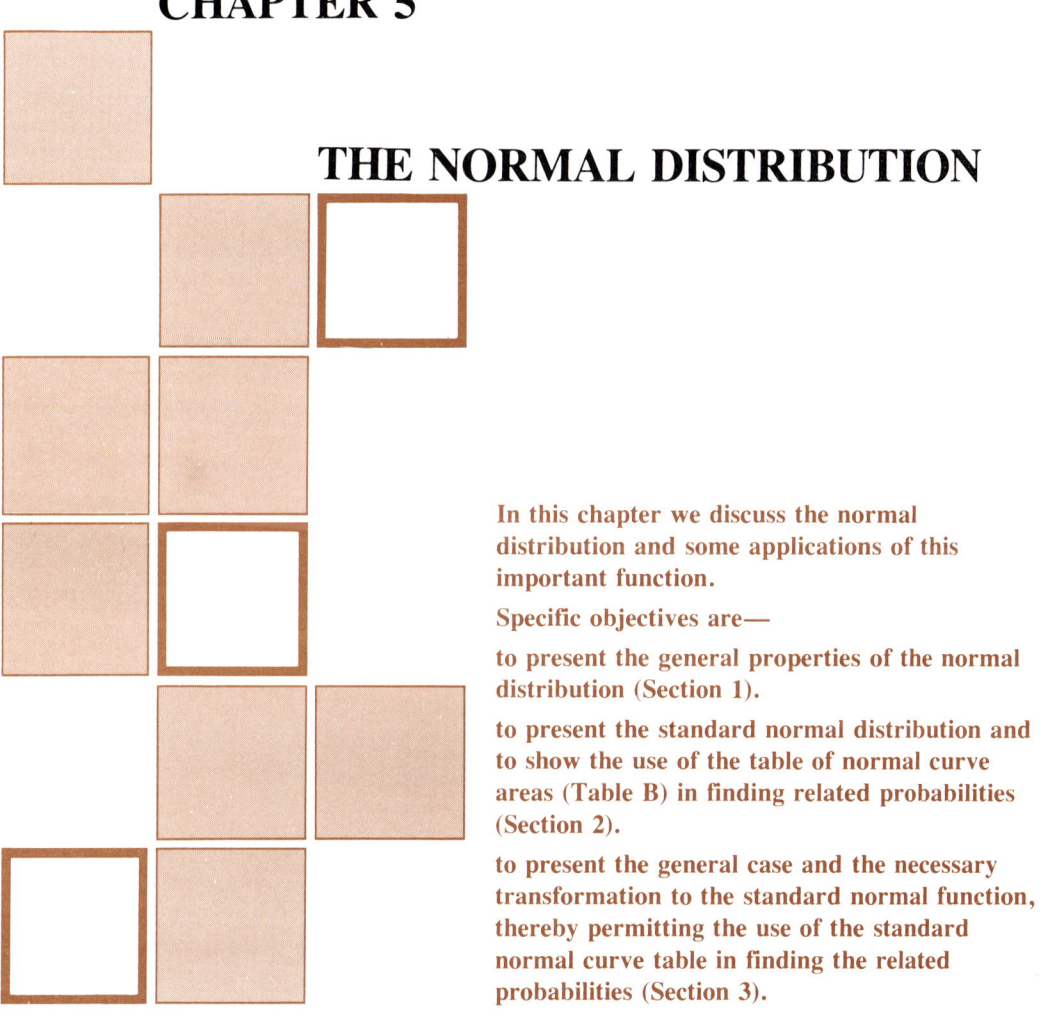

In this chapter we discuss the normal distribution and some applications of this important function.

Specific objectives are—

to present the general properties of the normal distribution (Section 1).

to present the standard normal distribution and to show the use of the table of normal curve areas (Table B) in finding related probabilities (Section 2).

to present the general case and the necessary transformation to the standard normal function, thereby permitting the use of the standard normal curve table in finding the related probabilities (Section 3).

The normal distribution is probably the most important distribution in applied statistics. Many populations of interest are normally or nearly normally distributed. More important, the Central Limit Theorem, which is discussed in the next chapter, tells us that the sample mean is approximately normally distributed for large sample sizes even if the population variable is not normally distributed. Finally, the normal distribution provides a useful approximation to the binomial distribution for a wide range of values of n and π. The reader will want to become thoroughly acquainted with the normal distribution.

5.1 General Properties of the Normal Distribution

The graph of the normal density function is given in Figure 5.1. Because of its shape when graphed, the normal density function is often referred to as the *bell-shaped curve*. It is a continuous function, unimodal and symmetrical about the mean. The mean, median, and mode are all equal and at the center of the distribution. The symmetry about the mean is important to note because it means that 1/2 the area under the curve is to the left of the mean and 1/2 is to the right. As pointed out in Chapter 4, the total area under a probability density function is equal to 1. The normal distribution is a 2-parameter distribution; the 2 parameters are the mean and the standard deviation. If the mean and standard deviation of a random variable are given and if, in addition, the variable is known to be normally distributed, then it is possible to answer any probability question that might be asked concerning the random variable.

Figure 5.1 Graph of the Normal Probability Density Function

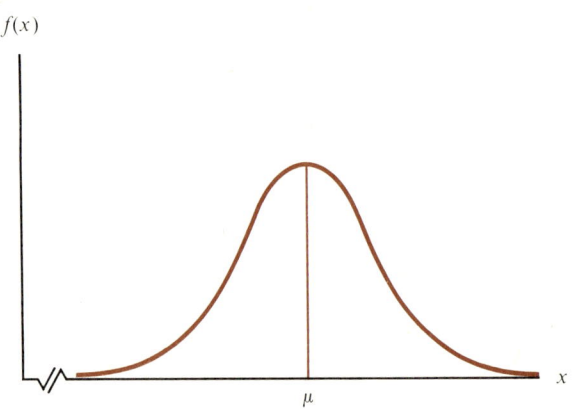

We should keep in mind that probability is determined by the area under the curve of a continuous distribution. In Figure 5.2 we have shown 3 intervals and, thus, 3 corresponding areas of general interest. The first is symmetrical about the mean and of 2 standard deviations in length. The area under the curve and within this interval is .6826; the probability that the normal random variable will take on a value in this interval is the same. That is, the probability that a normal random variable will fall in the interval $\mu - \sigma$ to $\mu + \sigma$ is .6826; or, alterna-

Figure 5.2 Normal Probabilities for Three Intervals

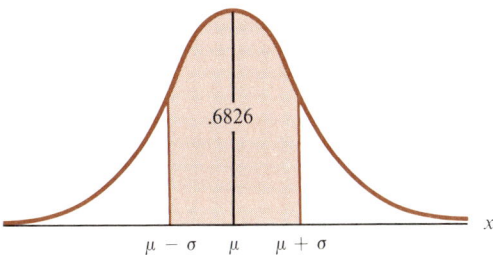

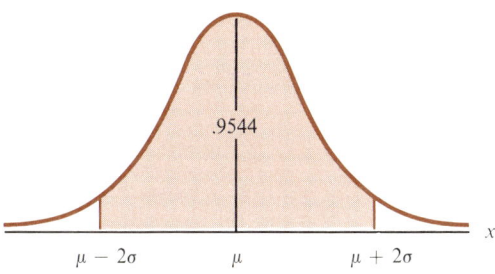

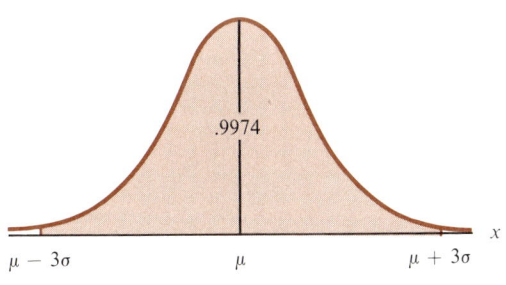

General Properties of the Normal Distribution

tively, 68.26 percent of all values in a normal population fall within 1 standard deviation of the mean. For example, assume IQ is normally distributed with a mean of 100 and a standard deviation of 15; then the probability is .6826 that a person chosen at random will have an IQ between 85 and 115; or 68.26 percent of the population have IQ's between these 2 values. Referring again to Figure 5.2, 95.44 percent of the values from a normal population fall within 2 standard deviations of the mean, and 99.74 percent fall within 3 standard deviations of the mean. A little over 95 percent of the population have IQ's between 70 and 130, and nearly everyone has an IQ of at least 55 and no greater than 145.

Beyond 3 standard deviations from the mean the curve gets very close to the x-axis. In formal terms the area is said to converge. For many practical applications we proceed as if the area beyond 3 standard deviations were 0, and it would indeed be a rare problem where the probability of a normal variable's falling beyond 4 standard deviations would be given serious consideration. The probability density function for the normal distribution is an exponential expression that need not concern the beginning student, but we present it so that you will at least be familiar with the nature of the function:

$$f(x) = \frac{1}{\sigma\sqrt{2\pi}} e^{\frac{-(x-\mu)^2}{2\sigma^2}}$$

where

μ = the population mean
σ = the population standard deviation.

At first glance this may appear to be a function, not only of x, but also of μ, σ, π, and e. The mean, μ, and standard deviation, σ, are as mentioned above the parameters that fix the specific distribution of the 2-parameter family. As you may recall $\pi = 3.1416$ and is the ratio of the circumference of a circle to its diameter; e, like π, is a constant and an irrational number with a value of approximately 2.7183. This very interesting number is the base for natural logarithms, and it appears frequently in applied work in business and economics as well as in the hard sciences. A business example using e with which you may be familiar is the formula for instantaneous compound interest, $P_t = P_o e^{rt}$, where P_o is the initial amount, P_t is the amount at time t, and r is the interest rate per unit of time.

5.2 The Standard Normal Distribution

Because of the nature of the normal density function, techniques beyond elementary calculus are needed to find the area under the curve over some specified interval. These techniques are very time consuming and, for a practical matter, the skilled mathematician, as well as you and I, will use the table of normal curve areas, Table B, to find normal curve probabilities. This is a table of areas for what we call the *standard normal random variable*. This is a normal random variable with a mean of 0 and a standard deviation of 1. Traditionally, the variable is designated by the letter z. We will first learn how to use this table for probability problems related to the standard normal variable. Then, with a simple algebraic transformation, we will learn to use this table for probabilities relating to any normal random variable.

The areas (or probabilities) that have been calculated for us and given in Table B are those from 0, the mean value of the standard normal distribution, to some fixed z, say z_0, for positive values of z_0 only. This is represented by the shaded area in Figure 5.3. The shaded area under the curve between 0 and z_0 is the probability that the random variable, z, will assume a value in this interval, i.e., the shaded area is equal to $P(0 \leq z \leq z_0)$. It is important to keep in mind that the values of the standard normal variable are represented along the z-axis while the probability that the assumed value will lie in a specified interval is the area under the curve over that particular interval.

Referring again to Table B, the left-hand column has values of z_0 to the first decimal place. The second decimal place is given as column headings for the remaining columns in the table. The numbers in the body of the table are the shaded areas from 0 to z_0 (as indicated in the diagram at the top of the table). Again, the number in the body of the table for a given value of z, z_0, is the $P(0 \leq z \leq z_0)$. (It should also be recalled at this point that for a continuous random variable it does not matter whether or not the endpoints are included in the interval—the probability is the same; i.e., $P(0 \leq z \leq z_0) = P(0 < z < z_0)$.) To find the probability that the standard normal random variable, z, will take on a value between 0 and 1.43, we select the row with 1.4 in the left-hand column and go across this row to the column headed .03 and read .4236

Figure 5.3 The Standard Normal Distribution

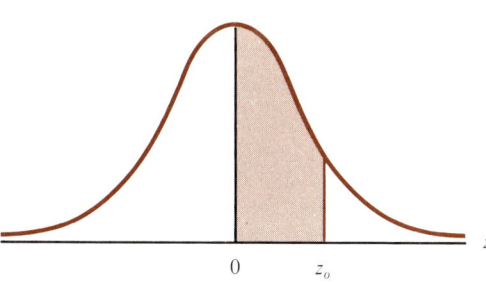

The Standard Normal Distribution

in the body of the table. This result can be stated as follows: $P(0 \leq z \leq 1.43) = .4236$.

Using only the given tabular areas, we can find the probability that z will fall within any closed interval, or the probability that z exceeds (or is less than) some fixed value of z. In order to do this we will find it helpful to keep the following in mind: (1) The total area under the curve is unity; (2) half of this area is to the right of 0 and half is to the left of 0; i.e., $P(z \geq 0) = .5$; and (3) the function is symmetrical about 0. Using these observations and Table B, we will be able to find any standard normal probability. In working probability problems using the standard normal table, the beginning student will find it helpful to sketch a diagram of the interval of interest and the corresponding area under the curve.

Example 5.2.1   Find $P(0 \leq z \leq 2.05)$.

The area over the interval from 0 to 2.05 will give us the desired probability. From Figure 5.4a we observe that this area matches the tabular area and is therefore read directly from the table.

$P(0 \leq z \leq 2.05) = .4798$.

Example 5.2.2   Find $P(-1.78 \leq z \leq 0)$.

The area of interest is indicated in Figure 5.4b. Because of the symmetry of the normal distribution, this is an equivalent amount of area as the area from 0 to 1.78. The area from 0 to 1.78 can be read directly from the table, giving us the desired probability.

$P(-1.78 \leq z \leq 0) = P(0 \leq z \leq 1.78) = 0.4625$.

Example 5.2.3   Find $P(-1.83 \leq z \leq 0.78)$.

As indicated in Figure 5.4c, the area over the interval from -1.83 to $.78$ represents the desired probability. This area is separated into two areas that can be determined from Table B. The area from -1.83 to 0 is equal to that from 0 to 1.83. This and the area from 0 to .78 are read directly from Table B. The tabular values are then summed to obtain the area from -1.83 to $.78$:

$P(-1.83 \leq z \leq .78) = .4664 + .2823 = .7487$.

Example 5.2.4   Find $P(z \leq 1.34)$.

As with the previous problem, this problem is solved by separating the

Figure 5.4 Using Table B to Find Standard Normal Probabilities

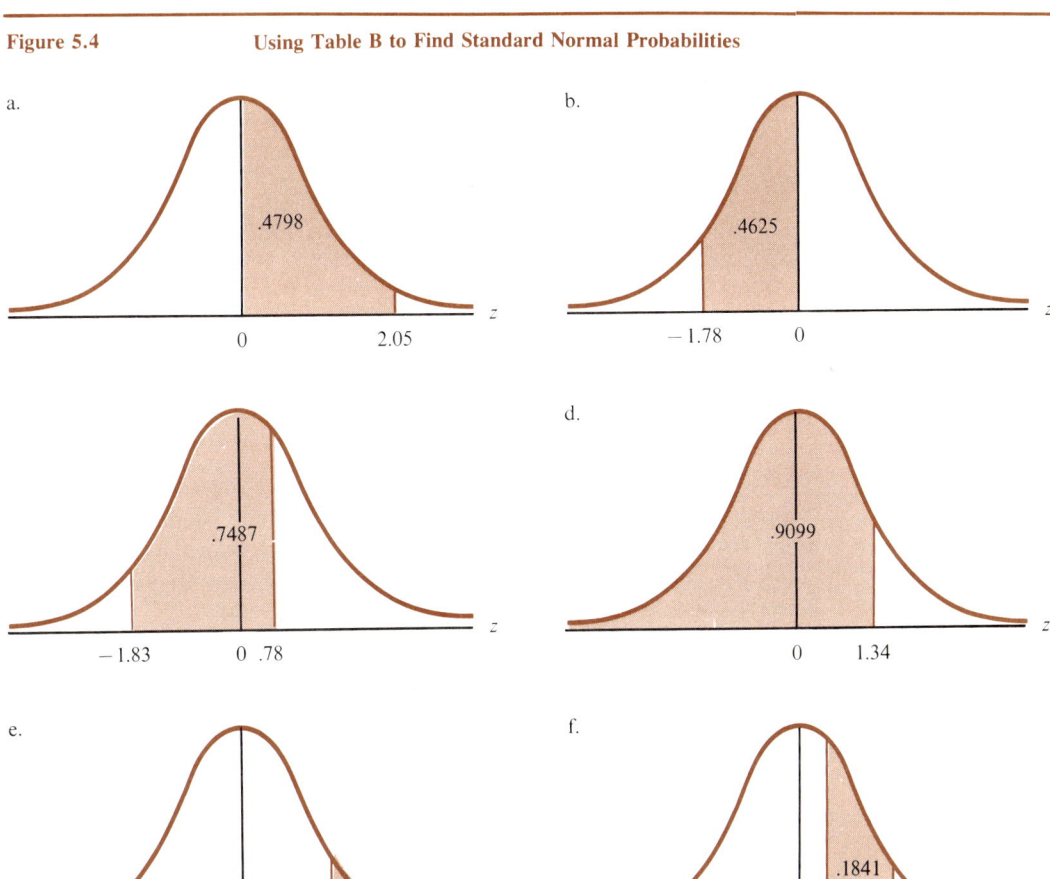

area to the left of 1.34 (see Figure 5.4d) into 2 areas. Because of symmetry, 1/2 the area under the curve lies to the left of 0. Therefore, the area from $-\infty$ to 0 is .5000. This plus the tabular area from 0 to 1.34 gives us our desired probability:

$P(z \leq 1.34) = .5000 + .4099 = .9099$.

Example 5.2.5 Find $P(z \geq 1.56)$.

As indicated in Figure 5.4e, the desired area is to the right of 1.56. This is what we often call a "tail" area. We observe that the area from 0 to 1.56, which can be determined from Table B, plus the area from 1.56 to $+\infty$ is equal to the area from 0 to $+\infty$. Since the area from 0 to $+\infty$ is

.5000, the desired area is the difference between this and the tabular area from 0 to 1.56:

$P(z \geq 1.56) = .5000 - .4406 = .0594.$

Example 5.2.6 Find $P(.78 \leq z \leq 1.83)$.

From Figure 5.4f we see that this problem is similar to the problem worked in the previous example. The desired probability is obtained by subtracting the area between 0 and .78 from the area between 0 and 1.83. (Note that both of these tabular areas were found in Example 5.2.3 above.)

$P(.78 \leq z \leq 1.83) = .4664 - .2823 = .1841.$

The reader might like to use the standard normal table to verify the probabilities given in Figure 5.2. Remember that the mean and standard deviation of the standard normal curve are 0 and 1 respectively. Thus, the $\mu - \sigma$ to $\mu + \sigma$ interval is the interval from $0 - 1$ to $0 + 1$ or from -1 to $+1$. You would be checking the following three probabilities:

1. $P(-1.00 \leq z \leq 1.00) = .3413 + .3413 = .6826.$
2. $P(-2.00 \leq z \leq 2.00) = .4772 + .4772 = .9544.$
3. $P(-3.00 \leq z \leq 3.00) = .4987 + .4987 = .9974.$

To this point we have been finding probabilities from the body of the table given the normal curve z-values. Many statistical inference problems discussed in later chapters will necessitate the reversal of this procedure; i.e., find a z-value or some z-values that satisfy a given probability statement. Consider the following two examples.

Example 5.2.7 Find z_o such that $P(z \geq z_o) = .04$.

We are asked to find a value of z along the z-axis such that the area to the right of z_o is equal to .04. The z_o we seek must be to the right of 0 as indicated in Figure 5.5a because the upper tail area is less than .5; i.e., if z_o is to the left of 0, then the area to the right of 0 must necessarily be greater than .5. The area from 0 to the desired z_o is .4600. This latter area corresponds with the area given in Table B. The closest tabular area to this is .4599 corresponding to a z value of 1.75. Without interpolating for additional accuracy, the answer to the nearest hundredth is $z_o = 1.75$.

Example 5.2.8 Find z_o such that $P(-z_o \leq z \leq z_o) = .98$.

Figure 5.5 Determining z_o Such That $P(z \geq z_o) = .04$ and $P(-z_o \leq z \leq z_o) = .98$

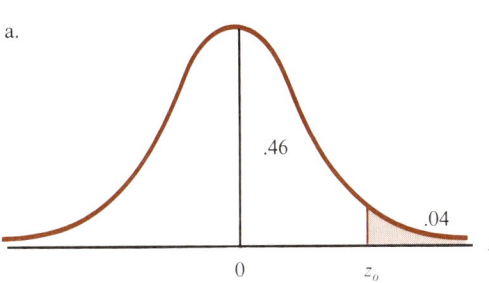
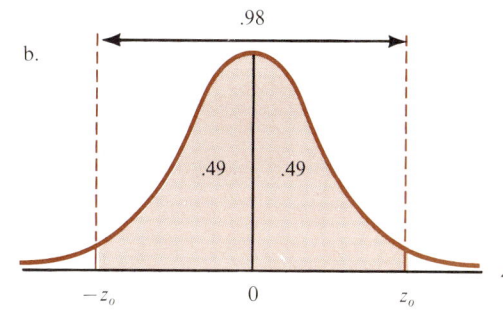

Because we are asking for a symmetrical area about the mean of 0, the answer is unique. If the one z-value were not the negative of the other, there would be infinitely many intervals with a probability of .98. See Figure 5.5b for a sketch of the area. Since the area is symmetrical, the area from 0 to z_o would be the same as that from $-z_o$ to 0, and each must then be .49. The closest tabular area to .49 is .4901, corresponding to a z-value of 2.33. Without interpolating for additional accuracy, $z_o = 2.33$ and, of course, $-z_o = -2.33$.

 ## 5.3 Normal Random Variables: The General Case

To this point, we have considered only the standard normal distribution. We would like to be able to find probabilities for any normal random variable. We do this by transforming the general probability problem to an equivalent problem in terms of the standard normal random variable. We can then use Table B to solve the latter problem. This process is called *standardizing* or *normalizing* the original variable, x, and the linear transformation that accomplishes this is given by the equation

$$z = \frac{x - \mu}{\sigma}.$$

This is a simple linear mapping (or change of linear scale) from the x-variable to the z-variable. Figure 5.6 has a sketch of a normal random variable with mean, μ, and standard deviation, σ. The corresponding standard normal curve that is obtained from transforming the x variable is drawn directly beneath the general normal distribution. Note that when x equals μ, z is 0; if $x = \mu + \sigma$, then $z = \{(\mu + \sigma) - \mu\}/\sigma = 1.0$; if

Figure 5.6 Mapping the General Normal Distribution into the Standard Normal Distribution

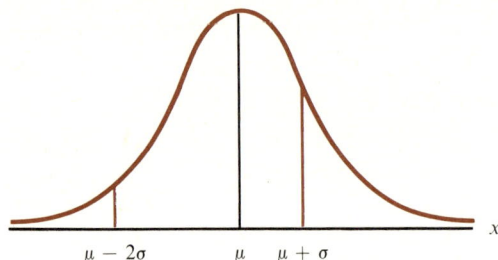

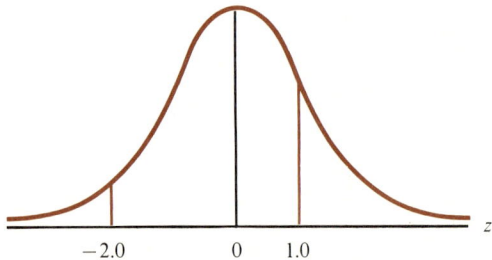

$x = \mu - 2\sigma$, then $z = -2.0$; etc. In this mapping, z measures the number of standard deviations the variable x is from its mean. With this transformation, it is possible to get normal probability estimates for any random variable if μ and σ are known. These probabilities will be exact only if we know that the random variable is normally distributed. Let us consider the following examples.

Example 5.3.1 The random variable x is normally distributed with the mean equal to 160 and the standard deviation equal to 20. Find the probability that x will take on a value between 135 and 175.

Figure 5.7 has a diagram of the normal random variable with $\mu = 160$ and $\sigma = 20$. Directly below is a diagram of the standard normal curve.

Because x is normally distributed, $P(135 \leq x \leq 175) = P(z_1 \leq z \leq z_2)$, where z_1 and z_2 are the normalized values of 135 and 175 respectively. These z-values are found as follows:

$$z_1 = \frac{x_1 - \mu}{\sigma} = \frac{135 - \mu}{\sigma} = \frac{135 - 160}{20} = -1.25$$

Figure 5.7 Using the Standard Normal Distribution to Find $P(135 \leq x \leq 175)$

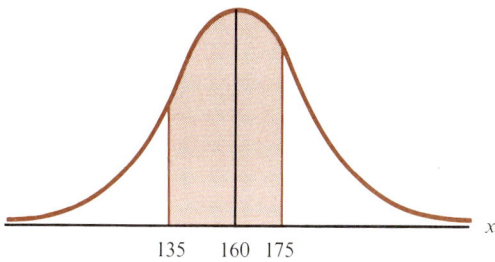

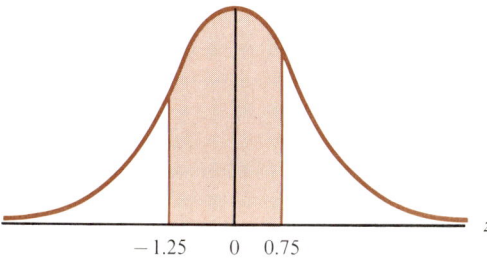

and

$$z_2 = \frac{x_2 - \mu}{\sigma} = \frac{175 - \mu}{\sigma} = \frac{175 - 160}{20} = 0.75.$$

(Note that 135 is 1.25 standard deviations below the mean, and 175 is .75 standard deviations above the mean.) The probability that the random variable will take on a value between 135 and 175 can now be found using the probabilities obtained from Table B:

$$P(135 \leq x \leq 175) = P(-1.25 \leq z \leq 0.75) = .3944 + .2734 = .6678.$$

Some students find the following formulation a helpful summary of the technique used to solve normal distribution probability problems for the general case. If the random variable x is normally distributed with mean μ and standard deviation σ, then the probability that x will assume a value greater than some value a can be found using the following formula:

$$P(x \geq a) = P\left(z \geq \frac{a - \mu}{\sigma}\right).$$

The probability that x will take on a value between two values a and b ($a < b$) is obtained in a similar manner:

$$P(a \leq x \leq b) = P\left(\frac{a - \mu}{\sigma} \leq z \leq \frac{b - \mu}{\sigma}\right).$$

Example 5.3.2 The mean weight of ABC's canned pears is 14.00 ounces with a standard deviation of .12 ounces. Assume the weights are normally distributed. What percent of the cans of pears will weigh more than 14.30 ounces?

The percent of cans that weigh more than 14.30 ounces is the same as the probability that a can selected at random will weigh more than 14.30 ounces. Therefore, we can find the desired percentage by finding the probability that x will be greater than 14.30 ounces. Figure 5.8 has a sketch of this problem.

$$P(x \geq 14.30) = P\left(z \geq \frac{14.30 - \mu}{\sigma}\right) = P\left(z \geq \frac{14.30 - 14.00}{.12}\right) =$$

$P(z \geq 2.50) = .5000 - .4938 = .0062.$

We estimate that .62 percent (or less than 1 percent) of the cans of pears weigh more than 14.30 ounces.

Earlier we considered several examples where we found a value of the standard normal variable, z_o, that would satisfy a given probability statement. The analogous problem arises with any normal variable;

Figure 5.8 $P(x \geq 14.30) = P(z \geq 2.50)$

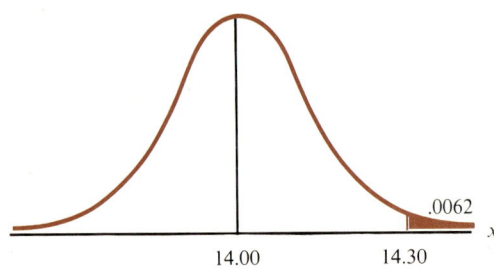

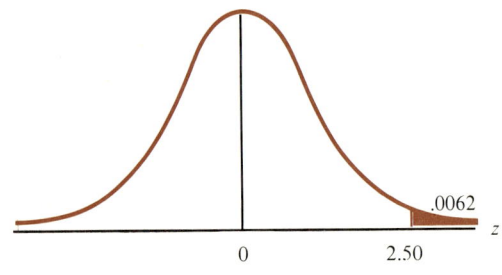

i.e., find a value of x, say x_0, that satisfies a probability statement. Consider the following example.

Example 5.3.3 For a certain college entrance exam, the national average is 70.00 and the standard deviation is 8.00. Assuming that the test scores follow a normal distribution, what should be the cutoff score if we wish to accept approximately 80 percent of the students taking the test?

Restating the problem in probability form we have: Find x_0 such that $P(x \geq x_0) = .80$. For this type of problem we first find z_0 such that $P(z \geq z_0) = .80$ and then use the inverse of the z-transformation, $x_0 = \mu + z_0\sigma$. This latter expression is obtained by solving the z-transformation,

$$z_0 = \frac{x_0 - \mu}{\sigma},$$

for x_0. From Figure 5.9 it can be seen that the value of z with an area of .80 to the right must be negative and in magnitude equal to that z-value

Figure 5.9 Determining x_0 Such That $P(x \geq x_0) = .80$

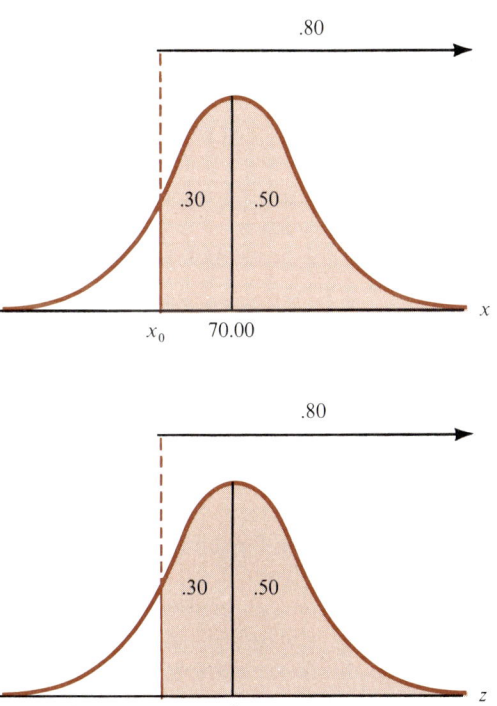

Normal Random Variables: The General Case

with an area of .30 from 0 to the z-value. The closest tabular area to .30 is .2995, and the corresponding z-value is .84. Again, because z_o is to the left of 0 it is negative, $z_o = -.84$. For this particular problem, the value of x_o obtained from the inverse transformation is

$$x_o = \mu + z_o \sigma = 70.00 + 8.00 z_o = 70.00 + 8.00(-.84) = 63.28.$$

If the cutoff score is set at 63.28, then approximately 80 percent of those taking the test will be accepted.

5A Appendix: The Normal Approximation to the Binomial Distribution

It will be recalled from the previous chapter that the binomial distribution is a 2-parameter family, the 2 parameters being the number of independent trials, n, and the probability of success on a single trial, π. If n is large and both $n\pi$ and $n(1 - \pi)$ are not too small, then the normal distribution can be used to get good approximations to binomial probabilities. The Central Limit Theorem, discussed in the next chapter, provides the theoretical basis for this. If $\pi = .5$, the binomial distribution has perfect symmetry, and good approximations will result with relatively small values of n. The more π deviates from .5, either above or below, the larger n must be in order to get a good approximation. As a rule of thumb, n should be at least 30 and both $n\pi$ and $n(1 - \pi)$ should be at least 5; but remember that the larger the sample size and the closer π is to .5 the better the approximation will be to the true values of the binomial probabilities.

In Chapter 4 the mean and standard deviation for the binomial distribution were given as

$$\mu = n\pi$$

and

$$\sigma = \sqrt{n\pi(1 - \pi)}.$$

These are the only 2 parameters needed for normal curve estimates of probabilities for any function. The binomial distribution is a discrete random variable with non-0 probabilities defined for $n + 1$ values of the variable (the integers from 0 to n). On the other hand, the normal distribution is a continuous probability distribution with non-0 probabilities defined for an infinite number of intervals along the real number line, and the probability assigned to any point is 0. The problem is to define an interval for which the probability calculated using the normal distribution is approximately equal to the binomial point probability (or sum of the probabilities of a number of points).

Example 5A.1 If 20 percent of all college students take a mathematics course in college, what is the probability that in a random sample of 100 students exactly 25 will take a mathematics course?

It is helpful in this type of problem to distinguish carefully between the original binomial problem and the normal distribution that is used to obtain the estimated probability. Let x_B represent the binomial random variable and use x_N to designate the normal random variable with the same mean and standard deviation as x_B. The question asked in this example can be stated, in terms of the original binomial variable, as

$$P(x_B = 25 | n = 100, \pi = .20) = ?$$

If we were to try to solve this problem using $P(x_N = 25)$, we would not obtain very useful results. $P(x_N = 25) = 0$ since x_N is a continuous random variable. It is necessary to find an interval for which the probability that x_N will take on a value in this interval is approximately equal to the probability that x_B is equal to exactly 25.

Figure 5.10 illustrates how the appropriate interval is selected. Just a part of the normal curve dealing directly with the problem is shown. The normal curve ordinate above 25 should be very close to the probability that $x_B = 25$, the actual height of the rectangle. The bottom of the rectangle runs from 24.5 to 25.5 and thus has length 1; because of this, the area of the rectangle is also equal to $P(x_B = 25)$. From Figure 5.10 it is apparent that the area under the normal curve from 24.5 to 25.5 could be a good approximation of $P(x_B = 25)$. We are using a trapezoidal-type area to approximate the true rectangular area. Thus, $P(x_B = 25) \cong P(24.5 \leq x_N \leq 25.5)$. The right-hand side of the equation is a straightforward normal distribution probability problem.

Figure 5.10 Using the Normal Distribution to Approximate a Binomial Probability

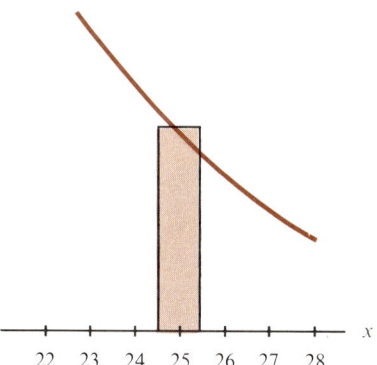

To solve the normal probability problem, it is necessary to calculate the mean and standard deviation of the binomial distribution, which are the parameters of the normal distribution used for the approximation:

$$\mu = n\pi = 100(.2) = 20$$

and

$$\sigma = \sqrt{n\pi(1-\pi)} = \sqrt{100(.2)(.8)} = 4.$$

We then have

$$P(24.5 \le x_N \le 25.5) = P\left(\frac{24.5 - 20}{4} \le z \le \frac{25.5 - 20}{4}\right) =$$

$$P(1.13 \le z \le 1.38) = .4162 - .3708 = .0454.$$

The estimated probability that exactly 25 of the 100 students sampled will take mathematics in college is .0454.

Example 5A.2   Refer to the previous example. What is the probability that at least 25 of the students sampled will take mathematics in college? Is this the same as the probability that more than 25 students sampled will take mathematics in college?

The answer to the latter question is no. Because we are dealing with a discrete random variable, these are 2 different events. The first question includes 25 in the probability event, and the second question excludes it. Whether or not the interval contains the endpoint is important for a discrete random variable, because the endpoint will have a non-0 probability of occurring. In this case, as we have seen, it is approximately equal to .0454. The 2 questions asked are

1. $P(x_B \ge 25) = ?$
2. $P(x_B > 25) = ?$

The first includes 25 and, thus, would be estimated as follows:

$$P(x_B \ge 25) \cong P(x_N \ge 24.5) = P(z \ge 1.13) = .1292.$$

In the second case, 25 is not included, and we have

$$P(x_B > 25) \cong P(x_N > 25.5) = P(z \ge 1.38) = .0838.$$

We observe that $P(x_B > 25)$ is equal to $P(x_B \ge 26)$ since 26 is the first value greater than 25 that has a non-0 probability of occurring.

Example 5A.3   Referring again to the first example in this appendix, find the probability that the number of students in the sample taking mathematics courses in college is greater than 18 but less than 27. Compare this with the probability that the number of students taking mathematics courses will be at least 18 but no more than 27.

The difference between these questions is similar to the distinction made in the previous example. The endpoints are not included in the first question. The requested probability is as follows:

$P(18 < x_B < 27) \cong P(18.5 \leq x_N \leq 26.5) = P(-.38 \leq z \leq 1.63) = .5964.$

The second question is stated so that the endpoints are included as part of the event for which the probability is desired. The inclusion of the endpoints does, as we would expect, result in a higher probability value.

$P(18 \leq x_B \leq 27) \cong P(17.5 \leq x_N \leq 27.5) = P(-.63 \leq z \leq 1.88) = .7056.$

Note that the intervals used to calculate $P(x_B = 18)$ and $P(x_B = 27)$, namely, 17.5 to 18.5 and 26.5 to 27.5, are completely included in the interval 17.5 to 27.5 used to find this probability.

It is interesting to compare estimated values using the normal approximation to values taken from the binomial table. In Table A, 25 is the largest n, which is slightly smaller than the rough-rule-of-thumb value of 30; but for π close to .5, say $\pi = .4$, the approximation should be fairly good. For $n = 25$ and $\pi = .4$, we have

$\mu = n\pi = 25(.4) = 10$
$\sigma = \sqrt{n\pi(1-\pi)} = \sqrt{25(.4)(.6)} = \sqrt{6} = 2.45.$

Calculating the normal approximation of the probability that the binomial random variable will assume a value of 11 or less:

$P(x_B \leq 11) \cong P(x_N \leq 11.5) = P\left(z \leq \frac{11.5 - 10}{2.45}\right) = P(z \leq .61)$

$= .5000 + .2291 = .7291.$

Taken from the binomial table, the correct probability to 3 places is .732. Our estimate is in error by less than 1/2 of 1 percent.

Problems

1. Find the indicated probabilities using the standard normal table.
    a. $P(0 < z < 1.83)$
    b. $P(-1.18 < z < 0)$
    c. $P(-1.26 \leq z < 2.11)$
    d. $P(-2.08 < z \leq .63)$
    e. $P(z \geq 1.47)$
    f. $P(z < -.95)$

g. $P(z > -1.09)$
h. $P(.86 \leq z \leq 1.72)$
i. $P(z < 1.56)$
j. $P(-1.96 < z < -1.28)$

2. Find a value, z_o, such that:
   a. $P(z \geq z_o) = .06$
   b. $P(z > z_o) = .85$
   c. $P(z < z_o) = .12$
   d. $P(z \leq z_o) = .92$
   e. $P(-z_o < z < z_o) = .95$
   f. $P(-z_o \leq z \leq z_o) = .88$

3. The random variable, x, has a normal distribution with mean $\mu = 160$, and standard deviation, $\sigma = 16$. Find the indicated probabilities.
   a. $P(156 \leq x < 188)$
   b. $P(124 < x < 168)$
   c. $P(x \geq 140)$
   d. $P(x < 176)$
   e. $P(x > 136)$
   f. $P(x \leq 148)$
   g. $P(180 \leq x < 200)$
   h. $P(148 < x < 152)$

4. For the random variable in Problem 3, find a value, x_o, such that:
   a. $P(x \leq x_o) = .75$;
   b. $P(x \geq x_o) = .08$;
   c. $P(x < x_o) = .16$.

5. The machine filling 18-ounce cereal boxes is set so that the average fill is 18.2 ounces with a standard deviation of .7 ounces. What percentage of the boxes is filled with less than the stated amount of cereal, 18 ounces? (Note that this is the same as asking the probability a box selected at random will weigh less than 18 ounces.)

6. A soft-drink machine discharges an average of 7.5 ounces per cup with a standard deviation of .4 ounces. The ounces of fill are normally distributed.
   a. What is the probability of overfilling an 8-ounce cup?
   b. How large a cup is required so that the probability the cup will overflow is only 3 percent?

7. The average life of color picture tubes manufactured by ACR Corporation is 33.6 months with a standard deviation of 5.7 months. The picture tube life is normally distributed.
   a. What percent of the picture tubes will not last 24 months or more?
   b. What percent will last more than 30 months?
   c. What percent will last more than 48 months?

d. If we wanted to set a warranty period, how many months would it have to be in order to only have to replace 1 percent of the picture tubes sold?

8. A government study determined that the number of people fishing on Lake Powell at any daylight hour is approximately normally distributed with a mean of 380 and standard deviation of 64.
   a. What is the probability that there will be no more than 500 people fishing at any given time?
   b. What is the probability that there will be between 300 and 450 people fishing at any given time?

9. A normally distributed, continuous random variable, x, has a mean of 38.2 and a standard deviation of 4.8. Answer each of the following.
   a. $P(30.0 \leq x \leq 35.0) = ?$
   b. $P(x \geq 32.5) = ?$
   c. 15 percent of the population values will exceed what value of x?
   d. For what value of k will the mean plus or minus k include 50 percent of the scores?

10. Hourly wage rates for the unskilled labor force in the Ogden area are normally distributed with a mean of $3.28 and a standard deviation of $.52. An individual is randomly selected from this population.
    a. What is the probability that the wage rate will be greater than $3.00 per hour?
    b. What is the probability that the wage rate will be less than $4.00 per hour?
    c. What percent of the unskilled labor force in the Ogden area have an hourly wage between $2.75 and $4.25?
    d. Ninety-five percent of the population have wage rates between what 2 symmetric limits about the mean?

11. A large food processor runs 2 restaurant chains. He has a small chain of expensive restaurants and a large chain of family restaurants. The weight of steaks he receives from his supplier follows a normal distribution with a mean weight of 8.72 ounces and a standard deviation of .16 ounces. At his expensive restaurants, he only serves steaks that weigh 9 ounces or more. What proportion of the steaks received from his supplier are large enough to serve in the expensive restaurant chain?

12. Refer to Problem 11. If you randomly select four steaks from the supplier, what is the probability they will all be too small to use in the expensive restaurants?

13. The diameter of a certain shaft must be within .627 ± .003 inches to be acceptable. The grinding process that produces the shafts is

known to vary, and the standard deviation of the diameters of the shafts it produces is .0015.
   a. What percent of the parts will be defective if the process average is centered at .627, i.e., $\mu = .627$?
   b. In a metal removal situation like this, an oversized part is usually less undesirable than an undersized part. If the process setting is moved to .628 (i.e., $\mu = .628$), what percent of the parts will be undersized? Compare this with the percent of undersized parts when $\mu = .627$.

14. The purchasing agent for a large firm is trying to select 1 of 2 suppliers of small generators. Supplier A's generators have an average output of 26.9 volts with a standard deviation of 1.2 volts. The average output of generators produced by supplier B is 28.1 volts, which is higher than supplier A's, but the variability of supplier B's product is greater. The standard deviation for supplier B is 2.2 volts. Which of the 2 suppliers will have the highest percentage of useable generators if the engineering specifications call for generators with an electrical output of at least 25 volts? What is the percentage?

15. If a part has an inside diameter greater than 2.68 inches, the part is defective. The average inside diameter of all parts produced by machine A is 2.52 inches with a standard deviation equal to .08 inches. What proportion of the parts produced by machine A are defective?

16. Refer to Problem 15. If you randomly select four parts produced by machine A, what is the probability all four parts will be good?

17. Balances on past due Disaster Charge Card accounts carried by the Last Security Bank were approximately normally distributed with $\mu = \$395$ and σ about 20 percent of μ. All accounts with balances over $250 were given to a hard-nosed collection agency.
   a. What percent of the past due accounts was the agency asked to collect?
   b. What is the probability that an account balance will exceed $500?

18. The XTZ Army Depot has some 6,000 employees. Ninety percent of the employees commute by auto, and 10 percent ride the commuter buses. The mean travel distance of the auto commuters is 16.4 miles. Commuter buses all come from a town which is 5.8 miles from the depot. Assume that auto travel distance is approximately normally distributed with a standard deviation of 3.6 miles.
   a. What percent of the auto commuters travel more than 5.8 miles?
   b. What percent of the auto commuters travel between 10 and 20 miles?

c. Only 1 percent of the commuters travel more than _____ miles.

19. In order to graduate at a certain large university, a student's GPA must be at least 2.00 and, of course, no GPA exceeds 4.00 because A = 4.00. In order to be eligible for admission to a graduate program, the GPA must be at least 3.60. What percent of the graduating seniors are eligible for graduate school if:
    a. GPA is approximately normally distributed with $\mu = 3.00$ and $\sigma = .33$;
    b. GPA has a uniform distribution between 2.00 and 4.00;
    c. GPA has an equilateral triangular distribution with $f(x) = 0$ for $x \leq 2$ and $x \geq 4$?
    (Hint: Refer to the material in Section 4.5 of Chapter 4 for parts b and c.)

20. Roland Bear, Inc., manufactures Item #G2007, a roller bearing, with a mean diameter of 2.475 centimeters. Even though the plant reflects the latest technology, the standard deviation is .048 centimeters. Specifications from Space Uttle & Co., one of the major buyers of #G2007, call for these bearings to be 2.475 ± .035 centimeters. The bearings are sorted on some Go, No-Go gauges; Space Uttle purchases the full production within their specs.
    a. What percent of the output of this particular bearing does Space Uttle purchase?
    b. Ram Rack & Sons purchase some of these bearings. Their specifications call for 2.475 ± 1.00 centimeters. Roland Bear, Inc., sorts the rejects from the Space Uttle sort to fill orders for Ram Rack. Sketch the distribution of diameters of bearings being purchased by Ram Rack. Assuming the sort for Ram Rack is a good one, i.e., all bearing diameters fall within their specifications, do you think they might change suppliers if they discovered the nature of this distribution?

■ The remaining problems require material from the appendix.

21. The random variable x has a binomial distribution with $n = 25$ and $\pi = .4$. Find the following probabilities using the normal approximation to the binomial distribution, and compare the answers with those obtained from the binomial table.
    a. $P(x \leq 9)$.
    b. $P(x = 9)$.
    c. $P(x < 13)$.
    d. $P(7 \leq x \leq 12)$.
    e. $P(7 < x < 12)$.
    f. $P(x > 14)$.

22. A sample of 400 persons is drawn from a large universe that includes 20 percent smokers. Use the normal curve estimate to find:
    a. the probability that the sample will contain 84 smokers;

b. the probability that there will be 75 or fewer smokers in the sample.

23. Fifty percent of the voters generally approve of the president's actions and policies to date. (Use binomial table for parts a and b.)
    a. In a random sample of 25, what is the probability that he will find a majority in his favor?
    b. In a random sample of 25, what is the probability that he will find at most 10 in his favor?
    c. Find the normal curve approximation of parts a and b.
    d. In a sample of 400, we could expect at least 52 percent to approve of the president with a probability of _____ .
    e. The odds are 19:1 that the number of voters who approve of the president (in the sample of 400) will be at least _____ and no more than _____ .

24. Consider the toss of 180 honest dice. Use the normal curve approximation to find:
    a. the probability that at least 28 sixes will occur;
    b. the probability that exactly 34 ones will occur.

25. Refer to the information in Problem 18. The commanding officer of XTZ Army Depot wants to organize a worker's committee of 60 persons. The commanding officer directed his statistical assistant to make a random selection. Let x be the number of bus commuters, and use the normal curve approximation to find:
    a. the probability that exactly 6 bus commuters will be on the committee;
    b. the probability that there will be 10 or more bus commuters on the committee;
    c. the probability there will be at least 3 but no more than 8 bus commuters on the committee.

CHAPTER 6

SAMPLING THEORY

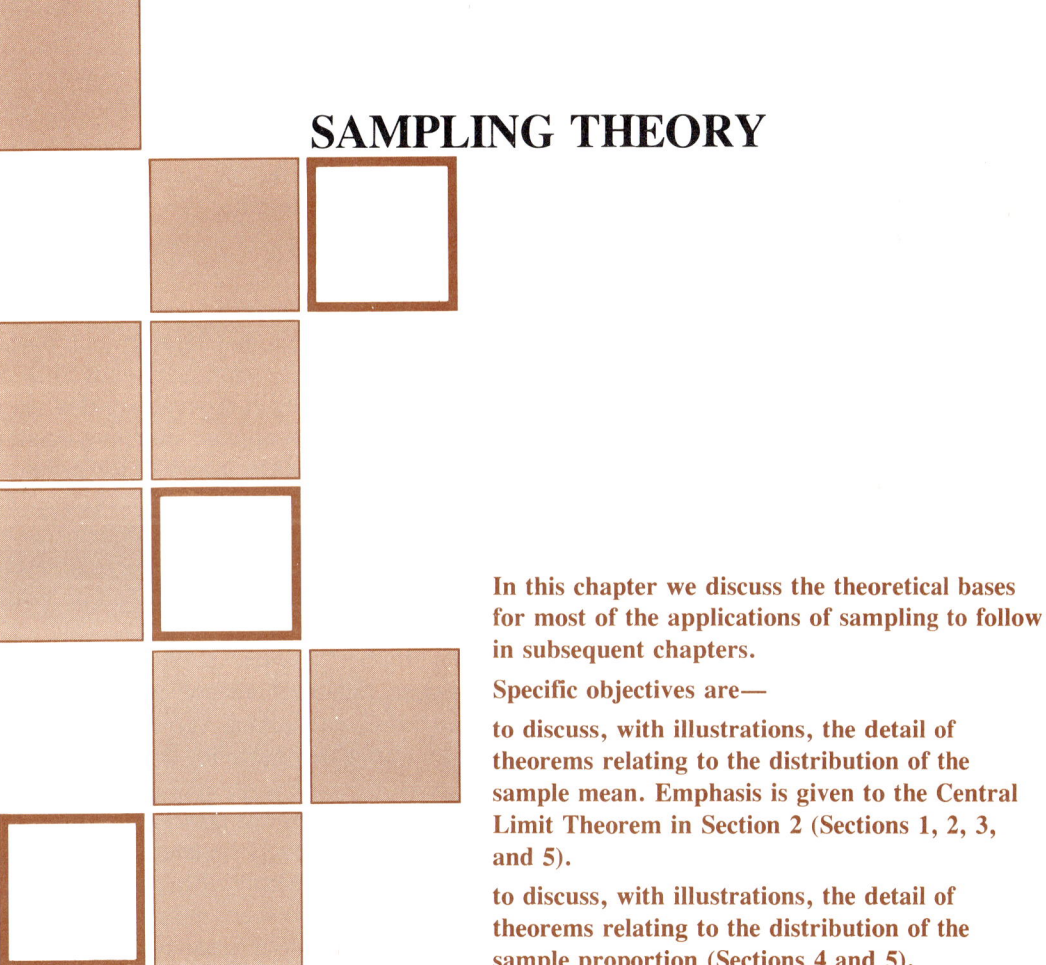

In this chapter we discuss the theoretical bases for most of the applications of sampling to follow in subsequent chapters.

Specific objectives are—

to discuss, with illustrations, the detail of theorems relating to the distribution of the sample mean. Emphasis is given to the Central Limit Theorem in Section 2 (Sections 1, 2, 3, and 5).

to discuss, with illustrations, the detail of theorems relating to the distribution of the sample proportion (Sections 4 and 5).

Many of the topics discussed in this chapter are theoretical and somewhat difficult to understand on first encounter. Your understanding and appreciation of this material will increase as the fundamental concepts presented here are applied and discussed throughout many of the chapters that follow. The subject matter in this chapter provides the linkage between probability theory and random variables, covered in the last three chapters, and statistical inference, the underlying theme in most of the remaining chapters. In Chapter 1 it was pointed out that one of the major problems of concern in statistics is that of making statements or inferences about the characteristics of a population from the information that is contained in a sample of observations selected from the population. In this chapter we are concerned with probability distributions of random variables that are a function of the sample data. One of the main concepts you should learn from discussion in this chapter is that a sample statistic is a function of sample data; i.e., a calculated statistic from the sample data, such as the sample mean or sample standard deviation, is itself a random variable.

Throughout the remainder of the text, when reference is made to a sample it will be assumed that the sample is a random one.

A sample selection is *random* if the trials are independent. For N finite, this implies that each element in the population has the same probability of being selected.

This is also called a *simple probability sample* and, for a finite universe of size N, the definition above is equivalent to saying that each of the possible samples (there are C_n^N of them) have an equal probability of being chosen. Implementing random sampling in the real world is often very difficult, and problems associated with applied sampling occupy volumes in the literature. As indicated in the discussion of the problem in Chapter 1, bias resulting from a nonrandom sampling procedure can give misleading results. When called upon to make a judgment as to whether or not a sample is in fact a random sample or an approximately random sample, you will find it helpful to refer to the definition given above.

When discussing sampling theory and/or statistical inference, confusion may arise if the student fails to note and understand the distinction between the two commonly encountered types of sampling. The terms *variable sampling* and *attribute sampling* are used to differentiate between the two. *Variable sampling* refers to taking a measurement on each of the items included in the sample. *Attribute sampling* refers to the procedure whereby the items sampled are classified into specifically defined categories. For example, if the members of a random sample of the Memphis labor force were asked, "What was your income last month?" this would be variable sampling. The measurement taken is the income of the respondent being sampled. In cases of variable sampling, we would probably be interested in calculating the mean of the sample incomes in order to say something about the population mean. On the other hand, if we had asked the respondents if they were currently employed, this would be an example of attribute sampling. For the attribute sampling case, we would be interested in calculating the proportion or percentage of respondents in our sample that are unemployed. We might use this sample proportion as an estimate of the unemployment rate for the entire Memphis labor force.

6.1 The Distribution of the Sample Mean

When a random sample of a given size is selected from a population, it usually represents one of a large number of samples of that size that could have been selected.

We will be concerned with sampling from large populations; but, even when sampling from small populations, there is a large number of possible samples that can be selected. For example, if a sample of 10 items is selected from a population of 30 items, there are over 60 million different samples that can be selected.[1] For each sample, a sample mean can be calculated; thus, the sample mean itself is a random variable. The value it assumes depends upon the measurements taken from the randomly selected elements of the population.

Because the sample mean is a random variable, it is natural to inquire about the average value of all possible sample means for samples of a given size selected from a specific population. Another question of interest would be concerned with the variability of the sample means.

Theorem 1   The expected value of the sample mean, i.e., the mean of all possible sample means, is equal to the mean of the population from which the sample is taken:

$$\mu_{\bar{x}} = E(\bar{x}) = \mu$$

where

$\mu_{\bar{x}}$ = the mean of the sample means
μ = the mean of the population from which the sample is drawn.

Note that there are no conditions on the nature of the population or on

---

1. The number of samples is the number of combinations that can be made from 30 items taken 10 at a time. This can be calculated using the Combination Formula in Appendix 3A of Chapter 3.

the population size. The expected value of the sample mean or the average value of all sample means is equal to the population mean regardless of the sample size or the population size. If the mean balance of Casey's Department Store's 500,000 charge accounts was $194.00, the expected value of the sample mean for a sample of 200 or a sample of 2 would be $194.00. If Charlie's Corner Grocery showed a mean balance of $194.00 in 500 charge accounts, this would also be the expected value of the sample mean for any sample size. In the next theorem we see for the first time a specific measure of the variability of the sample mean, the *statistical error,* as mentioned in Chapter 1.

Theorem 2

If the sample is drawn from an infinite population or if the sample size is small relative to the population size, then the standard deviation of the sample mean is equal to the standard deviation of the population sampled divided by the square root of the sample size:

$$\sigma_{\bar{x}} = \frac{\sigma}{\sqrt{n}}$$

where

$\sigma_{\bar{x}}$ = the standard deviation of the sample mean
σ = the standard deviation of the population
  from which the sample is drawn
n = the sample size.

The standard deviation of the sample mean is a measure of the variability of all possible values the sample mean could assume. We would expect the variability of the possible sample mean values to be a function of the variability of the population from which the sample is being selected. The latter is, of course, measured by σ, the population standard deviation. Based upon our intuition and/or our personal experience, most of us would say that a large sample is "better" than a small sample. What do we mean by "better"? We would feel better about using the sample mean from a large sample as an estimate of a population mean than we would about using the mean from a small sample because we feel that the sample mean of a large sample will be closer to the population mean than that of a small sample. In Theorem 2 above we see that our intuition has been vindicated—at least in part. The variability of the sample mean decreases as the sample size increases and, quite specifically, the standard deviation of the sample mean (also known as *the standard error of the mean*) varies inversely with the sample size. As an example of this important concept, let us assume that the population standard deviation of Casey's Department Store's 500,000 account balances is $72.00; i.e., σ = $72.00. For selected sample sizes, note how $\sigma_{\bar{x}}$ decreases as the sample size increases. Note also that n must be quadrupled in order to halve $\sigma_{\bar{x}}$.

n	4	16	36	144	576
$\sigma_{\bar{x}}$	$36	$18	$12	$6	$3

The following theorem permits us to make probability statements about the sample mean when sampling from normally distributed populations.

Theorem 3   If the population being sampled is normally distributed with mean, μ, and standard deviation, σ, then the sample mean will be normally distributed with

$$\mu_{\bar{x}} = \mu \text{ and } \sigma_{\bar{x}} = \frac{\sigma}{\sqrt{n}}.$$

(Note: See the two previous theorems.)

This is equivalent to saying that the ratio

$$\frac{\bar{x} - \mu}{\sigma_{\bar{x}}} \text{ or } \frac{\bar{x} - \mu}{\frac{\sigma}{\sqrt{n}}}$$

is normally distributed with mean of 0 and standard deviation of 1.

Using this transformation and the standard normal curve table, we can find probabilities concerning the sample mean, $\bar{x}$. You will recall from Chapter 5 that to find the probability that a normally distributed random variable, x, is greater than some value, a, find

$$P(z > z_1)$$

where

$$z_1 = \frac{a - \mu}{\sigma}.$$

Probabilities concerning the sample mean are found in a similar manner.

To find the probability that the mean of a random sample from a normal population will be greater than some value, a, find

$$P(z > z_1)$$

where

$$z_1 = \frac{a - \mu}{\frac{\sigma}{\sqrt{n}}}$$

μ = the population mean
σ = the population standard deviation
n = the sample size
z = the standard normal variable.

The Distribution of the Sample Mean

The two transformation formulas can also be written using the alternative format introduced in Chapter 5:

$$P(x > a) = P\left(z > \frac{a - \mu}{\sigma}\right)$$

$$P(\bar{x} > a) = P\left(z > \frac{a - \mu}{\frac{\sigma}{\sqrt{n}}}\right).$$

In this same format, the probability that the sample mean, $\bar{x}$, will fall in the interval from a to b is written

$$P(a < \bar{x} < b) = P\left(\frac{a - \mu}{\frac{\sigma}{\sqrt{n}}} < z < \frac{b - \mu}{\frac{\sigma}{\sqrt{n}}}\right).$$

We note again that the standard deviation of the sample mean is smaller than the population standard deviation by a reciprocal factor of the square root of the sample size. This is important. We illustrate this point in the following example.

Example 6.1.1

A normally distributed population has a mean of 100.00 and a standard deviation of 20.00. (a) If 1 item is selected randomly from this population, what is the probability that it will have a value greater than 110.00? (b) If 16 items are randomly selected, what is the probability that the sample mean will exceed 110.00?

Figure 6.1, parts a and b, illustrates graphically the solution to the respective parts of the above problem. The solutions are calculated as follows:

a. $P(x > 110) = P\left(z > \frac{110 - \mu}{\sigma}\right) = P\left(z > \frac{110 - 100}{20}\right)$

   $= P(z > .5) = .5 - .1915 = .3085.$

b. $P(\bar{x} > 110) = P\left(z > \frac{110 - \mu}{\frac{\sigma}{\sqrt{n}}}\right) = P\left(z > \frac{110 - 100}{\frac{20}{\sqrt{16}}}\right)$

   $= P(z > 2.0) = .5 - .4772 = .0228.$

As we should have expected, the second probability is considerably smaller than the first. This is due to the fact that the sample mean has less variability than the population itself; i.e., on the average the sample mean will be closer to the population mean than a randomly selected single value. Refer again to Figure 6.1. The scales for x and $\bar{x}$ in the first parts of a and b are the same. The sample means are more tightly clustered about the population mean than the individual values

Figure 6.1 **Comparison: Distribution of Normal Variable with That of Sample Mean**

a.

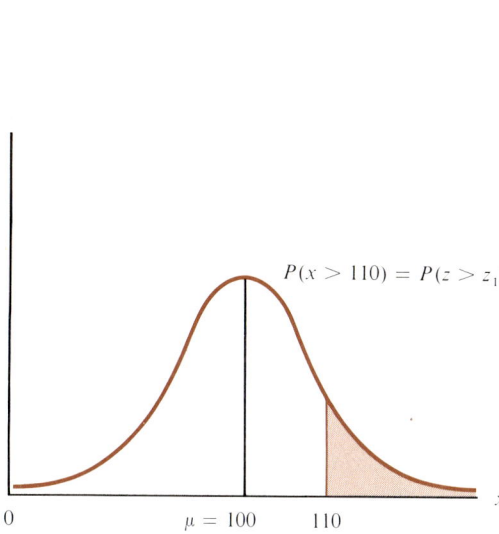

b.

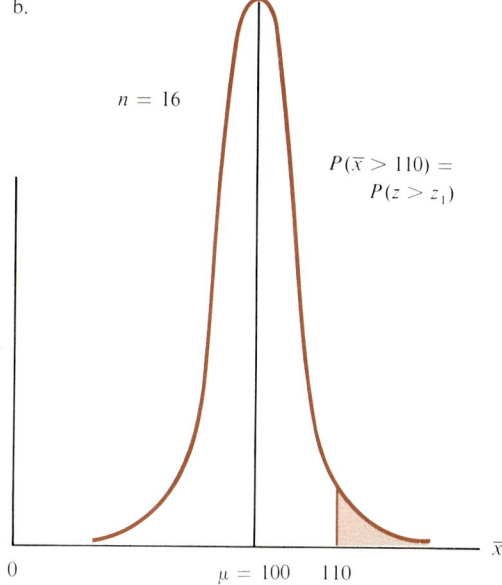

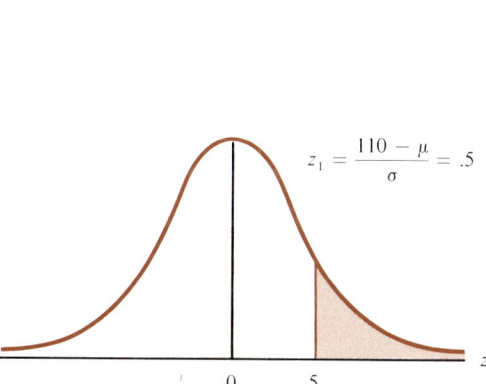

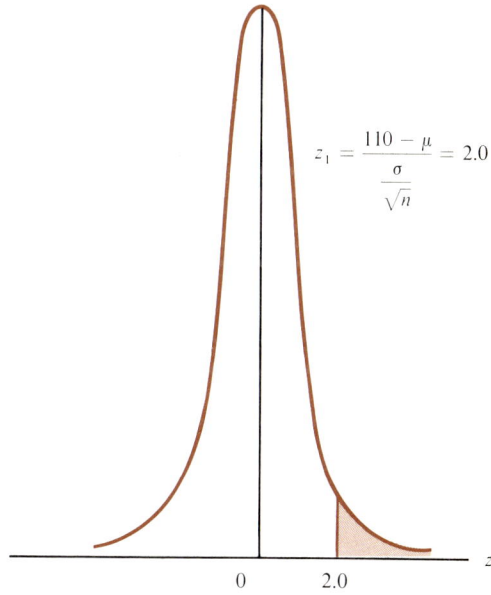

x is a normally distributed random variable with mean, μ, and standard deviation, σ; $\bar{x}$ represents the mean of a random sample of 16 values.

The Distribution of the Sample Mean

of x. The narrower and higher curve in part b of Figure 6.1 compared to that in part a illustrates the fact that the standard deviation of the sample mean ($\sigma_{\bar{x}} = 20/\sqrt{16} = 5$) is much smaller than the population standard deviation. Consider a second example that will further illustrate the inverse relationship between the standard deviation of the mean and the sample size.

Example 6.1.2

Aluminum castings manufactured by Reyco Company have a mean weight of 25.0 pounds and a standard deviation of .36 pounds. The distribution of the weights of the individual castings follows an approximately normal distribution. Joe's Job Shop buys 4 of these castings, and Fred's Foil, Inc., buys 100 castings. Assume that both purchases are random samples from a potentially infinite production. What is the probability that the mean weight of Joe's purchase will fall within $\pm 1/10$ pound of the 25.0-pound mean weight of all aluminum castings produced by Reyco? Answer the same question for Fred's purchase.

Joe's purchase represents a random sample of 4 castings from a normal population with mean, $\mu = 25$, and standard deviation, $\sigma = .36$, and we are to find the probability that the sample mean will lie within the interval 25.0 ± 0.1. This probability can be found with the standard normal table after using the transformation given above:

$$P(24.9 < \bar{x} < 25.1) = P\left(\frac{24.9 - \mu}{\frac{\sigma}{\sqrt{n}}} < z < \frac{25.1 - \mu}{\frac{\sigma}{\sqrt{n}}}\right)$$

$$= P\left(\frac{24.9 - 25}{\frac{.36}{\sqrt{4}}} < z < \frac{25.1 - 25}{\frac{.36}{\sqrt{4}}}\right)$$

$$= P(-.56 < z < .56)$$

$$= .2123 + .2123 = .4246.$$

The standard deviation of the mean, $\sigma_{\bar{x}}$, is

$$\frac{0.36}{\sqrt{4}} = 0.18,$$

and the symmetric interval is .56 standard deviation from the mean of 25.0.

This central area is .4246, and from this we conclude that there is a 42 percent chance that the average weight of the 4 castings purchased by Joe will be within 1/10 pound of the 25.0-pound mean weight of all castings.

The parameters of the problem are the same for Fred's purchase with the exception that the sample size is 100. We now want to find the probability that the sample mean of a sample of size 100 will fall within the interval 24.9 to 25.1. The probability is calculated in the same manner as above:

$$P(24.9 < \bar{x} < 25.1) = P\left(\frac{24.9 - \mu}{\frac{\sigma}{\sqrt{n}}} < z < \frac{25.1 - \mu}{\frac{\sigma}{\sqrt{n}}}\right)$$

$$= P\left(\frac{24.9 - 25}{\frac{.36}{\sqrt{100}}} < z < \frac{25.1 - 25}{\frac{.36}{\sqrt{100}}}\right)$$

$$= P(-2.78 < z < 2.78)$$

$$= .4973 + .4973 = .9946.$$

Here we see the power of a large sample. Fred is almost certain (there is a 99.5 percent chance) that the mean weight of his purchase will differ by no more than 1/10 pound from the population mean of 25.0 pounds. Figure 6.2 has the distribution of the individual castings as they are produced by Reyco (part a), the distribution of the mean for samples of size 4 (part b), and the distribution of the mean for samples of 100 castings (part c). These results illustrate and, we hope, add credence to the important fundamental proposition we have been emphasizing: *As the sample size increases, the standard deviation of the sample mean decreases.* The standard error of the mean for $n = 100$ is .036 as compared to .18 for $n = 4$. The larger the sample size, the higher the probability that the sample mean will fall within a fixed interval of the population mean.

Example 6.1.3

A component for a consumer product consists of 10 laminated disks. If the total weight of the disks is less than 23.00 ounces, the component is considered to be defective. The average weight of disks coming from the stamping machine is 2.43 ounces. The standard deviation is .18 ounce. The distribution of the weights of the disks is approximately normal. The 10 disks selected for any given component are essentially randomly selected. What percent of the components will be defective with respect to the weight criterion?

The proportion of defective components is the same as the probability that the sum of the weights of the 10 disks making up a component will be less than 23.00 ounces. The 10 disks for a given component represent a random sample from a normally distributed population with mean, $\mu = 2.43$ ounces, and standard deviation, $\sigma = .18$ ounce. Therefore, if we can state the problem in terms of a probability question concerning the sample mean, then we can find the desired probability.

Figure 6.2 Distribution of Weights and Mean Weights (Reyco Example)

a.

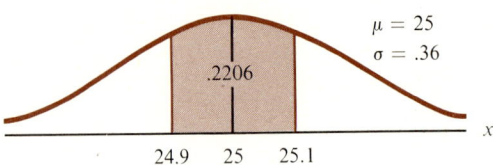

b.

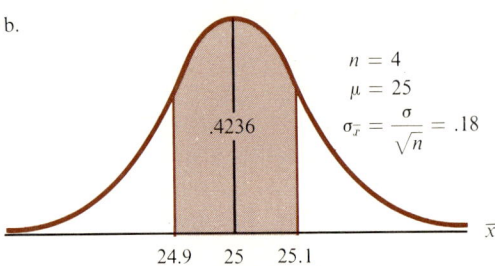

c.

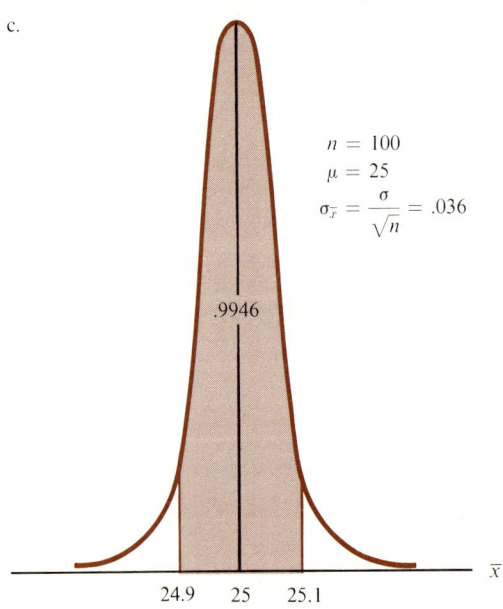

We observe that if the sum of weights of the 10 disks were equal to 23 ounces, then the average weight of the 10 disks would be 2.3 ounces; i.e.,

$$\bar{x} = \frac{\sum_{i=1}^{n} x_i}{n} = \frac{23}{10} = 2.3 \text{ ounces.}$$

Multiplying both sides of the definition of $\bar{x}$ by n, we also have

$$\sum_{i=1}^{n} x_i = n\bar{x}.$$

If the mean of the sample 10 disks making up the component is less than 2.3 ounces, then the sum of the weights of the 10 disks,

$$\sum_{i=1}^{n} x_i,$$

will be less than 23 ounces: If

$$\bar{x} < 2.3,$$

then

$$\Sigma x_i = n\bar{x} = 10\bar{x} < 10(2.3) = 23.$$

The desired probability is the probability that the sample mean is less than 2.3 ounces:

$$P(\Sigma x_i < 23) = P(\bar{x} < 2.3).$$

The latter probability can be found by transforming it to the equivalent standard normal curve probability statement:

$$P(\bar{x} < 2.30) = P\left(z < \frac{2.30 - \mu}{\frac{\sigma}{\sqrt{n}}}\right)$$

$$= P\left(z < \frac{2.30 - 2.43}{\frac{.18}{\sqrt{10}}}\right) = P(z < -2.28)$$

$$= .5 - .4887 = .0113.$$

The probability that the average weight of the 10 disks will be less than 2.3 ounces is .0113. This is the same as the probability that any given component will weigh less than 23 ounces and, therefore, the probability of a defective component. In other words, 1.13 percent of the components will be defective with respect to weight.

6.2 The Central Limit Theorem

The previous theorem concerning the distribution of the sample mean depended upon the assumption that the population sampled was normally distributed. We would indeed be quite limited in the scope of applied statistics if it were necessary to know the nature of the population distribution and even more limited if it were necessary to assume normalcy. The Central Limit Theorem is one of the most powerful theorems in applied statistics because it frees us from the assumption of a normally distributed population.

**Theorem 4
The Central
Limit Theorem[2]**

If n is large, the sample mean is approximately normally distributed regardless of the nature of the population distribution; or

$$\frac{\bar{x} - \mu}{\sigma_{\bar{x}}} = \frac{\bar{x} - \mu}{\frac{\sigma}{\sqrt{n}}} \cong z,$$

if n is large.
(Recall from the early part of the chapter that

$$\mu_{\bar{x}} = \mu \text{ and } \sigma_{\bar{x}} = \frac{\sigma}{\sqrt{n}}.)$$

The Central Limit Theorem is the fundamental sampling theorem. It is because of this theorem (and variations thereof), and not because of nature's questionable tendency to normalcy, that the normal distribution plays such a key role in our work. Based upon experimental sampling from a variety of distributions, it has been found that, for a sample size of 30 or more, the distribution of the sample mean is very close to normal. Throughout the text this will be our somewhat arbitrary rule: "A sample size of 30 or more is a *large* sample." You could argue, "If 30 is a large sample, then 29 is also large; if 29 is a large sample, then surely 28 must also be; etc." (Plato addressed a similar problem several hundred years B.C. when the question arose as to how many hairs constitute a beard.) If we were aware that a population distribution was characterized by a somewhat extreme nonnormal attribute, such as skewness, multimodality, flatness, etc. (see Figure 6.3), then $n = 30$ would clearly be a lower bound for a "large" sample; on the other hand, if we knew that the population distribution was unimodal and not highly skewed, sample means for n considerably less than 30 would be very close to being normally distributed.

---

2. This is a "limit" theorem and correctly stated we would have: As $n \to \infty$,

$$\frac{\bar{x} - \mu}{\frac{\sigma}{\sqrt{n}}} \to z,$$

the standard normal curve variable with mean of 0 and standard deviation of 1.

Figure 6.3 Examples of Some Continuous Distributions

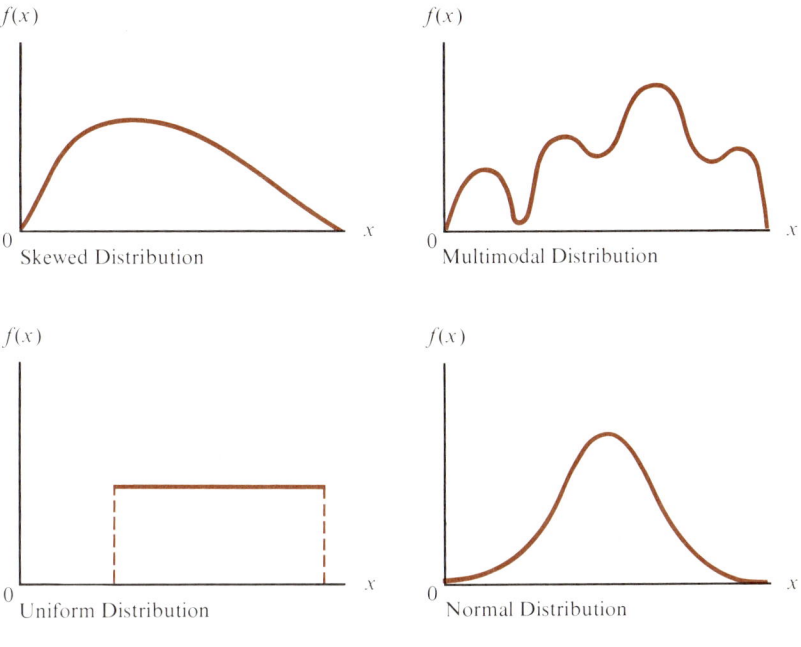

An important corollary to the Central Limit Theorem permits the substitution of the sample standard deviation, s, for the population standard deviation, σ, in the normal transformation equation.

Corollary 1

As $n \to \infty$, $s \to \sigma$ and it follows that

$$\frac{\bar{x} - \mu}{s/\sqrt{n}} \to \frac{\bar{x} - \mu}{\sigma/\sqrt{n}} \cong z.$$

(This is read, "The ratio,

$$\frac{\bar{x} - \mu}{s/\sqrt{n}},$$

approaches the standard normal distribution as the sample size, n, goes to infinity, i.e., as n gets large.")

Because the population standard deviation is generally not known, this corollary to the Central Limit Theorem provides us with one of the key statistics to be used in statistical estimation and hypothesis testing, the major topics of the next several chapters.

The following example will help clarify the implications of the Central Limit Theorem.

Example 6.2.1 For a certain metropolitan area, the U.S. Internal Revenue Service took a complete census of personal income tax paid in 1979. The mean and standard deviation of this variable were $\mu = \$2,180$ and $\sigma = \$1,816$.

a. Would we expect sample means to be normally distributed for $n = 9$? for $n = 64$?
b. Find the standard error of the mean for each of the above sample sizes. What is the expected value of the mean in each case?
c. What is the probability that the sample mean for $n = 64$ is no more than $300 from the universe mean?

Income distributions are typically highly skewed, and we would expect income tax distributions also to be positively skewed. The facts that σ is of the same magnitude as μ and 0 is a lower bound on tax payments add credence to the skewness notion. Assuming this is the case, means for $n = 64$ would be approximately normally distributed, but those for samples of size 9 would not be. The standard error of the mean for $n = 9$,

$$\sigma_{\bar{x}} = \frac{\sigma}{\sqrt{n}} = \frac{1{,}816}{\sqrt{9}} \cong \$605,$$

and for $n = 64$,

$$\sigma_{\bar{x}} = \frac{1{,}816}{\sqrt{64}} = 227.$$

In both cases the expected value of the sample mean is the universe mean of $2,180.

For $n = 64$, we want to find the probability that $\bar{x}$ is within $300 of the population mean. Written in terms of the normal curve transformation,

$$P(1{,}880 \leq \bar{x} \leq 2{,}480) = P\left(\frac{1{,}880 - 2{,}180}{227} \leq z \leq \frac{2{,}480 - 2{,}180}{227}\right)$$

$$= P(-1.32 \leq z \leq 1.32) = .8132.$$

There is an 81 percent chance that a mean from a sample of size 64 will be within $300 of the universe mean.

6.3 The Student-*t* Distribution

Probability statements about the sample mean in the above examples depended upon having a large sample and/or knowing the population standard deviation. If the population standard deviation is not known, it is still possible to make probability statements concerning the sample mean. We introduce a new distribution, the Student-*t* Distribution, with the following theorem.

Theorem 5   If the population variable, *x*, is normally distributed, then the ratio,

$$\frac{\bar{x} - \mu}{\frac{s}{\sqrt{n}}},$$

has a Student-*t* distribution with (*n*-1) degrees of freedom.[3]

In lieu of writing Student-*t*, we will frequently use *t*; i.e.,

$$t = \frac{\bar{x} - \mu}{\frac{s}{\sqrt{n}}}$$

where

t is a random variable that has a Student-*t* distribution with $n - 1$ degrees of freedom

n = the sample size
$\bar{x}$ = the sample mean
s = the sample standard deviation.

Note the parallel form of this new ratio,

$$t = \frac{\bar{x} - \mu}{\frac{s}{\sqrt{n}}},$$

with the one we have been using,

$$z = \frac{\bar{x} - \mu}{\frac{\sigma}{\sqrt{n}}}.$$

Also note the important difference: the *t* variable is a function of two sample statistics, $\bar{x}$ and *s*, whereas the normal curve variable is a function of $\bar{x}$ only.

As implied above, the Student-*t* refers to a family of distributions with the specific member distribution determined by the degrees of freedom, *df*. In the theorem introduced above, the *df* equal sample size less 1, $n - 1$. The Student-*t* Distributions are symmetrical and unimodal with mean and median both 0; in these respects they are similar to the normal curve. Because of the variability of both $\bar{x}$ and *s*, the

---

3. Around the turn of the century, W. S. Gosset, an English accountant-statistician, discovered and developed this distribution. Because of problems related to his employment, he published his paper in 1908 under the pen name of Student.

The Student-*t* Distribution

Student-*t* Distribution with fixed *df* is flatter than the normal curve; i.e., the dispersion is greater. See the comparison in Figure 6.4 between a *t* distribution with $df = 4$ and the standard normal curve. The smaller the sample size (and, therefore, the *df*), the greater the variability of the *t*-distribution; and, as the sample size gets very large, the *t*-distribution approaches the normal curve.

To have the flexibility accorded us by the standard normal table would require a different table for each degree of freedom. In Table C, tail areas or probabilities for several useful *t*-values and for integral degrees of freedom from 1 to 30 are presented. The Student-*t* Distribution is an important distribution for statistical inference problems when the sample size is small and the population sampled is normally distributed. Examples and applications of the Student-*t* Distribution are much more meaningful in the context of statistical inference; therefore, we will defer further discussion of this distribution and the use of Table C until statistical inference is discussed in the next two chapters.

6.4 The Distribution of the Sample Proportion

At the beginning of this chapter, a distinction was made between variable sampling and attribute sampling. To this point in the chapter, theorems and examples have dealt with variable sampling only; i.e., a measurement was attached to the sample element. In the case of attribute sampling, the item sampled is classified into one of two or more categories. We will be concerned only with the dichotomy, attribute *A* or not-*A*, even though there could well be more than two categories in the original classification. If the sampling trials are independent, you should recognize this as the binomial problem from Chapter 4. The variable used in our previous discussion was the sample number of successes. We find it more convenient from the arithmetic standpoint

Figure 6.4   Comparison: Normal and Student-*t* Distributions

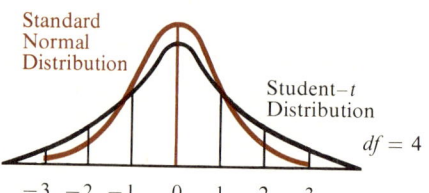

and more consistent with our previous discussions to use p, the sample proportion of successes, rather than the sample number of successes. The probability of success or the population proportion of the attribute of interest will be denoted by π. For example, the stockholders of a corporation may be divided into two groups: those who approve of the chairman of the board's new emphasis on social awareness and those who don't. If the characteristic of interest is "approve the chairman's new policy," then π would be the proportion of all stockholders that approve of the new emphasis on social awareness. Other examples would be the proportion of all customers who prefer the new package design, the proportion of parts coming off an assembly line that are defective, the proportion of Indiana high school graduates that attend a university in Indiana, etc.

If, as in the earlier discussion of the binomial distribution, we let x be the sample number of items having the specified attribute, then the sample proportion will be the ratio of x to the sample size, n; i.e., $p = x/n$. Note that the possible values of x are nonnegative integral values from 0 to n and, therefore, the variable p may take on the discrete values, $0, 1/n, 2/n, \ldots n - 1/n$ and n/n (or 1).

The sample proportion, when multiplied by 100, will, of course, be the percentage of sample items having the characteristic of interest. See the following example for clarification.

Example 6.4.1

The manager of a supermarket is interested in whether or not his customers like the new checkout facilities. Of the 25 randomly selected customers, 14 said they liked the new facilities. The other 11 customers in the sample said they preferred the old checkout facilities. Calculate the sample proportion.

The manager has indicated that the characteristic of interest is "the customer likes the new facility"; thus the sample proportion is

$$p = \frac{x}{n} = \frac{14}{25} = .56.$$

Fifty-six percent of the customers sampled preferred the new facilities.

As is the case with the sample mean, the sample proportion, p, is itself a random variable. It varies from sample to sample.

Theorem 6

$\mu_p = E(p) = \pi,$

where

μ_p = the mean of the population of all possible sample proportions for a given sample size
π = the population proportion.

If the sample size is small relative to the population size, if we are sampling from an infinite population or a process, or if we are sampling from a small population with replacement, then

$$\sigma_p = \sqrt{\frac{\pi(1-\pi)}{n}}$$

where

σ_p = the standard deviation of the sample proportion, p, (or the standard error of the proportion)
π = the population proportion
n = the sample size.

The expected value of the sample proportion is the population proportion, π. This is analogous to the case for the sample mean.

We are not too surprised at the analogous results because, as was pointed out in Chapter 2, the sample proportion is a special case of the sample mean. In Example 6.4.1 25 customers were asked if they liked the new checkout facilities installed by the supermarket. With a value of 1 assigned to those customers who liked the new facility and a value of 0 assigned to those customers who preferred the old facilities, the following sample results were obtained:

1, 1, 0, 1, 0, 0, 1, 0, 1, 1, 0, 1, 0, 1, 0, 1, 1, 0, 0, 1, 0, 1, 1, 0, 1.

The mean of this sample of data can be calculated using the formula for the sample mean:

$$\bar{x} = \frac{\sum_{i=1}^{n} x_i}{n} = \frac{1 + 1 + 0 + \cdots + 1 + 0 + 1}{25} = \frac{14}{25} = .56.$$

As expected, the mean of the 0–1 values is equal to the sample proportion calculated above.

For a given sample size, n, the standard deviation of the sample proportion, σ_p, is a maximum for $\pi = 1/2$; i.e., when the population is evenly divided between the 2 categories. For a given population proportion, π, the standard deviation of the sample proportion decreases as n increases. This is what we would intuitively expect, as was the case with the sample mean.

To calculate probabilities concerning the outcome of the sample proportion for small sample sizes, it is necessary to convert the problem to a binomial distribution problem and use the binomial probability function discussed in Chapter 4. For large sample sizes, the Central Limit Theorem rescues us from the tedious calculations required by the binomial formula.

Theorem 7
Special Case of the Central Limit Theorem

The sample proportion, p, tends to be approximately normally distributed as the sample size, n, approaches infinity (gets very large); i.e.,

$$\frac{p - \pi}{\sigma_p} = \frac{p - \pi}{\sqrt{\frac{\pi(1 - \pi)}{n}}} \to z \text{ as } n \to \infty.$$

In addition we have the following special case of Corollary 1, Corollary 2.

Corollary 2

$$\frac{p - \pi}{\sqrt{\frac{p(1 - p)}{n}}} \to z \text{ as } n \to \infty.$$

The corollary is used more frequently than the theorem in statistical inference; this will become evident in the next two chapters. It is a stronger statement in that the sample proportion is substituted for the typically unknown population proportion in the denominator.

Again, the same question that occurred with the sample mean has to be dealt with here. When is the sample size large enough to use the normal approximation? It is generally agreed that n is sufficiently large if the following 3 conditions are satisfied: $n \geq 30$, $n\pi > 5$, and $n(1 - \pi) > 5$. The latter 2 conditions are added because the skewness of the distribution varies as the population proportion, π, varies. If π is equal to .5, then the distribution of the sample proportion, p, is symmetrical about the mean, $\mu_p = \pi = .5$. The distribution of p becomes more asymmetrical or skewed as π gets smaller than .5 or as π gets larger than .5.

Example 6.4.2

A cellophane packaging process is supposed to package cookies so that no more than 5 percent of the packages have broken cookies that can be observed by the customer through visual inspection. The marketing manager is upset with the production manager because a random sample of 200 packages had 14 defective packages, or 7 percent defective. If the packaging process is in fact operating at a rate of only 5 percent defective, what is the probability that a random sample of 200 packages will have 7 percent or more defective?

The question asked deals with the sample proportion, p; i.e., what is the probability that the sample proportion of a sample of 200 items selected from a population with population proportion, $\pi = .05$, will be greater than or equal to .07 or $P(p \geq .07) = ?$ The conditions for using the normal distribution to calculate the probability are satisfied for this problem; the sample size is greater than 30 and $n\pi = 200(.05) = 10$ is greater than 5. Once it is assumed that p is approximately normally distributed, then the probability can be found in the same manner as for any other normally distributed random variable:

$$P(p \geq .07) = P\left(z \geq \frac{.07 - \pi}{\sqrt{\frac{\pi(1-\pi)}{n}}}\right)$$

$$= P\left(z \geq \frac{.07 - .05}{\sqrt{\frac{(.05)(.95)}{200}}}\right) = P(z \geq 1.30)$$

$$= .5000 - .4032 = .0968.$$

There is a 9.7 percent chance that in a random sample of 200 packages there will be 7 percent or more defective packages. If you studied the material on the normal approximation to the binomial distribution in the appendix to Chapter 5, then you may have observed the similarity between these two types of problems. The Central Limit Theorem provides the justification for the technique covered in the Chapter 5 appendix: the normal approximation of the binomial distribution.

6.5 Sampling from a Finite Population

Up to this point it has been assumed that the population sampled was infinite in size or at least very large relative to the sample size. If a significant proportion of the population is being sampled, then the formulas for the standard deviation of the sample mean, $\sigma_{\bar{x}}$, and the standard deviation of the sample proportion, σ_p, should be modified as follows:

Theorem 8   If the sample consists of a significant part of the population, then

$$\sigma_{\bar{x}} = \frac{\sigma}{\sqrt{n}} \sqrt{\frac{N-n}{N-1}}$$

and

$$\sigma_p = \sqrt{\frac{\pi(1-\pi)}{n}} \sqrt{\frac{N-n}{N-1}}$$

where

n = the sample size
N = the population size.

Except for the factor,

$$\sqrt{\frac{N-n}{N-1}},$$

the formulas for $\sigma_{\bar{x}}$ and σ_p are the same as those given earlier for an infinite or large population. The expression,

$$\sqrt{\frac{N-n}{N-1}},$$

is called the *finite population correction factor*—abbreviated as the *fpc*. Except for the uninteresting case when the sample size is equal to 1, the finite population correction factor is always less than 1; thus, when sampling from a finite population, the standard deviation of the sample mean is less than that obtained when sampling from an infinite population. While sampling a significant proportion of the population does have an effect on the standard deviation of the mean (or the sampling error), it is important to realize that it is the sample size itself that is the main determinant of the variability of the sample mean and not the proportion sampled. It is erroneously assumed by many people that a fixed-percent sample yields the same sampling error regardless of population size. This is a naive notion, and the following example will help clarify the differing effects of sample size versus percent sampled.

Example 6.5.1 In all of the parts below, assume that the population standard deviation, σ, is the same, namely, $\sigma = 50$. Find the standard deviation of the sample mean, $\sigma_{\bar{x}}$, for each of the following cases.

a. The population is infinite; the sample size is 4.
b. The population is infinite; the sample size is 100.
c. The population size is 80; the sample size is 4 (a 5 percent sample).
d. The population size is 2,000; the sample size is 100 (a 5 percent sample).
e. The population size is 10,000; the sample size is 100 (a 1 percent sample).
f. The population size is 100,000; the sample size is 100 (a .1 percent sample).
g. The population size is 1 million; the sample size is 100 (a .01 percent sample).

The answers to each part can readily be calculated using the formulas for the standard deviation of the sample mean. Remember that $\sigma = 50$ for all parts of the problem.

a. $\sigma_{\bar{x}} = \dfrac{\sigma}{\sqrt{n}} = \dfrac{50}{\sqrt{4}} = 25.$

b. $\sigma_{\bar{x}} = \dfrac{\sigma}{\sqrt{n}} = \dfrac{50}{\sqrt{100}} = 5.$

c. $\sigma_{\bar{x}} = \dfrac{\sigma}{\sqrt{n}} \sqrt{\dfrac{N-n}{N-1}} = \dfrac{50}{\sqrt{4}} \sqrt{\dfrac{80-4}{80-1}} = 25(.98083) = 24.52075.$

d. $\sigma_{\bar{x}} = \dfrac{\sigma}{\sqrt{n}} \sqrt{\dfrac{N-n}{N-1}} = \dfrac{50}{\sqrt{100}} \sqrt{\dfrac{2{,}000-100}{12{,}000-1}} = 5(.97492)$
   $= 4.87460.$

e. $\sigma_{\bar{x}} = \dfrac{\sigma}{\sqrt{n}} \sqrt{\dfrac{N-n}{N-1}} = \dfrac{50}{\sqrt{100}} \sqrt{\dfrac{10{,}000-100}{10{,}000-1}} = 5(.99504)$
   $= 4.9752.$

f. $\sigma_{\bar{x}} = \dfrac{\sigma}{\sqrt{n}} \sqrt{\dfrac{N-n}{N-1}} = \dfrac{50}{\sqrt{100}} \sqrt{\dfrac{100{,}000-100}{100{,}000-1}} = 5(.99950)$
   $= 4.99750.$

g. $\sigma_{\bar{x}} = \dfrac{\sigma}{\sqrt{n}} \sqrt{\dfrac{N-n}{N-1}} = \dfrac{50}{\sqrt{100}} \sqrt{\dfrac{1{,}000{,}000-100}{1{,}000{,}000-1}} = 5(.99995)$
   $= 4.99975.$

There is a significant reduction in part b compared to part a. Increasing sample size does have a significant effect on the sampling error. However, the gain is not linear. The sample size was increased by a factor of 25 with a reduction in $\sigma_{\bar{x}}$ of only

$$\sqrt{\dfrac{1}{25}} = \dfrac{1}{5}.$$

In parts c and d, the calculation of $\sigma_{\bar{x}}$ was adjusted because 5 percent of a finite population was sampled. Here again, the sample size makes a tremendous difference in the sampling error, $\sigma_{\bar{x}}$, even though the percentage sampled is the same for both cases.

For parts d, e, f, and g, the sample size of 100 is the same, but the percentage sampled changes. The effect on $\sigma_{\bar{x}}$ is minimal when the percentage sampled is 1 percent or less. Even for a 5 percent sample the effect is quite small. It is particularly interesting to compare parts e and g. In part g the percent sampled is .01 percent. In part e the percent sampled is increased a hundredfold to 1 percent. But the sampling error is decreased only slightly from 4.99975 to 4.9752.

Our general rule of thumb would be to use the finite population correction factor if more than 5 percent of the population is being sampled; otherwise treat the data as if it were a sample from an infinite population. It could be that you are concerned about the effect of the finite population correction factor even though the percent sampled is less than 5 percent. An estimate of the *fpc* can quickly be calculated for larger populations by observing that $N - 1$ is approximately equal to N for larger populations. Noting this, we get

$$fpc = \sqrt{\frac{N-n}{N-1}} \cong \sqrt{\frac{N-n}{N}} = \sqrt{1 - \frac{n}{N}}.$$

The finite population correction factor is approximately equal to the square root of 1 minus the proportion sampled, which is the same as the proportion not sampled. For example, if 3 percent of the population is sampled,

$$fpc \cong \sqrt{1 - .03} = \sqrt{.97} = .9849.$$

Similarly, for a 2 percent sample the finite population correction factor is approximately $\sqrt{.98}$; for a 10 percent sample it is approximately $\sqrt{.90}$; etc.

6A Appendix: Selected Proofs and Additional Theorems

The expected value and variance theorems covered in Appendix 4A can be used to demonstrate the validity of the formulas for the expected value and standard deviation of the sample mean:

$$\mu_{\bar{x}} = \mu$$

and

$$\sigma_{\bar{x}} = \frac{\sigma}{\sqrt{n}}$$

if sampling from an infinite population (or from a process or sampling with replacement from a finite population). For this case each individual item sampled is independent of any other item sampled and can be treated as 1 value randomly selected from a population with mean, μ, and standard deviation, σ. Thus for the ith sample observation, x_i, we have

$E(x_i) = \mu$, the mean of the population being sampled

and

$\sigma_x = \sigma$, the standard deviation of the population being sampled.

The sample mean is by definition the sum of n of these x_i variables divided by n; i.e.,

$$\bar{x} = \frac{\sum_{i=1}^{n} x_i}{n} = \frac{x_1 + x_2 + \cdots x_n}{n} = \frac{1}{n}(x_1 + x_2 + \cdots + x_n).$$

The mean of $\bar{x}$, $\mu_{\bar{x}}$, is equal to the $E(\bar{x})$, and in Appendix 4A it was proved that the expected value operator, E, can be taken across the summation sign and that a constant can be factored out; thus we can write

$$\mu_{\bar{x}} = E(\bar{x}) = E\left[\frac{1}{n}(x_1 + x_2 + \cdots + x_n)\right]$$

$$= \frac{1}{n} E(x_1 + x_2 + \cdots + x_n) = \frac{1}{n} E(x_1) + E(x_2) + \cdots + E(x_n)$$

$$= \frac{1}{n}(\mu + \mu + \cdots + \mu) = \frac{1}{n}(n\mu) = \mu.$$

The theorem in Appendix 4A proved that the variance of the sum of 2 independent random variables is equal to the sum of the variances: $\sigma_{x+y}^2 = \sigma_x^2 + \sigma_y^2$; this extends to the sum of any number of independent random variables. It was also demonstrated that the variance of a constant times a random variable is equal to the square of the constant times the variance of the random variable, $\sigma_{kx}^2 = k^2\sigma_x^2$. These 2 propositions can be used to derive the formula for the standard deviation of the sample mean, $\sigma_{\bar{x}}$:

$$\sigma_{\bar{x}}^2 = \sigma_{\Sigma x_i/n}^2 = \sigma_{(1/n)\Sigma x_i}^2 = \left(\frac{1}{n}\right)^2 \sigma_{\Sigma x_i}^2.$$

The variance of the mean is the same as the variance of a constant, $(1/n)^2$, times the variance of another random variable, Σx_i. The variance of the Σx_i is equal to the sum of the individual variances of each sample observation:

$$\sigma_{\bar{x}}^2 = \left(\frac{1}{n}\right)^2 \sigma_{\Sigma x_i}^2 = \left(\frac{1}{n}\right)^2 \left[\sigma_{x_1}^2 + \sigma_{x_2}^2 + \cdots + \sigma_{x_n}^2\right]$$

$$= \frac{1}{n^2}(\sigma^2 + \sigma^2 + \cdots + \sigma^2) = \frac{1}{n}(n\sigma^2) = \frac{\sigma^2}{n}$$

$$\sigma_{\bar{x}} = \sqrt{\sigma_{\bar{x}}^2} = \sqrt{\frac{\sigma^2}{n}} = \frac{\sigma}{\sqrt{n}}.$$

It was observed above that $\sigma_{x_i} = \sigma$; therefore, $\sigma_{x_i}^2 = \sigma^2$, the population variance.

The following theorem concerning linear functions of normal random variables has important applications in industry and can also be used to demonstrate the validity of the standard normal transformation used in Chapter 5 and this chapter for finding probabilities.

Theorem 9   A linear function of independent normally distributed random variables has a normal distribution.

In other words, if the random variable w is a linear function of independent, normally distributed random variables, then w is a normally distributed random variable. The mean and variance of w would be found using the expected value and variance theorems in Appendix 4A.

In Chapter 5 the following statement was made:

If the random variable x is normally distributed with mean, μ, and standard deviation, σ, then the random variable z has a normal distribution with $\mu = 0$ and $\sigma = 1$; z is defined as

$$z = \frac{x - \mu}{\sigma}.$$

A normal random variable with 0 mean and a standard deviation equal to 1 was arbitrarily defined as a standard normal variable. The expression for z can be rewritten as follows:

$$z = \frac{x - \mu}{\sigma} = \frac{x}{\sigma} - \frac{\mu}{\sigma} = \frac{1}{\sigma} x - \frac{\mu}{\sigma}.$$

Since μ and σ are the population mean and standard deviation, they are constants for a specific population, and z is clearly a linear function of the normally distributed random variable x. Therefore, z itself is normally distributed.

The facts that $\mu_z = 0$ and $\sigma_z = 1$ can be demonstrated using the theorems from Appendix 4A:

$$\mu_z = E(z) = E\left(\frac{x - \mu}{\sigma}\right) = E\left(\frac{1x}{\sigma} - \frac{\mu}{\sigma}\right)$$

$$= \frac{1}{\sigma} E(x) - E\left(\frac{\mu}{\sigma}\right) = \frac{1}{\sigma} \mu - \frac{\mu}{\sigma} = 0.$$

The above proof used the fact that by definition $E(x) = \mu$. For finding σ_z we again employ the theorems from Appendix 4A and observe that $\sigma_x = \sigma$:

$$\sigma_z^2 = \sigma_{(1/\sigma x - \mu/\sigma)}^2 = \sigma_{1/\sigma x}^2 + \sigma_{\mu/\sigma}^2$$

$$= \left(\frac{1}{\sigma}\right)^2 \sigma_x^2 + 0 = \frac{1}{\sigma^2}(\sigma^2) = 1$$

$$\sigma_z = \sqrt{\sigma_z^2} = \sqrt{1} = 1.$$

The variance of μ/σ is 0 because it is a constant; there is no variability in a constant term.

A theorem referred to as Chebyshev's Inequality might possibly have been used more in applied statistics if the Central Limit Theorem were not true. This theorem puts bounds on probabilities for any random variable.

Chebyshev's Inequality

If the random variable, x, has mean, μ, and standard deviation, σ, then

$$P(\mu - k\sigma \leq x \leq \mu + k\sigma) \geq 1 - \frac{1}{k^2}$$

where k is an arbitrary constant greater than 0.

This puts a lower bound on the percentage of items falling within an interval about the population mean. For example, if $k = 2$, then the interval of interest is from $\mu - 2\sigma$ to $\mu + 2\sigma$, or the mean ± 2 standard deviations. Chebyshev's Inequality states that, regardless of the population distribution, in any population at least

$$1 - \frac{1}{k^2} = 1 - \frac{1}{(2)^2} = 1 - \frac{1}{4} = \frac{3}{4} = .75 = 75 \text{ percent}$$

of the items fall within ± 2 standard deviations from the mean. This is a weaker probability statement than can be made for a normal population (you will recall from Chapter 5 that about 95 percent of the items fall within $\mu \pm 2\sigma$) because it applies to any population.

Problems

1. If a sample of 25 values is randomly selected from a normal population with $\mu = 128$ and $\sigma = 20$, find the following probabilities ($\bar{x}$ = the sample mean).
    a. $P(123 < \bar{x} < 134)$.
    b. $P(\bar{x} > 137)$.
    c. $P(\bar{x} < 125)$.
    d. $P(129 < \bar{x} < 138)$.
    e. $P(119 < \bar{x} < 121)$.

2. Refer to the previous problem. Double the size of the sample to 50, and find the indicated probabilities.

3. Twelve electric mixers selected from a population with mean weight of 16 pounds each are packed in a carton weighing 5 pounds. The shipping clerk puts in lightweight packing material weighing 2 pounds. He wants to pay less than 200 pounds shipping rate; there is a lump sum jump in the shipping cost for packages weighing 200 pounds or more. If the carton and materials do in fact

equal exactly 7 pounds, what is the probability that a package of 12 mixers will weigh more than 200 pounds if the weight of electric mixers produced by the company follows a normal distribution with standard deviation equal to .15 pound?

4. One thousand two hundred honest dice are tossed. Let p be the sample proportion of sixes.
   a. What is the expected value of p? The standard deviation of p?
   b. Find the $Pr(.14 \leq p \leq .175)$.
   c. The probability is .75 that there will be at least _____ sixes. (How many?)

5. Twenty percent of the students in a statistics class of size 150 also take mathematics. From a random sample of size 60, find:
   a. the probability that p (the sample proportion that take mathematics) will be at least .15;
   b. $Pr(.18 \leq p \leq .24)$;
   c. k, such that the probability is .80 that p will be no more than k.

6. Let x be monthly rental paid by married students at a large university. Given $\mu = \$118.00$, $\sigma = \$24.00$, and $N = 7,800$. For samples of size 144, what is:
   a. $E(\bar{x})$; b. $\sigma_{\bar{x}}$; c. $P(\bar{x} \leq 120)$?
   d. What assumption, if any, would be necessary to answer part c if n were only 4? Given this assumption, answer questions a through c for $n = 4$.
   e. Answer these same questions for $n = 144$ and for another small university where $N = 300$ (same μ and σ).

7. Consider the variable, x, the time between arrivals at a university hospital. This variable has a mean of 20.00 minutes and a standard deviation of 8.00 minutes. Assume it is also normally distributed. Find the following.
   a. The probability that x will exceed 2/3 hour.
   b. The odds are 2 to 1 that x will lie within what symmetric interval about the mean?
   Consider now the mean of a random sample of 81 arrivals. Find:
   c. $E(\bar{x})$; d. $\sigma_{\bar{x}}$; e. $P(\bar{x} \leq 22.00)$;
   f. A number k such that $P(20.00 - k \leq \bar{x} \leq 20.00 + k) = .97$.

8. Ten 5-pound aluminum ingots are melted and used to coat 500 casings. If the total weight of the aluminum is less than 49.8 pounds, the batch of 500 casings will have an unacceptable number of bad casings. If the average weight of the ingots is 5 pounds and the weight follows a normal distribution with $\sigma = 0.3$ pound, what percent of the batches of 500 casings coated by this process will have an excessive number of defects?

9. Given $\sigma = 24$, find $\sigma_{\bar{x}}$ for each of the following.
   a. $n = 36, N = \infty$

b. $n = 100, N = 200$
c. $n = N$
d. $n = 144, N = 14,400$
e. A 1 percent sample from $N = 1,000$
f. A 1 percent sample from $N = 10,000$
g. A 1 percent sample from $N = 100,000$

10. Complete the following table of values for σ_p. (Assume $N = \infty$.)

$\downarrow \pi \quad n \rightarrow$	25	100	400
.1(.9)	.06		
.2(8)		.04	
.5			

11. A large herd of Texas steers is known to have a mean weight of 914 pounds and a standard deviation of 120 pounds.
    a. Assuming that the weight of individual steers is normally distributed, what is the probability that a steer picked at random weighs at least 1,000 pounds?
    b. 100 steers are chosen at random.
        (i). We would expect the mean weight of this sample to be close to _____, and
        (ii). the standard deviation of this mean weight would be _____.
        (iii). In fact, it is an even-money wager that the sample mean weight would differ from 914 pounds by no more than _____.
        (iv). The probability is _____ that the total weight of the 100 steers would be at least 90,000 pounds.
        (v). What assumptions (if any) are necessary regarding the population distribution in order to make the statements in parts 3 and 4?
        (vi). The theorem used in this problem (b) is known as the _____.

12. Mean daily beer consumption by adult residents of a large city in West Germany is known to be 2.8 liters with a standard deviation of 3.6 liters. (With a lower bound of 0, could this possibly be a normal distribution?) What is the probability, in a random sample of 81 adult residents, that the sample mean will not exceed 3.5 liters?

CHAPTER 7

STATISTICAL ESTIMATION

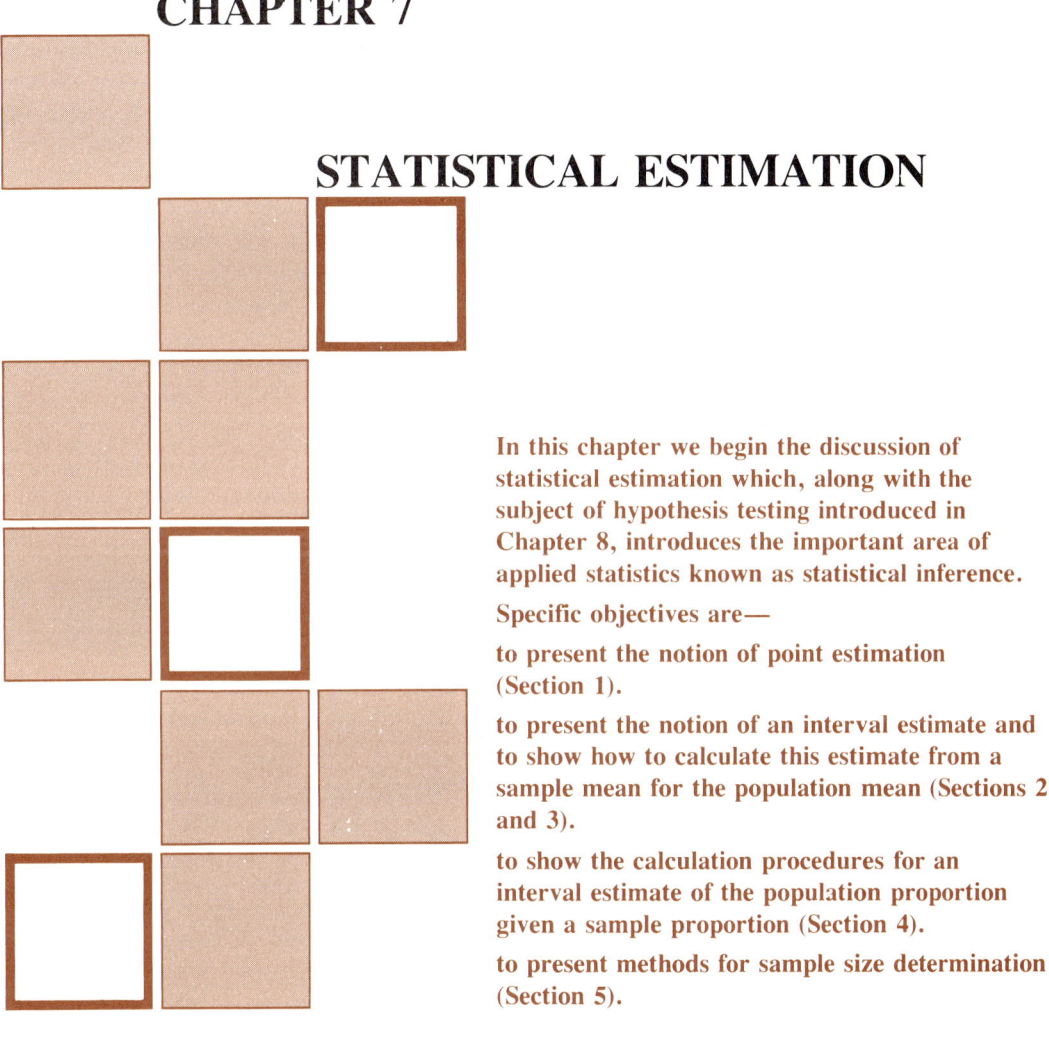

In this chapter we begin the discussion of statistical estimation which, along with the subject of hypothesis testing introduced in Chapter 8, introduces the important area of applied statistics known as statistical inference.

Specific objectives are—

to present the notion of point estimation (Section 1).

to present the notion of an interval estimate and to show how to calculate this estimate from a sample mean for the population mean (Sections 2 and 3).

to show the calculation procedures for an interval estimate of the population proportion given a sample proportion (Section 4).

to present methods for sample size determination (Section 5).

Up to this point we have assumed that the parameters of the population of interest are known. When working with the normal distribution, the population mean, μ, and the standard deviation, σ, were given. When dealing with an attribute variable, the proportion of the items in the population, π, having the attribute of interest was a known quantity. The questions asked dealt with the problem of finding the probability of specified events occurring given a specific value of μ and σ or given a specific value of π.

In this chapter we make an abrupt change in the type of problem with which we are concerned. Sometimes students miss the abruptness of the change and unnecessary confusion results. Even though the theory of probability and distribution of random variables covered in the previous chapters provide the theoretical underpinning for the methods of estimation and hypothesis testing (Chapter 8), the type of problem considered is entirely different from those with which we have been dealing. In this chapter we are concerned with problems that arise when the population parameters, μ, σ, or π, are *not* known. As you work through this chapter, be very much aware of the fact that the population parameters are not known.

What do we do when the population parameter of interest—say for example the population mean, μ—is not known? We can either calculate it exactly or estimate it. To calculate it exactly means enumerating the whole population. To measure every item in the population costs money and takes time. The cost and time requirements may be such that it is economically infeasible or physically impossible to enumerate all of the items in the population of interest. In this case, we are limited to the information obtained from a sample of items randomly selected from the population.

The value of the unknown population parameter is estimated or inferred from the information that can be obtained from the sample data. As long as we are limited to sample data and are unable to enumerate the whole population, all we have is an estimate: the population parameter remains unknown. However, we are not completely in the dark. We are able to make some statements about the probability that the estimate obtained from the sample is a "satisfactory" estimate—i.e., it is "reasonably" close to the value of the parameter being estimated. This chapter is concerned with the problem of obtaining a "good" estimate from sample data and making some assessment of the accuracy of the particular estimate.

7.1 Point Estimation

In Chapter 6 it was pointed out that the sample mean is itself a random variable; i.e., two samples of the same size from the same population will generally have different means. There is, however, a high probability that large sample means will be closer in value to the population mean than those from small samples. We choose the mean of our sample data as the estimate of the unknown population mean because the expected value or the average of all sample means of a given size is equal to the population mean. A sample statistic is an *unbiased estimator* if the expected value of this statistic is equal to the population parameter. Thus, for example, $\bar{x}$ is an unbiased estimator of μ because $E(\bar{x}) = \mu$. Because of this property, the sample mean, $\bar{x}$, is referred to as a *point estimate* of μ. A common notation used for the point estimate of μ is $\hat{\mu}$, read as "μ-hat." If we have an attribute random variable and want to estimate the population proportion, π, we would select the sample proportion, p, as the estimator of π. The expected value of the sample proportion is equal to the population proportion; i.e., $E(p) = \pi$, or, using the "hat" notation, we would write $\hat{\pi} = p$.

If your only exposure to statistical reporting is to that which appears in newspapers and magazines, you are familiar only with the point estimate because this is the only result typically reported. Unfortunately, the general public often interprets this sample result to be a population value. Almost daily you will find newspaper reports with items such as, "63.8 percent of the voters approve of the president's position . . .", or "the average per person-day tourist expenditure in Nevada is $48.60." If we read on, we find it is usually the case that these most certainly would have been results of a sample survey; and they would truly be point estimates if the sample taken were a probability (random) sample. From your experience and your work in Chapter 6, you probably feel that estimates of this type are more reliable if the sample size is large and that they could be unreliable if the sample were small.

For more formal statistical reporting, we need to develop the notion of an interval estimate—a confidence interval—that will reflect the power of the sample size and the precision obtained therefrom. That is, we want some method that gives us an indication of how good the point estimate is. A level of confidence is specified and a corresponding interval estimate of the population mean is then obtained from the sample data. If this interval estimate is relatively large, then we would conclude that there is a considerable amount of variability in the estimator, and we would be concerned that there is a relatively high probability that the point estimate could be considerably different from the population mean. On the other hand, if the confidence interval is relatively small, we consider our point estimate to be fairly precise with a small likelihood that the point estimate is an extreme value.

7.2 Interval Estimation for μ: The Large-Sample Case

From Chapter 6, we know that

$$\frac{\bar{x} - \mu}{\sigma/\sqrt{n}} = z$$

if the sample comes from a normally distributed population. However, the Central Limit Theorem states that this is a valid relationship regardless of the nature of the population distribution if the sample size, n, is large. The following probability statement can be derived using the above relationship:[1]

$$P\left(\mu - z_o \frac{\sigma}{\sqrt{n}} \leq \bar{x} \leq \mu + z_o \frac{\sigma}{\sqrt{n}}\right) = K$$

where K is an arbitrarily specified probability level and z_o is determined from the normal curve table by finding a value of the standard normal random variable, z, such that $P(-z_o \leq z \leq z_o) = K$. With simple operations on the two inequalities above, μ instead of $\bar{x}$ can be made the subject of the statement; i.e.,

$$P\left(\bar{x} - z_o \frac{\sigma}{\sqrt{n}} \leq \mu \leq \bar{x} + z_o \frac{\sigma}{\sqrt{n}}\right) = K.$$

This is a probability statement because $\bar{x}$ represents a random variable. After the sample is taken, $\bar{x}$ represents the actual sample mean. When this calculated number is represented by $\bar{x}$, then we no longer have a probability statement, and we prefer to write

$$\text{Conf}\left\{\bar{x} - z_o \frac{\sigma}{\sqrt{n}} \leq \mu \leq \bar{x} + z_o \frac{\sigma}{\sqrt{n}}\right\} = K.$$

In most applications, σ is unknown, as is μ. Recall that, for large sample sizes, the sample standard deviation is a good approximation for σ. If the sample size is large, we may substitute s for σ in the confidence interval expression; thus, the formula for the confidence interval becomes

$$\text{Conf}\left\{\bar{x} - z_o \frac{s}{\sqrt{n}} \leq \mu \leq \bar{x} + z_o \frac{s}{\sqrt{n}}\right\} = K.$$

The expression in brackets is the K percent confidence interval estimate of the population mean. The confidence interval could be written

$$\text{Conf}\{\bar{x} - \epsilon \leq \mu \leq \bar{x} + \epsilon\} = K$$

where

---

1. The details of the algebraic manipulations can be found in Appendix 7A of this chapter.

$$\epsilon = z_o \frac{s}{\sqrt{n}}.$$

The latter expression, ϵ, is referred to as the error or precision term of the confidence interval estimate.

The notation for the confidence interval used in the preceding paragraph is preferable because it emphasizes the fact that an interval is calculated to estimate the population mean. Some students find the following shorthand formulation helpful for summarizing the calculation of a confidence interval estimate.

Confidence Interval Estimate of the Population Mean

To find the K percent confidence interval estimate of the population mean when the sample size, n, is large:

1. First find a value z_o such that

    $$P(-z_o < z < z_o) = K.$$

2. Then calculate the confidence interval using the following formula:

    $$\bar{x} \pm z_o \frac{s}{\sqrt{n}}$$

    where $\bar{x}$ = the sample mean
    s = the sample standard deviation
    n = the sample size.

The notion of a confidence interval can also be illustrated graphically. For $K = .95$, z_o is equal to 1.96. (This value of z_o was obtained by finding z_o such that $P(-z_o < z < z_o) = .95$ using the method discussed in Chapter 5, Section 5.2.) Thus, for $K = .95$,

$$\epsilon = 1.96 \frac{\sigma}{\sqrt{n}}$$

and 95 percent of all sample means would fall within ϵ of μ as indicated by the shaded area in Figure 7.1.

Figure 7.1 An Interval Including 95 Percent of All Sample Means

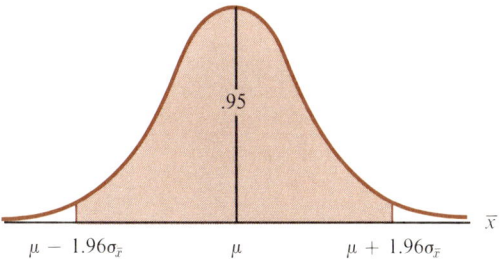

Interval Estimation for μ: The Large-Sample Case

For any value of the sample mean, $\bar{x}_o$, falling within this interval, the interval from $\bar{x}_o - \epsilon$ to $\bar{x}_o + \epsilon$ would contain μ; so, for 95 percent of all samples taken, the sample mean would fall within ϵ of the population mean, μ. Figure 7.2 illustrates a confidence interval constructed about an observed sample mean, $\bar{x}_o$.

Example 7.2.1 A department store carries 100,000 accounts on its books. The fiscal year closed June 30 and, for tax purposes, it was necessary to estimate the mean balance of all accounts at the close of business on this day. Because of the time constraint, it was not possible to check all the accounts. However, it was possible to take a random sample of 625 accounts; this was adequate to satisfy government regulations. From the sample, $\bar{x}$ and s were found to be $53.80 and $29.25 respectively. As representatives of a statistical consultant firm, what kind of report should we prepare? Management has specified a desired confidence level of 95 percent.

We would, of course, first report the best estimate of the mean balance of all accounts:

$$\hat{\mu} = \bar{x} = \$53.80.$$

We would also report a confidence interval for μ. This would be of the form $\bar{x} \pm \epsilon$, where

$$\epsilon = z_o \frac{s}{\sqrt{n}}.$$

Since management has selected the 95 percent confidence level, we select z_o such that $P(-z_o < z < z_o) = .95$. This value is obtained by selecting the value of z that corresponds to an area of .4750 in Table B, the standard normal table. Figure 7.3 illustrates how the area referred to in the table was determined. For a 95 percent confidence interval,

Figure 7.2 A Confidence Interval Constructed About an Observed Sample Mean

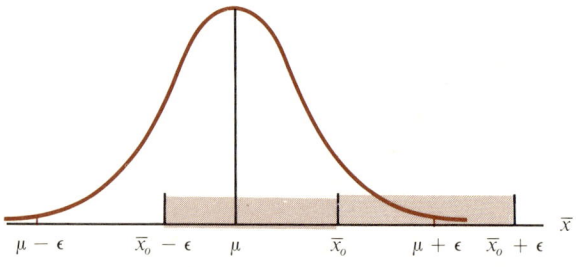

Statistical Estimation

Figure 7.3 Finding the Value of z for a 95 Percent Confidence Interval

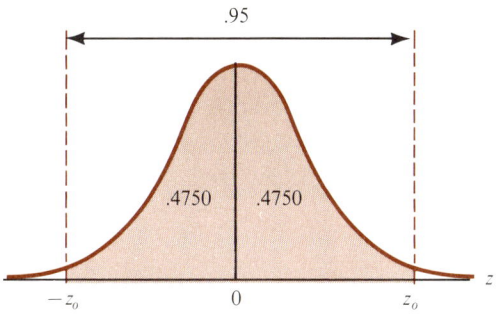

the value of z_0 is 1.96. The error or precision term for the confidence interval is

$$\epsilon = z_0 \frac{s}{\sqrt{n}} = 1.96 \frac{29.25}{\sqrt{625}} = \$2.29.$$

Thus, the confidence interval is

$$\bar{x} \pm \epsilon = 53.80 \pm 2.29.$$

For the second and perhaps most important part of our report, we would make the following statement:

We are 95 percent confident that the mean balance of all accounts is at least $51.51 and no more than $56.09.

In a formal mathematical statement, this would appear as

$$\text{Conf}\{\$51.51 \leq \mu \leq \$56.09\} = .95.$$

It is possible that the nonstatistician would better understand the following statement:

The best estimate of the average balance of all accounts is $53.80, and we are 95 percent confident (or we are willing to lay odds of 19 to 1) that this is in error by no more than $2.29.

Just a note as to why the term *confidence* is used instead of *probability*. In the strict classical sense, before the sample is taken, there is a .95 probability that the sample mean $\pm 1.96\sigma_{\bar{x}}$ will contain the population mean. After the sample is taken, the classical statistician would say that this is no longer a probability problem; i.e., the experiment (obtaining an interval estimate) has been performed, and either the interval contains the population mean or it does not. Most applied statisticians are not too concerned about holding to this strict position. To illustrate this less rigorous position: The probability that heads will occur if an honest coin is flipped is .5; if, however, a coin is flipped and covered on the wrist, most of us would still wager even money on heads showing. The classicial statistician would say there is no probability of heads because the probability experiment has been performed and the outcome is determined. However, we would not see this as 2 different games.

Estimating the Total of a Finite Population

In many problems, such as the one above, the aggregate or total estimate is also very meaningful. Inasmuch as the total, T, is the product of the population size and the population mean,

$$\hat{T} = N\hat{\mu}$$

and

$$\sigma_{\hat{T}} = N\sigma_{\bar{x}} = N\frac{\sigma}{\sqrt{n}}$$

where

$\hat{T}$ = the estimate of the population total
N = the total number of items in the population.

(The population size must, of course, be finite.)

The confidence interval for estimating the total accounts receivable in the above example is obtained by multiplying the confidence interval for the mean by N, the number of items in the population. In the above example there were 100,000 accounts. Our best estimate of the total accounts receivable is

$$\hat{T} = N\bar{x} = 100{,}000(\$53.80) = \$5{,}380{,}000.$$

We are 95 percent confident that this is in error by no more than

$$N\epsilon = 100{,}000(\$2.29) = \$229{,}000$$

or

$$\text{Conf}\{\$5{,}151{,}000 \le \hat{T} \le \$5{,}609{,}000\} = .95.$$

Suppose that the department store manager is shocked at the size of the error in the estimate. A lower confidence level would give a smaller interval, but then it wouldn't be a solution to the same problem. The manager has specified a 95 percent confidence level; he has determined this based on his feeling about risk, apart from the statistical problem. The only real alternative is to make more time and money available for a larger sample. But the return to sampling is not linear; we should point out to the manager that doubling the sample size will reduce ϵ by only 29 percent. We must quadruple the sample size in order to cut the error in half. Section 7.5 in this chapter has more on sample size.

7.3 Interval Estimation for μ: The Small-Sample Case

If the sample size is small, then the Central Limit Theorem doesn't apply, and we can't use the z-distribution to calculate a confidence interval when the population standard deviation, σ, is unknown (which is usually the case). Traditionally, the rule of thumb is that the sample size must be at least 30 for the Central Limit Theorem to be applicable. If the sample size is less than 30 and we are sampling from a normally distributed population, then we use the t-statistic in place of the z-statistic in the calculation of the confidence interval estimate of the population mean. You will recall from Chapter 6 that the ratio

$$\frac{\bar{x} - \mu}{s/\sqrt{n}}$$

has a t-distribution with $n - 1$ degrees of freedom (df) if the population sampled has a *normal* distribution; i.e.,

$$\frac{\bar{x} - \mu}{s/\sqrt{n}} = t$$

with

$df = n - 1.$

You will also recall from Chapter 6 that the Student-t Distribution is a family of distributions; the number of degrees of freedom, in this case a function of the sample size, determines the specific member of the t-distribution family.

The term *degrees of freedom* is one that even experts often have trouble defining, and it is always difficult for the beginner to understand. This refers to the number of free variables used in the calculation of a particular statistic. With a sample size, n, we have n free variables to begin with, and hence the sample mean has n degrees of freedom. The mean is a necessary ingredient in the formula for the sample standard deviation; i.e.,

$$s = \sqrt{\frac{\Sigma(x_i - \bar{x})^2}{(n - 1)}}.$$

If we knew the population mean, we could use it to calculate the sample standard deviation. But we don't know what the population mean is, and so we have to use an estimate of it calculated from the sample data; we thereby lose a degree of freedom. Note that if we know the sample mean and $(n - 1)$ of the sample numbers, the nth number is no longer "free"; it is fixed. This is because the sample mean has the property that $\Sigma(x_i - \bar{x}) = 0$. Note also that if $n = 1$, $\bar{x}$ exists, but the sample standard deviation, the measure of variability, cannot be calculated. The statistic, s, is said to have $(n - 1)$ degrees of freedom as does the Student-t statistic, since s is an integral part of the t-ratio.

The t-distribution, like the normal, is unimodal and is symmetric about the mean, which is also 0, but the dispersion varies inversely with the degrees of freedom. Figure 7.4 below illustrates the area given in the t-table (Table C). The t-values are given for upper-tail areas (or probabilities) of .10, .05, .025, .01, and .005; they are also given for degrees of freedom, df, running sequentially from 1 through 29 and the last row with df equal to infinity. In this last row, t-values are identical with normal curve values with the same upper tail area. Referring to the table, for an upper-tail area of .025 and degrees of freedom equal to 29, the critical t-value, t_o, is 2.045. Translating this to probability, for 29 degrees of freedom, $P(t \geq 2.045) = 0.025$; also $P(t < 2.045) = .975$. Inasmuch as most confidence intervals are 2-sided, we should also note that $P(-2.045 \leq t \leq 2.045) = .95$.

The point estimate of μ is, of course, still $\bar{x}$, and the interval estimate is of the same form as that for the large sample size with z_o replaced by the proper t-value. The K percent confidence interval can be written as

$$\text{Conf}\left\{\bar{x} - t_o \frac{s}{\sqrt{n}} < \mu < \bar{x} + t_o \frac{s}{\sqrt{n}}\right\} = K.$$

The following formulation summarizes the calculation of the confidence interval estimate.

Figure 7.4 The Value of t Given in Table C

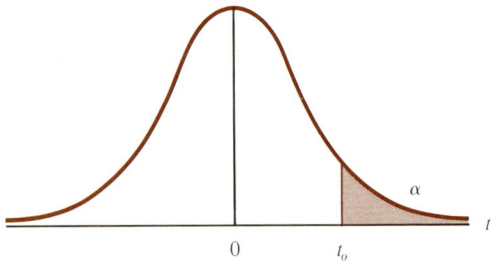

Confidence Interval Estimate Using the t-Distribution

To find the K percent confidence interval estimate of the population mean when the population is normally distributed,

1. Find a value t_o such that

$$P(-t_o \leq t \leq t_o) = K$$

with

$$df = n - 1.$$

2. Calculate the confidence interval using the following formula:

$$\bar{x} \pm t_o \frac{s}{\sqrt{n}}$$

where

$\bar{x}$ = the sample mean
s = the sample standard deviation
n = the sample size.

This confidence interval formula is valid for any size sample if the population distribution is a normal distribution. For large sample size, we will just use the z-statistic. As already indicated above, as the degrees of freedom get large, the value of t_o is very nearly the same as z_o for a given level of confidence. From our table we see that t_o with $df = \infty$ is equal to z_o.

Example 7.3.1

Myron's Meat Market bought 16 turkeys from Sudwurst Turkey Packaging Corporation. The sample mean weight turned out to be 18.7 pounds and the sample standard deviation was 2.8 pounds. Assume that the 16 turkeys represent a random sample from Sudwurst's large turkey inventory and that the individual turkey weights in the inventory are normally distributed. Give the best point estimate and the 90 percent confidence interval estimate of the mean weight of all Sudwurst's turkeys.

The best point estimate of μ is the sample mean:

$$\hat{\mu} = \bar{x} = 18.7 \text{ pounds.}$$

The 90 percent confidence interval is calculated by first finding a value of the t-distribution with $n - 1 = 16 - 1 = 15$ degrees of freedom such that

$$P(-t_o \leq t \leq t_o) = .90.$$

From Figure 7.5 we see that this is equivalent to a t_o such that the area in the upper tail is .05. We look under the column for $\alpha = .05$ and along

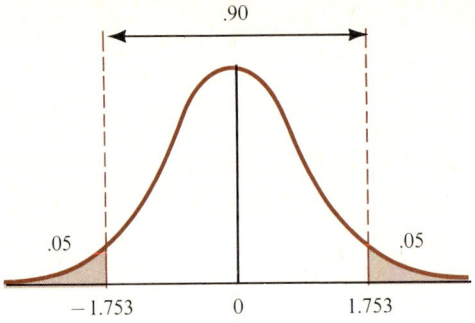

Figure 7.5 The t-Values for a 90 Percent Confidence Interval with $df = 15$

the row for 15 degrees in the t-table (Table C) to obtain the value for t_o, 1.753. The 90 percent confidence interval can now be calculated using the formula given above:

$$\bar{x} \pm t_o \frac{s}{\sqrt{n}} = 18.7 \pm 1.753 \frac{2.8}{\sqrt{16}}$$

$$= 18.7 \pm 1.23$$

or $\text{Conf}\{17.47 \leq \mu \leq 19.93\} = .90$.

At the 90 percent level of confidence, we would say the population mean, μ, has a value between 17.47 and 19.93. As an alternative statement, we could say that our best estimate of the mean turkey weight is 18.7 pounds and we are willing to lay odds of 9 to 1 that this is in error by no more than 1.23 pounds.

Example 7.3.2 Professor Smith of Examples 2.1.1 and 2.2.1 took 6 measurements of his driving time from his home to the university. We calculated these measurements to have a mean of 13.2 minutes and a standard deviation of 2.64 minutes. Assume that the parent distribution is normal and estimate μ, the population mean driving time.

The best point estimate is the mean value of the sample:

$$\hat{\mu} = \bar{x} = 13.2 \text{ minutes}.$$

To find the 95 percent confidence interval, we would use the t-table much as we did for the previous example. For $n = 6$ we have 5 degrees of freedom, and the area in each tail of the distribution is .025, i.e., the result of half the equation $1 - .95 = .05$. The intersection of the row for 5 degrees of freedom and the column headed .025 gives us a t-value of

$t_o = 2.571$. The 95 percent confidence interval for estimating the population mean driving time can now be calculated using this t-value:

$$\bar{x} \pm t_o \frac{s}{\sqrt{n}} = 13.2 \pm 2.571 \frac{2.64}{\sqrt{6}} = 13.2 \pm 2.8.$$

We can tell the professor that his mean driving time, with 95 percent confidence, is at least 10.4 minutes and no more than 16.0 minutes.

7.4 Estimating the Population Proportion

Another important problem is that of estimating the proportion of a population having a particular attribute. For example, we may want to estimate the proportion of people in a given region who support fluoridation of drinking water; or a manufacturer may want to know the proportion of defective flashcubes produced by a new production process. It could be very costly to interview every citizen in the region in order to determine the proportion favoring fluoridated drinking water. The manufacturer doesn't gain much if he must field-test all of his flashcubes in order to determine the proportion defective. In both of these examples, the use of sample information would seem to be the only reasonable approach.

The sample proportion, p, is an unbiased estimator of the population proportion, π, i.e., $E(p) = \pi$. Therefore, we use p as the point estimate of π, or $\hat{\pi} = p$. The sample proportion, p, is the proportion of items in the sample that have the attribute of interest.

The Sample Proportion

Formally, p is defined as follows:

$$p = \frac{x}{n}$$

where

p = the sample proportion
x = the number of items in the sample having the attribute of interest
n = the sample size.

The sample proportion, p, is a random variable; i.e., a different random sample of the same size yields a different sample proportion. The standard deviation of the sample proportion, σ_p, is a function of the population proportion:

$$\sigma_p = \sqrt{\frac{\pi(1-\pi)}{n}}.$$

A corollary to the Central Limit Theorem discussed in Chapter 6 makes it possible to provide a confidence interval estimate of the population proportion, π, when the sample size is large. The corollary states that for large sample sizes we can write

$$\frac{p - \pi}{\sqrt{\frac{p(1-p)}{n}}} = z.$$

The expression

$$\sqrt{\frac{p(1-p)}{n}}$$

is an estimate of the standard deviation of p,

$$\sqrt{\frac{\pi(1-\pi)}{n}}.$$

Using algebraic operations similar to those used in Section 7.1 to derive the confidence interval of the mean results in the following confidence interval estimate for the population proportion.

$$\text{Conf}\left\{p - z_o\sqrt{\frac{p(1-p)}{n}} \leq \pi \leq p + z_o\sqrt{\frac{p(1-p)}{n}}\right\} = K.$$

The calculation of the confidence interval estimate of the population proportion can be summarized as follows.

Confidence Interval Estimate of the Population Proportion

To find the K percent confidence interval estimate of the population proportion, π, when the sample size, n, is large,

1. Find a value, z_o, such that

$$P(-z_o \leq z \leq z_o) = K.$$

2. Calculate the confidence interval using the following formula:

$$p \pm z_o\sqrt{\frac{p(1-p)}{n}}.$$

We observe that the latter formula can also be written as $p \pm z_o\hat{\sigma}_p$; this is analogous to the confidence interval estimate of the mean which can be written $\bar{x} \pm z_o\hat{\sigma}_{\bar{x}}$.

Example 7.4.1 If 400 citizens were randomly selected from a region with a large population, and 220 of those selected favored fluoridation of the drinking water, what is the best estimate of the population proportion, π? Find a 98 percent confidence interval estimate of the population proportion.

The best point estimate of the population proportion is the sample proportion:

$$\hat{\pi} = p = \frac{x}{n} = \frac{220}{400} = .55.$$

Our point estimate is that 55 percent of all citizens in the region favor fluoridation. The value of z_o used in the confidence interval estimate is determined in exactly the same manner as for estimating the population mean when the sample size is large. From Figure 7.6 we see that we want a z_o value consistent with an area in the body of Table B equal to .4900. The value of z_o, 2.33, thus obtained is then used to calculate the 98 percent confidence interval:

$$p \pm z_o \sqrt{\frac{p(1-p)}{n}} = .55 \pm 2.33 \sqrt{\frac{(.55)(1-.55)}{400}} = .55 \pm .058.$$

We could also state this interval estimate as follows:

$$\text{Conf}\{.492 \leq \pi \leq .608\} = .98.$$

We are 98 percent confident that π is at least .492 and no more than .608. It seems likely that the majority of the citizens favor fluoridation, but we are not absolutely certain.

Example 7.4.2 Refer to Example 7.2.1. It was observed that 75 of the 625 accounts receivable sampled were inactive. Find the best point estimate and the 90 percent confidence interval estimate for the proportion of all charge accounts that are inactive.

For the best point estimate, we have

$$\hat{\pi} = p = \frac{x}{n} = \frac{75}{625} = .12.$$

Figure 7.6 Finding the Value of z for a 98 Percent Confidence Interval

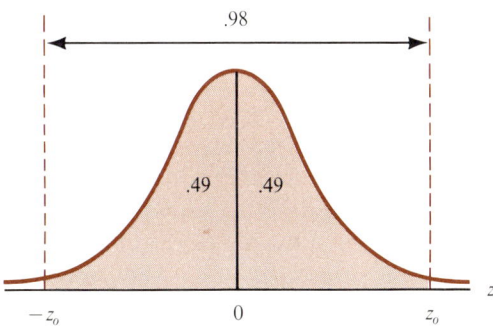

Estimating the Population Proportion

The z_o value that corresponds to an area of .4500 (half of .90) in Table B is 1.65 (to the nearest hundredth). Using this value, the 90 percent confidence interval estimate of the proportion of all accounts receivable that are inactive can be calculated:

$$p \pm z_o \sqrt{\frac{p(1-p)}{n}} = .12 \pm 1.65 \sqrt{\frac{(.12)(.88)}{625}} = .12 \pm .021.$$

We might report this result as follows: "The best estimate of the percentage of inactive accounts is 12 percent, and we are willing to lay odds of 9 to 1 that this is in error by no more than 2.1 percentage points."

Example 7.4.3

Statistical reporting in the newspapers is becoming more sophisticated. The following is similar to a recent news release: "The latest poll indicates that a thin majority of Americans believe the president is doing a less-than-adequate job after 39 months in office. The poll showed that 51 percent of 1,600 adults polled gave the president a negative overall rating and 45 percent approved of his performance. . . . The margin of error . . . is 'a very rough plus or minus 3 percent,' the poll continued." "Roughly" what level of confidence was the spokesperson giving to the press?

The K percent level of confidence interval for those pleased with the president was given as $.45 \pm .03$. K is not given; however, we can determine the z_o value used (there will be some rounding error) by solving for z_o in the formula for ϵ. We have

$$\epsilon = z_o \sqrt{\frac{p(1-p)}{n}}$$

or

$$.03 = z_o \sqrt{\frac{(.45)(.55)}{1,600}}.$$

Solving for z_o, we get $z_o = 2.41$. Referring to the normal curve table, we see that this z-value is just over the value for the 98 percent confidence level. Do you think the public would have been better informed if the poll had used the term *98 percent confident*?

The confidence interval estimate given in this section requires that the sample size be large. If the sample size is small, then we should not attempt an interval estimate of the population proportion using the standard normal distribution, z. Unfortunately, there is no simple small-sample theorem we may use to develop a small-sample interval estimate for the population proportion. We can't use the t-distribution as was done for small-sample estimation of the mean because the t-

distribution requires a normally distributed population; the sample proportion is, of course, binomially distributed.

 ## 7.5 Sample Size Determination

Preliminary to the whole problem of interval estimation is that of determining what size sample to take. It is evident from the confidence interval estimation formulas that the sample size is the controlling factor in the precision of the estimate. Before the appropriate sample size can be determined, it is necessary for the statistical consultant, the manager, the researcher, the graduate student, or whoever is planning to do statistical estimation to make an exact statement of his or her desired precision. The following kinds of statements are of no help. "I just want a sample size large enough to be respectable"; "So-and-so always takes a 10 percent sample. He seems to have his feet on the ground; so let's appropriate enough money to go with a 10 percent sample." The idea that, regardless of population size, a fixed percentage sample always gives the same precision is statistically naive.

In order to determine sample size, we must be able to make the following kind of statement: "We wish to estimate μ (or π) within a maximum error of $\pm \epsilon$ at the K percent confidence level." For example, we may want to estimate the average life of a certain brand of light bulbs with an error of no more than ± 50 hours at the 95 percent level of confidence. That is, when we calculate the 95 percent confidence interval after the sample has been taken, we want the interval to be

$\bar{x} \pm 50$.

The maximum error allowable of 50 hours is the ϵ-term when the confidence interval is written as

$\bar{x} \pm \epsilon$.

Sample Size for Estimating the Population Mean

The formula for the sample size, n, can be solved for algebraically from the equation for ϵ:

$$\epsilon = z_o \sigma_{\bar{x}} = z_o \frac{\sigma}{\sqrt{n}}.$$

Squaring both sides and solving for n,

$$n = \frac{z_o^2 \sigma^2}{\epsilon^2}$$

where

n = the sample size
σ = the population standard deviation

ϵ = the error term
z_o = the z-value determined according to the specified confidence level.

The formula for sample size requires the population standard deviation, which will not be known most of the time. Before the sample size can be determined, it is necessary to have some estimate or guess of σ. There are several possibilities, all of which require some subjective judgment on the part of the user or statistician:

1. Take a pilot sample, and use the resulting sample standard deviation, s, as an estimate of σ. A *pilot sample* is a relatively small preliminary sample of the population.
2. Use the sample standard deviation obtained from a previous and similiar type of study.
3. Use σ from a previous census.
4. Make an educated guess of σ. For example, if one is already experienced in the area of study, one might have a good idea of the maximum and minimum values that the variable, x, takes on, and therefore an estimate of the range of x. If the population distribution were thought to be normal or near normal, dividing the range by 5 or 6 would give a good guess of σ; if the population distribution were known to be flat or skewed, dividing the range by 3 or 4 could give a good guess of σ.
5. Some combination of the above. For example, from the 1970 U.S. Census, we may have σ for family income in our region, and we may wish to estimate the mean family income for the current year. Based on our expertise in this area, we know that mean family income has increased considerably during the period, and we feel very strongly that dispersion has increased but not nearly so much as the mean value. In our judgment the mean has increased by about 5 percent per year, but our experience indicates that dispersion has increased by about only 1 percent per year. As an estimate for σ_{78}, the standard deviation of family income in 1978, we would have $\sigma_{78} = \sigma_{70}(1.01)^8$.

The determination of the population standard deviation usually requires some subjective judgment; nevertheless, it does provide us with at least some idea of how large the sample size has to be in order to achieve a desired level of precision in estimating the population mean.

Example 7.5.1

We wish to estimate the mean hourly wage rate of the unskilled labor force of the Bay Area within ±$.05 at the 95 percent confidence level. A 1974 study (large sample size) by a government agency resulted in a sample standard deviation, s = $.835. We are willing to use this as an approximation for σ, the population standard deviation. How large must the sample size be in order to attain the desired level of precision?

The maximum allowable error, ϵ, equals .05. The z_o value for a 95 percent confidence interval is determined from the standard normal table: $z_o = 1.96$. Our best estimate of the population standard deviation is $\sigma = .835$. Substituting these values in the formula gives us the required sample size:

$$n = \frac{z_o^2 \sigma^2}{\epsilon^2} = \frac{(1.96)^2(.835)^2}{(.05)^2} = 1,071.38.$$

(Use $n = 1,072$, because n must be an integer.)

We round up to 1,072 because we want the sample size to be large enough. If our estimate of the population standard deviation is good, the sample standard deviation should be fairly close to this value, and a sample of size 1,072 will result in a 95 percent confidence interval on μ of approximately $\bar{x} \pm .05$, where $\bar{x}$ is the mean of the sample of 1,072 unskilled laborers.

The procedure and formulas for determining the sample size for an interval estimate of π, the population proportion, parallel, with a few exceptions, those used above for the interval estimate of μ, the population mean. The level of precision or error term, ϵ, must be prescribed by the manager or the statistician: "I wish to estimate π within $\pm \epsilon$ at the K percent confidence level."

Sample Size for Estimating the Population Proportion

As was the case for the mean, we obtain a formula for the sample size by solving for n using the formula for ϵ, the error term of the interval estimate, $p \pm \epsilon$. We have

$$\epsilon = z_o \sqrt{\frac{\pi(1-\pi)}{n}}$$

$$n = \frac{z_o^2 \pi(1-\pi)}{\epsilon^2}$$

where

n = the sample size
ϵ = the error term
π = the population proportion
z_o = z-value determined according to the specified confidence level.

As before, the numerator is the problem. Note that we need to know or have some estimate of π, the population proportion, the very parameter we are trying to estimate with our sample! The five possibilities for getting an estimate of σ, the population standard deviation, listed

Sample Size Determination

above are also possibilities for getting an estimate of π; but we do have an additional alternative for the initial estimate of π. For the largest possible sample size needed, regardless of the true value of π, we can let $\pi = .5$. (The proof of this is very simple for those who know the basic elements of calculus.) Using n^* to indicate this maximum value of n, we have

$$n^* = \frac{z_0^2(.5)(.5)}{\epsilon^2}.$$

For all values of π different from .5, the required sample size is less than n^*. The sample size, n^*, calculated using $\pi = .5$ guarantees that the sample size is adequate, and we could use it if we have no information on the magnitude of the population proportion. Using the maximum sample size, n^*, could lead to considerable waste of time and money on sampling. For example, if we didn't know π but were quite certain it was at most .10, we would use $\pi = .10$ in the sample size determination formula; n determined this way would give, at the very least, the desired precision for $\pi \leq .10$. This is clearly evident from the table of sample sizes presented in Table 7.1. For $\epsilon = .02$ and $\pi = .50$, $n^* = 2,401$; for the same value of ϵ and $\pi = .10$, $n = 865$.

Table 7.1 Sample size, n, for Selected Values of π and ϵ at the 95 Percent Confidence Level

ϵ \ π	.1	.3	.5	.7	.9
.01	3,458	8,068	9,604	8,068	3,458
.02	865	2,017	2,401	2,017	865
.05	139	323	385	323	139

Example 7.5.2   As part of his thesis, Mike needs an estimate of the proportion of full-time undergraduate students who, in addition to their schoolwork, hold a wage-paying job of at least 20 hours per week. He wishes to estimate π within $\pm .025$ at the 90 percent confidence level. What size sample should he interview if he has no idea as to what π might be? What should the sample size be if, being well acquainted with large numbers of undergrads, he believes very strongly that π is no more than .25?

For this problem, the specified level of precision is $\epsilon = .025$. The value of z_o for a 90 percent confidence interval, determined from the standard normal table, is 1.65 (to the nearest hundredth). The first question asked states that Mike has no knowledge of what the population proportion is. In this case we use $\pi = .5$:

$$n = \frac{z_o^2 \pi(1 - \pi)}{\epsilon^2} = \frac{(1.65)^2(.5)(.5)}{(.025)^2} = 1,089.$$

The answer to the first question is that he should sample 1,089 undergraduate students.

The second question assumes that he has enough experience with the population to make a reasonable judgment about the population proportion. If he is confident that π is no more than .25, he should use this value in calculating the required sample size:

$$n = \frac{z_o^2 \pi(1 - \pi)}{\epsilon^2} = \frac{(1.65)^2(.25)(.75)}{(.025)^2} = 816.75 = 817.$$

If his assumption about π is correct, then he can achieve the desired precision level with a sample of 817 undergraduate students.

We know from our previous work that, if the sample size is large, both $\bar{x}$ and p are approximately normally distributed. Inasmuch as the standard normal curve value, z_o, was used in the sample size determination formulas, a resulting large value of n will surely give the desired precision. If the sample size calculated using the formula is small, it is very probable that the stated precision will not be accomplished. The sample proportion, p, is not normally distributed for small sample sizes; therefore, an interval estimate of π would be unreliable. If μ is the parameter to be estimated and we can argue that x is normally distributed, then the interval estimate requires the use of the t-distribution, which results in larger interval estimates. If, as is more often than not the case, we really don't know much about the nature of the parent distribution, it is highly recommended that, if at all possible, the sample size be arbitrarily increased to the point where the Central Limit Theorem becomes effective.

7A Appendix: The Derivation of a Confidence Interval

If we sample from a normal population, the sample mean has a normal distribution, and we can write

$$z = \frac{\bar{x} - \mu}{\sigma/\sqrt{n}}.$$

The Central Limit Theorem permits us to make this statement for any sample mean regardless of the nature of the population distribution if the sample size, n, is large. A corollary to the Central Limit Theorem permits us to substitute the sample standard deviation, s, for the population standard deviation, σ, when the sample size is large. Thus, for large n the following statement is approximately true:

$$z = \frac{\bar{x} - \mu}{s/\sqrt{n}}.$$

If z_0 is a number such that

$$P(-z_0 < z < z_0) = K$$

for some probability K, then the following statement is also true for large sample sizes:

$$P(-z_0 < \frac{\bar{x} - \mu}{s/\sqrt{n}} < z_0) = K.$$

In the latter expression we have simply substituted an equivalent expression for z. The following equivalent probability statements are obtained by performing a series of simple algebraic manipulations on the two inequalities. Start with

$$P\left(-z_0 < \frac{\bar{x} - \mu}{s/\sqrt{n}} < z_0\right) = K.$$

Multiplying through by $s/\sqrt{n}$ gives us

$$P\left(-z_0 \frac{s}{\sqrt{n}} < \bar{x} - \mu < z_0 \frac{s}{\sqrt{n}}\right) = K.$$

Adding μ to the three parts results in

$$P\left(\mu - z_0 \frac{s}{\sqrt{n}} < \bar{x} < \mu + z_0 \frac{s}{\sqrt{n}}\right) = K.$$

This gives us a probability statement about the likelihood that the sample mean has a value that falls in an interval centered about the population mean. To obtain a confidence interval estimate of the mean, we would like an equivalent probability statement for an interval centered around the sample mean.

The desired probability statement is obtained by taking the next to last of the three equivalent probability statements given above,

$$P\left(-z_0 \frac{s}{\sqrt{n}} < \bar{x} - \mu < z_0 \frac{s}{\sqrt{n}}\right) = K,$$

and subtracting $\bar{x}$ from all parts to get

$$P\left(-\bar{x} - z_0 \frac{s}{\sqrt{n}} < -\mu < -\bar{x} + z_0 \frac{s}{\sqrt{n}}\right) = K.$$

The minus sign in front of the population mean is removed by multiplying through by -1:

$$P\left(\bar{x} + z_o \frac{s}{\sqrt{n}} > \mu > \bar{x} - z_o \frac{s}{\sqrt{n}}\right) = K.$$

The sense of the inequalities is changed because we multiplied by a negative number. Generally, we prefer to arrange the expression so that the lower value of the interval is written first. This is done simply by changing the order in which we write the terms while maintaining the same inequalities:

$$P\left(\bar{x} - z_o \frac{s}{\sqrt{n}} < \mu < \bar{x} + z_o \frac{s}{\sqrt{n}}\right) = K.$$

The latter expression is a statement of the probability that an interval constructed around the sample mean will contain the unknown population mean. The lower bound of the interval is

$$\bar{x} - z_o \frac{s}{\sqrt{n}}$$

and the upper bound is

$$\bar{x} + z_o \frac{s}{\sqrt{n}}.$$

As indicated in Section 7.2, before the sample is taken the above is a probability statement; i.e., $\bar{x}$ represents a random variable. After the sample is taken, $\bar{x}$ becomes a number and, technically, we no longer have a probability statement. We now have a K percent confidence interval estimate designated in Section 7.2 as

$$\text{Conf}\left\{\bar{x} - z_o \frac{s}{\sqrt{n}} < \mu < \bar{x} + z_o \frac{s}{\sqrt{n}}\right\} = K.$$

7B Appendix: Finite Population Interval Estimation and Sample Size

In Chapter 6 it was pointed out that, if a significant proportion of a finite population is being sampled, then we might want to adjust our estimate of the standard deviation by applying the finite population correction factor (usually denoted as fpc). The formula for the finite population was given as

$$fpc = \sqrt{\frac{N-n}{N-1}}$$

where

fpc = the finite population correction factor
n = the sample size
N = the population size.

It was suggested in Chapter 6 that a good rule for beginning students is to use the finite population correction factor whenever more than 5 percent of the population is being sampled, i.e., whenever $n/N > .05$ or $n > .05N$.

The finite population correction factor is used in interval estimation by simply replacing the estimation of $\sigma_{\bar{x}}$, $s/\sqrt{n}$, by

$$\frac{s}{\sqrt{n}} \sqrt{\frac{N-n}{N-1}}$$

in the formula for the confidence interval estimate, resulting in the following confidence interval formula:

$$\bar{x} \pm z_o \frac{s}{\sqrt{n}} \sqrt{\frac{N-n}{N-1}}$$

or

$$\text{Conf} \left\{ \bar{x} - z_o \frac{s}{\sqrt{n}} \sqrt{\frac{N-n}{N-1}} \leq \mu \leq \bar{x} + z_o \frac{s}{\sqrt{n}} \sqrt{\frac{N-n}{N-1}} \right\} = K.$$

It may not be a serious offense to omit the finite population correction factor when sampling from a finite population (unless the percent sampled is quite large) because the finite population correction factor is always less than 1. You are being conservative when you don't use it; i.e., you are understating your case. You are giving an interval estimate that is larger than necessary.

We observe that if the population size, N, is reasonably large, there is little difference between N and $N - 1$. Replacing $N - 1$ by N in the finite population correction factor formula changes it only slightly[2] but results in a more manageable formula:

$$fpc = \sqrt{\frac{N-n}{N-1}} \cong \sqrt{\frac{N-n}{N}} \cong \sqrt{1 - \frac{n}{N}}.$$

The ratio n/N is simply the fraction or proportion of the population sampled. If 10 percent of the population is being sampled, then

$$fpc = \sqrt{1 - \frac{n}{N}} = \sqrt{1 - .10} = \sqrt{.90} = .94868.$$

---

2. In many examples the population size may be known with some accuracy but not precisely. For example, the number of students enrolled at your institution is changing from day to day. The beginning-of-the-term head count may be precise, but N usually decreases through the term and is not normally known. This is another reason why the difference between N and $N - 1$ is frequently irrelevant to the accuracy of the problem.

Using the *fpc* in this case reduces the size of the interval estimate by 5 percent. It is a matter of judgment whether or not you consider this as significant; possibly you would. If only 1 percent of the population is sampled, the following results are obtained:[3]

$$fpc = \sqrt{1 - \frac{n}{N}} = \sqrt{1 - .01} = \sqrt{.99} = .99499.$$

For this case the *fpc* is very nearly 1, and whether or not it is used wouldn't change the interval estimate appreciably. You would probably choose not to use it. The *fpc* for the strictly arbitrary 5 percent rule given in Chapter 6 is .97468. Is this a good rule? You decide. It's strictly a matter of judgment; almost all would agree that, if the sample is a very small percent of the total population, then the *fpc* can be ignored.[4]

In the development of the formula for sample size used in this chapter, we assumed that the population size was infinite or, for all practical purposes, very large relative to the sample size. If the sample size calculated from the formula is an appreciable percentage of the population size, then, as you have seen, the *fpc* should be used when making the confidence interval estimate. The sample size n obtained from the formula will be adequate, but the required precision can be achieved by using a sample size smaller than that obtained from the original formula when n/N is large.

The formula for the sample size when the finite population correction factor is used can be found from

$$\epsilon = z_0 \frac{\sigma}{\sqrt{n}} \sqrt{1 - \frac{n}{N}}$$

and solving for n:

$$n_f = \frac{\left(\frac{z_0^2 \sigma^2}{\epsilon^2}\right) N}{N + \left(\frac{z_0^2 \sigma^2}{\epsilon^2}\right)}.$$

Note that our original

$$n = \frac{z_0^2 \sigma^2}{\epsilon^2},$$

and so

---

[3]. A quick and quite accurate estimate of the square root of a number close to 1.00 is the average of the number and 1.00—halfway between 1.00 and the original number. This gives an efficient estimate of the error introduced by ignoring the *fpc*.

[4]. The best rule is, of course, to consider all sources of error and the relative accuracy that is possible in the statement of results. Use the *fpc* if it has any effect upon the relative accuracy.

$$n_f = \frac{nN}{N+n}.$$

Dividing by N, we get

$$n_f = \frac{n}{1 + n/N}$$

where

n_f = the sample size when the finite population correction factor is used
N = the population size
ϵ = the desired precision
z_o = the value of z determined by the specified confidence level
σ = the population standard deviation.

The denominator, $1 + n/N$, may be thought of as the *fpc* factor for sample size estimation and, as with interval estimation, we recommend its use if n exceeds $.05N$.

Example 7B.1

Example 7.5.1 asked for the sample size required in order to estimate the mean hourly wage rate of the unskilled labor force of the Bay Area within ± $.05 at the 95 percent confidence level. The population standard deviation, σ, was estimated from a previous government study to be $.835. Assume the same data and desired precision as in example 7.5.1 except, instead of the Bay Area, we want to make this estimate for the Battle Mountain Area, where the unskilled labor force consists of 1,937 workers as compared to 150,000 for the Bay Area.

Since the confidence level and precision level are the same, the sample size obtained for this example using the sample size formula for large or infinite populations will be the same as the sample size obtained for Example 7.5.1:

$$n = \frac{z_o^2 \sigma^2}{\epsilon^2} = \frac{(1.96)^2(.835)^2}{(.05)^2} = 1{,}072.$$

This sample size is a fraction of 1 percent of the unskilled labor force of the Bay Area; the sample size *fpc* would correct this downward to 1,065—hardly worth the trouble. For the Battle Mountain Area, a sample of size 1,072 is 55.3 percent of the population being sampled, (1,072/1,937 = .553), and so we would definitely want to use the finite population correction factor.

Using the formula for sample size when the population is small (developed above) we have

$$n_f = \frac{1{,}072}{1 + (1{,}072/1{,}937)} = 691.$$

A sample size of 691 would give the same precision in the Battle Mountain Area as that of 1,072 in the Bay Area (assuming $\sigma = .835$ is correct for both areas). It is also interesting to note that this is approximately a 35 percent sample of the Battle Mountain Area labor force while only .7 percent (less than 1 percent) of the population in the Bay Area is sampled; even if N were infinite, a sample size of 1,072 would give the desired precision (again assuming $\sigma = .835$).

Problems

1. A random sample test of 40 Eastinghaus 60-watt bulbs yielded a mean burning time of 680 hours and a standard deviation of 96 hours. What is the best estimate of the mean burning time for all bulbs of this type? Give the 99 percent confidence interval for mean burning time.

2. A week before the season opened, 150 fish were taken from a lake by the State Fish and Game Department and were found to have an average weight of 1.2 pounds with a standard deviation of .4 pounds. Give the 95 percent confidence estimate of the average weight of fish in the lake.

3. Estimates of home values were sought in a random sample of 400 homeowners, with the following results:

   $\bar{x} = \$68{,}300$
   $s = \$8{,}400.$

   What is the best point estimate of μ, the mean value of all homes? What is the 95 percent confidence interval estimate for this parameter?

4. A sample of 220 families showed that 180 of them own more than 1 automobile. What is the best point estimate of the percent of all families that own more than 1 automobile? Give the 90 percent confidence interval for this estimate.

5. A sample of 500 families resulted in 460 with health insurance. Give the 95 percent confidence interval for the proportion of all families with health insurance.

6. From a random sample of 380 households it was recorded that 320 of the families shop on Saturdays.
   a. The best point estimate of π, the universe proportion of families who shop on Saturdays, is _____.
   b. We are 98 percent confident that π is at least _____ and no more than _____.

7. A sample of 16 accounting majors showed a mean GPA of 2.92 and a standard deviation of .44. Give the 98 percent confidence interval for the mean GPA of all accounting majors. What assumption did you have to make regarding the distribution of GPA?

8. The following impact strengths in foot-pounds were determined for 5 randomly selected electrical insulators:

    7.01, 7.38, 7.59, 7.93, and 7.17 foot pounds.

    Assume impact strength is a normally distributed variable, and find the 99 percent confidence interval estimate of the average impact strength of all electrical insulators produced by the firm.

9. A random sample of 26 steers was taken from a large herd. The sample mean weight was 843 pounds, and sample standard deviation, s, was 72 pounds. What are the best point estimate and the 95 percent confidence interval estimate of the average weight of all steers in the herd?

10. A random sample of 1,600 voters showed 864 favored candidate Wow for U.S. senator. What is the best estimate for the universe proportion favoring Mr. Wow? Find the 95 percent confidence interval estimate.

11. In a random sample of 265 homeowners in the county, the average mortgage balance was $20,800 and the standard deviation was $6,300. What is the best estimate of μ, the mean mortgage balance of all 250,000 homeowners? What is the 98 percent confidence interval estimate for μ? What is the best estimate of countywide total outstanding mortgage debt?

12. A local restaurant chain wants to estimate the daily demand for pickles within ± 1.5 jars at the 95 percent confidence level. A similar study performed by another restaurant chain estimated the standard deviation to be 6.8 jars. What sample size should be used to attain the desired level of precision?

13. We wish to estimate the proportion of all women in the city between the ages of 16 and 65 who are in the labor force ($N = 250{,}000$). This estimate should be in error by no more than ± .01 at the 98 percent confidence level. If we have no idea of the value of the universe proportion, π, what should the sample size be? If the 1970 census indicated π to be approximately .20, what sample size might be more appropriate?

14. What sample size would you recommend in order to estimate the proportion of students in a certain large university who have taken at least 1 university level mathematics course, if this estimate is to be within ± .03 at the 90 percent confidence level?
    a. If you have no idea of the value of π, $n =$ _____.
    b. If a similar study in 1975 resulted in a p of .80, and you are

willing to use this as a first estimate of the current value of π, $n =$ _____.

15. What sample size should be used to estimate the proportion unemployed in the labor force within $\pm .02$ at the 95 percent confidence level? This proportion is believed to be approximately .10.

16. Your production process requires the use of high-grade steel rods. You receive the rods in lots of 10,000. You want to estimate the average weight of steel per rod in a lot from a new supplier with an error of no more than $\pm .3$ pound with a confidence level of 99 percent. A sample of 35 similar rods taken for a study performed 5 months ago had a mean of 45.8 pounds and standard deviation of 1.2 pounds. How large a sample size should be taken to obtain a satisfactory estimate of the average weight of steel per rod?

17. The following times to assemble a power drill were randomly selected:

    4.8, 2.6, 10.7, 6.5, and 7.1 minutes.

    Find the 90 percent confidence interval estimate of the mean assembly time. (Assume the time to assemble a drill follows a normal distribution.)

18. If the length of a motor shaft is greater than 6.45 inches, it is considered defective. In a random sample of 60 shafts taken from a large lot, 8 are found to be longer than 6.45 inches. Find the 97 percent confidence interval for estimating the proportion of defective motor shafts in the lot.

19. A random sample of 85 customers of a large department store had 48 customers who indicated they preferred to shop on the second floor because they like the decor and the background music. The average purchase of those sampled was $27.53 with a standard deviation of $8.17. Find the 98 percent confidence interval estimate of the proportion of all customers who prefer shopping on the second floor.

20. You want to estimate the proportion of customers who prefer your new package design over the company's current package design. You want an error of no more than ± 3 percentage points at the 98 percent level of confidence, and, from previous experience with similar problems, you feel that at least 70 percent of the customers will prefer your new package design. How large should the sample size be to attain the desired accuracy?

21. The materials manager wants to estimate the average weight of parts made out of an expensive steel alloy. The following weights in ounces were obtained from 12 randomly selected parts:

    7.3, 6.8, 6.4, 8.2, 6.7, 7.5, 6.9, 7.1, 7.4, 6.6, 7.2, and 7.6.

Find the 90 percent confidence interval estimate of the mean weight of all components made out of the steel alloy.

22. Your firm has purchased a large shipment of light bulbs to be used in its factories. You need a good estimate of the average length of life of this shipment of light bulbs so that you can work out a replacement plan for your factories. You would like to get a 98 percent confidence interval estimate that is no more than 20 hours in length, i.e., the point estimate ± 10 hours. Past statistical studies indicate that this type of light bulb will have a standard deviation of approximately 60 hours. How large a sample size is necessary to obtain the desired precision?

23. You want to estimate the proportion of overdue accounts for a large retail chain. You would like to estimate this proportion with an error of no more than ± 2 percentage points at the 96 percent level of confidence. You feel confident that no more than 15 percent of the accounts are overdue. How large should your sample size be in order to attain the desired level of accuracy?

24. The production scheduler for a large manufacturing concern was concerned about the accuracy of the assembly time per component estimated by the industrial engineering department. He decided to obtain his own estimate by randomly selecting time of day and work station; unobserved, he recorded the time various workers required to assemble 1 component. The mean value of a sample of 115 assembly times was 4.35 minutes with a sample standard deviation of 1.42 minutes.
   a. Find the 97 percent confidence interval estimate of the average assembly time per component for all components produced by the firm.
   b. Find the estimated assembly time in hours required to produce 2,480 components. What is the 97 percent confidence interval for the estimate?

25. If the outside diameter of a pump seal is greater than 2.75 inches, the seal is considered to be defective. A random sample of 80 parts from a large lot of pump seals has 10 defective seals. Find the 93 percent confidence interval estimate of the proportion of bad seals in the lot.

26. The vice-president of marketing is making an economic feasibility study for constructing a new store in a newly developed suburban area. The growth in the area has taken place so fast that the census income data for the area would be outdated. He is able to learn from the local realtors' association that 14,280 families currently live in the area. A sample of 95 families randomly selected from the area yielded a mean family income of $16,280 with a standard deviation of $1,286.

a. Find the 98 percent confidence interval for estimating the average family income of all the families in the potential marketing area.
b. Find the 98 percent confidence interval estimate of the total family income of the 14,280 families in the marketing area.

27. In the above example, the marketing vice-president would like to have some idea of the proportion of families that would want to shop at his type of store if it were introduced in the area. In a random sample of 150 families, 87 indicated a desire to patronize the type of store under consideration. Find the 95 percent confidence interval estimate of the proportion of all families in the area interested in this type of store.

28. The personnel manager wants to estimate the average length of the morning coffee break of clerical employees on the fourth floor. During the past week he has randomly observed and timed the clerical help and obtained the following coffee break times in minutes:

    17.2, 12.8, 21.1, 16.5, 25.3, 23.7, 14.9, 17.1, and 19.4.

    Find the 95 percent confidence interval estimate of the average coffee break time for clerical employees on the fourth floor. (Assume the times are normally distributed.)

29. The inventory manager for a large manufacturing firm wants to obtain a quick estimate of the current value of the 3,423 inventory items in the $10-to-$50 class. A random sample of 275 items has a mean value of $38.17 and a standard deviation equal to $3.96.
    a. Find the 96 percent confidence interval estimate of the average value per item of all the items in this inventory class.
    b. Find the 96 percent confidence interval estimate of the total inventory value of the 3,423 items in the $10-to-$50 class.

■ The remaining problems require material from the appendices.

30. We wish to estimate the mean monthly household grocery expenditures in the state (N is 350,000). The standard deviation of this variable is known to be approximately $42.00. Our estimate should be in error by no more than $2.00 at the 95 percent confidence level.
    a. What sample size is required?
    b. If we wish the same precision for a small community of 1,000 households (assume the same standard deviation), what sample size is required?

31. A random sample of 125 bearings is selected from a lot of 1,000 bearings. The average diameter of the sample of 125 bearings is 2.73 inches with a standard deviation of .83 inches. Find the 95

percent confidence interval estimate of the average diameter of the lot of bearings.

32. Refer to the previous problem. It is desired to estimate the mean bearing size of the lot of 1,000 bearings with an error of no more than ± .06 inches at the 95 percent level of confidence. How large should the sample size be?

CHAPTER 8

HYPOTHESIS TESTING

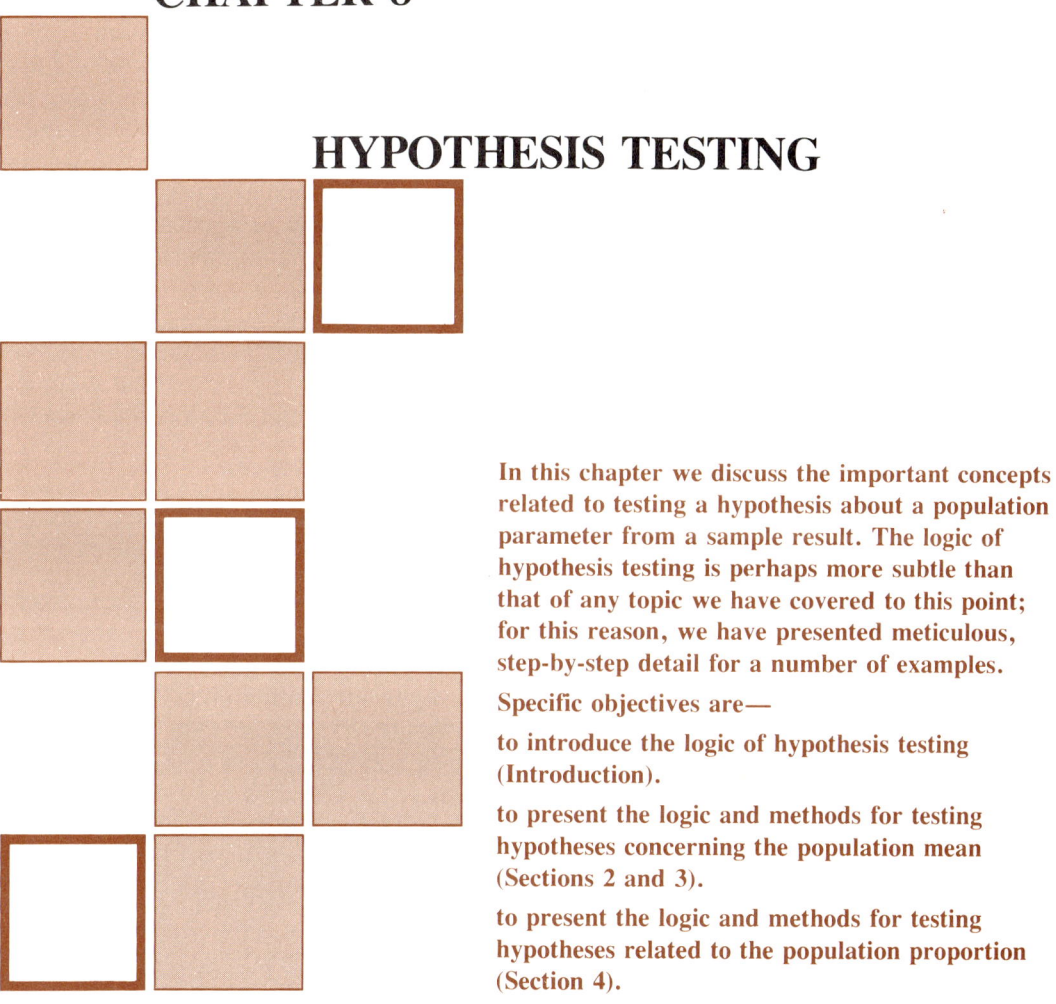

In this chapter we discuss the important concepts related to testing a hypothesis about a population parameter from a sample result. The logic of hypothesis testing is perhaps more subtle than that of any topic we have covered to this point; for this reason, we have presented meticulous, step-by-step detail for a number of examples.

Specific objectives are—

to introduce the logic of hypothesis testing (Introduction).

to present the logic and methods for testing hypotheses concerning the population mean (Sections 2 and 3).

to present the logic and methods for testing hypotheses related to the population proportion (Section 4).

The arithmetic, the algebra, and the statistical models used for hypothesis testing are similar to those used for statistical estimation, but the concept and logic are different. We are still forced to make a statement about the population parameter values based on the limited information contained in sample data. A hypothesis—a thesis or conjecture—regarding a population parameter is suggested for consideration. A random sample is then taken from the population, and the hypothesis is "tested" with the appropriate sample statistic; i.e., the sample result is the basis for making a decision concerning the validity of the hypothesis.

For example, it may be hypothesized that the average life of batteries produced by a company is 30 months. It is decided to select randomly 81 batteries and determine the average life of these batteries under normal use. A decision will be made about the hypothesized value for the mean life of all batteries produced by the company based on the average life of the 81 batteries in the sample; the hypothesis that the population mean is equal to 30 will be either accepted or rejected. Even if the population mean is equal to 30, it is highly unlikely that the mean of a random sample is going to be exactly equal to 30. Assume a sample result of $\bar{x} = 28.7$ months. The hypothesis $\mu = 30$ will be accepted if we decide that this value of the sample mean is attributable to a chance deviation, or the hypothesis will be rejected if we decide that 28.7 is sufficiently different from 30 to conclude that the sample data came from a population with a mean different from 30. We can never know the correct decision with certainty as long as our information is limited to sample data. Using the statistical theory developed in previous chapters, it is possible, however, to say something about the probability of the rightness or wrongness of the decision that is made.

8.1 Two Types of Error and Classical Hypothesis Testing

It is traditional to refer to the hypothesis consisting of the initial hypothesized value of the population parameter as the null hypothesis. The symbol H_o will be used to designate the null hypothesis. For the example above, the hypothesis that the average life of batteries is equal to 30 is written

$H_o: \mu = 30.$

Two Types of Error

Inasmuch as we are asked to either reject or accept this hypothesis, there are two kinds of mistakes we can make. Statisticians have arbitrarily labeled these two types of error as Type I and Type II.

- Type I error: Reject H_o when H_o is true.
- Type II error: Accept H_o when H_o is false.

The Type I error is also referred to as the α-error and the Type II error is called the β-error. This is a result of the convention to let α designate the probability of a Type I error and let β be the probability of a Type II error; i.e.,

P(Type I error) = α.
P(Type II error) = β.

The two types of hypothesis-testing errors are summarized in the table below.

State of Nature	Decision	
	Reject the Null Hypothesis	Accept the Null Hypothesis
The Null Hypothesis Is True	Type I Error	Correct Decision
The Null Hypothesis Is False	Correct Decision	Type II Error

A decision is made based on the sample data obtained and, because the whole population is not enumerated, the best we can do is make a statement about the probability that a mistake will be made; we cannot guarantee a correct decision.

There are various approaches to hypothesis testing. One distinguishing feature among these approaches is the treatment of the two types of errors. In this chapter we concentrate on the classical approach to hypothesis testing. The classical method of hypothesis testing is concerned primarily with controlling the probability of Type I error. It is the approach most frequently encountered by business and economics students. The classical approach is used in the hypothesis testing that is an integral part of such important statistical techniques as regression analysis, analysis of variance, and chi-square tests.

Classical hypothesis testing requires the specification of an explicit value for the probability of a Type I error. On the other hand, the probability that a Type II error will occur when the hypothesis test is performed is neither calculated nor specified. As you study hypothesis testing in the remainder of this chapter, you should not assume that Type II error is being ignored. If this were the case, then we could conclude this chapter at this point by suggesting the decision rule, "Always accept the null hypothesis." Inasmuch as a Type I error can only be committed when the null hypothesis is rejected, this rule assures a 0 probability of committing a Type I error. This decision rule is, of course, uninteresting because it means that the probability of a Type II error is 1 or certainty; thus, when performing hypothesis tests, we are asked to specify the *maximum* acceptable probability of a Type I error. For a given set of circumstances, a larger specified probability of a Type I error results in a smaller probability of a Type II error. Concern for the probability of a Type II error is also reflected in the fact that a so-called one-tail test is used when appropriate. A one-tail test may be preferred for some types of problems because it could mean a lower probability of committing a Type II error.

8.2 Hypothesis Tests Concerning the Mean—Large Sample Size

Four steps that summarize classical hypothesis testing are given below and then discussed in detail. A researcher reporting his results in the literature may not explicitly state the four steps developed in this chapter, but they are implicitly used whenever classical hypothesis testing is employed. If you have a good understanding of these four steps, you will be able to understand reported hypothesis-testing results regardless of the format in which the results are presented. Also, once you have mastered the basic concepts of hypothesis testing for one type of problem, such as hypothesis tests concerning the mean, you can readily do hypothesis testing for other types of problems. The statistical distributions or the models may be different, but the basic approach to classical hypothesis testing is always the same. In regression analysis (covered in a later chapter), hypothesis testing is a topic of major importance. Analysis of variance is a topic primarily concerned with hypothesis testing.

Four Steps to Test a Hypothesis

The four steps that will be followed throughout the text when hypothesis testing is required are:

1. State the null and alternate hypotheses.
2. Specify the significance level and a test statistic.
3. Determine a decision rule.
4. Obtain sample data, calculate the necessary statistics, and make a decision based upon these results.

It should be emphasized that the first three steps are completed before the sample is taken. Determining the proper test statistic for Step 2 could require a knowledge of the sample size; however, this is something that is determined before the sample is taken. For textbook examples and problems, all of the information must necessarily be given at the same time. But it will help to avoid confusion if the reader will keep in mind that, strictly speaking, the first three steps are performed before the sample is taken; i.e., the sample data are not available until Step 4. A detailed explanation of each of the steps will be given while working the following example.

Example 8.2.1 A machine used to fill cans of soup is properly adjusted if the average net weight of soup is 11.2 ounces. A random sample of 64 cans had an average fill of 11.12 ounces and standard deviation of .18 ounces. Is the machine adjusted properly? Test at the 5 percent significance level.

The null hypothesis is, "The machine is properly adjusted"; i.e., "The mean fill of all cans is 11.2 ounces." The alternate hypothesis is, "The mean fill is not equal to 11.2 ounces." If it is greater than 11.2 ounces, the accountant is unhappy. If the mean fill is less than 11.2 ounces, the legal department is unhappy. The null hypothesis always contains the equality expression and, as indicated above, is designated by H_o. The symbol H_a is used to label the alternate hypothesis. These two hypotheses are stated in Step 1.

1. $H_o: \mu = 11.2$
   $H_a: \mu \neq 11.2$.

The significance level called for in Step 2 has already been given in the problem as 5 percent or .05. This is the maximum probability of making a Type I error (reject H_o when H_o is true) that the user of the statistical results is willing to accept. Since α is used to designate the probability of Type I error, the significance level for this example is written as $\alpha = .05$.

If the reader has the feeling that the .05 value was simply pulled out of thin air, you're not too far wrong. In classical hypothesis testing, the statistical theory does *not* tell how the significance level is to be determined. The investigator decides this based on his or her feelings about the desirability or undesirability of making a Type I error. The investigator selects the maximum probability that seems acceptable because of the trade-off between the Type I and Type II errors referred to above—the larger the Type I error probability, the smaller the Type II error probability and vice versa. Traditionally, researchers reporting their results in journals use as typical significance levels .01, .05, and .10. This is strictly traditional; there is nothing in the theory of classical hypothesis testing that indicates the "goodness" or "badness" of these significance levels. In business applications, the cost of making a particular type of error will influence the determination of the significance level.

For a test statistic, we want to select a function of the population parameter and the corresponding sample statistic of interest that will permit us to make a statement about the probability of the sample statistic having come from a population with the hypothesized parameter.

Test Statistic for Large Sample Sizes

For large sample sizes, the following statistic has an approximately standard normal distribution:

$$z = \frac{\bar{x} - \mu}{s/\sqrt{n}}$$

where

$\bar{x}$ = the sample mean
s = the sample standard deviation
n = the sample size
μ = the population mean.

This test statistic is a function of both the sample mean and the population mean. In addition, we can use the standard normal table (Table B) to make statements about the probability of this statistic's assuming specified values.

Looking ahead to Step 4, we will want to obtain a value of z calculated from the sample data. At that point in the hypothesis-testing process, a sample is taken from the population to obtain values of $\bar{x}$ and s. In order to complete the calculation of the test statistic in Step 4, it is necessary to answer the question, What should be used for the value of μ, the population mean? (The population mean is not known; that is why we have to resort to hypothesis testing.) The answer to this question is determined by the basic approach to classical hypothesis testing: set up the test assuming that the null hypothesis is true, and then observe whether or not the sample data are consistent with this assumption. For this problem, the hypothesized value for μ is 11.2, which is the value for μ used in the test statistic. Thus for Step 2 we have

2. $\alpha = .05$

$$z = \frac{\bar{x} - \mu}{s/\sqrt{n}} = \frac{\bar{x} - 11.2}{s/\sqrt{n}}.$$

The decision rule called for in Step 3 is an important part of hypothesis testing. Once the decision rule has been determined and the sample taken, it is simply a clerical matter to make the necessary calculations using the sample results and then compare the calculated value of the test statistic with the value specified by the decision rule and make the corresponding decision. The decision rule is determined by first applying the basic approach of classical hypothesis testing referred to above, which is to set up the test assuming the null hypothesis is true. If the null hypothesis is true, then the test statistic in Step 2,

$$\frac{\bar{x} - 11.2}{s/\sqrt{n}},$$

will have a standard normal distribution centered around 0 as indicated in Figure 8.1. We want to state a decision rule such that the probability of a Type I error (reject H_o when H_o is true) will be equal to the significance level (for this problem $\alpha = .05$) specified in Step 2.

In classical hypothesis testing, the validity of the null hypothesis is questioned when the sample results tend strongly to favor the alternate hypothesis. The alternate hypothesis for this problem, stated in Step 1, is $H_a: \mu \neq 11.2$. A sample mean greater than 11.2 as well as a sample mean less than 11.2 are consistent with the alternate hypothesis. Due to the randomness of the sample, we expect $\bar{x}$ to be different from 11.2; however, if the sample mean were considerably larger than 11.2, then we might question the validity of the null hypothesis; i.e., "Could these sample data have come from a population with $\mu = 11.2$?" As the value of $\bar{x}$ gets larger than 11.2, the calculated value of the test statistic,

$$z = \frac{\bar{x} - 11.2}{s/\sqrt{n}},$$

gets larger. (The sample size n is fixed for a given problem, and the sample standard deviation s will not vary appreciably from the population standard deviation.) The null hypothesis is rejected when large positive values of the test statistic are obtained from the sample results as is indicated by the rejection region on the right side of Figure 8.1. On the other hand, if the sample mean is considerably less than 11.2, the calculated value of the test statistic will assume a large negative value.

Figure 8.1 Determination of the Value of z for a Two-Tail Test with $\alpha = .05$

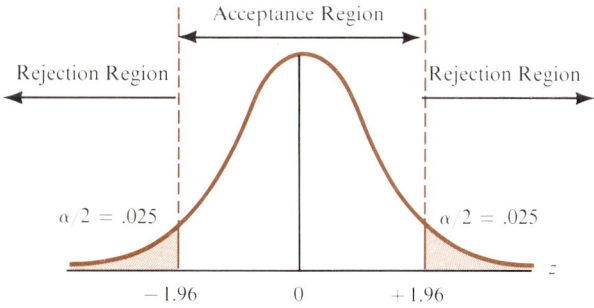

It follows that the null hypothesis is also rejected for large negative values of z as indicated on the left side of Figure 8.1.

The values of the test statistic, z, to be used for the decision rule, are found by dividing the significance level, $\alpha = .05$, equally into both tails of the z-distribution and then finding the corresponding values of z from the standard normal table as already discussed in Chapter 5. (In this case, the z_o value corresponding to $.5000 - .0250 = .4750$ in the body of the table is selected; i.e., $P(z > 1.96) = .025$.) If we let z_c denote the calculated value of the test statistic (i.e., the value that is obtained from "plugging in" the sample mean and standard deviation), then the decision rule called for in Step 3 can be written

3. Reject H_o if and only if $z_c > 1.96$ or $z_c < -1.96$.

This is referred to as a "2-tail test" because the significance level, $\alpha = .05$, is split between the 2 tails. If the null hypothesis is true, there is a .025 probability that the mean of a sample drawn from the population will be so large that the calculated value of z will be greater than 1.96, causing us to reject the null hypothesis and thereby commit a Type I error. Similarly, if the null hypothesis is true, there is also a .025 probability that the sample mean will be so small that the calculated value of the sample statistic will be less than -1.96, which would also mean that the null hypothesis is rejected. (The reader is reminded that the sample is not taken until after the decision rule has been determined.) There are two mutually exclusive ways in which a Type I error can be committed. (These are also exhaustive—i.e., there is no other way.) The probability of committing a Type I error if the decision rule in Step 3 is applied is the sum of the two probabilities $.025 + .025 = .05$.

Once the decision rule has been completed in Step 3, all one has left to do is take a sample and use the sample data to obtain a calculated value of the test statistic. This calculated value is then applied to the decision rule which automatically specifies the decision:

4. $z_c = \dfrac{\bar{x} - 11.2}{s/\sqrt{n}} = \dfrac{11.12 - 11.2}{.18/\sqrt{64}} = -3.56.$

    Reject H_o.

The calculated value of the test statistic, -3.56, is less than the -1.96 critical value in the decision rule; so the null hypothesis is rejected. This is the same as rejecting the hypothesis that the machine is properly adjusted. The test was set up so that there would be at most a .05 probability of making a mistake if the sample results indicate that the null hypothesis should be rejected. It was determined before performing the hypothesis test that a 5 percent probability of a Type I error is acceptable; thus, when the sample data indicate that the null hypothesis should be rejected, we feel confident and are quite positive in making the statement that the machine needs to be adjusted. We realize

Two-Tail Hypothesis Test

The four hypothesis-testing steps for the example just worked are summarized below without the detailed explanation.

1. $H_o: \mu = 11.2$
   $H_a: \mu \neq 11.2$
2. $\alpha = .05$

$$z = \frac{\bar{x} - \mu}{s/\sqrt{n}} = \frac{\bar{x} - 11.2}{s/\sqrt{n}}.$$

3. Reject H_o if and only if $z_c > 1.96$ or $z_c < -1.96$.
4. $z_c = \dfrac{\bar{x} - 11.2}{s/\sqrt{n}} = \dfrac{11.12 - 11.2}{.18/\sqrt{64}} = -3.56.$

Reject H_o.

The hypothesis test just performed was of the type often referred to as a "two-tail" test. This is because the significance level was split between the two tails as indicated in Figure 8.1. The next example deals with a situation in which a so-called one-tail test is appropriate. A one-tail test is usually appropriate in situations where there is a bone of contention. For example, we may prove our point only if the population mean is less than the hypothesized value. In classical hypothesis testing, the null and alternate hypotheses are set up in such a way that we demonstrate the validity of our claim by rejecting the null hypothesis in favor of the alternate hypothesis. This is done because we control the probability of Type I error. Thus, if we demonstrate our claim or theory by rejecting the null hypothesis, we know that our probability of being wrong is no greater than the significance level specified before the sample was taken. On the other hand, if we accept the null hypothesis, we usually are unable to specify the probability of making a mistake; i.e., the probability of a Type II error is not usually known because there are an infinite number of values of μ consistent with the alternate hypothesis. It isn't always possible to set up a hypothesis test so that we prove our point by rejecting the null hypothesis, but whenever possible it is definitely the preferred approach.

Example 8.2.2

A manufacturer claims that the average life of batteries produced by his firm is 30 months. You disagree, contending that the average life of the batteries is less than 30 months. A random sample of 81 batteries has a mean of 28.7 months and a standard deviation of 8 months. Perform the appropriate hypothesis test. Use a significance level of .05.

This example is different from the first example as to what alternate values of the population mean we are interested in. In the first example, if the machine is out of adjustment either way—i.e., if its average fill is either too great or too small—we want to reject the hypothesized value. In the present example, we are only interested in rejecting the manufacturer's claim or the hypothesized value of the mean, $\mu = 30$ months, if the mean is less than the hypothesized value. (If the average life of the batteries produced by the manufacturer is greater than 30, then this just strengthens his claim.) The difference in what we are trying to prove in these two problems is reflected in the different statements of the alternative hypothesis.

A good rule to follow is to always have equality in the null hypothesis and let the alternate hypothesis reflect the values for which it is desired to reject the null hypothesis. Then the decision rule is formulated so that we reject the null hypothesis in favor of the alternate hypothesis, i.e., whenever the sample data are consistent with the alternate hypothesis, then we consider the possibility of rejecting the null hypothesis. In this problem we prove our point only if the mean battery life is less than 30 months. The appropriate alternate hypothesis is $H_a: \mu < 30$. We have set up the alternate hypothesis so that we prove our claim by rejecting the null hypothesis and accepting the alternate hypothesis. The complete hypothesis test is given below followed by some additional discussion concerning the various hypothesis-testing steps.

One-Tail Hypothesis Test

The one-tail hypothesis test is as follows:

1. $H_o: \mu = 30$
   $H_a: \mu < 30$.

2. $\alpha = .05$

   $$z = \frac{\bar{x} - \mu}{s/\sqrt{n}} = \frac{\bar{x} - 30}{s/\sqrt{n}}.$$

3. Reject H_o if and only if $z_c < -1.65$.

4. $z_c = \dfrac{\bar{x} - 30}{s/\sqrt{n}} = \dfrac{28.7 - 30}{8/\sqrt{81}} = -1.46.$

   Do *not* reject H_o.

The sample size is large (the traditional rule of thumb is $n > 30$), allowing us to take advantage of the Central Limit Theorem and use the z-statistic for the test statistic in Step 2. Note that, even though the significance level is the same as in the first example, the decision rule in Step 3 is quite different. Not only is the decision rule stated differently, but the critical value selected from the standard normal table is also different. This difference is due to the fact that the alternate hypothesis is different. The alternate hypothesis determines the form of the decision rule, i.e., determines whether it is a two-tail or a one-tail test.

The null hypothesis is rejected when the sample results strongly favor the alternate hypothesis. The alternate hypothesis for this problem is $\mu < 30$; so we only consider rejecting the null hypothesis if the sample mean, $\bar{x}$, is less than 30. (As indicated above, it would be difficult to argue that the average life of batteries produced by the manufacturer is less than 30 if our sample of 81 batteries should happen to produce a sample mean larger than 30.) It is realized that a population with a mean of 30 is going to generate a sample mean less than 30 for 1/2, on the average, of the random samples taken from that population. However, if the sample mean is significantly less than 30, the possibility that it (the sample mean) came from a population with mean less than 30 becomes greater. As the sample mean, $\bar{x}$, gets much smaller than 30, the test statistic,

$$\frac{\bar{x} - 30}{s/\sqrt{n}},$$

also gets much smaller and assumes large negative values; thus, the null hypothesis is rejected for sample results leading to large negative values of the test statistic as indicated in Figure 8.2.

Note that the total value of α is in the lower tail of the z-distribution. (The critical value is found by finding .4500 in the body of the standard normal table and selecting the corresponding value of z to the nearest hundredth.) This is done because there is only one way that a Type I error can occur: the sample mean is much smaller than the hypothesized value when the hypothesized value is true. For sample means larger than 30, the null hypothesis is never rejected; so there is no possibility of making a Type I error if positive values of z are attained. If in the question under consideration it is only of interest to

Figure 8.2 Determination of the Value of z for a One-Tail Test with $\alpha = .05$

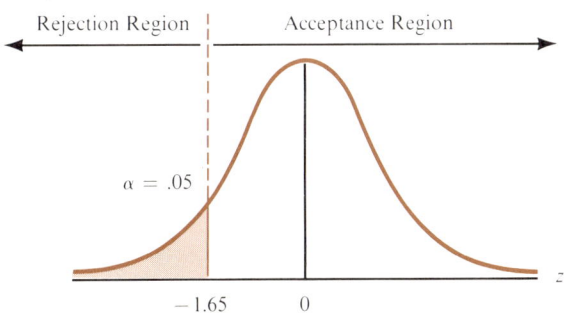

Hypothesis Tests Concerning the Mean—Large Sample Size

reject the null hypothesis one way, as is the case in this example, then a one-tail test is used because this minimizes the probability of a Type II error for the given probability of a Type I error.

One-tail tests are often referred to as "lower-tail" or "upper-tail" tests depending on whether the null hypothesis is rejected only for small values of the test statistic or only for large values of the test statistic. The case just cited is an example of a lower-tail test. Example 8.4.2 in the latter part of this chapter is an example of an upper-tail test.

In Step 4, the wording used to indicate the decision made was "do not reject H_o" as opposed to "accept H_o." Often, in a practical situation we are forced to either accept or reject the null hypothesis; but the statistician prefers the expression "do not reject H_o" because it expresses a hesitancy to accept wholeheartedly the null hypothesis. This reluctance on the part of the statistician is due, in part, to the fact that the probability of a Type II error is not known; i.e., one doesn't know the probability of making a mistake when the null hypothesis is accepted. When the sample data indicate that the null hypothesis, H_o, should be rejected, we feel very confident about rejecting H_o because the probability of being wrong is no more than the significance level that had already been determined to be an acceptable risk before the sample was taken. In this example, the sample results are consistent with the null hypothesis at the specified significance level. This hedging about accepting H_o is evident in the following frequently used expressions: "The sample data do not indicate that the null hypothesis should be rejected"; or "Based on the sample results, we cannot reject the null hypothesis." These types of statements may also indicate that the statistician wishes to reserve judgment until more data are obtained.

The probability of a Type II error is not known because the statement of the alternate hypothesis doesn't specify a value for the population mean, but presents a range of possible values, $\mu < 30$. For each of these possible values of μ, there is a different probability of making a Type II error. The probability of a Type II error can be as large as $1 - \alpha$ (for this example $1 - \alpha = .95$), but it can also be very small. The greater the difference between the true mean (which is of course unknown) and the hypothesized mean, the smaller is the probability of accepting the hypothesized value and thereby committing a Type II error. The appendix at the end of this chapter shows how the probability of a Type II error can be calculated if specific alternate values for μ are specified. The above problem will be used to illustrate the method of calculation. From Appendix 8A, we see that the probability of accepting the null hypothesis (H_o: $\mu = 30$) if the mean is in fact 29.5 is calculated to be .8621; there is an 86.21 percent chance of accepting $\mu = 30$ when μ is actually 29.5. If the true population mean were 26, then the probability of Type II error would be only .0022.

# 8.3 Hypothesis Tests Concerning the Mean—Small Sample Size

In the above examples, the z-statistic was the proper statistic because the sample sizes were large. If the sample size is small, then the Central Limit Theorem is not applicable, and it is necessary to cast about for another test statistic. In Chapter 6 it was noted that, if the sample is drawn from a normal population, then the expression $(\bar{x} - \mu)/(s/\sqrt{n})$ has a Student-t distribution with $n - 1$ degrees of freedom.

Test Statistic when Sampling from a Normal Population

This provides us with a test statistic for testing hypotheses about the population mean when the population has a normal distribution:

$$t = \frac{\bar{x} - \mu}{\frac{s}{\sqrt{n}}} \text{ with } df = n - 1.$$

This statistic is not dependent on sample size; however, it does require that the population under investigation have a normal distribution. If the sample size, n, is large, the z-statistic can be used regardless of the population distribution by virtue of the Central Limit Theorem; but if the sample size is small and the t-statistic is used for the test statistic, then it is necessary to be sampling from a population with a normal distribution.

Example 8.3.1

The following sample was drawn randomly from a normally distributed population:

$-1.3, 2.7, 4.1,$ and $2.5.$

Test the hypothesis that the mean of the population from which this sample was taken is 3.6. Test at the 10 percent significance level.

This problem simply asks us to test the hypothesis that the population mean is equal to 3.6; so the alternate hypothesis is $\mu \neq 3.6$. The sample size is small; so the Central Limit Theorem is not applicable. However, the sample is from a normal population. This permits the use of the t-statistic as the test statistic. The solution to this problem is as follows.

1. $H_o: \mu = 3.6$
    $H_a: \mu \neq 3.6.$
2. $\alpha = .10$

$$t = \frac{\bar{x} - \mu}{s/\sqrt{n}} = \frac{\bar{x} - 3.6}{s/\sqrt{n}}$$

with $df = n - 1 = 4 - 1 = 3.$

3. Reject H_o if and only if $t_c < -2.353$ or $t_c > 2.353$.

4. $\bar{x} = \dfrac{\Sigma x_i}{n} = \dfrac{-1.3 + 2.7 + 4.1 + 2.5}{4} = 2.0$

$s^2 = \dfrac{\Sigma(x_i - \bar{x})^2}{n - 1}$

$= \dfrac{(-1.3 - 2.0)^2 + (2.7 - 2.0)^2 + (4.1 - 2.0)^2 + (2.5 - 2.0)^2}{4 - 1}$

$= 5.35.$

$s = \sqrt{5.35} = 2.31$

$t_c = \dfrac{\bar{x} - 3.6}{s/\sqrt{n}} = \dfrac{2.0 - 3.6}{2.31/\sqrt{4}} = \dfrac{-1.6}{1.155} = -1.39.$

Do *not* reject H_o.

The alternate hypothesis for this problem is the not-equal-to hypothesis ($H_a: \mu \neq 3.6$); so the null hypothesis is rejected for both large and small values of the test statistic. Thus, the significance level, $\alpha = .10$, is divided between the two tails (see Figure 8.3). The degrees of freedom for this example are: $df = n - 1 = 4 - 1 = 3$; 2.353 is obtained from the t-table (Table C) under the column headed .05 (the area in the tail) and in the row for 3 degrees of freedom.

Where appropriate, a one-tail test using the t-statistic can be implemented through a method much like that employed with the z-statistic. If the alternate hypothesis for the problem just worked were $\mu > 3.6$, the total value of α would be put in the upper tail of the t-distribution. The resulting decision rule would be to reject H_o if the value of the t-statistic calculated from the sample data is larger than 1.638. (This value was found in the t-table under the column headed .10

Figure 8.3 Two-Tail Test with $df = 3$ and $\alpha = .10$

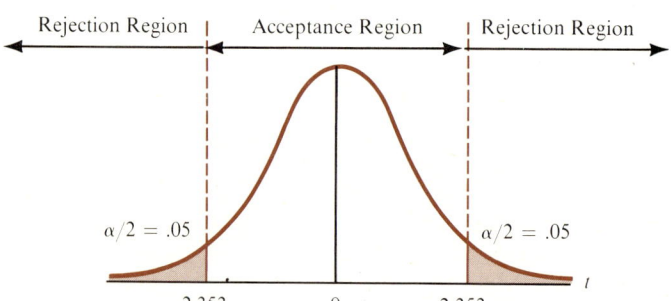

Hypothesis Testing

and along the row for 3 degrees of freedom.) A complete solution for a one-tail t-test is given for the following example.

Example 8.3.2

The manager feels that the laboratory supervisor is budgeting too much time for performing an analysis of processed food. He feels that the average time per analysis should be much less than the 15 minutes budgeted by the supervisor. He decides to test his hypothesis by observing the time it takes to perform 16 randomly selected analyses. He wants to have a strong case when he confronts the laboratory supervisor; so he decides that a significance level of .01 will be required. The average time for the 16 observations turns out to be 13.4 minutes, with a standard deviation equal to 1.3 minutes.

The manager is interested in demonstrating that the average time per analysis is less than the amount budgeted by the supervisor. Consistent with this desire, he sets up the null hypothesis as $\mu = 15$ (the value he is hoping to reject) and the alternate hypothesis as $\mu < 15$ (that which he is hoping to demonstrate). The sample size is small; so the z-statistic would not be appropriate.

It is, however, plausible to assume a normal or near-normal distribution for a random variable such as the time to perform an analysis of processed food. This permits the use of the t-statistic as the test statistic for this problem. The solution is given below.

1. $H_o: \mu = 15$
   $H_a: \mu < 15$.
2. $\alpha = .01$

   $$t = \frac{\bar{x} - \mu}{s/\sqrt{n}} = \frac{\bar{x} - 15}{s/\sqrt{n}},$$

   with

   $df = n - 1 = 15$.

3. Reject H_o if and only if $t_c < -2.602$.
4. $t_c = \dfrac{\bar{x} - 15}{s/\sqrt{n}} = \dfrac{13.4 - 15}{1.3/\sqrt{16}} = -4.92$.

   Reject H_o.

Based on the sample results, the manager rejects the hypothesis that the average time per analysis is 15 minutes and accepts the alternate hypothesis that it is less than 15 minutes. He set up the hypothesis test so that there would be at most a probability of .01 that he would make the mistake of rejecting the null hypothesis if it were true. The total value of the significance level, $\alpha = .01$, is in the lower tail of the t-distribution (see Figure 8.4) because the null hypothesis will only be

Figure 8.4 One-Tail Test with $df = 15$ and $\alpha = .01$

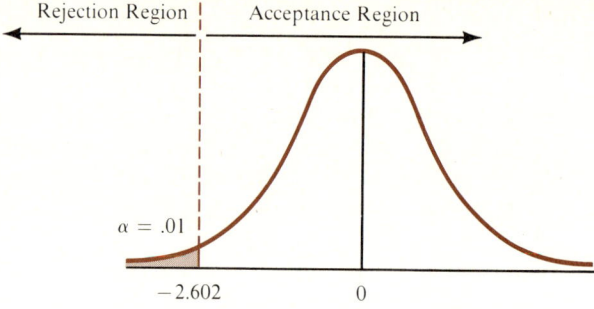

rejected when the sample mean, $\bar{x}$, is consistent with the alternate hypothesis, i.e., only for large negative values of the t-statistic. (The t-value of 2.602 is found in the t-table in the row for 15 degrees of freedom and in the column with the .01 heading.)

 ## 8.4 Hypothesis Tests Concerning the Population Proportion

Test Statistic for the Population Proportion

For large sample sizes, the test statistic used for testing a hypothesis concerning the proportion of a population having a specified attribute comes from a corollary of the Central Limit Theorem discussed in Chapter 6:

$$z = \frac{p - \pi}{\sqrt{\frac{\pi(1 - \pi)}{n}}}$$

where

z = the standard normal random variable
π = the population proportion
p = the sample proportion.

You will recall from a previous chapter the rule of thumb that states, "If the sample size is greater than 30 and $n\pi > 5$ and $n(1 - \pi) > 5$, then the normal distribution provides a good approximation of the binomial distribution." You will also recall from the last chapter that p was defined as the proportion of observations in the sample having the characteristic of interest. If the attribute of interest were "a person approves of the president's economic policy" and if in a sample of

2,000 people 1,200 approve the president's economic policy, the sample proportion would be $p = 1{,}200/2{,}000 = .60$.

Example 8.4.1 A manufacturer claims that no more than 3 percent of the parts he supplies your company are defective. You use large quantities of this part in your manufacturing process and are concerned that, while his claim may have been true in the past, it may not be the case now. You decide to test the manufacturer's claim using a significance level of .02. A random sample of 300 parts has 12 defective parts. Perform the hypothesis test.

The attribute of interest for this problem is whether or not a part is defective. The population proportion, π, is the proportion of all parts supplied by the manufacturer that are defective. The null hypothesis is the equality hypothesis; for this problem, $\pi = .03$. The alternative hypothesis is $\pi > .03$ because you are interested in rejecting the null hypothesis only if the state of affairs is worse than what the manufacturer is claiming. If the proportion of defective parts is less than 3 percent, then the manufacturer's claim is valid. The solution to this problem is presented below.

1. $H_o: \pi = .03$
   $H_a: \pi > .03$.

2. $\alpha = .02$

$$z = \frac{p - \pi}{\sqrt{\frac{\pi(1-\pi)}{n}}} = \frac{p - .03}{\sqrt{\frac{(.03)(.97)}{n}}}.$$

3. Reject H_o if and only if $z_c > 2.05$.

4. $p = \dfrac{\text{number of successes}}{\text{sample size}} = \dfrac{12}{300} = .04$

$$z_c = \frac{.04 - .03}{\sqrt{\frac{(.03)(.97)}{300}}} = 1.015.$$

Do *not* reject H_o.

The sample result indicates that you should not reject the manufacturer's claim.

Note that the hypothesized value of the population proportion, $\pi = .03$, is used in the denominator of the test statistic. This is consistent with the basic method of classical hypothesis testing, which is to set up the hypothesis test assuming the null hypothesis is true. The decision rule is determined in the same manner as that for the case of hypothesis testing about the population mean when the sample size is large. The null hypothesis is rejected only if the sample proportion, p, is consistent with the alternative hypothesis, $\pi > .03$. In addition it must be significantly larger than .03. This means that the calculated value of z is considerably larger than 0. The null hypothesis is rejected only for large values of z, and the total significance level, $\alpha = .02$, is put in the upper tail of the standard normal distribution as indicated in Figure 8.5. This is referred to as an "upper-tail test." The value for z is obtained from the standard normal table.

The z-statistic was appropriate as a test statistic for this problem because the sample size was large. If the sample size were small, the z-statistic could not have been used. Unfortunately, there is no test statistic that can be used for hypothesis tests about the population proportion when the sample size is small. The t-statistic is not appropriate. The t-statistic requires that the sample statistic be normally distributed, which is the case for the sample mean, $\bar{x}$, if the sample is drawn from a normal population. This is *not* the case for the sample proportion, p, when the sample size is small. For the small-sample case, it is necessary to calculate the error probabilities using the binomial probability function (or use binomial tables).

8.5 Additional Comments

The reader should appreciate by now the statement made at the beginning of this chapter: Once the basic methodology of classical hypothesis testing is mastered, then you can do a variety of hypothesis-testing problems. The steps were basically the same for all examples. The parameters of interest and the statement of the alternative hypothesis varied from problem to problem, and different test statistics were used; but all the examples were worked in essentially the same manner.

Figure 8.5 Determination of the Value of z for a One-Tail Test with $\alpha = .02$

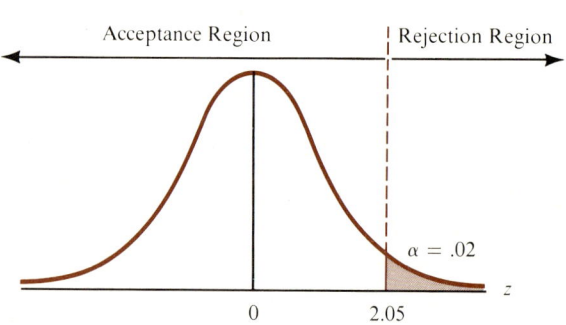

When performing a one-tail test, some authors prefer to include an inequality sign along with the equality sign in the statement of the null hypothesis. They would use $H_o: \mu \geq 30$ and $H_a: \mu < 30$ in Example 8.2.2 for the statement of hypotheses, instead of $H_o: \mu = 30$ and $H_a: \mu < 30$ as used in this text. As long as the equality sign is included in the null hypothesis, the ideas are essentially the same. The approach used in this text helps the student by emphasizing that (1) the equality sign belongs in the null hypothesis and (2) whenever possible, it is desirable to set up the null and alternative hypotheses so that the researcher demonstrates a theory or conjecture by rejecting the null hypothesis and accepting the alternate hypothesis. Again, this is done because the probability of Type I error is specified and controlled by the researcher when using the classical method of statistical hypothesis testing.

Appendix 8A shows how the probability of a Type II error can be calculated for a specific value selected from the infinite possible values that satisfy the alternative hypothesis. As already discussed in this chapter, the probability of a Type II error is different for each of these possible values. However, there are some things that can be said in general about the probability of Type II error. For a given problem, the larger the sample size, the smaller the probability of a Type II error. If the sample size is large and the null hypothesis is accepted, then the investigator might make some judgment about whether or not the probability of a Type II error is acceptably small because the sample size is large. This is, of course, strictly judgmental. Also, another factor to be considered is that the probability of Type II error decreases as the difference between the true population mean and the hypothesized mean increases. It is quite unlikely that the null hypothesis will be accepted for a reasonably large sample size, if the population mean is considerably different from the hypothesized mean.

8A Appendix: Type II Error in Hypothesis Testing

For most applications, the alternate hypothesis is an open-ended interval. For example, the alternate hypothesis in Example 8.2.2 is $H_a: \mu < 30$. For this case, the alternate hypothesis is true for an infinite number of values of μ. For each of these values, there is a different probability of Type II error, i.e., a different probability of accepting the null hypothesis for each population mean less than 30. Three of the infinite number of values consistent with the alternate hypothesis are selected arbitrarily, and the probability of Type II error is calculated for these values. The values selected are 29.5, 28, and 26. You should stop at this point and make sure it is clear in your mind that, if the population mean should be any one of these three values, it would be a mistake, a Type II error, to accept the null hypothesis, $\mu = 30$.

As researchers limited to the information obtained from sample data, we do not know the value of the population mean. If we perform the hypothesis test in the solution to Example 8.2.2, we are simply asking, "What is the probability that we will accept the null hypothesis if the population mean is in fact 29.5?" The solution to Example 8.2.2 is repeated below.

1. $H_o: \mu = 30$
   $H_a: \mu < 30$.
2. $\alpha = .05$

$$z = \frac{\bar{x} - \mu}{s/\sqrt{n}} = \frac{\bar{x} - 30}{s/\sqrt{n}}.$$

3. Reject H_o if and only if $z_c < -1.65$.

4. $z_c = \dfrac{28.7 - 30}{8/\sqrt{81}} = -1.46$

Do *not* reject H_o.

The conditional value notation from Chapter 3 can be used as a shorthand notation for writing the definition of the probability of a Type II error:

P(Type II error) $= P$(accept $H_o | H_a$ is true).

You will recall from Chapter 3 that the right-hand expression is read, "The probability that H_o will be accepted *given* that H_a is true." This notation can be used to state the probability problem that has to be solved to find the probability of a Type II error if the population mean is the arbitrarily selected value, 29.5. The probability of a Type II error for this value of the mean can be obtained by finding P(accept $H_o | \mu = 29.5$).

The calculation of the probability of a Type II error requires a value for the population standard deviation, σ. For this example, the sample size, $n = 81$, is large; therefore, the sample standard deviation provides us with a good estimate of σ. Accordingly, we will assume σ is approximately equal to 8. If σ is known, we use σ instead of the sample standard deviation, s, in the test statistic in Step 2,

$$z = \frac{\bar{x} - \mu}{\sigma/\sqrt{n}} = \frac{\bar{x} - 30}{8/\sqrt{n}}.$$

Substituting this expression for z_c in the statement of the decision rule results in the following decision rule. (The sample size, 81, is also substituted for n.)

4. Reject H_o if and only if $\dfrac{\bar{x} - 30}{8/\sqrt{81}} < -1.65$.

The only unknown in the decision rule expression is the sample mean $\bar{x}$. Solving for $\bar{x}$, we get

$$\bar{x} < 30 - 1.65(8/\sqrt{81}) \text{ or } \bar{x} < 28.53.$$

Thus, if the sample mean, $\bar{x}$, is less than 28.53 when a sample of size 81 is taken, the calculated value of the test statistic, z_c, will be less than -1.65. The decision rule in Step 4 calls for rejecting the null hypothesis whenever the sample result is such that the calculated value of the test statistic is less than -1.65. Using this information, an equivalent decision rule can be written in terms of $\bar{x}$:

4. Reject H_o if and only if $\bar{x} < 28.53$.

It follows that, for sample mean values greater than or equal to 28.53, the null hypothesis is accepted. The rejection and acceptance values for $\bar{x}$ are illustrated in Figure 8.6.

Now the probability of a Type II error for a specified value of the population mean consistent with the alternate hypothesis can readily be calculated. It is simply a matter of calculating the probability that the sample mean obtained will be greater than 28.53 when the sample is taken from a population with the mean equal to the specified value and standard deviation equal to the estimated value, $\sigma = 8$. The probability of a Type II error when the population mean is equal to 29.5 is obtained by finding $P(\bar{x} \geq 28.53 | \mu = 29.5)$. Using the methods developed in Chapter 6, we find:

$$P(\bar{x} \geq 28.53 | \mu = 29.5) = P\left(z \geq \dfrac{28.53 - 29.5}{8/\sqrt{81}}\right) = P(z \geq -1.09)$$
$$= .8621.$$

The probability of Type II error is very large when $\mu = 29.5$. But 29.5 is quite close to the hypothesized value, $\mu = 30$. Part a of Figure 8.7

Figure 8.6 Values of $\bar{x}$ for Which the Null Hypothesis is Rejected or Accepted

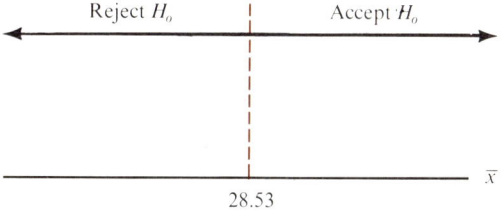

Figure 8.7 Probability of a Type II Error for Different Values of μ

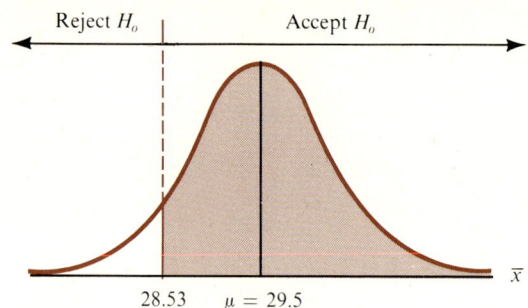

$P(\bar{x} \geq 28.53 \mid \mu = 29.5) = .8621$

b.

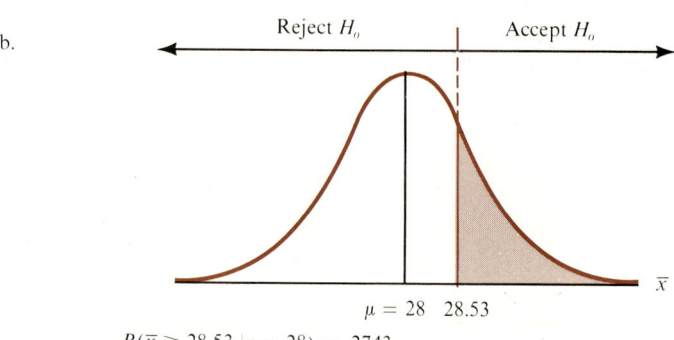

$P(\bar{x} \geq 28.53 \mid \mu = 28) = .2743$

c.

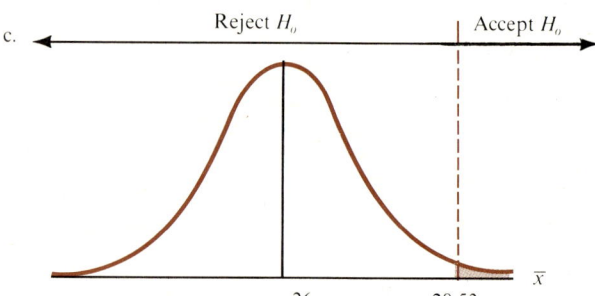

$P(\bar{x} \geq 28.53 \mid \mu = 26) = .0022$

illustrates graphically the above calculation. Parts b and c of Figure 8.7 illustrate how the probability of Type II error is found for specified values of the population mean equal to 28 and 26, respectively. The calculations are given as follows.

$P(\text{Type II error}|\mu = 28) = P(\bar{x} \geq 28.53|\mu = 28) =$

$P\left(z \geq \dfrac{28.53 - 28}{8/\sqrt{81}}\right) = P(z \geq .60) = .2743$

$P(\text{Type II error}|\mu = 26) = P(\bar{x} \geq 28.53|\mu = 26) =$

$P\left(z \geq \dfrac{28.53 - 26}{8/\sqrt{81}}\right) = P(z \geq 2.85) = .0022.$

The probability of accepting $H_o: \mu = 30$, β, decreases as μ decreases. There is a range of values of μ for which, for all practical purposes, $\beta = 0$. The .2743 probability for $\mu = 28$ is considerably smaller than the .8621 value for $\mu = 29.5$. For $\mu = 26$ the probability is less than 1 percent; thus, if the true population mean for this example were less than 26, the probability of accepting the hypothesized value, $\mu = 30$, would be less than .0022.

The sample size has an effect on the probability of a Type II error. If the sample size were increased in this example, then the probability of Type II error would be less for each of the respective alternate values of μ. For example, if $n = 400$, then the rejection region with respect to $\bar{x}$ for this example is calculated as follows:

$\bar{x} < 30 - 1.65 (8/\sqrt{400}) = 30 - .66 = 29.34.$

Thus the null hypothesis is accepted whenever $\bar{x} \geq 29.34$. For $n = 400$ with everything else the same as before, the probability of Type II error, given that $\mu = 29.5$, is

$P(\bar{x} \geq 29.34|\mu = 29.5) = P\left(z \geq \dfrac{29.34 - 29.5}{8/\sqrt{400}}\right)$

$= P(z \geq -.40) = .6554.$

This is considerably less than the .8621 probability for $n = 81$. The reader can verify that the probability of Type II error for $\mu = 28$ is very small, less than .001, for $n = 400$. For $n = 81$ the probability was .2743 when $\mu = 28$.

Problems

1. The pump seals produced by a production line are supposed to have an average outside diameter of 2.76 inches. A sample of 64 seals has a mean of 2.81 inches and standard deviation equal to .6 inches. On the basis of this sample, can you conclude that the average diameter of the seals produced by the production line is different from 2.76 inches? (Use $\alpha = .05$.)

2. A manufacturer of refrigerators claims that his product possesses an average defect-free life of 30 months. You have heard some complaints about his product and therefore feel that the average defect-free life claimed by the manufacturer is somewhat high. In a sample of 85 refrigerators, the average life until repairs were needed was 28.3 months with a standard deviation of 5.0 months. Perform a hypothesis test, testing the manufacturer's claim against your claim. (Use $\alpha = .01$.)

3. A recent survey showed that the average car mileage obtained by car owners in your city is 15.6 miles per gallon. You believe that the gasoline sold by the station in your neighborhood is inferior. You randomly select 80 customers of this station and find that the average mileage for this sample is 14.8 miles per gallon with standard deviation equal to 5.4. Do these sample results substantiate your claim? (Use $\alpha = .03$.)

4. Two years ago an extensive government survey found that 12.6 percent of all secretaries in a large metropolitan area were males. Some are arguing that the percentage of male secretaries is higher today; others argue that it is lower. A local official decides to determine if the proportion is still 12.6 percent by randomly sampling 125 secretaries. Thirteen of the 125 secretaries sampled were male secretaries. Perform the statistical hypothesis test for the local official, using a 6 percent significance level.

5. You feel that you've come up with a package design that customers would prefer to the packaging currently being used by the company. However, top management disagrees with you. They think that the public is indifferent to the two designs and therefore it would not be worth the investment required to change designs. You decide to perform a statistical analysis to demonstrate that more than half of the company's customers would prefer your design. In a sample of 100 customers, 68 selected your package design and 32 selected the current design. Can you conclude that your design would be preferred by the majority of the company's customers? (Use $\alpha = .04$.)

6. You feel that the large retail chain you are working for is a little lax in collection policy. You believe that the percent of overdue accounts is considerably more than the 10 percent claimed by your supervisor. You decide to perform a statistical analysis to support your claims. A random sample of 200 accounts proved to have 32 overdue accounts. Do the statistical results support your claim? (Use $\alpha = .06$.)

7. The following data were randomly selected from a normally distributed population: 5.1, −4.7, 2.3, 1.5. Test the hypothesis that the population mean is equal to 0. (Use $\alpha = .01$.)

8. A cereal-filling machine is properly adjusted if it fills cereal boxes with an average of 20.4 ounces per box. The accountant is concerned that the machine may be overfilling the cereal boxes. A random sample of 16 cereal boxes has a mean of 21.2 ounces and standard deviation equal to .6 ounces. Test the hypothesis that the machine is properly adjusted against the alternative proposed by the accountant. (Use $\alpha = .01$.)

9. As the design engineer, you want to perform a statistical hypothesis test to demonstrate that the new electrical insulators produced by the firm do have an average impact strength greater than the 4.92 average impact strength of the old insulators. A random sample of 17 of the new insulators has a mean impact strength of 5.23 foot-pounds with a standard deviation of 1.18 foot-pounds. Perform the hypothesis test. (Use $\alpha = .05$.)

10. A coin-operated peanut machine was designed to discharge, on the average, 1.5 ounces of peanuts per purchase. To test the machine, 81 purchases of peanuts were made from the machine and weighed. The mean and standard deviation of the 81 measurements were 1.3 and .27 ounces, respectively. Do these data present sufficient evidence to indicate that the mean weight per purchase differs from 1.5 ounces? (Use $\alpha = .01$.)

11. A manufacturer of refrigerators claims that no more than 10 percent of the refrigerators he produces will need repairs before the warranty expires. You think the figure is higher. In a random sample of 400 refrigerators, 58 need repairs before the warranty expires. Test the manufacturer's claim against the alternative you have proposed. (Use $\alpha = .05$.)

12. The MBI computer representative claims that the average time to decode messages between corporate computer centers sent in computer code to avoid espionage is no more than 12 minutes. You want to test his claim with a random sample of 70 messages. The average time for the sample is 13.6 minutes with a standard deviation of 3.1 minutes. (Use $\alpha = .02$.)

13. A filling machine is properly adjusted if it discharges an average of 11.5 ounces each time a can is filled. A sample of 56 cans has a mean of 11.35 ounces and standard deviation of .6 ounces. Is the machine properly adjusted? (Use $\alpha = .02$.)

14. A hospital administrator suggests that 30 percent of the patients admitted to the hospital stay more than 4 days. No recent studies have been done on this subject; so you have been asked to randomly sample 210 recent patients and determine if the sample results are consistent with the administrator's estimated percentage. You find that 70 of the patients sampled stayed longer than 4 days. Perform the indicated hypothesis test. (Use $\alpha = .01$.)

15. A sample poll of 120 members of the AFL-CIO chosen at random reveals that 66 are against trade with Red China. Test the hypothesis that the true proportion opposing trade with Red China is 50 percent, against the alternate hypothesis that it is greater than 50 percent. (Use $\alpha = .10$.)

16. You do not agree with the manufacturer's claim that the average life of the light bulbs he produces is at least 950 hours. You hope to demonstrate your assertion with a statistical test. A random sample of 23 light bulbs has an average life of 927 hours with a standard deviation of 28 hours. Perform the indicated hypothesis test. (Use $\alpha = .05$.)

17. You are concerned that the mean time between failures of the radar system used at the municipal airport is less than the 150 hours claimed by the manufacturer of the system. A random sample of 78 times between failures had a mean value of 143.8 hours and a standard deviation of 12.6 hours. Test the manufacturer's claim against the alternative proposed by yourself at the 1 percent significance level.

18. Last year a survey of all adult members of your city showed that 40 percent of them wanted fluoridated water. A random sample of 150 adults has 72 who want fluoridated water. Based on this sample, can you conclude that the percent of adults who would like fluoridated water is different than it was last year? (Use $\alpha = .05$.)

19. You are concerned that bolts made on a certain screw machine are on the average too small. Too many of the bolts are loose when used in the production line. The average diameter of the bolts is supposed to be 1.2 inches. A sample of 121 bolts has a mean diameter of 1.15 inches with a standard deviation of .22 inches. Do these sample results support your belief that the average diameter of bolts produced by the screw machine is less than 1.2 inches? (Use $\alpha = .03$.)

20. An insurance industry safety-research group wants the government to take action against companies making allegedly unsafe wheels and rims on school buses and medium-sized trucks. The manufacturer's trade association claims that at most 1 percent are defective. The safety-research group disputes this claim and randomly selects 500 wheels and rims. Nine of the 500 sampled prove to be defective. Perform the indicated hypothesis test for the safety-research group. (Use $\alpha = .04$.)

21. The manager of an exclusive department store suggests that 20 percent of the customers coming into the store make purchases totaling more than $1,000.00. No recent studies have been done on this subject; so you have been asked to randomly sample 200 customers and determine if the sample results are consistent with the

manager's estimated percentage. You find that 48 of the customers sampled made purchases totaling more than $1,000.00. Perform the indicated hypothesis test. (Use $\alpha = .07$.)

22. The motor shafts produced by a production line are supposed to have an average length of 6.45 inches. A random sample of 81 shafts has a mean equal to 6.60 inches and a standard deviation of .60 inches. Is the average length of shafts produced by this production line equal to 6.45 inches? (Use $\alpha = .06$.)

23. A marketing executive wants to test the effect of background music. The average customer purchase was $23.82 for the last few months. The executive changes the sound system to softer music with brief periodic silent periods. The average of 125 purchases made during this period was $25.16 with a standard deviation of $6.88. Can we conclude that the experimental background music has a positive effect on the amount purchased per customer at the .05 significance level?

24. A special electronics component required an average of 5.7 minutes assembly time in the past. The design engineer made a slight change that he thinks will reduce assembly time. A random sample of 75 components has a mean value, $\bar{x} = 5.48$ minutes, with a standard deviation, $s = .73$. Based on this sample, can we conclude that average assembly time has been reduced by the change? (Use $\alpha = .02$.)

25. The internal auditor is concerned about the accuracy of bills sent out to customers. She randomly samples 15 invoices ready to be mailed out and obtains the following error amounts: $0, 2.68, -1.41, 0, .25, .67, 5.12, 0, -6.10, -2.05, 6.83, 0, -.91, 0, .06$. A positive discrepancy means the billing clerk calculated the bill in excess of what should have been charged. A negative amount indicates an undercharge. Use these sample results to test the hypothesis that the clerical billing errors balance each other out: i.e., test the hypothesis that the mean of all billing errors is equal to 0. (Use $\alpha = .01$.)

26. In the previous problem the invoices were checked for errors before they were sent out. The auditor is now concerned that the actual amount collected may not be equal to the amount billed because the customer may recalculate the bill. The auditor's concern is that a majority of customers may only make the correction one way, i.e., pay the amount of the invoice when it is understated, but pay the recalculated correct amount for the overstated invoices. She wants to perform a hypothesis test to prove her claim. An average underpayment for all invoices would be consistent with her claim. She randomly samples 65 invoices and for each invoice calculates the difference between the actual payment made

and what the correct payment should have been. The average of the 65 differences was -1.63 with a standard deviation of $2.17. Can the auditor conclude that the clerical billing errors result in an average underpayment for all invoices? (Use $\alpha = .05$.)

27. One policy of a large retail outlet is that defective merchandise will be replaced. The manager is particularly perturbed with the manufacturer of an electric car-racing set that was a popular Christmas item. The manufacturer assured the retailer that his quality control was so good that no more than 1 percent of the electric car-racing sets would have operational defects. However, 19 of the 568 sets sold by the large retail outlet during the Christmas season were returned with serious operational defects right after the holiday. If the 568 sets sold are treated as a random sample, can it be concluded, at the 1 percent significance level, that the manufacturer's quality control methods are not as good as claimed?

28. A new branch manager for a large bank would like to spot-check the accuracy of her tellers' calculations at the end of the day. She realizes that because she only has time to take a small sample, there is a high probability of a Type II error (she will accept the tellers' calculations even if they should be wrong); however, for her own satisfaction she would like to do some spot checking in case there are some flagrant errors. Because she is a new boss and also because it isn't her job to audit the tellers' calculations, she doesn't want to make any accusations unless she feels pretty good about her sample results; thus, she wants to set a 1 percent significance level. Based on the number of checks handled and the total dollar amount given, the average amount per check of one of the tellers should be $43.82. The manager feels that this teller is easily distracted from his work and may, therefore, make some errors. A random sample of 7 checks produced the following dollar amounts: $36.48, 2.75, 161.06, 58.49, 16.17, 28.67, and 63.43. Perform the hypothesis test for the manager.

29. A union organizer wants to determine if the average wage of blue collar employees for a large nonunion company is less than $7.00 per hour. The company considers this confidential information; so the organizer is forced to use sampling methods. He is worried that a sample of those workers who readily cooperate in giving him the desired information will be downward biased. He chooses to select a smaller random sample and concentrate his efforts on obtaining the desired information from each of those randomly selected. The 10 employees selected had the following hourly wages: $5.83, 7.61, 6.43, 5.94, 6.56, 7.11, 6.23, 6.75, 5.98, and 6.05. Can the union organizer conclude at the 5 percent level of significance that the average hourly wage of blue collar workers employed by the company is less than $7.00 per hour?

30. A telephone sales manager claims that, if you use his introductory sentence, at least 80 percent of the people approached via the telephone will listen to the remainder of your sales pitch. You believe that this percentage is too high. In a random sample of 136 people contacted by phone, 41 hung up before you got into the body of your sales pitch even though you used the manager's introductory sentence. Perform the hypothesis test to test the manager's claim against the alternate proposed by you. (Use $\alpha = .03$.)

31. The total inventory value for 7,138 items is reported to be $453,834.04. If this latter figure is accurate, then the average value per item in inventory should be $63.58. A random sample of 80 inventory items has a mean value of $61.23 with a standard deviation equal to $16.13. Based on these sample results, can the auditor conclude that the mean value of all items in inventory is different from the reported value of $63.58 per item at the 3 percent significance level?

32. You want to determine if the average mileage per gallon your new car attains is equal to 18.4 miles per gallon. You arrived at the 18.4 mile figure by discounting the EPA estimate, 23 mpg, by 20 percent. The miles per gallon from 5 different tankfuls of gas were as follows: 16.2, 18.6, 17.5, 15.3, and 17.9 miles per gallon. Based on these sample results, can you conclude that your new car is performing at less than the expected level of 18.4 miles per gallon? (Use $\alpha = .10$.)

33. With the price of copper at an all-time high, the materials manager is concerned that the 50-pound rolls of copper wire he buys on a continuous basis for use in the manufacture of electric motors may weigh on the average less than the 50 pounds the firm is paying for. A sample of 78 rolls randomly selected from the previous month's purchases has a mean weight of 49.6 pounds and a standard deviation of .8 pound. Perform the appropriate hypothesis test for the materials manager, using a significance level of 1 percent.

34. The mayor of Jonesburg estimates that the average convention attendee spends $30.00 per day in Jonesburg. You feel that his figure is too low. A random sample of 49 convention attendees has a mean of $28.00 and a standard deviation of $14.00. Set up the hypothesis test to test the mayor's claim against the alternative hypothesis consistent with your feeling about the mayor's estimate. (Use $\alpha = .08$.)

■ The remaining problems require material from the appendix.

35. The mean lifetime of a sample of 100 light bulbs is computed to be 1,630 hours with a sample standard deviation of 100 hours. The

hypothesis to be tested is $\mu = 1,600$ hours against the alternate hypothesis, $\mu > 1,600$ hours. (Use $\alpha = .05$.)
   a. What is the probability of a Type II error if the true population mean is 1,640 hours?
   b. Consider the above problem with the alternate hypothesis changed to $\mu \neq 1,600$; everything else remains the same. What is the probability of a Type II error if the true population mean is 1,605 hours?

36. A plastic injection mold is working properly if the average amount of plastic used per component produced is equal to 15 ounces. A sample of 100 components has an average weight of 15.3 ounces and a standard deviation of 2.0 ounces.
   a. If the mold is injecting 15.4 ounces, what is the probability of a Type II error if the hypothesis test is performed with the alternate hypothesis, $H_a: \mu > 15$, and $\alpha = .02$?
   b. If everything is the same as in part a, except that the alternate hypothesis is $H_a: \mu \neq 15$, what is the probability of a Type II error?

37. Refer to Problem 3. Find the probability of a Type II error if the mean mileage of all customers buying gas from the station in your neighborhood is 15.0 miles per gallon.

38. Refer to Problem 31. If the value of the total inventory is in fact equal to $442,556.00, i.e., average cost per item is $62.00,
   a. what is the probability that the hypothesis test in the solution to Problem 31 will result in accepting the hypothesis that the mean value per item is $63.58;
   b. what is the probability of a Type II error if the true inventory value is $64.00 per item;
   c. what is the probability of a Type II error if the true inventory value is $68.50 per item?

39. If the sample size in Problem 37 is doubled to 160, what is the probability of a Type II error? What is the probability of a Type II error in Problem 37 if the sample size is quadrupled to 320?

CHAPTER 9

STATISTICAL INFERENCE: TWO POPULATIONS

In this chapter we continue the discussion of statistical inference, extending the concepts of estimation and hypothesis testing to include the estimation of differences between population parameters and also hypothesis testing of possible differences.

Specific objectives are—

to present concepts and techniques of testing hypotheses related to the comparison of two population means (Sections 1 and 2).

to present methods of testing hypotheses about two population proportions (Section 3).

to present procedures for estimating the difference between two population means given the corresponding sample means, and to present the parallel procedures for two population proportions (Section 4).

The topics covered in this chapter could have been included in the chapters on statistical estimation and hypothesis testing (Chapters 7 and 8). There is a fundamental difference, however, between the material in this chapter and that in the last two chapters. This chapter deals with the problem of comparing the parameters of two populations where sample data are the only information available. Up to now we have been concerned with making statements about one population. Because we are dealing with two populations and therefore two different samples, one from each population, the test statistics and the confidence interval formulas are more complex and somewhat more cumbersome than those for the single-population case. The basic concepts of hypothesis testing and statistical estimation are the same for the two-population case as those for the single-population case.

The primary emphasis in dealing with two populations is on hypothesis testing; therefore, hypothesis testing is discussed first in this chapter. There are business and economic applications, however, where it is necessary to estimate the difference between two population parameters. Confidence interval formulas for estimating the difference between two population parameters are readily derived from the test statistics used for hypothesis testing. The latter part of this chapter is concerned with estimation.

9.1 Testing Hypotheses about Two Population Means—Large Sample Size

In this section we are concerned with testing the hypothesis that the population means of two different populations have the same value. The null hypothesis for this problem is stated as follows:

$H_o: \mu_1 = \mu_2$

where

μ_1 = the mean of the first population of interest
μ_2 = the mean of the second population of interest.

Note that it is not necessary to specify particular values for the population means. The concern here is whether or not the population means are equal.

In Chapter 8 we considered the problem of determining whether or not the average life of batteries produced by a particular firm was equal to a specified value, 30 months, as claimed by the manufacturer. The population of interest was all the batteries produced by the particular firm. In this chapter we are concerned with a different type of problem. The problem of interest is whether or not the average life of batteries produced by two different firms is the same for both firms.

Example 9.1.1

We wish to determine if the average life of batteries produced by the NEB Battery Company and the average life of batteries produced by the OK Battery Company are the same. A random sample of 56 batteries produced by NEB has an average life of 26.4 months with a standard deviation of 6.4 months. The average life of a sample of 68 OK batteries was 28.9 months with a standard deviation of 5.3 months. Test the hypothesis that the average life is the same for the two makes of batteries. Use a significance level of 5 percent.

The first population of interest is all the batteries produced by the NEB Battery Company and the second population of interest is all the batteries produced by the OK Battery Company. If we label the mean of the first population, μ_1, and the second, μ_2, then the null hypothesis for this problem can be written as indicated above:

$H_o: \mu_1 = \mu_2$.

We observe that the following is an equivalent hypothesis:

$H_o: \mu_1 - \mu_2 = 0$.

If the means are equal, then their difference must be equal to 0. This latter observation suggests a sample statistic that could be used to test the above hypothesis, namely, the difference of the two values of the two sample means obtained from the respective populations:

$D = \bar{x}_1 - \bar{x}_2$

where

$\bar{x}_1$ = the mean of the sample randomly selected from the first population
$\bar{x}_2$ = the mean of the sample randomly selected from the second population.

It is important that the two samples be taken independently of each other. If they are not independent samples, then the test statistic used in this section would not be applicable. (If the two samples are not independent, it is possible we might be able to use a "paired difference test." This is discussed in Appendix 9A.) The difference of the two sample means, $D = \bar{x}_1 - \bar{x}_2$, is a linear function of the two random variables $\bar{x}_1$ and $\bar{x}_2$. The expected value theorems in Appendix 4A can be used to show that the mean of the difference of the sample means,

μ_D or $\mu_{\bar{x}_1 - \bar{x}_2}$,

is equal to the difference of the means of the two populations from which the two samples were drawn:

$$\mu_{\bar{x}_1 - \bar{x}_2} = \mu_1 - \mu_2.$$

If the samples are selected independently of each other, as assumed here, then the variance theorems in Appendix 4A can be used to obtain

$$\sigma^2_{\bar{x}_1 - \bar{x}_2},$$

the variance of the difference of the two sample means:

$$\sigma^2_{\bar{x}_1 - \bar{x}_2} = \sigma^2_{\bar{x}_1} + \sigma^2_{\bar{x}_2}.$$

Recalling that

$$\sigma^2_{\bar{x}} = \frac{\sigma^2}{n},$$

we can write the above as

$$\sigma^2_{\bar{x}_1 - \bar{x}_2} = \frac{\sigma_1^2}{n_1} + \frac{\sigma_2^2}{n_2}.$$

The standard deviation is defined as the square root of the variance, resulting in

$$\sigma_{\bar{x}_1 - \bar{x}_2} = \sqrt{\frac{\sigma_1^2}{n_1} + \frac{\sigma_2^2}{n_2}}$$

as the formula for the standard deviation of the difference of the two sample means.

As was the case in Chapter 8, the population standard deviation is generally not known when the population mean is not known. It is necessary to use the sample standard deviation, s, as an estimate of the population standard deviation, σ; thus the estimated standard deviation of the difference of the sample means,

$$\hat{\sigma}_{\bar{x}_1 - \bar{x}_2},$$

is a function of the sample standard deviations:

$$\hat{\sigma}_{\bar{x}_1 - \bar{x}_2} = \sqrt{\frac{s_1^2}{n_1} + \frac{s_2^2}{n_2}}$$

where

s_1 = the standard deviation of the first sample
s_2 = the standard deviation of the second sample
n_1 = the number of observations in the first sample

n_2 = the number of observations in the second sample.

The sample statistics described above can be used to develop a test statistic for hypothesis testing. You will recall from Chapter 8 that the test statistic was equal to the sample statistic, $\bar{x}$, minus its expected value, μ, divided by the estimate standard deviation of $\bar{x}$, $s/\sqrt{n}$; i.e.,

$$z = \frac{\bar{x} - \mu}{\frac{s}{\sqrt{n}}}.$$

A corollary to the Central Limit Theorem permitted the use of the standard normal distribution when the sample size n was large. A similar corollary applies in the case of $\bar{x}_1 - \bar{x}_2$ when both sample sizes n_1 and n_2 are large. As was done in Chapter 8, the desired test statistic is obtained by taking the sample statistic, $\bar{x}_1 - \bar{x}_2$, subtracting its mean or expected value,

$$\mu_{\bar{x}_1 - \bar{x}_2} = \mu_1 - \mu_2,$$

and dividing the result by the estimated standard deviation of

$$\bar{x}_1 - \bar{x}_2,$$
$$\hat{\sigma}_{\bar{x}_1 - \bar{x}_2}.$$

Test Statistic for Large Sample Size

This results in the following test statistic for testing the hypothesis of the equality of two population means when both sample sizes are large:

$$z = \frac{(\bar{x}_1 - \bar{x}_2) - (\mu_1 - \mu_2)}{\sqrt{\frac{s_1^2}{n_1} + \frac{s_2^2}{n_2}}}.$$

You will recall from the previous chapter that the hypothesis test is set up assuming that the null hypothesis is true and then observing whether or not the sample results are consistent with this assumption. If the null hypothesis, $H_0: \mu_1 = \mu_2$, is true, then $\mu_1 - \mu_2 = 0$, as has already been noted above. Substituting this value in the above expression gives us the final test statistic:

$$z = \frac{\bar{x}_1 - \bar{x}_2}{\sqrt{\frac{s_1^2}{n_1} + \frac{s_2^2}{n_2}}}.$$

We are now ready to proceed with the hypothesis test requested in Example 9.1.1 above. The hypothesis test is performed following the four steps outlined in Chapter 8:

1. $H_o: \mu_1 = \mu_2$
    $H_a: \mu_1 \neq \mu_2.$

2. $\alpha = .05$

$$z = \frac{\bar{x}_1 - \bar{x}_2}{\sqrt{\frac{s_1^2}{n_1} + \frac{s_2^2}{n_2}}}.$$

3. Reject H_o if and only if $z_c > 1.96$ or $z_c < -1.96$.

4. $z_c = \dfrac{\bar{x}_1 - \bar{x}_2}{\sqrt{\dfrac{s_1^2}{n_1} + \dfrac{s_2^2}{n_2}}} = \dfrac{26.4 - 28.9}{\sqrt{\dfrac{(6.4)^2}{56} + \dfrac{(5.3)^2}{68}}} = -2.34.$

Reject H_o.

We conclude that the batteries produced by the two firms are different with respect to length of life.

The alternate hypothesis in the above example was the not-equal-to hypothesis because we were simply asked to determine if the two brands were the same with respect to length of life. We did not specify one of the firms as having a superior product, in which case we would have had a different alternate hypothesis. The decision rule was determined from the standard normal table in exactly the same manner as in Chapter 8. The alternate hypothesis indicates this is a two-tail test; so the significance level, $\alpha = .05$, was split into two tails, with the result that we looked for the z-value corresponding to a value of .4750 in the body of the standard normal table as indicated in Figure 9.1. The calculations in Step 4 were made using the values calculated from the two samples as given in the example.

A one-tail hypothesis test is also possible when comparing two population parameters. It may be desired to demonstrate that the mean of a specified population is greater than the mean of another population.

Example 9.1.2 A credit manager trainee feels that the customers from the well-to-do north end of the city may actually take longer to pay off overdue

Figure 9.1 Determining the value of z for a Two-Tail Test with $\alpha = .05$

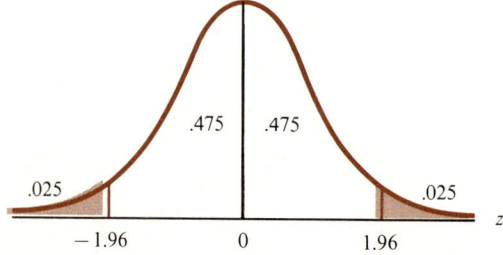

accounts than the customers from the less affluent south section of the city. A random sample of 45 customers from the north end of the city who paid up on overdue accounts had an average time of 14.6 days until the overdue account was paid with a standard deviation of 4.7 days. A random sample of 70 of the same type of customers from the south end of the city had an average time of 11.7 days and a standard deviation of 5.1 days. Test the hypothesis that the north-end customers take longer than the south-end customers to pay off overdue accounts. Use $\alpha = .02$.

It should be evident to the reader that it is strictly arbitrary which of the populations is labeled number one; but, when performing a one-tail test as required by this particular example, it is important to keep track of which population was arbitrarily labeled number one. For this example, we arbitrarily let the north-end customers be labeled as population number one. Thus, we have

μ_1 = the mean time to pay off an overdue account for all customers from the north end of the city
μ_2 = the mean time to pay off an overdue account for all customers from the south end of the city.

The credit manager trainee is performing the hypothesis test in an effort to demonstrate her claim that the average time for settling overdue accounts is greater for north-end customers than for south-end customers. As was the case in the previous chapter, if the person performing the statistical study has a bone to pick, it is reflected in the alternate hypothesis. For this example we have

$H_a: \mu_1 > \mu_2$

for the alternate hypothesis.

The decision rule is determined by putting the entire value of the significance level, $\alpha = .02$, in the upper tail as indicated in Figure 9.2. The argument for doing this is the same as that given in Chapter 8. The

Figure 9.2 Determining the Value of z for a One-Tail Test with $\alpha = .02$

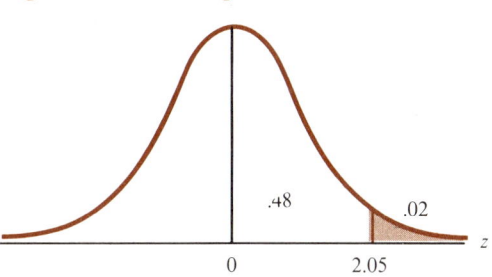

researcher will only consider rejecting the null hypothesis if the sample results are consistent with the alternate hypothesis; in this case, if $\bar{x}_1$ is greater than $\bar{x}_2$. He or she will reject H_o only for positive values of z_c (the numerator of the test statistic, $\bar{x}_1 - \bar{x}_2$, is positive when $\bar{x}_1$ is greater than $\bar{x}_2$); so the total value of α is included in the upper tail. The complete hypothesis test is as follows:

1. $H_o: \mu_1 = \mu_2$
   $H_a: \mu_1 > \mu_2$.

2. $\alpha = .02$

$$z = \frac{\bar{x}_1 - \bar{x}_2}{\sqrt{\dfrac{s_1^2}{n_1} + \dfrac{s_2^2}{n_2}}}.$$

3. Reject H_o if and only if $z_c > 2.05$.

4. $z_c = \dfrac{\bar{x}_1 - \bar{x}_2}{\sqrt{\dfrac{s_1^2}{n_1} + \dfrac{s_2^2}{n_2}}} = \dfrac{14.6 - 11.7}{\sqrt{\dfrac{(4.7)^2}{45} + \dfrac{(5.1)^2}{70}}} = 3.12.$

Reject H_o.

The credit manager concludes that the average time for the more affluent customers from the north end of the city to pay overdue accounts is greater than that for the south-end customers.

9.2 Testing Hypotheses about Two Population Means—Small Sample Size

The test statistic in the previous section depended upon the requirement that both sample sizes be large. A large sample size generally means a sample size greater than 30 (as was suggested for a rule of thumb in previous chapters). If one of the sample sizes is less than 30, then the test statistic given above would not be appropriate for performing a hypothesis test concerning the equality of two population means. The t-statistic can be used if the following assumptions are satisfied:

1. Both populations are normally distributed.
2. The population variances are the same.
3. The respective samples are selected independently of each other.

Test Statistic when Sampling from Normal Populations

If the above assumptions are satisfied then the test statistic,

$$t = \frac{\bar{x}_1 - \bar{x}_2}{\sqrt{\frac{(n_1 - 1)s_1^2 + (n_2 - 1)s_2^2}{n_1 + n_2 - 2}}\sqrt{\frac{1}{n_1} + \frac{1}{n_2}}}$$

with

$df = n_1 + n_2 - 2,$

can be used for testing the hypothesis that two population means are equal. Note that the degrees of freedom are a function of both sample sizes and are equal to $n_1 + n_2 - 2$.

The test statistic given here appears to be slightly more complex than that for the large sample case. The derivation depends upon the three assumptions given above, and they should be kept in mind. If the first two assumptions are only approximately true, then this test statistic will still give satisfactory results. If the population variances are not equal but their ratio is known, it is possible to use a modification of the above test statistic (not covered in this text). Fortunately, there are many applications where it is not unreasonable to assume that the variances of the two populations of interest are approximately equal.

Example 9.2.2

A purchasing agent wants to determine whether the fuel provided by two suppliers is the same with regard to performance in engines that the firm manufactures. Eight engines were randomly selected, and each was run on the same amount of fuel obtained from supplier A. The following running times, in minutes, were recorded for the eight engines:

18.1, 22.6, 19.2, 19.3, 21.5, 18.8, 20.4, and 19.7 minutes.

Nine engines were randomly selected and run on equivalent amounts of fuel from supplier B. The following running times, in minutes, were recorded for the nine engines:

20.3, 18.9, 22.4, 24.7, 19.8, 21.6, 22.2, 21.3, and 23.2 minutes.

Are the two fuels equivalent with regard to engine running time? Use a significance level of 10 percent.

At least 1 of the sample sizes (in this problem both of them) is small, less than 30. Therefore, the z-statistic is not applicable for this problem. The t-statistic can be used if the 3 conditions given for its use are satisfied or approximately satisfied: normal populations, equal variances, and independent samples. The last condition is satisfied because the 8 engines in the first sample were selected randomly and independently of the selection of the 9 engines used in the second example. Differences that arise in mechanical components, such as an engine, do tend to follow a normal distribution. Inasmuch as most of the variability for the 2 populations, engines run on fuel A and engines run on fuel B, is due to the engines themselves, it is reasonable to assume that the population variances are approximately equal.

The question asked is whether or not the two means are equal; therefore, the alternate hypothesis is the not-equal-to hypothesis. The solution to this example is given below. Step 4, the calculation step, is somewhat lengthy because it is necessary to calculate the sample means and sample variances from the raw sample data.

1. $H_o: \mu_1 = \mu_2$
   $H_a: \mu_1 \neq \mu_2$.

2. $\alpha = .10$

$$t = \frac{\bar{x}_1 - \bar{x}_2}{\sqrt{\frac{(n_1 - 1)s_1^2 + (n_2 - 1)s_2^2}{n_1 + n_2 - 2}} \sqrt{\frac{1}{n_1} + \frac{1}{n_2}}}$$

with

$$df = n_1 + n_2 - 2.$$

3. Reject H_o if and only if $t_c < -1.753$ or $t_c > 1.753$.

4. $\bar{x}_1 = \frac{\Sigma x_i}{n} = \frac{18.1 + 22.6 + \cdots + 19.7}{8} = \frac{159.6}{8} = 19.95$

$$s_1^2 = \frac{\Sigma(x_i - \bar{x})^2}{n - 1}$$

$$= \frac{(18.1 - 19.95)^2 + (22.6 - 19.95)^2 + \cdots + (19.7 - 19.95)^2}{8 - 1}$$

$$= \frac{15.42}{7} = 2.2029$$

$\bar{x}_2 = \frac{\Sigma x_i}{n} = \frac{20.3 + 18.9 + \cdots + 23.2}{9} = \frac{194.4}{9} = 21.6$

$$s_2^2 = \frac{\Sigma(x_i - \bar{x})^2}{n - 1}$$

$$= \frac{(20.3 - 21.6)^2 + (18.9 - 21.6)^2 + \cdots + (23.2 - 21.6)^2}{9 - 1}$$

$$= \frac{25.48}{8} = 3.185$$

$$t_c = \frac{\bar{x}_1 - \bar{x}_2}{\sqrt{\frac{(n_1 - 1)s_1^2 + (n_2 - 1)s_2^2}{n_1 + n_2 - 2}} \sqrt{\frac{1}{n_1} + \frac{1}{n_2}}}$$

$$= \frac{19.95 - 21.6}{\sqrt{\frac{(8 - 1)(2.2029) + (9 - 1)3.185}{8 + 9 - 2}} \sqrt{\frac{1}{8} + \frac{1}{9}}}$$

$$= \frac{-1.65}{\sqrt{2.727} \sqrt{.236}} = \frac{-1.65}{(1.651)(.486)} = -2.056.$$

Reject H_o.

Based on the sample results, the purchasing agent concludes that the two fuels perform differently in the engines manufactured by his firm.

When using the above test statistic, it is important to remember that the degrees of freedom are a function of both sample sizes. For the above example, the degrees of freedom total 15: $n_1 + n_2 - 2 = 8 + 9 - 2 = 15$. This is a two-tail test, so the α-value, $\alpha = .10$, is split between the two tails as indicated in Figure 9.3. The t-table (Table C) at the back of this text is a one-tail table, and the 1.753 value used in Step 3 above was obtained by looking under the column headed .05 and across the row for 15 degrees of freedom. This is the same procedure followed in Chapter 8 when the t-statistic was used for single-population hypothesis testing with small sample size.

The test statistic used in this section was obtained using a theorem from mathematical statistics concerning the ratio of a standard normal distribution to the square root of a chi-square distribution divided by its degrees of freedom. Nothing would be gained by presenting the derivation of the test statistic here; but the discussion of the pooled variance concept in Appendix 9C may help make the test statistic used above somewhat more tractable and understandable to you.

Figure 9.3 Two-Tail Test with $df = 15$ and $\alpha = .10$

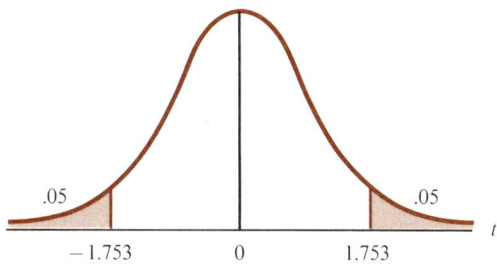

9.3 Testing Hypotheses about Two Population Proportions

Testing hypotheses concerning two population proportions is analogous to testing hypotheses concerning two population means, with the exception that the t-statistic cannot be used when sample sizes are small because the t-statistic requires sampling from a normal population. As in Chapter 8 for the single population case, hypothesis testing for the two-population case will be discussed only for the large-sample-size case; i.e., both sample sizes must be large. The null hypothesis for testing the equality of two population proportions is written as:

$H_o: \pi_1 = \pi_2$.

The symbol π_1 represents the first population proportion or the proportion of items in the first population having the characteristic of interest. Similarly, π_2 is the proportion of items in the second population having the characteristic of interest. We observe, as was the case for the means, that

$H_o: \pi_1 - \pi_2 = 0$

is an equivalent hypothesis.

This latter observation suggests a sample statistic similar to the one used above for the two-population-means hypothesis tests. The sample statistic used for proportion problems is

$$D = p_1 - p_2$$

where

p_1 = the proportion of items having the characteristic of interest in the sample randomly selected from the first population, and

p_2 = the proportion of items having the characteristic of interest in the sample randomly selected from the second population.

The expected value of the expression $p_1 - p_2$, $\mu_{p_1 - p_2}$, can be derived using the expected value theorems in Appendix 4A of Chapter 4:

$\mu_{p_1 - p_2} = \pi_1 - \pi_2$.

This is of course analogous to the case where $\bar{x}_1 - \bar{x}_2$ was the sample statistic. The analogy continues on with the standard deviation, $\sigma_{p_1 - p_2}$, of the sample statistic $p_1 - p_2$:

$$\sigma_{p_1 - p_2} = \sqrt{\sigma_{p_1}^2 + \sigma_{p_2}^2} = \sqrt{\frac{\pi_1(1 - \pi_1)}{n_1} + \frac{\pi_2(1 - \pi_2)}{n_2}}.$$

In Chapter 6 the variance of the sample proportion, σ_p^2, was stated to be $\sigma_p^2 = \pi(1-\pi)/n$. This formula was used to determine $\sigma_{p_1}^2$ and $\sigma_{p_1}^2$ in the right-hand side of the above expression for $\sigma_{p_1-p_2}$.

The Central Limit Theorem permits us to use the sample estimate of $\sigma_{p_1-p_2}$, as shown below:

$$\hat{\sigma}_{p_1-p_2} = \sqrt{\frac{p_1(1-p_1)}{n_1} + \frac{p_2(1-p_2)}{n_2}}.$$

This is obtained by replacing the population proportions, π_1 and π_2, with their respective sample estimates, p_1 and p_2, in the above formula.

Test Statistic for Two Population Proportions

Thus, a corollary to the Central Limit Theorem states that the following expression has an approximately standard normal distribution or z-distribution when both sample sizes are large:

$$z = \frac{(p_1 - p_2) - (\pi_1 - \pi_2)}{\sqrt{\frac{p_1(1-p_1)}{n_1} + \frac{p_2(1-p_2)}{n_2}}}.$$

This provides us with a test statistic to be used for testing hypotheses about two population proportions when both sample sizes are large. Note that this is really no different from any other z-type test statistic used in this chapter and in Chapter 8. It is equal to the sample statistic, $p_1 - p_2$, minus its expected value, the difference being divided by the estimated standard deviation of the sample statistic.

Example 9.3.1

A random sample of 200 workers in France had 152 workers that supported the Common Market concept. In West Germany, 207 workers in a random sample of 300 were in favor of the Common Market concept. Is the proportion of workers favoring the Common Market concept the same for both countries? Test at the 3 percent significance level.

The solution to this problem is given below. In Step 2 the test statistic does not have the expression $\pi_1 - \pi_2$ in the numerator because, if the null hypothesis is true, the population proportions are equal and their difference is equal to 0.

1. $H_o: \pi_1 = \pi_2$
   $H_a: \pi_1 \neq \pi_2$.

Testing Hypotheses about Two Population Proportions

2. $\alpha = .03$

$$z = \frac{p_1 - p_2}{\sqrt{\frac{p_1(1-p_1)}{n_1} + \frac{p_2(1-p_2)}{n_2}}}.$$

3. Reject H_o if and only if $z_c < -2.17$ or $z_c > 2.17$.

4. $p_1 = \frac{x_1}{n_1} = \frac{152}{200} = .76$

$p_2 = \frac{x_2}{n_2} = \frac{207}{300} = .69$

$$z_c = \frac{p_1 - p_2}{\sqrt{\frac{p_1(1-p_1)}{n_1} + \frac{p_2(1-p_2)}{n_2}}} = \frac{.76 - .69}{\sqrt{\frac{(.76)(.24)}{200} + \frac{(.69)(.31)}{300}}}$$

$$= \frac{.07}{.0403} = 1.74.$$

Do *not* reject H_o.

Based on these sample results, we cannot conclude that there is a difference in the attitude of all French workers and all West German workers with regard to the Common Market. The decision rule is found in exactly the same manner as before. Once it has been determined that the test statistic has an approximate z-distribution, then the standard normal table can be used to determine that 2.17 should be used in the decision rule by looking for the z-value most closely corresponding to .4850 in the body of Table B as indicated in Figure 9.4. The significance level, $\alpha = .03$, was split between the 2 tails because the alternate hypothesis calls for a two-tail test.

As with two population means, it is also possible to perform a one-tail test with two population proportions. If the researcher is interested in whether or not a specific population proportion is greater

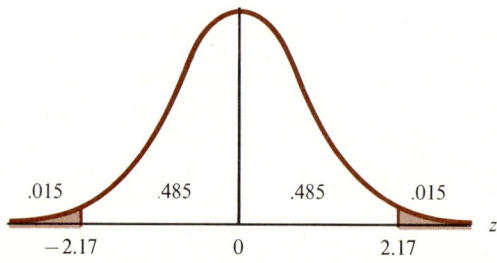

Figure 9.4 Determining the Value of z for a Two-Tail Test with $\alpha = .03$

(or less) than another population proportion, then a one-tail test would be appropriate.

Example 9.3.2 A marketing manager is interested in cooking demonstrations that are being offered in a large metropolitan area. In one of the two-hour "free to the public" cooking classes, certain brand-name products were specifically used in the recipes, and the instructor made off-the-cuff promotional comments concerning these brand-name products. The manager feels that this might have had a negative effect on the audience. She wants to compare the audience reaction to this class with response to another cooking class that was essentially the same but had no sly promotional pitches for brand-name food products. A random sample of 145 people attending the session promoting brand-name products had 70 people saying it was enjoyable. At the second cooking class, where no brand names were promoted, 215 people were randomly selected; 143 of those sampled said they enjoyed the session. Do these data support the marketing manager's contention that the promotion of brand-name food products during a cooking class makes the class less enjoyable? Perform the hypothesis test at a 5 percent significance level.

Let π_1 be the proportion of all people attending classes where brand-name products were promoted who found the class enjoyable. Let π_2 be the proportion of all people attending similar cooking classes without brand-name promotional pitches who found the class enjoyable. Then the set of hypotheses for this example would be stated as follows:

$H_o: \pi_1 = \pi_2$
$H_a: \pi_1 < \pi_2$.

The marketing manager is requesting the statistical analysis, and the alternate hypothesis reflects the point she is trying to prove. The complete hypothesis test is presented below.

1. $H_o: \pi_1 = \pi_2$
   $H_a: \pi_1 < \pi_2$.
2. $\alpha = .05$

$$z = \frac{p_1 - p_2}{\sqrt{\frac{p_1(1 - p_1)}{n_1} + \frac{p_2(1 - p_2)}{n_2}}}.$$

3. Reject H_o if and only if $z_c < -1.65$.
4. $p_1 = \frac{x_1}{n_1} = \frac{70}{145} = .483$

   $p_2 = \frac{x_2}{n_2} = \frac{143}{215} = .665$

$$z = \frac{p_1 - p_2}{\sqrt{\frac{p_1(1-p_1)}{n_1} + \frac{p_2(1-p_2)}{n_2}}} = \frac{.483 - .665}{\sqrt{\frac{(.483)(.517)}{145} + \frac{(.665)(.335)}{215}}}$$

$$= \frac{-.182}{.053} = -3.43.$$

Reject H_o.

The null hypothesis is rejected in favor of the alternate hypothesis that π_1 is less than π_2. The marketing manager concludes that the brand-name promotion during a cooking class does have a detrimental effect on the participants' enjoyment of the class.

The decision rule in Step 3 was determined in the same manner as that for other one-tail hypothesis tests discussed in this and the previous chapter. The sample result is consistent with the alternate hypothesis only when p_1 is less than p_2 or when $p_1 - p_2$ is a negative number. When $p_1 - p_2$ is negative, then the calculated value of z is negative. Inasmuch as we consider rejecting the null hypothesis only when the sample results are consistent with the alternate hypothesis, we will only reject H_o for relatively large negative calculated values of z. The entire value of α, $\alpha = .05$, is placed in the lower tail as indicated in Figure 9.5, and the appropriate value of z is found from the standard normal table.

9.4 Estimating the Difference between Two Population Parameters

There are applications in business and government that call for an estimate of the difference between two population means; i.e., they call for an answer to the question, How much greater is one population mean than another population mean? It may also be necessary to estimate the difference between two population proportions. The method

Figure 9.5 Determining the Value of z for a One-Tail Test with $\alpha = .05$

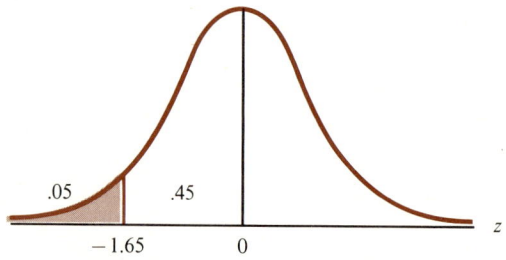

of estimation is the same as that for the single-population case except that the confidence interval formulas are a bit more involved because we are dealing with a sample from each of two populations as opposed to a sample from one population.

You will recall from Chapter 7 that, when estimating the mean of a single population, we first found a point estimate and then calculated an interval estimate centered around the point estimate. The length of the interval was a function of the specified level of confidence and the variability of the point estimator. The point estimate used for the population mean was the sample mean $\bar{x}$. The confidence interval estimate was of the form

$$\bar{x} \pm \epsilon$$

where ϵ is referred to as the statistical error term or precision term. For a large sample size ϵ was defined as

$$\epsilon = z_o \frac{s}{\sqrt{n}}$$

where z_o is a specific value of the standard normal random variable, z, and a function of the specified level of confidence, s is the sample standard deviation, and n is the sample size. Note that $s/\sqrt{n}$ is an estimate of $\sigma_{\bar{x}} = \sigma/\sqrt{n}$, the standard deviation of the point estimate, $\bar{x}$.

To estimate the difference between 2 population means, $D = \mu_1 - \mu_2$, we use the difference between the 2 respective sample means, $\hat{D} = \bar{x}_1 - \bar{x}_2$, as the point estimate of $\mu_1 - \mu_2$. The confidence interval estimate of $\mu_1 - \mu_2$ is calculated in a manner analogous to the confidence interval estimate of μ. The confidence interval is equal to the point estimate plus and minus the error or precision term:

$$(\bar{x}_1 - \bar{x}_2) \pm \epsilon$$

where

$$\epsilon = z_o \sqrt{\frac{s_1^2}{n_1} + \frac{s_2^2}{n_2}}.$$

The above formula requires both sample sizes to be large and the samples to be independent of each other. These are the same conditions for the test statistic discussed at the beginning of this chapter. In that discussion it was pointed out that the formula,

$$\sqrt{\frac{s_1^2}{n_1} + \frac{s_2^2}{n_2}},$$

is the estimated standard deviation of the expression $\bar{x}_1 - \bar{x}_2$. The error term for the interval estimate of 2 means is equal to the standard normal variable value, z_o, times the estimated standard deviation of the point estimate.

The calculation of the confidence interval estimate of the difference of two population means is summarized in the following formulation.

Confidence Interval Estimate of the Difference Between Two Population Means

To find the K percent confidence interval estimate of the difference between 2 population means when both sample sizes are large:

1. First find a value, z_o, of the standard normal random variable, z, such that

   $P(-z_o < z < z_o) = K.$

2. Then calculate the confidence interval using the following formula:

$$(\bar{x}_1 - \bar{x}_2) \pm z_o \sqrt{\frac{s_1^2}{n_1} + \frac{s_2^2}{n_2}}$$

where

$\bar{x}_1$ = the mean of the sample drawn from the first population
$\bar{x}_2$ = the mean of the sample drawn from the second population
s_1 = the standard deviation of the first population sample
s_2 = the standard deviation of the second population sample
n_1 = the size of the first sample
n_2 = the size of the second sample.

This formulation can be used to obtain the solution to the following problem.

Example 9.4.1

The operations manager wants to estimate the difference between the average size of the orders received by the firm's eastern branch and the average size of the orders received by the firm's western branch. The random sample of 87 orders from the eastern branch had a mean value of $13.16 with a standard deviation equal to $2.58. Orders received by the western branch of the firm were also randomly sampled. The average of the 53 orders sampled was $11.57 and the standard deviation was $3.23. Find the 98 percent confidence interval estimate of the difference between the average orders of the 2 branches.

The value for z_o to be used in the confidence interval formula is found in the same manner as in Chapter 7: the confidence level is divided equally on both sides of the mean of the z-distribution (see Figure 9.6), and the appropriate value is then obtained from the standard normal table. This value, along with the sample data, is then put into the confidence interval formula given above:

Figure 9.6 Determining the Value of z for a 98 Percent Confidence Interval

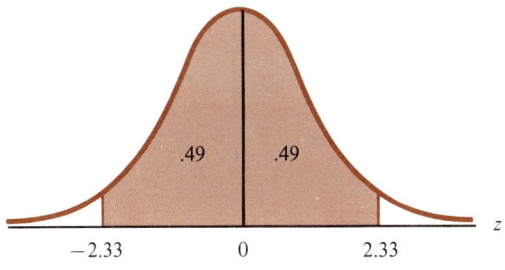

$$(\bar{x}_1 - \bar{x}_2) \pm z_o \sqrt{\frac{s_1^2}{n_1} + \frac{s_2^2}{n_2}}$$

$$(13.16 - 11.57) \pm 2.33 \sqrt{\frac{(2.58)^2}{87} + \frac{(3.23)^2}{53}}$$

$\$1.59 \pm 1.22$,

or, using the notation from Chapter 7, we can write

Conf $\{\$.37 < \mu_1 - \mu_2 < \$2.81\} = 98$ percent.

The point estimate of the difference in average order size between the 2 branches is $1.59. The operations manager is 98 percent confident that the true difference in the average order size is within $1.22 of this point estimate or falls in the interval between $.37 and $2.81.

The difference between 2 population proportions is estimated in the same manner as the difference of 2 population means. The point estimate of the difference between 2 population proportions, $\pi_1 - \pi_2$, is the difference between the respective sample proportions, $p_1 - p_2$. The estimated standard deviation of the difference in sample proportions,

$$\hat{\sigma}_{p_1 - p_2} = \sqrt{\frac{p_1(1 - p_1)}{n_1} + \frac{p_2(1 - p_2)}{n_2}},$$

derived above for the test statistic, is used in the summary formulation given below for calculating the confidence interval estimate of $\pi_1 - \pi_2$.

Confidence Interval Estimate of the Difference Between Two Population Proportions

To find the K percent confidence interval estimate of the difference between 2 population proportions when both sample sizes are large:

1. First find a value, z_o, of the standard normal random variable, z, such that

    $P(-z_0 < z < z_0) = K$.

Estimating the Difference between Two Population Parameters

2. Then calculate the confidence interval using the following formula:

$$(p_1 - p_2) \pm z_o \sqrt{\frac{p_1(1-p_1)}{n_1} + \frac{p_2(1-p_2)}{n_2}}$$

where

p_1 = the proportion of items having the characteristic of interest in the sample randomly selected from the first population

p_2 = the proportion of items having the characteristic of interest in the sample randomly selected from the second population

n_1 = the number of observations in the first sample

n_2 = the number of observations in the second sample.

This confidence interval formula can be used to find the solution to the following problem.

Example 9.4.2 A political analyst is confident that the proportion of voters claiming to belong to the Democratic Party who favor the current Democratic president's energy policy is larger than the proportion of Republican voters who are satisfied with the current president's energy policy. A random sample of 358 Democratic voters has 219 voters in favor of the current energy policy. A random sample of 412 Republican voters has 161 in favor of the president's energy policy. Find the 95 percent confidence interval estimate of the difference in the proportion of Democratic voters and proportion of Republican voters who are satisfied with the current president's energy policy.

The value of z_o is found from the standard normal table as indicated in Figure 9.7. This value, along with the sample data, is used to find the desired confidence interval:

Figure 9.7 Determining the Value of z for a 95 Percent Confidence Interval

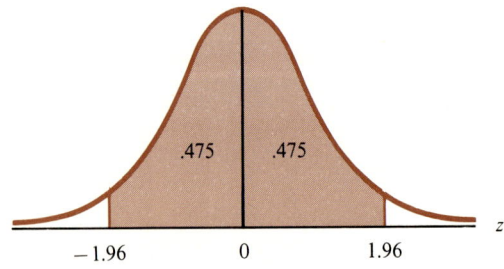

$$p_1 = \frac{x_1}{n_1} = \frac{219}{358} = .612$$

$$p_2 = \frac{x_2}{n_2} = \frac{161}{412} = .391$$

$$(p_1 - p_2) \pm z_o \sqrt{\frac{p_1(1-p_1)}{n_1} + \frac{p_2(1-p_2)}{n_2}}$$

$$(.612 - .391) \pm 1.96 \sqrt{\frac{(.612)(.388)}{358} + \frac{(.391)(.609)}{412}}$$

$.221 \pm .069$.

Or, using the notation from Chapter 7, we have

Conf $\{.152 < \pi_1 - \pi_2 < .290\} = 95$ percent.

The point estimate of the difference between the proportion of Democratic voters approving of the president's energy policy and the proportion of Republican voters favoring the president's energy policy is .221, or 22.1 percentage points. The political analyst is 95 percent confident that the true difference in the proportion of voters favoring the president's energy policy is between .152 and .290 or, in other words, this difference falls within ± 6.9 percentage points of 22.1 percent.

There is no simple formula for finding a confidence interval estimate of the difference between two population proportions if one or both of the sample sizes are small. A confidence interval estimate formula can be derived for estimating the difference between two population means from the test statistic for small sample sizes used for hypothesis testing in Section 9.2. It is necessary that the same conditions be satisfied for confidence interval estimation as were required for hypothesis testing when the small-sample-size test statistic was employed.

Confidence Interval Estimate when Sampling from Normal Populations

In order to find the K percent confidence interval estimate of the difference between 2 population means when the populations are normally distributed and the 2 population variances are equal:

1. First find a value, t_o, of the t-distribution with $n_1 + n_2 - 2$ degrees of freedom such that,

   $P(-t_o < t < t_o) = K$.

2. Then calculate the confidence interval using the following formula:

$$(\bar{x}_1 - \bar{x}_2) \pm t_o \sqrt{\frac{(n_1 - 1)s_1^2 + (n_2 - 1)s_2^2}{n_1 + n_2 - 2}} \sqrt{\frac{1}{n_1} + \frac{1}{n_2}}$$

where

$\bar{x}_1$ = the mean of the sample drawn from the first population
$\bar{x}_2$ = the mean of the sample drawn from the second population
s_1 = the standard deviation of the first sample
s_2 = the standard deviation of the second sample
n_1 = the number of observations in the first sample
n_2 = the number of observations in the second sample.

The solution to the following problem can be calculated using the above formula.

Example 9.4.3

A restaurant has a new addition and an older section. The meal prices and service are the same throughout the restaurant; however, the new section is definitely better appointed than the old section. A random sample of 15 tips received in the new section had an average value of $.2.35 with a standard deviation of $.68. A random sample of 13 tips from the old section had a mean value of $1.98 with a standard deviation of $.62. Find the 95 percent confidence estimate of the difference in the average tip received in the new section and the average tip received in the old section.

The value of t_o is found as was done for confidence interval estimation in Chapter 7 when the t-distribution was employed. For this particular example, the degrees of freedom are $n_1 + n_2 - 2 = 15 + 13 - 2 = 26$. The value in the tails, $1 - .95 = .05$, is split between the 2 tails as indicated in Figure 9.8. The 2.056 value was obtained under the .025 column in the t-table and along the row for 26 degrees of freedom. Using this value and the sample data given in the problem, the desired confidence interval can be calculated:

Figure 9.8 Determining the Value of t for a 95 Percent Confidence Interval with $df = 26$

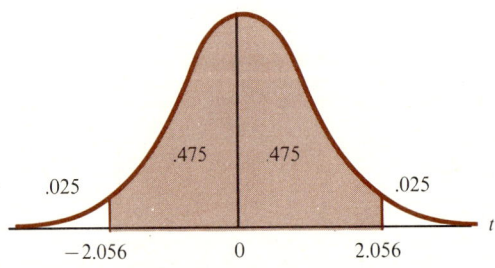

$$(\bar{x}_1 - \bar{x}_2) \pm t_o \sqrt{\frac{(n_1 - 1)s_1 + (n_2 - 1)s_2^2}{n_1 + n_2 - 2}} \sqrt{\frac{1}{n_1} + \frac{1}{n_2}}$$

$$(2.35 - 1.98) \pm 2.056 \sqrt{\frac{(15 - 1)(.68)^2 + (13 - 1)(.62)^2}{15 + 13 - 2}} \sqrt{\frac{1}{15} + \frac{1}{13}}$$

$.37 \pm .51$.

Or, using the notation from Chapter 7, we can write

Conf $\{\$-.14 < \mu_1 - \mu_2 < \$.88\} = 95$ percent.

At the 95 percent level of confidence, we can say that the difference between average tips in the new and old sections is somewhere between $-\$.14$ and $\$.88$. This relatively large interval is primarily a result of the small sample sizes.

If you haven't already noticed, we would like to call your attention to the near-perfunctory hypothesis tests that follow from interval estimates calculated in the last 3 examples. If an interval estimate of $(\mu_1 - \mu_2)$ or $(\pi_1 - \pi_2)$ contains 0—i.e., if the interval goes from a negative number to a positive—we would not be able to reject the hypothesis, H_o: $\mu_1 - \mu_2 = 0$ (or $\pi_1 - \pi_2 = 0$), at the $1 - k$, two-tailed significance level (where k is the confidence level).

In Example 9.4.1, the 98 percent confidence interval extends from $.37 to $2.81; this does not include 0. If we were testing the hypothesis H_o: $\mu_1 - \mu_2 = 0$ against H_a: $\mu_1 - \mu_2 \neq 0$ with $\alpha = .02$, we would reject the null hypothesis. In Example 9.4.3, the 95 percent confidence interval contains 0 so we would *not* reject the null hypothesis at the 5 percent significance level.

9A Appendix: The Paired Difference Test for Equality of Two Population Means

The methodology used throughout this chapter assumed that the samples taken from the respective populations of interest were independent of each other. This simply means that the sample values selected for the first sample were in no way related to or influenced by the sample values selected for the second sample and vice versa. In this appendix we want to look at a method of sampling in which paired values sampled from the two populations are very strongly identified with each other; for such samples, the test statistics used so far in this chapter would not be applicable. If the sampling is performed in the manner described below, there is a test statistic similar to that used in the previous chapter that can be used to test the hypothesis that two population means are equal.

The type of hypothesis test discussed here is referred to as a *paired difference test* because of the way in which the sample data are selected. A classic example is comparing two brands of typewriters with respect to the typing speed of the users measured in words per minute. If we want to determine if the design features of the two brands of typewriters are different enough so as to affect typing speed, we can use one of the following two approaches for selecting the sample data. The first approach would be to select two independent samples, as we have been doing. We could randomly select seven secretaries and direct them to type on seven randomly selected brand A typewriters and then select another seven secretaries to type on randomly selected brand B typewriters. The second approach, and the one we wish to discuss in this section, would be to randomly select seven secretaries and assign each to type on both brands of typewriters. The typewriters each secretary uses would, of course, be randomly selected.

The second approach does *not* lead to independent sample values. The same secretary types on both brands of typewriters: if he or she is an unusually fast typist, then both typewriter brands will show a relatively high score, and if the typist happens to be slow, then both brands will each receive a low typing score. With this approach, there are 7 secretaries producing 14 sample values as opposed to the first approach wherein the 14 sample values were produced by 14 different secretaries. The 14 sample values produced by the 7 secretaries are naturally paired by secretary to generate 7 pairs of sample values.

Example 9A.1

It is desired to determine if the design differences in two brands of typewriters affect the typing speed of the user. Seven brand A typewriters were randomly selected and randomly assigned to seven randomly selected secretaries. The same seven secretaries were also randomly assigned to seven randomly selected brand B typewriters. The words per minute achieved by each of the secretaries on the respective typewriters are recorded below.

	Words per Minute	
Secretary	Brand A Typewriter	Brand B Typewriter
A	69	63
B	75	72
C	46	48
D	61	57
E	58	56
F	83	78
G	57	53

Can we conclude that the different design features do in fact influence the typing speed of the user? Use a significance level of 5 percent.

Table 9.1 Paired Differences for Example 9A.1

Secretary	Brand A	Brand B	Brand A Score Minus Brand B Score
A	69	63	6
B	75	72	3
C	46	48	-2
D	61	57	4
E	58	56	2
F	83	78	5
G	57	53	4

Clearly, the sampling for this problem was done according to the second approach discussed above. It would be wrong to treat the 2 columns of numbers given as data as if they represented 2 independent samples. What this sampling procedure results in is 7 pairs of numbers. The difference between each pair of values can be readily calculated resulting in a sample of 7 *paired differences*. This is what has been done in Table 9.1. It is important that the subtraction be done in the same direction for each pair; i.e., in Table 9.1 the brand B score is subtracted from the brand A score for each pair. Once it is recognized that this type of sampling results in 7 sample values, then we can use the methods developed in Chapter 8 to test the hypothesis that these 7 paired sample values came from a population with the mean equal to 0.

Inasmuch as the sample size is small and we feel it is not unreasonable to assume that the different typing speeds among secretaries follow a normal distribution, we will use the t-statistic discussed in Section 3 of Chapter 8. The calculation of the mean and sample standard deviation of the paired differences is shown in Table 9.2. We let d_i

Table 9.2 Calculations for Example 9A.1

Secretary	d_i	d_i^2
A	6	36
B	3	9
C	-2	4
D	4	16
E	2	4
F	5	25
G	4	16
Totals	22	110

$$\bar{d} = \frac{\Sigma d_i}{n} = \frac{6 + 3 + (-2) + 4 + 2 + 5 + 4}{7} = \frac{22}{7} = 3.14$$

$$s_d^2 = \frac{\Sigma d_i^2 - \frac{(\Sigma d_i)^2}{n}}{n - 1} = \frac{110 - \frac{(22)^2}{7}}{7 - 1} = \frac{40.86}{6} = 6.81$$

$$s_d = \sqrt{s_d^2} = \sqrt{6.81} = 2.61$$

d_i = the difference between the brand A score and the brand B score as calculated in Table 9.1.

represent the ith difference; therefore, $\bar{d}$ is the mean of the paired differences and s_d is the standard deviation. The test statistic for this problem is

$$t = \frac{\bar{d} - (\mu_1 - \mu_2)}{s_d/\sqrt{n}}$$

where n is equal to the number of paired differences and $\bar{d}$ is equal to the mean of the population of paired differences. As has already been noted in this chapter, the hypothesis test that $\mu_1 = \mu_2$ is the same as the hypothesis that the difference, $\mu_1 - \mu_2$, is equal to 0. Thus, the mean value of the population of paired differences is hypothesized to equal 0. Substituting the hypothesized value of $\mu_1 - \mu_2$ in the above test statistic gives us

$$t = \frac{\bar{d} - (\mu_1 - \mu_2)}{s_d/\sqrt{n}} = \frac{\bar{d}}{s_d/\sqrt{n}}$$

with $df = n - 1$.

This latter test statistic is used in the solution to Example 9A.1 given below.

1. $H_o: \mu_1 = \mu_2$
   $H_a: \mu_1 \neq \mu_2$.

2. $\alpha = .05$

   $$t = \frac{\bar{d}}{s_d/\sqrt{n}} \quad \text{with } df = n - 1 = 7 - 1 = 6.$$

3. Reject H_o if and only if $t_c < -2.447$ or $t_c > 2.447$.

4. $t_c = \dfrac{\bar{d}}{s_d/\sqrt{n}} = \dfrac{3.14}{2.61/\sqrt{7}} = 3.18$

   Reject H_o.

Figure 9.9 Determining the Value of t with $df = 6$ and $\alpha = .05$

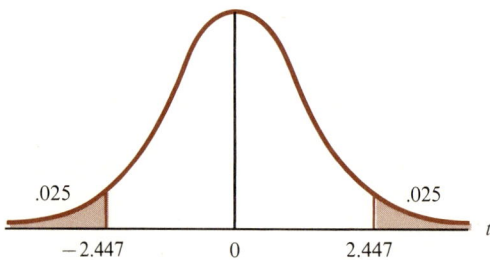

On the basis of this sample result, we conclude that there is a difference in typing efficiency between the 2 brands. The decision rule was determined by observing that the degrees of freedom for this problem are $7 - 1 = 6$ and looking up the appropriate value for a two-tail test as indicated in Figure 9.9.

9B Appendix: More On Hypothesis Tests Concerning Two Population Means

In the hypothesis tests concerning 2 population means discussed in this chapter, the null hypothesis was always assumed to be the equality hypothesis, i.e., $\mu_1 = \mu_2$ or $\mu_1 - \mu_2 = 0$. It is possible, with very little modification, to perform hypothesis tests where it is hypothesized that the population means differ by a specified amount. For example, it may be desirable to test the hypothesis that one of the population means is 10 units greater than the other; this could be stated as $H_o: \mu_1 - \mu_2 = 10$. For large sample sizes, the test statistic originally discussed in Section 9.1 of this chapter can be used to test this hypothesis:

$$z = \frac{(\bar{x}_1 - \bar{x}_2) - (\mu_1 - \mu_2)}{\sqrt{\frac{s_1^2}{n_1} + \frac{s_2^2}{n_2}}}.$$

The hypothesized value of $\mu_1 - \mu_2$ is simply substituted in the test statistic. Consider the following modification of Example 9.1.1.

Example 9B.1

The production engineer for the NEB Battery Company believes that the batteries produced by his firm have an average life that is more than 1.5 months greater than the average life of the batteries produced by a competitor, the OK Battery Company. To test this hypothesis, a sample of 79 NEB batteries was randomly selected and tested. The average life of batteries in this sample was 29.7 months with a standard deviation of 5.8 months. A random sample of 112 batteries produced by the OK Battery Company had a sample mean of 26.1 months with a standard deviation of 6.2 months. Are these sample results consistent with the production engineer's claim? Use a 2 percent significance level.

If we let

μ_1 = the mean life of all NEB batteries

and

μ_2 = the mean life of all OK batteries,

then the null hypothesis can be written as

$H_o: \mu_1 - \mu_2 = 1.5$.

The alternate hypothesis for this problem,

$H_a: \mu_1 - \mu_2 > 1.5$,

reflects the production engineer's claim that the difference in the 2 average battery lives is greater than 1.5 months. The complete hypothesis test for this example is given below.

1. $H_o: \mu_1 - \mu_2 = 1.5$
   $H_a: \mu_1 - \mu_2 > 1.5$.

2. $\alpha = .02$

$$z = \frac{\bar{x}_1 - \bar{x}_2 - (\mu_1 - \mu_2)}{\sqrt{\frac{s_1^2}{n_1} + \frac{s_2^2}{n_2}}} = \frac{\bar{x}_1 - \bar{x}_2 - 1.5}{\sqrt{\frac{s_1^2}{n_1} + \frac{s_2^2}{n_2}}}.$$

3. Reject H_o if and only if $z_c > 2.05$.

4. $z_c = \dfrac{\bar{x}_1 - \bar{x}_2 - 1.5}{\sqrt{\dfrac{s_1^2}{n_1} + \dfrac{s_2^2}{n_2}}} = \dfrac{29.7 - 26.1 - 1.5}{\sqrt{\dfrac{(5.8)^2}{79} + \dfrac{(6.2)^2}{112}}} = 2.39.$

Reject H_o.

The sample results are consistent with the production manager's claim that the NEB batteries outlast OK batteries by more than 1.5 months on the average. The decision rule is one-tailed because of the alternate hypothesis, $\mu_1 - \mu_2 > 1.5$. If the sample data are consistent with the alternate hypothesis—i.e., $\bar{x}_1 - \bar{x}_2 > 1.5$—then the numerator of the test statistic, $\bar{x}_1 - \bar{x}_2 - 1.5$, is greater than 0; thus, the null hypothesis is rejected only for relatively large positive values of z, and the entire value of α is placed in the upper tail of the z-distribution as indicated in Figure 9.10.

Figure 9.10 Determining the Value of z for a One-Tail Test with $\alpha = .02$

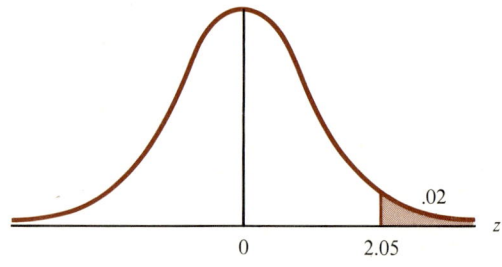

Statistical Inference: Two Populations

The approach used above is also applicable to hypothesis tests concerning a specified difference in 2 population proportions. The only difference is in the test statistic used. For example, if it were desired to test the hypothesis that 2 population proportions differ by 10 percentage points, the null hypothesis would be

$H_o: \pi_1 - \pi_2 = .10,$

and the appropriate test statistic would be

$$z = \frac{(p_1 - p_2) - (\pi_1 - \pi_2)}{\sqrt{\frac{p_1(1-p_1)}{n_1} + \frac{p_2(1-p_2)}{n_2}}} = \frac{p_1 - p_2 - .10}{\sqrt{\frac{p_1(1-p_1)}{n_1} + \frac{p_2(1-p_2)}{n_2}}}$$

for this particular problem. (It is assumed, of course, that both sample sizes are large.) Given this, solving the problem is simply a matter of obtaining the sample data and following the 4 basic steps of hypothesis testing.

9C Appendix: Pooled Variance

The test statistic used in Section 9.2,

$$t = \frac{\bar{x}_1 - \bar{x}_2}{\sqrt{\frac{(n_1 - 1)s_1^2 + (n_2 - 1)s_2^2}{n_1 + n_2 - 2}} \sqrt{\frac{1}{n_1} + \frac{1}{n_2}}},$$

is derived theoretically by taking the ratio of a standard normal distribution to the square root of a chi-square distribution divided by its degrees of freedom. (The chi-square distribution is discussed in Chapter 14.) As indicated in Section 9.2, it is assumed that the 2 samples are taken independently of each other and the two sampled populations are normally distributed with equal variances.

The function of the 2 sample variances in the denominator of the above test statistic,

$$\frac{(n_1 - 1)s_1^2 + (n_2 - 1)s_2^2}{n_1 + n_2 - 2},$$

is a weighted average (the weights are $n_1 - 1$ and $n_2 - 1$, respectively) of the two sample variances s_1^2 and s_2^2. Some writers refer to this as the *pooled variance*, meaning that the 2 sample variances are pooled together to estimate the equal variance of the 2 populations. If we let s_p^2 equal the pooled variance and s_p equal the square root of the pooled variance, the denominator in the above test statistic can be written as

$$s_p \sqrt{\frac{1}{n_1} + \frac{1}{n_2}}.$$

In Section 9.1 of this chapter, the formula for the standard deviation of the difference of 2 sample means,

$$\sigma_{\bar{x}_1 - \bar{x}_2} = \sqrt{\frac{\sigma_1^2}{n_1} + \frac{\sigma_2^2}{n_2}},$$

was discussed. If, as is assumed in the small-sample case, the 2 population variances are equal—i.e., $\sigma_1^2 = \sigma_2^2 = \sigma^2$—then

$$\sigma_{\bar{x}_1 - \bar{x}_2} = \sqrt{\frac{\sigma^2}{n_1} + \frac{\sigma^2}{n_2}} = \sigma \sqrt{\frac{1}{n_1} + \frac{1}{n_2}}.$$

It is now evident that, under the assumption of equal population variances, the estimated standard deviation of the difference of 2 sample means is

$$\hat{\sigma}_{\bar{x}_1 - \bar{x}_2} = s_p \sqrt{\frac{1}{n_1} + \frac{1}{n_2}}.$$

Thus, the above test statistic can be written in a simpler form,

$$t = \frac{\bar{x}_1 - \bar{x}_2}{s_p \sqrt{\frac{1}{n_1} + \frac{1}{n_2}}},$$

and it can be viewed in the same light as the test statistics used in Chapter 8, i.e., the sample statistic $\bar{x}_1 - \bar{x}_2$, minus its expected value (0 if the null hypothesis is true) divided by the estimated standard deviation of the test statistic. This latter statement should not be construed to mean that this appendix outlines a proof that the above test statistic has a t-distribution. The proof is based on the theoretical proposition referred to in the first sentence of this appendix.

Problems

1. An auditor is trying to determine if finished goods inventory is in fact evenly distributed with regard to cost between 2 warehouses as claimed by management. A random sample of 79 items from the first warehouse has an average cost of $27.86 with a standard deviation of $7.63. The 93 items sampled from the second warehouse had an average cost of $25.73 with a standard deviation of $6.88. Can the auditor conclude at the 2 percent significance level that the allocation of inventory between the 2 warehouses is the same with respect to the average cost per item?

2. The production manager wants to determine if a change in the alloy, made on the basis of cost consideration, of the filaments for light bulbs produced by the company has a detrimental effect on

the life of the light bulb. A random sample of 125 bulbs produced with the old alloy had a mean life of 756.3 hours with a standard deviation of 31.2 hours. The 90 bulbs randomly selected from the production run using the new alloy had an average life of 711.8 hours with a standard deviation of 42.6 hours. Can the production manager conclude that the new alloy is reducing the average life of light bulbs produced by the company? Test at the 4 percent significance level.

3. A retail and wholesale plumbing supply store has a large walk-in trade. The owner wants to try to determine whether the sales clerk tending the store on Saturdays, when the do-it-yourselfers make up the bulk of the customers, affects the amount sold per purchase. Fifty randomly selected sales from Saturdays when Darrel was tending the store had an average of $21.31 with a standard deviation of $6.79. The 60 randomly selected sales when Wayne was tending the store had an average of $24.58 with a standard deviation of $7.93. Do these sample results indicate that there is a difference in the average amount sold per customer for the 2 sales clerks? (Use $\alpha = .03$.)

4. A manufacturer producing 110-volt motors judges the quality of the motors produced by the minimum voltage required to start the motors running. The new energy-efficiency requirements have tended to raise the minimum voltage required to start the motors. The design engineer wants to determine whether there is any difference in the minimum voltage requirements for 2 different energy-efficient motor designs he is considering for mass production. A sample of 9 prototypes of the first design required an average voltage of 91.2 volts to start with a standard deviation of 8.7 volts. The 11 randomly selected prototypes of the second design required 95.8 volts with a standard deviation of 9.3 volts. Is there a difference in the 2 designs with respect to the minimum amount of voltage required to start the motors? (Use $\alpha = .10$.)

5. A marketing analyst for a vending machine distributor feels that the location of the vending machines is more important than the price that is charged. Her boss does not agree with her and asks her to test his assertion that even slightly higher prices will result in less sales per machine. She randomly selects 7 vending machines for which the prices are held constant, and she randomly selects 8 vending machines for which the prices of all items are increased by $.05. The daily sales for the 7 vending machines with constant prices are $68.05, $51.25, $75.20, $46.50, $58.15, $63.80, and $60.40. The 8 vending machines for which the price has been raised a nickel per item have the following daily sales: $44.90, $65.35, $70.65, $49.15, $60.45, $55.70, $57.80, and $59.05. Based on these sample results, can the marketing analyst conclude that

her boss is right in his assertion about the effect of a price increase? (Use $\alpha = .05$.)

6. The materials manager for a large firm wants to determine whether there is a difference in the average weight of vials produced by 2 of the firm's suppliers. Seventeen vials randomly selected from the first supplier's shipment had the following weights: 7.5, 8.4, 13.7, 14.8, 12.8, 15.5, 9.2, 9.4, 11.6, 9.5, 10.8, 8.1, 9.4, 11.1, 12.9, 13.2, and 11.9 grams. The weights of 11 vials randomly selected from the second firm's shipments were as follows: 7.2, 7.5, 10.1, 8.7, 7.6, 9.1, 10.3, 9.4, 9.0, 7.1, and 7.4 grams. Are the average weights of vials produced by the 2 suppliers the same? (Use $\alpha = .01$.)

7. A political analyst wants to determine whether the proportion of people in favor of the president's energy policy is the same for both Democrats and Republicans. The 118 Democrats sampled had 53 who favored the president's energy policy. One hundred and thirty Republicans were sampled, with 68 indicating satisfaction with the president's energy program. Is the proportion of voters favoring the president's energy policy the same for the 2 groups of voters—Democrats and Republicans? (Use $\alpha = .03$.)

8. A marketing manufacturer for a large bakery chain argues that his company should use a different package design for its retail outlets according to whether or not the retail outlets are in high-income or low-income areas. He considers 2 package designs, A and B, and argues that a higher percentage of customers will prefer design A in the low-income areas than in the high-income areas. A random sample of 85 customers of a low-income retail outlet showed 67 preferring design A. Thirty-two of the 58 customers sampled at a high-income retail outlet preferred design A. Perform the hypothesis test for the marketing manager at the 5 percent significance level.

9. A small manufacturer of glass bottles has 2 machines producing the bottles. The percent of defective bottles produced by the firm is unacceptable to the manager. He feels that before he tries to remedy the problem it is necessary to first determine whether or not both machines are producing the same proportion of defective bottles. A random sample of 90 bottles from the first machine contained 12 defective bottles. A random sample of 120 bottles from the second machine contained 7 defectives. Are the 2 machines the same with regard to proportion of defective bottles produced? (Use $\alpha = .08$.)

10. A credit manager is certain that the amount per account of the accounts receivable at the southwest store is larger than the amount at the east-side store. He would like an estimate of the average difference. A random sample of 80 accounts at the southwest store had an average amount per account equal to $78.21 with

a standard deviation of $23.69. The 110 accounts randomly selected from the east-side store had a mean of $36.54 with a standard deviation of $11.06. Find the 95 percent confidence interval estimate of the difference in the averages of the accounts receivable at the 2 stores.

11. A firm marketing 2 brands of disposable baby diapers wants to get an estimate of consumers' perception of improved quality in its diapers with respect to absorbency. Both of the brands have "new improved—more absorbent" printed on the boxes in which they are sold in a selected market. Only 1 of the brands has been significantly altered in its construction and materials used. After the diapers had been on the market for a month, 150 randomly selected customers of the physically altered brand were asked if they did perceive a noticeable improvement in the quality of the diapers. Ninety-seven of those sampled indicated that they did. Of the 130 customers sampled who used the brand with only the box changed, 23 indicated that they perceived a noticeable improvement in the quality of the diapers. Find the 98 percent confidence interval estimate of the difference in the proportion of customers perceiving a noticeable difference in quality for the 2 brands of diapers.

12. The dollar amounts of meat purchased per customer for 7 customers randomly selected from customers picking up their groceries in luxury cars were as follows: $18.75, $20.25, $15.67, $23.85, $31.23, $23.24, and $21.48. Ten randomly selected customers driving compact cars had the following dollar amount of meat purchases per customer: $6.25, $10.79, $18.61, $12.13, $9.88, $7.63, $8.37, $9.10, $11.05, and $8.03. Find the 95 percent confidence interval estimate of the difference between average meat expenditures of luxury car customers and compact car customers.

13. A keypunch supervisor wants to determine whether the number of errors per card is the same for buff cards and yellow cards. She sets up cards of the 2 colors in alternating order at the beginning of the day. At the end of the day, from the thousands of cards that have been keypunched she randomly selects 175 buff cards and 190 yellow cards. The average number of errors (a card can have more than 1 error) for the buff cards is .053 per card with a standard deviation of .019. The yellow cards have an average of .068 errors per card with a standard deviation of .026. Is there a difference in the average errors per card for buff cards and yellow cards keypunched by the keypunch operators working under the supervisor? Test at the 6 percent significance level.

14. In a random sample of 105 large business and financial institutions, 63 reported that the chief personnel executive's pay was equal to or more than the pay of top legal, administration, and manufac-

turing executives. Fifty-eight of 120 randomly selected medium-sized business and financial institutions reported that their chief personnel executive received pay equal to or greater than the pay of top executives in other classifications. Is there a difference in the proportion of large business and financial institutions that pay top salary dollar to their chief personnel executive and the corresponding proportion for medium-sized institutions? Test at the 5 percent significance level.

15. A produce manager feels that his relative prices for 2 grades of potatoes are O.K. if sales for both grades of potatoes are about the same during a given day. He is continuously varying prices according to the day of the week, sales on other products, and special store promotions. On 65 randomly selected days, an average of 258.7 pounds of the top grade of potatoes was sold with a standard deviation of 25.3 pounds. The average number of pounds of the inferior grade sold on 50 other randomly selected days was 245.1 pounds with a standard deviation of 31.3 pounds. Are the average pounds of potatoes sold the same for the 2 grades? Test at the 2 percent significance level.

16. A supermarket manager wants to determine whether people notice unadvertised price differences. She randomly selects 35 of the next 75 days, and on these 35 days she puts a higher price on tomato soup than during the other 40 days. The average number of cans sold on the 35 days of higher prices is 53.4 with a standard deviation of 7.8 cans. For the 40 days of lower prices, the average number of cans sold is 61.7 with a standard deviation of 6.8 cans. Are customers alert to the lower prices even though they aren't advertised? (Use $\alpha = .03$.)

17. The president has not made any significant changes in his economic policy during the last 2 months. One of his aides is interested in knowing if the mood of the people has deteriorated with regard to the president's current economic policy. The aide feels that the change in the economic environment is sufficient to cause more dissatisfaction with the president's economic policy. A random sample of 350 voters taken 2 months ago had 183 voters indicating approval of the president's economic policy. A random sample of 420 voters recently interviewed had 197 expressing approval of the president's economic policy. Has the mood of the people deteriorated with regard to the president's economic policy? Test at the 6 percent significance level.

18. A drug manufacturer has been working on an appetite depressant. The depressant was given to 15 newborn laboratory animals which were fed and housed with 12 other laboratory animals. After a specified time period, the weight gain for the 27 animals was recorded. The 15 animals receiving the appetite depressant gained an

average of 17.8 grams with a standard deviation of 3.4 grams. The 12 control animals experienced a weight gain of 29.3 grams with a standard deviation of 4.1 grams. Find the 95 percent confidence interval estimate of the difference in average weight gain for animals given the appetite depressant and for animals not given a drug of this nature.

19. The industrial engineer for a color television manufacturer has come up with a new molding process for making picture tube glass stems. The new process is considerably less costly then the old process but has the disadvantage that it produces more defective stems. Before a decision can be made about the new process, it is necessary to estimate the difference in the percentage of defectives for the 2 processes. A random sample of 130 glass stems from the new process has 15 defective stems. Only 6 of the 155 glass stems sampled from the old process were defective. Find the 90 percent confidence interval of the difference in the proportion of defectives for the 2 processes.

20. A large executive-recruiting firm randomly sampled 90 applicants rejected for top-level management jobs in large corporations and found that 37 were rejected because the "personal-chemistry factor" wasn't right; i.e., they were not "our kind of person." Twenty-one of the 75 randomly selected job applicants for top-level positions in medium-sized corporations were rejected because of the "personal-chemistry factor." On the basis of these sample data, can the executive-recruiting firm conclude at the 10 percent significance level that there is a difference in the importance of the "personal-chemistry factor" for the two sizes of firms?

21. A frozen foods product line manager wants to determine the difference between single-person households and 2-or-more-person households in the proportion of meals eaten at home that are precooked. A random sample of 350 meals eaten in single-person households had 209 meals that consisted primarily of precooked items. A random sample of 265 meals eaten in 2-or-more-person households had 96 meals that were for the most part precooked. Find the 95 percent confidence interval estimate of the difference in the proportions of precooked meals for the 2 types of households.

22. The owner of a large retail grocery store in a small city is certain that his rural customers purchase more per visit than his city customers. He would like to determine the difference in the average dollar purchase for the 2 types of customers. He decides to randomly select customers, determine if they are from outside the city or live in the city, and then unobtrusively record the total dollar amount of the purchase. This procedure results in a sample of 72

rural customers and 67 city customers. The average amount purchased per visit by the rural customers is $78.27. The average for the sample of city customers is $43.91. The grocer also calculates the sample standard deviation for each of the samples. The sample standard deviation for the rural customers is $19.68 and for the city customers it is $13.49. Find the 98 percent confidence interval estimate of the difference between the average purchase amounts for the 2 types of customers.

■ The remaining problems require material covered in the appendices.

23. A national distributor of antifreeze for cars wants to determine if there is a difference, measured in sales, in the effectiveness of 2 different promotional strategies under consideration. The sales department has identified 8 different types of markets in which it sells the product. The markets are classified according to such factors as population density, per capita income, severity of the weather, etc. For each market type, a retail location is randomly selected, and the first promotional strategy is employed. Similarly, for testing the second promotional strategy, 8 retail outlets are randomly selected, 1 in each type of market. Sales for the fall quarter (October, November, and December) are realized for the 16 outlets as follows:

Type of Market in Which Store Is Located	Sales Promotional Strategy Used A	B
1	$256.83	$278.19
2	656.18	829.20
3	427.89	388.77
4	313.22	401.10
5	527.61	575.80
6	186.20	171.39
7	678.51	731.48
8	277.83	290.00

Can the distributor conclude there is a difference in the 2 strategies with regard to their effectiveness in generating sales throughout the country? Test at the 5 percent significance level.

24. A research chemist for a car battery manufacturer is trying to convince the cost accountant that a new, improved, higher-quality acid adds at least 6 months to the average life of batteries produced by the company. An experiment using the 2 acids of different quality in similarly constructed batteries had the following results. The 85 batteries using the new, improved, high-quality acid had an average life of 38.7 months with a standard deviation of 6.3 months. The 70 batteries containing the unimproved acid had an average life of 30.9 months with a standard deviation of 5.8 months. Based on these sample results, can the company conclude

that the improved-quality acid adds more than 6 months to the life of the batteries produced by the company? Test at the 2 percent significance level.

25. A drill press operator has made a complaint to his union steward concerning the equipment he has been assigned. He claims that his productivity is lower than average because of faulty equipment and not because of his ability. The steward asks 7 other drill press operators to work 15 minutes on another piece of equipment and then to work 15 minutes on the equipment in question. The 15-minute work periods are randomly assigned during the day, and the operators are randomly assigned to pieces of equipment other than those which they normally work with. Their respective outputs on the allegedly faulty equipment and the randomly selected pieces of equipment are given below.

	Completed Parts	
Operator	"Faulty" Equipment	Other Equipment
1	21	25
2	17	19
3	18	16
4	20	23
5	19	20
6	16	22
7	15	21

Can the union steward conclude at the 5 percent significance level that the piece of equipment in question is defective?

26. A Washington apple grower has decided to use a more costly fertilizer on half of her orchard. If the new fertilizer results in more than 5 percentage points more top-grade apples, then it is worth the extra cost. A random sample of 135 apples from the half of the orchard using the standard fertilizer had 56 apples of top-grade quality. Eighty-three of the 160 apples randomly selected from the half of the orchard using the more costly fertilizer could qualify as top-grade apples. Can the apple grower conclude at the 10 percent significance level that the more costly fertilizer improves the percentage yield of top-grade apples by more than 5 percentage points?

27. A shoe manufacturer wants to compare the "new" type of leather offered by a competitor of his current supplier. Twelve pairs of shoes are produced with leather soles. Six pairs have the new leather on the right shoe and the other 6 have the new leather on the left shoe. The shoes are randomly assigned to 12 junior high school students. After 5 months the amount of wear on each sole is measured and converted to a measure of equivalent miles of wear

that could be obtained from each sole. The following sample results are obtained:

Student	Equivalent Miles of Wear	
	New Leather	Regular Supplier
1	343.7	298.3
2	302.1	315.4
3	401.8	369.7
4	389.2	356.7
5	297.1	325.1
6	412.3	368.0
7	386.1	378.8
8	371.4	349.5
9	393.6	399.2
10	411.8	388.9
11	367.4	341.1
12	406.0	402.2

Do these data substantiate the new supplier's claim that his leather is superior? (Use $\alpha = .01$.)

CHAPTER 10

REGRESSION ANALYSIS

In this chapter we begin a discussion of statistical methods applied to the relationships that might exist between or among two or more variables. The science of estimating, in functional form, the dependence of one variable upon one or more other variables is called *regression analysis*; this is the topic addressed in this chapter.

Specific objectives are—

to present in great detail and with illustrations the two-variable linear regression model (having one dependent and one independent variable), and to show how this model is used for both statistical estimation and hypothesis testing (Introduction and Sections 1, 2, 3, 4, and 6).

to present the linear correlation coefficient and its relationship to regression analysis (Section 5).

to discuss the use of the computer in regression analysis (Section 7).

to present the multivariable regression analysis model, the computer output for this model, and the use of this computer output in the analysis of the regression relationship (Section 8).

The search for relationships that exist between and among variables is never ending in the world of business and the related social sciences. The expression "there is a correlation between this and that" is frequently used by nonstatisticians with the implication that the two variables are somehow related. As you might expect, we will want to formalize such associations as functional relationships whenever possible. The science of estimating, in functional form, the dependence of one variable upon another (or several others) is called *regression analysis*.[1] As we shall see, the term *correlation* is also a precisely defined professional term and not unrelated to regression analysis.

The first problem in regression analysis is to determine the nature of the underlying functional relationship. Is it linear or nonlinear; and if nonlinear, what is the specific form of the nonlinearity? This is usually a difficult problem for the social scientist. Altogether too often, we are forced to one of two nonscientific alternatives wherein the data gathered dictate the form to be used. A visual inspection of the sample scattergram plot for a two-variable problem is one guide to a choice of the functional form; or alternatively, if a computer is available, the data may be run for a variety of possible functions and the "best" one chosen according to certain criteria to be discussed later. It is, of course, significantly better from the scientific viewpoint if we are able to argue the nature of the function a priori—that is, based upon realistic assumptions within our own science. In this respect, physical scientists typically have an edge on the social scientist. For example, assuming that the force of gravity is constant and neglecting air friction, the distance traveled by a falling body is a second-degree function of time. As a second example, the exponential growth (or decay) function, $y = Ae^{bt}$, is derived by methods of calculus from the basic assumption that the growth (or decay) is proportional to the amount of y present at any time, t. This function describes many different kinds of problems in the physical and biological sciences and is also the form for instantaneous compound interest ($P_t = P_o e^{rt}$, where r is the simple interest rate per period of time, t is time, P_o is initial amount, and P_t the amount at end of time, t).

Perhaps more often than can be justified, we in business and the social sciences use a linear model. See Appendix 10A for a review of the analytics of a straight line.

Aside from assumptions that lead to a linear model, some of the other arguments for its use are:

1. In the absence of evidence to the contrary, the linear model is certainly the simplest, the most efficient, and the easiest to explain and understand of all possible models. There is also a historical

---

[1]. In a paper published in the *Processing of the Royal Society* in 1886, Sir Francis Galton reported the results of a study on the relationship of heights of children and their parents. Usually, tall people produced tall children and short people short children, but he found that children's heights tend to regress toward the mean. Hence, the term *regression* became a part of statistical literature and developed as the accepted term in referring to the study of functional relationships.

and philosophical basis for not complicating matters more than necessary; this preference for the simplest solution is known as Occam's razor.[2]

2. If the nonlinearity of a relationship is mild, a linear function will be a good approximation *over a limited range*. (In many applications, piecewise linear functions are used to approximate nonlinearity—especially if the nature of the nonlinearity is not known.)

Some assumptions that appear on first consideration to require a nonlinear function may be satisfied by a linear function. For example, an economist may wish to assume that, as real income in the community increases, total consumption increases, but the proportion of income spent on consumption decreases—in his terms, the *average propensity* to consume decreases with increasing income. There are, of course, many functions that satisfy this assumption, but it is interesting to note that a linear function with both *positive y-intercept* and *positive slope* will satisfy these assumptions. These terms are discussed in the next section.

10.1 Linear Regression

We study and work with the linear function in the form, $\hat{y} = a + bx$. The hat (ˆ) is used to emphasize that the paired data points (x, y) are from a sample and the function, $\hat{y}$, is an estimated expectation of y for a given x. From your previous work in mathematics (see Appendix 10A for a review discussion) you may recall that a is the y-intercept or the value of $\hat{y}$ when x equals 0; and b is the slope (or slope-coefficient, the change in $\hat{y}$ per unit change in x). Let us begin with a very simple hypothetical example in order to illustrate the idea of *fitting a function* to a set of data points.

Assume that the manager of a bank decided to experiment with the number of tellers during the noon hour rush in order to study the length of the queue, or waiting line, that developed. A single queue fed all tellers. The manager chose a random time during lunch hour for the queue-length count. The results of his study (and our problem) are presented in Table 10.1, where x is the number of tellers and y is the number of customers in the waiting line.

Inasmuch as the manager knows very little about formal statistics, we should help him. We, in agreement with the manager, would expect the queue length to decrease as the number of tellers increases. We also feel that a queue length may be quite linear over a limited reasonable range. For large values of x, we would expect y to be 0 (or very close to 0), and for $x = 0$ the queue length would, of course, get very

---

2. Named for William of Ockham (or Occam), a 14th-century Englishman, who wrote that "entities should not be multiplied unnecessarily."

Table 10.1 Data on Queue Length and Number of Tellers

Number of Tellers x	Length of Queue y
2	8
2	11
3	4
4	6
5	3
5	1

large. Regardless of the above considerations, however, we want to find the linear equation that best describes the data.

A plot of the data in the rectangular coordinates, a *scattergram*, is a first step and offers the researcher a visual image of the possible relationship. See Figure 10.1 for this scattergram. An inspection of the scattergram supports most of our presuppositions. Several methods, based upon quite specific criteria, exist for finding the line of best fit. For several reasons, which will be discussed later in the chapter, we

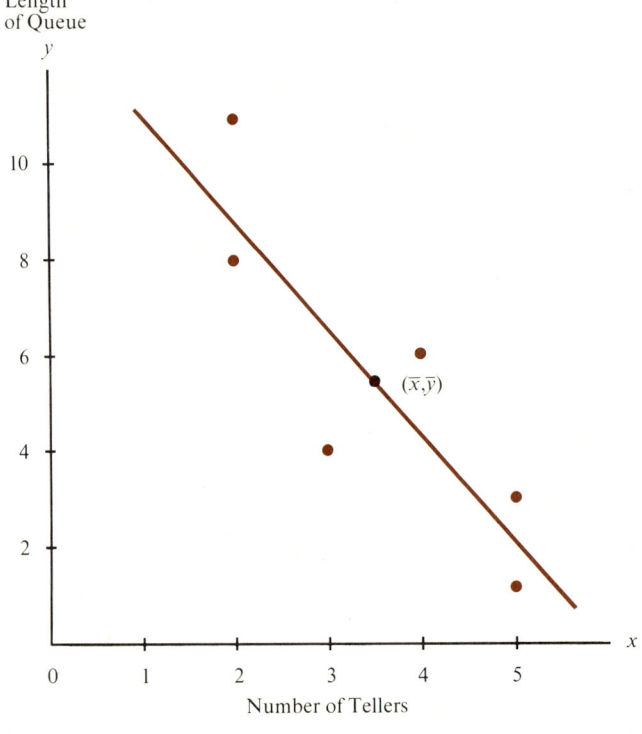

Figure 10.1 Queue Length and Number of Tellers: Scattergram and Linear Function

Table 10.2 Queue Length and Number of Tellers: Least Squares Calculations

	x	y	x^2	xy
	2	8	4	16
	2	11	4	22
	3	4	9	12
	4	6	16	24
	5	3	25	15
	5	1	25	5
Totals	21	33	83	94
	Σx	Σy	Σx^2	Σxy

x = number of tellers in bank
y = mean time in minutes for customer service

will use here the *least squares criterion*, that of finding the line such that the sum of the squared y-deviations (or vertical distances from the points to the line) is minimal. More formally, minimize

$$\sum_{i=1}^{n} (y_i - \hat{y}_i)^2.$$

This criterion, using relatively easy methods of calculus (see Appendix 10B), yields two equations that may be solved simultaneously to determine the values of a and b in the equation, $\hat{y} = a + bx$.

Normal Equations   These two equations are called the *normal equations*:

$\Sigma y = na + b\Sigma x$

$\Sigma xy = a\Sigma x + b\Sigma x^2.$

In order to find a and b we see that we need the following sums:[3] Σy, Σx, Σxy, and Σx^2. Refer to Table 10.2 for a suggested format and the calculations required to get the above sums. Substituting the actual sums into the two general equations, we get

$33 = 6a + 21b$
$94 = 21a + 83b.$

When the above equations are solved simultaneously, we get $b = 2.26$ and $a = 13.41$. The equation for the least squares line of best fit is

$\hat{y} = 13.41 - 2.26x.$

The $\hat{y}$-values from the equation are compared with the actual y-values in Table 10.3. The differences, y-$\hat{y}$, and the squares of these differences

---

[3]. You are already familiar with the expressions Σx, Σy, and Σx^2. These expressions were used in the formula for the sample mean and the computation formula for the standard deviation in Chapter 2. The expression Σxy is found by summing the products of each of the paired values; i.e., $\Sigma xy = x_1y_1 + x_2y_2 + \cdots + x_ny_n$. You should clearly distinguish in your mind the difference between "the sum of the cross-products," Σxy, and "the product of the sums," $\Sigma x \Sigma y$.

Table 10.3 Queue Length and Number of Tellers: The Squared Deviations

x	y	$\hat{y}$	$y - \hat{y}$	$(y - \hat{y})^2$
2	8	8.89	−.89	.79
2	11	8.89	2.11	4.45
3	4	6.63	−2.63	6.92
4	6	4.37	1.63	2.66
5	3	2.11	.89	.79
5	1	2.11	−1.11	1.23
Totals 21	33	33.00	0	16.84

are also shown. To emphasize a point, the summation of the squared $(y_i - \hat{y}_i)$ deviations, 16.840, is smaller than that for any other line in the 2-dimensional space. We might also note that $\Sigma y = \Sigma \hat{y}$ and $\Sigma(y_i - \hat{y}_i) = 0$. These propositions will always hold and, of course, one implies the other. As a further point of interest, $\Sigma(y_i - \hat{y}_i) = 0$ for any straight line that passes through the point of means. To assure ourselves that the least squares line goes through the point of means, $(\bar{x}, \bar{y})$, refer to the first of the 2 normal equations, $\Sigma y = na + b\Sigma x$. Dividing by n, we get $\bar{y} = a + b\bar{x}$; in a sense, this is 1 of the 2 conditions imposed by the least squares criterion.

10.2 Linear Regression—A More Comprehensive Example

General Solution of Normal Equations

If we solve the normal equations,

$$\Sigma y = na + b\Sigma x$$
$$\Sigma xy = a\Sigma x + b\Sigma x^2,$$

for a and b, we get

$$b = \frac{n\Sigma xy - \Sigma x \Sigma y}{n\Sigma x^2 - (\Sigma x)^2};$$

and, after finding b, solving the first equation for a yields $a = \bar{y} - b\bar{x}$.[4]

Without the use of a computer, we would employ these formulas rather than solve the normal equations for each and every problem. (As an exercise, use the summation from the previous example to check the coefficients derived from the simultaneous equation solution.)

---

4. This is the most efficient way to find a. The intercept coefficient can, of course, be found in terms of the summations; i.e.,

$$a = \frac{\Sigma x^2 \Sigma y - \Sigma x \Sigma xy}{n\Sigma x^2 - (\Sigma x)^2}.$$

To illustrate the use of these formulas, we move to a second example, one that appears to be more realistic and appropriate. In order that the firm will not be recognized, the actual data and the story have been considerably altered in what we present below. We will refer to this example and its results throughout the chapter.

Example 10.2.1 The manager of Casey's Department Store decided he should experiment with the weekly dollar expenditures on newspaper advertising in an effort to measure the effectiveness of the funds spent in this manner. The firm had been spending about $2,000 per week on newspaper ads, and total sales had been running about $400,000 per week. A program of variable newspaper advertising expense was set up for the next 12 weeks. Records were kept, with the results shown in Table 10.4; x and y represent newspaper advertising expense and total sales, respectively, in thousands of dollars per week.

As a first step we plotted the scattergram of these data in Figure 10.2. An inspection of the scattergram does not seem to contradict the possibility of a positive linear relationship between the two variables. The science of marketing dictates that sales be the dependent variable. We won't dwell upon the reasons for this, because the major objective for now and through the next several sections is to illustrate the technical and inferential aspects of linear regression.

The following summations were calculated from the above data: $\Sigma x = 25.8$; $\Sigma y = 5,090$; $\Sigma x^2 = 60.88$; $\Sigma y^2 = 2,214,100$; and $\Sigma xy = 11,386$. (See Table 10.5 for the display of these calculations.) The sum

Table 10.4 Casey's Department Store Data

Total Sales ($000) y	Newspaper Advertising Expense ($000) x
420	2.7
480	2.6
450	3.0
320	1.5
300	.5
540	3.0
510	2.1
410	2.3
380	1.7
390	1.9
440	2.2
450	2.3

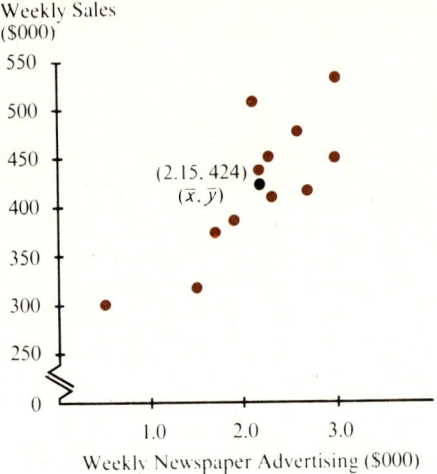

Figure 10.2 Scattergram: Casey's Department Store Sales as a Function of Advertising

of y^2 is not needed for the regression function calculation; it is given now and will be used later for other important calculations.

For the above formulas,

$$b = \frac{n\Sigma xy - \Sigma x \Sigma y}{n\Sigma x^2 - (\Sigma x)^2}$$

$$= \frac{12(11{,}386) - (25.8)(5{,}090)}{12(60.88) - (25.8)^2}$$

$$= 81.79,$$

and

$$a = \bar{y} - b\bar{x} = \frac{5{,}090}{12} - 81.79\left(\frac{25.8}{12}\right) = 424.17 - (81.79)(2.15) = 248.3.$$

The equation $\hat{y} = 248.3 + 81.8x$ is the line of best fit according to the least squares criterion.

This equation is graphed along with the scattergram in Figure 10.3. (Note that it passes through the point of means [2.15, 424].) The line appears to be a reasonable estimate of the linear functional relationship; i.e., major calculation errors would become obvious at this point. Mistakes in calculations, if any, are probably minor.

Before proceeding with further technical analysis, let us take a brief look at our function and interpret the coefficients. Both a and b are estimated expectations; i.e., we estimate that Casey's Department Store may expect $81.79 increment in sales per dollar of increased

Table 10.5 Casey's Department Store Weekly Data in ($000)

Sales y	Expenditures on Newspaper Advertising x	y^2	x^2	xy
420	2.7	176,400	7.29	1,134
480	2.6	230,400	6.76	1,248
450	3.0	202,500	9.00	1,350
320	1.5	102,400	2.25	480
300	.5	90,000	.25	150
540	3.0	291,600	9.00	1,620
510	2.1	260,100	4.41	1,071
410	2.3	168,100	5.29	943
380	1.7	144,400	2.89	646
390	1.9	152,100	3.61	741
440	2.2	193,600	4.84	968
450	2.3	202,500	5.29	1,035
5,090	25.8	2,214,100	60.88	11,386

newspaper advertising.[5] *Both x and y are in thousands of dollars;* therefore, the slope coefficient may be interpreted as dollars per dollar. The intercept of $248,300 is an estimate of sales if newspaper advertising were cut to 0. A regression function may be a poor estimate of actual results if we move too far from the data cluster. This is a word of

Figure 10.3 Scattergram with Plot of Line: Casey's Department Store Sales as a Function of Advertising

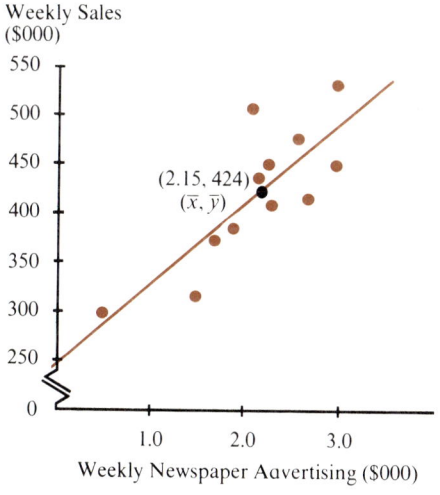

[5]. The slope, you will recall, is defined as $\Delta y/\Delta x$. If we set the estimated slope b equal to $\Delta y/\Delta x$ and solve for Δy, we get $\Delta y = b\Delta x$; thus, if Δx increases by 1 unit—i.e., $\Delta x = 1$—then Δy increases by an amount equal to b.

warning. Problems related to extrapolating beyond the domain of x will be discussed in a later section.

 ## 10.3 The Classical Linear Regression Model

In order to fully appreciate the problems of statistical inference and to intelligently address the question, How good is this equation? some consideration should be given to the assumptions underlying the Classical Regression Model. It is perhaps infrequent when a real world population truly satisfies all of the following assumptions. We recommend that you don't spend a lot of time on these now but, as you work your problems, return frequently to this section and ask yourself which of the assumptions seem realistic in terms of the population being studied.

The Classical Regression Model Assumptions

1. The independent variable, x, is considered to be *fixed*; i.e., repeated samples may be drawn for given values of x.
2. The dependent variable, y, is a random, or *stochastic*, variable.
3. The mean value of y is a linear function of x; $\mu_y = A + Bx$ where A and B are real numbers.
4. Each individual y-value is a linear function of x plus a random error or disturbance term, ϵ_i: $y_i = A + Bx_i + \epsilon_i$.
5. The ϵ_i are normally distributed for each x with $E(\epsilon_i) = 0$ (implied by 1 and 2), and the variance of the random error term, σ_ϵ^2, is constant for each x. (This is the assumption of *homoscedasticity*— versus that of *heteroscedasticity*, or unequal variance.)
6. The random error terms are independent or uncorrelated; the covariance[6] of ϵ_i with ϵ_j is 0.

The visual presentation in Figure 10.4 illustrates Assumptions 1 through 5. Consider several fixed values of x—say x_1, x_2, and x_3. The mean values of y, μ_1, μ_2, and μ_3 lie on the line $\mu_y = A + Bx$. For a given x-value, say x_1, the y-values are normally distributed about μ_1. Define the random error term as $\epsilon_{1i} = y_{1i} - \mu_1$. The ϵ_1 is normally distributed with a mean of 0 and a standard deviation, σ_ϵ; ϵ_2, ϵ_3, etc. all have the same distribution as ϵ_1. This is shown graphically. To emphasize this

---

6. The universe covariance, $cov(x,y)$, is defined as $E[(x - \bar{x})(y - \bar{y})]$; the sample covariance, $cov(x,y)$, equals
$$\frac{\Sigma(x - \bar{x})(y - \bar{y})}{n - 1}.$$
This is a measure of association or relationship between x and y. If x and y are independent, then $cov(x,y) = 0$. (The converse is not true.) If y is a perfect linear function of x, then $cov(x,y) = \pm s_x s_y$, plus or minus according as the relationship is direct or inverse.

Figure 10.4 Classical Regression Model

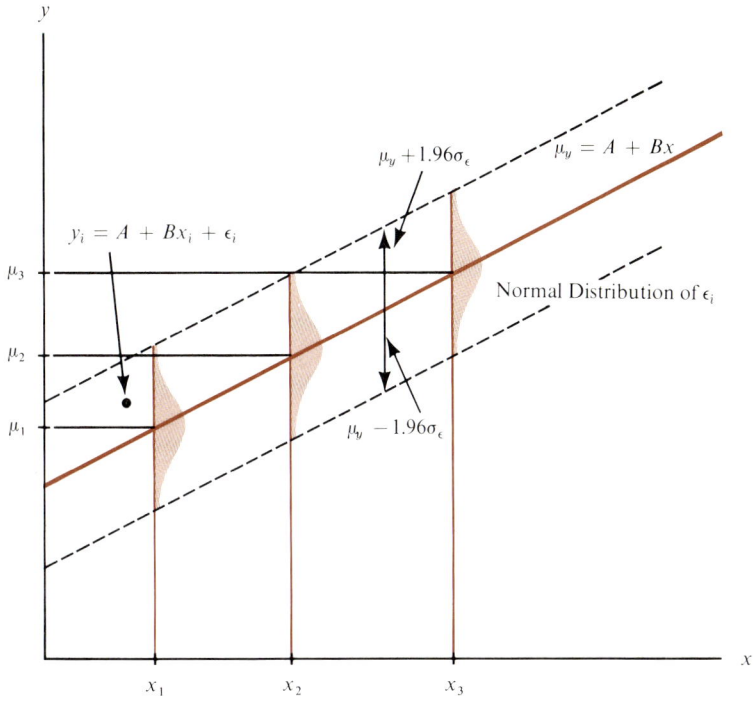

point, we have sketched two lines parallel to $\mu_y = A + Bx$, one $1.96\,\sigma_\epsilon$ below the line and the other $1.96\,\sigma_\epsilon$ above; 95 percent of all y-values fall within these parallel bands.

Nearly all important statistical inferences drawn from the sample regression results depend upon the above assumptions. Reference will be made to these as we proceed with the discussion.

10.4 Inference from Regression

The Standard Error of Regression

Recall that the sum of the squared deviations, $\Sigma(y_i - \hat{y}_i)^2$, was minimized in deriving the 2 conditions needed for finding a and b in the equation $\hat{y} = a + bx$. It can be shown that a and b are unbiased estimators of A and B respectively in the model $y_i = A + Bx + \epsilon_i$; i.e., $E(a) = A$ and $E(b) = B$. Each y-value of our data set could be viewed as $\hat{y}_i = a + bx_i$ plus some error term, e_i equal to $(y_i - \hat{y}_i)$; i.e., $\hat{y}_i = a + bx_i + e_i$. (Note that, as pointed out earlier, $\Sigma e_i = 0$.)

The Standard Error of the Regression

The e_i are estimates of the ϵ_i in the classical model, and the $\Sigma(y_i - \hat{y}_i)^2 = \Sigma e_i^2$, which, when divided by $(n - 2)$, defines an unbiased estimation of

$$\sigma_\epsilon^2 : \hat{\sigma}_\epsilon^2 = s_e^2 = \frac{\Sigma(y_i - \hat{y}_i)^2}{n - 2}.$$

This is the definition of the sample variance of the regression. The square root of this, s_e, is the standard error of the regression. The denominator is $(n - 2)$ because 2 degrees of freedom, 2 free variables, were lost in determining the coefficients of the regression function. (See Appendix 10A.) For example, if we had only 2 sample points, there would be a perfect fit; there would be no free variables available for estimating s_e^2. Only for sample sizes greater than 2 will there be "free" data points.

A Calculation Formula for the Standard Error of the Regression

To do hand calculations, the following is a more convenient formula:

$$s_e = \sqrt{\frac{\Sigma y_i^2 - a\Sigma y_i - b\Sigma x_i y_i}{n - 2}}.$$

(See Appendix 10C where we show that $\Sigma(y_i - \hat{y}_i)^2 = \Sigma y_i^2 - a\Sigma y_i - b\Sigma x_i y_i$.)

Statistical Inference on the Slope Coefficient

The formula above provides us with an estimate of the variability of the y-values about the universe line or hypothesized line, $A + Bx$. It should be clear that the sample estimate, b, of the hypothesized slope parameter, B, will vary from sample to sample. As we would expect, the variability of the slope parameter estimate is a function of the variability of the y-values about the hypothesized line.

The Standard Error of b, the Slope Coefficient

The estimated standard deviation, s_b, of the slope parameter estimator, b, can be calculated from the following formula:

$$s_b = \frac{s_e}{\sqrt{\Sigma(x - \bar{x})^2}} = \frac{s_e}{\sqrt{\Sigma x^2 - \frac{(\Sigma x)^2}{n}}}.$$

If the assumptions of the classical linear regression model discussed in Section 10.2 are satisfied, then the slope parameter, b, has a normal distribution and the following statistic can be used for making inferences about B, the hypothesized slope parameter:

$$\frac{b - B}{s_b} = t \text{ with } df = n - 2.$$

Note that this is analogous to the sample statistics having a *t*-distribution (or standard normal distribution for large sample size) used in the previous chapters: the numerator is equal to the sample estimator, b, minus the expected value of the estimator, B, and the denominator is the estimated standard deviation of the sample estimator. Also observe that the degrees of freedom are $n - 2$; the reason for $n - 2$ degrees of freedom has already been discussed above.

The application of the simple linear regression model is generally motivated by theoretical analysis that leads to the conclusion that one variable, y, is a function of the other variable, x. If the functional relationship is linear, we observe that a slope of 0 would imply that y is not a function of x; i.e., the graph of the linear function $y = A + Bx$ is a horizontal line, $y = A$, when $B = 0$, which means that the value of y is the same for every value of x. If we were to randomly select values of 2 random variables, x and y, then pair the values and calculate the least squares regression line, we would most probably obtain a calculated value of b different from 0 (positive or negative) even though the slope of the universe line is 0; i.e., $B = 0$. The *t*-statistic given above can be used to determine whether the non-0 value of b resulted from chance or because the sample data were generated by a universe line with a non-0 slope.

Example 10.4.1

Test the hypothesis that $B = 0$ at the 5 percent significance level using the sales and advertising expenditures data for Example 10.2.1 given in Table 10.4.

To perform the hypothesis test, it is necessary to calculate s_b, the estimated standard deviation of b from the sample data. The following summations were calculated from the sample data: $\Sigma x = 25.8$; $\Sigma y = 5{,}090$; $\Sigma xy = 11{,}386$; $\Sigma x^2 = 60.88$; and $\Sigma y^2 = 2{,}214{,}100$ (see Table 10.5). The least squares estimators of A and B calculated in the solution to Example 10.2.1 were $a = 248.3$ and $b = 81.8$. The standard deviation of the y-values about the hypothesized regression line is calculated using the formula given above:

$$s_e = \sqrt{\frac{\Sigma y^2 - a\Sigma y - b\Sigma xy}{n - 2}}$$

$$= \sqrt{\frac{2{,}214{,}100 - (248.3)5{,}090 - (81.8)11{,}386}{12 - 2}}$$

$$= \sqrt{\frac{18{,}900}{10}} = \sqrt{1{,}890} = 43.5.$$

The estimated standard deviation of b, s_b, can now be calculated using the formula given above:

$$s_b = \frac{s_e}{\sqrt{\Sigma x^2 - \frac{(\Sigma x)^2}{n}}} = \frac{43.5}{\sqrt{60.88 - \frac{(25.8)^2}{12}}} = 18.7$$

Hypothesis testing for regression analysis is no different from what you have learned in Chapter 8.

1. $H_o: B = 0$
   $H_a: B \neq 0$.

2. $\alpha = .05$
   $$t = \frac{b - B}{s_b} = \frac{b}{s_b}$$
   with degrees of freedom $= n - 2 = 12 - 2 = 10$.

3. Reject H_o if and only if $t_c < -2.228$ or $t_c > 2.228$.

4. $t_c = \frac{b}{s_b} = \frac{81.8}{18.7} = 4.37;$
   Reject H_o.

The null hypothesis $B = 0$ is rejected, and we conclude that there is a relationship between sales and advertising expenditures. The value of t used in the decision rule (Step 3) was obtained from Table C.

As discussed in Section 10.2, the slope of the regression line, B, is the amount by which y will change when x is increased by one unit. You will recall from your economics courses that incremental analysis plays a critical role in decision making. Management may require a confidence interval estimate of the hypothesized slope parameter.

Example 10.4.2 Find the 90 percent confidence interval estimate of the slope of the hypothesized regression line in Example 10.2.1.

The confidence interval estimate of B is calculated in the same manner as the confidence interval estimate of the population mean μ:

$b \pm t_o s_b$.

For this example, t_o is a value of the random variable t with $n - 2$ degrees of freedom such that $P(-t_o < t < t_o) = .90$. From Table C under the column headed .05 and along the row for $12 - 2 = 10$ degrees of freedom, we obtain the value of t_o, 2.228. The 90 percent confidence interval estimate of b is

$b \pm t_o s_b = 81.8 \pm 2.228(18.7) = 81.8 \pm 41.7$.

Using the notation from Chapter 7 we could state this result as follows:

Conf $\{40.10 \leq B \leq 123.50\} = .90$.

The report to management on this point would read something like this:

Our best estimate is that total sales can be expected to increase (decrease) by $81.80 per dollar of increase (decrease) in newspaper advertising; we are 90 percent confident that this is in error by no more than $41.70; or, the odds are 9:1 that the change in sales per dollar change in newspaper advertising is at least $40.10 and no more than $123.50; i.e., there is only a 5 percent chance that the true expected change is less than $41.70 and a 5 percent chance that it will exceed $123.50.

Estimating the Dependent Variable, y

Given the discussion and the derivation of s_e earlier in this section, you may now feel that s_e is an estimate of the standard deviation of the y-values about the line of best fit. Not so! Recall that s_e is an estimate of σ_ϵ, so s_e is an estimate of the variability of the y-values about the *universe straight line*.

The Standard Deviation of the Dependent Variable: A Function of the Independent Variable

Unfortunately, we must add another tree to the forest: because of the variability of b, the slope coefficient, the variability of y about the sample function is itself a divergent hyperbolic function of the fixed x value:

$$s_{y \cdot x} = s_e \sqrt{1 + \frac{1}{n} + \frac{(x - \bar{x})^2}{\Sigma(x_i - \bar{x})^2}}.$$

Several observations are in order:

1. If n is large (or as $n \to \infty$) $1/n$ is close to 0 (or $1/n \to 0$), and at $x = \bar{x}$ the last term under the radical is 0. We can conclude that at the mean value of x, $\bar{x}$, the variability of y_i about $\bar{y}$ is, for all practical purposes, measured by s_e even if n is of just modest magnitude.
2. If n is large and x is kept within the domain of the x variable, then, even though s_e underestimates $s_{y \cdot x}$, it could still be an acceptable estimate in many practical problems.
3. If n is small and/or x is taken outside the domain of the actual x-values, $s_{y \cdot x}$ will need to be calculated for each and every value of x.

Before proceeding with the consideration of other important standard errors, let us return to the newspaper advertising example introduced in Section 10.2. The major application of $s_{y \cdot x}$ would be that of estimation.

Example 10.4.3

The manager of Casey's decided to spend $2,150 on newspaper advertising next week and requested that we estimate sales for him. (Recall that $\bar{x} = \$2.15$ thousands.)

We would proceed as follows: For $x = 2.15 = \bar{x}, \hat{y} = 248.3 + 81.8(2.15) = 424.2 = \bar{y}$. We report $\hat{y}$, our point estimate of sales, to be \$424,200. The interval estimate would be of the form, $424,200 \pm t\, s_{y \cdot x}$, where t would be determined by the desired confidence level (assume 90 percent) and the degrees of freedom, $n - 2$—10 in our example. From the t-table we find $t = 2.228$, and $s_{y \cdot x}$ is calculated as follows:

$$s_{y \cdot x} = s_e \sqrt{1 + \frac{1}{n} + \frac{(x - \bar{x})^2}{\Sigma(x_i - \bar{x})^2}} = 43.5 \sqrt{1 + \frac{1}{12} + 0} = 45.3.$$

(The last term is 0 because $x = \bar{x}$.)

The 90 percent confidence interval calculated on the estimated sales would be

$$424,200 \pm 2.228(45.3)$$
$$= 424,200 \pm 101,900.$$

Our memo to the manager could read, "We estimate sales to be \$424,200, and we are 90 percent confident that this will be in error by no more than \$101,900."

Note that we have used a Student-t in arriving at our interval estimate. The Student-t distribution was derived assuming the parent distribution to be normal. Statistical inferences relating to individual estimates are truly valid only with the full set of assumptions from the Classical Regression Model.

Example 10.4.4

Assume now that the manager decides to buy \$500 worth of advertising next week. Find both the point and interval estimates of sales.

For $x = 500, \hat{y} = 248.3 + 81.8(.5) = 289.2$.

$$s_{y \cdot x} = 43.5 \sqrt{1 + \frac{1}{12} + \frac{2.723}{5.410}} = 54.8.$$

The 90 percent confidence interval is: $289.2 + 2.228(54.8) = 289.2 \pm 122.1$.

Note particularly the increase in $s_{y \cdot x}$ as we move away from $\bar{x}$ and also that the contributing term to this divergence is the numerator, $(x - \bar{x})^2$, of the third term under the radical. All other terms and the denominator of this term are constant. The denominator of the third term, $\Sigma(x_i - \bar{x})^2$, was calculated using the identity developed in Chapter 2, $\Sigma(x_i - \bar{x})^2 = \Sigma x^2 - n\bar{x}^2$.

Estimating the Mean Value of y

More closely related to our previous work on estimation in Chapter 7 is the problem of estimating the mean value of y or, if you please, the universe $\mu_y = A + Bx$, as it depends upon x. The standard error of this estimate is relatively more sensitive to sample size and the magnitude of $(x - \bar{x})^2$ than the standard error of the individual y discussed above.

The Standard Error of the Mean Value of the Dependent Variable

The formula for this standard error is

$$s_{\hat{\mu} \cdot x} = s_e \sqrt{\frac{1}{n} + \frac{(x - \bar{x})^2}{\Sigma(x_i - \bar{x})^2}}.$$

Note that $s_{\hat{\mu} \cdot x}$ is also a divergent function of $(x - \bar{x})^2$. Following along with our example, let us estimate expected sales or mean sales for $x = 2.15$ and for $x = .50$. The point estimates for mean sales are the same as those for the individual and are calculated from the regression function:

for $x = \bar{x} = 2.15$, $\hat{\mu}_y = 424.2$

and

for $x = .50$, $\hat{\mu}_y = 289.2$.

Calculating the standard errors, we have

for $x = \bar{x} = 2.15$, $s_{\hat{\mu} \cdot x} = 43.5 \sqrt{\frac{1}{12} + 0} = 12.5$

for $x = .50$, $s_{\hat{\mu} \cdot x} = 43.5 \sqrt{\frac{1}{12} + \frac{2.723}{5.410}} = 33.3$.

In both cases, the 90 percent confidence interval would be of the form,

$\hat{\mu} \pm 2.228 \, s_{\hat{\mu} \cdot x}$.

Interpreting the first case, if management decided to fix advertising expenses at $2,150 per week, the point estimate of expected sales would be $424,200, and we would be willing to lay odds of 9:1 that this estimate of expected sales would be in error by no more than $27,850 [(2.228)(12.5) thousands of dollars]. Make a similar statement for the second case where $x = .50$.

Two special cases of $s_{\hat{\mu} \cdot x}$ are worth noting:

1. At $x = 0$, $s_{\hat{\mu} \cdot x} = s_a = s_e \sqrt{\frac{1}{n} + \frac{\bar{x}^2}{\Sigma(x_i - \bar{x})^2}}$

2. At $x = \bar{x}$, $s_{\hat{\mu} \cdot x} = \frac{s_e}{\sqrt{n}}$

Summary of Statistical Inference in the Two-Variable Linear Model

Let us summarize the important material covered to this point.

1. The coefficients, a and b, of the linear regression function, $\hat{y} = a + bx$, are determined by

$$b = \frac{n\Sigma xy - \Sigma x \Sigma y}{n\Sigma x^2 - (\Sigma x)^2}$$

and

$a = \bar{y} - b\bar{x}$.

Inference from Regression

The criterion used to derive these formulas is known as the "least squares" criterion; the sum of the $(y_i - \hat{y}_i)^2$ was minimized.

2. The sample standard error of regression, s_e, is defined as

$$s_e = \sqrt{\frac{\Sigma(y_i - \hat{y})^2}{n - 2}}.$$

It is usually more convenient to find s_e from an equivalent formula:

$$s_e = \sqrt{\frac{\Sigma y_i^2 - a\Sigma y_i - b\Sigma x_i y_i}{n - 2}}.$$

Three important standard deviations are each a function of s_e.

3. The standard error of the slope-coefficient, b, is

$$s_b = \frac{s_e}{\sqrt{\Sigma(x_i - \bar{x})^2}}.$$

(From Chapter 2, recall that $\Sigma(x_i - \bar{x})^2 = \Sigma x_i^2 - n\bar{x}^2$.) For statistical estimation of the population slope-coefficient, B, the point estimate of B is b and the interval estimate of B is $b \pm ts_b$ where t depends upon the degrees of freedom, $(n - 2)$, and the desired confidence level. For hypothesis testing, the test statistic is

$$t = \frac{b - B}{s_b} \text{ with } df = n - 2.$$

If $H_o: B = 0$, then $t = \dfrac{b}{s_b}$.

4. The standard deviation of the individual y-value for any given x is

$$s_{y \cdot x} = s_e \sqrt{1 + \frac{1}{n} + \frac{(x - \bar{x})^2}{\Sigma(x_i - \bar{x})^2}}.$$

We noted that for $x = \bar{x}$ the last term under the radical is 0, and as x departs from $\bar{x}$, $s_{y \cdot x}$ is a divergent function of this departure; also for large n and x reasonably close to $\bar{x}$, $s_{y \cdot x} \cong s_e$. For the problem of estimating an individual y given x: the point estimate is $\hat{y} = a + bx$ and the interval estimate is $\hat{y} \pm ts_{y \cdot x}$.

5. The standard deviation of the mean y-value for a given x is

$$s_{\hat{\mu} \cdot x} = s_e \sqrt{\frac{1}{n} + \frac{(x - \bar{x})^2}{\Sigma(x_i - \bar{x})^2}}.$$

Special cases of this are

At $x = \bar{x}$, $s_{\hat{\mu} \cdot x} = \dfrac{s_e}{\sqrt{n}}$,

and at $x = 0$ we have the standard error of the intercept coefficient,

$$s_a = s_e \sqrt{\frac{1}{n} + \frac{\bar{x}^2}{\Sigma(x_i - \bar{x})^2}}.$$

If we are estimating the mean value of y for a given x: the point estimate is $\hat{\mu} = a + bx$ and the interval estimate is $\hat{\mu} + ts_{\hat{\mu} \cdot x}$.

10.5 The Coefficient of Correlation and the Coefficient of Determination

The coefficient of linear correlation, ρ for the population and r for the sample, can be more adequately addressed and interpreted in terms of another model, the bivariate normal model. In this model both x and y are random variables, with y normally distributed for every x and vice versa. The square of this coefficient, either ρ^2 or r^2, is called the coefficient of determination and does have an important interpretation in the Classical Regression Model.

The sample coefficients, either r or r^2, may be defined and calculated in several ways. Before going to the formulas let us look at some of the properties and the interpretation of these coefficients.

Properties of the Coefficient of Linear Correlation

Properties of the coefficient of linear correlation, r, are:

1. r has the same sign as b, the slope coefficient.
2. r may take on any value from -1 to $+1$.
3. If $r = \pm 1$, all data points fit the regression function, $\hat{y} = a + bx$; i.e., we have a perfect fit.
4. If $r = 0$ then $b = 0$ and $\hat{y} = a = \bar{y}$. There is no relationship between y and x, and the best estimate of y given any x is $\bar{y}$.

Properties of the Coefficient of Determination

Properties of r^2, the coefficient of determination are:

1. r^2 is always nonnegative and takes on values from 0 to 1.
2. r^2 is an estimate of ρ^2, and ρ^2 measures the proportion of the variability in the dependent variable, y, explained by the variability in x. If the sample size is large, then r^2 will be a reasonably good estimate of ρ^2; at one extreme, $r^2 = 1$, and 100 percent of the variability in y is explained by x; at the other extreme, $r^2 = 0$, and x is simply not an explanatory variable.

Formula for the Coefficient of Correlation

The basic formula for r is

$$r = \frac{n\Sigma xy - \Sigma x \Sigma y}{\sqrt{[n\Sigma x^2 - (\Sigma x)^2][n\Sigma y^2 - (\Sigma y)^2]}}.$$

Note that the numerator is the same as that for b and, as with b, the denominator is always positive; therefore r will have the same sign as b.

There are two interesting relationships involving r—one between r and b and the second between r and s_e. Recall from Chapter 2 that the standard deviation of x, s_x, equals

$$\sqrt{\frac{\Sigma(x_i - \bar{x})^2}{n - 1}}.$$

(We used the subscript on s_x because the standard deviation of y, s_y, calculated similarly, is also going to be used.)

Relationship of r to b

With mild algebraic manipulations on the formulas for b and r, it can be shown that

$$r = b \frac{s_x}{s_y}.$$

From an altogether different approach (see Chapter 14 for a basic analysis of variance on the regression model), it can be shown that

$$s_e^2 = s_y^2(1 - r^2)\left(\frac{n - 1}{n - 2}\right)$$

and, for large n,

$$s_e^2 \cong s_y^2(1 - r^2).$$

Relationship of r to s_e

Solving this for r^2, we get

$$r^2 = 1 - \frac{s_e^2}{s_y^2}.$$

From our original discussion it was clear that both r^2 and s_e were, in one sense, measures of the "goodness" of the functional relationship of y to x. It is not too surprising to find out that they are related. Inspection of the last equation shows that $r^2 \to 1$, $s_e \to 0$, and as $r^2 \to 0$, $s_e \to s_y$. This is entirely consistent with our previous discussion. If $r^2 = 1$, there is a perfect fit and all the variability in y is explained; therefore $\Sigma(y_i - \hat{y}_i)^2 = 0$ and $s_e = 0$. Also, if $r^2 = 0$, $b = 0$, $\hat{y} = \bar{y}$ for all x, and $\Sigma(y_i - \hat{y}_i)^2 = \Sigma(y_i - \bar{y})^2$; therefore $s_e \cong s_y$.

Let us return to our newspaper advertising example from the previous section and calculate r and r^2 for this problem. The following summations were given earlier: $\Sigma x = 25.8$, $\Sigma y = 5,090$, $\Sigma x^2 = 60.88$, $\Sigma y^2 = 2,214,100$, and $\Sigma xy = 11,386$. From the first definition of r above,

$$r = \frac{12(11{,}386) - 25.8(5{,}090)}{\sqrt{[12(60.88) - (25.8)^2][12(2{,}214{,}100) - (5{,}090)^2]}}$$

$$= \frac{5{,}310}{\sqrt{(64.92)(661{,}100)}} = \frac{5{,}310}{6{,}551} = .81.$$

We found r to be .81 and $r^2 = .66$. A sample size of 12 is too small to interpret r^2 as if it were ρ^2. If the sample size were several times 12, we could say that 66 percent of the variability in sales is determined by variability in newspaper advertising and the other 34 percent is left unexplained—perhaps to be explained by other independent variables.

10.6 An Alternative Set of Computational Formulas (Optional)

We have found the following set of formulas and sequence of calculations to be useful when a nonprogrammable hand calculator is the most powerful tool available. In addition to increased efficiency and probable accuracy, important relationships among and between the sample statistics are emphasized.

The formulas and sequence of calculations are listed below. The Casey's Department Store example will be used subsequently in a step-by-step illustration.

1. Calculate Σx_i, Σy_i, Σy_i^2, Σx_i^2, and $\Sigma x_i y_i$.
2. Find $\bar{y}$, $\bar{x}$, s_y, s_x, and $cov(x,y)$: where

$$\bar{y} = \frac{\Sigma y_i}{n},$$

$\bar{x}$ similarly;

$$s_y = \sqrt{\frac{\Sigma y_i^2 - n\bar{y}^2}{n - 1}},$$

s_x similarly;
and

$$cov(x,y) = \frac{\Sigma x_i y_i - n\bar{x}\bar{y}}{n - 1}.$$

3. Find the regression coefficients, a and b, where

$$b = \frac{cov(x,y)}{s_x^2}$$

and $a = \bar{y} - b\bar{x}$.

4. Find r and r^2 where

$$r = b \frac{s_x}{s_y}$$

or $r = \frac{cov(x,y)}{s_x s_y}$.

5. Find s_e where

$$s_e = s_y \sqrt{1 - r^2} \sqrt{\frac{n-1}{n-2}}.$$

Note: If n is large, $s_e \cong s_y \sqrt{1 - r^2}$.

6. Find s_b where

$$s_b = \frac{s_e}{\sqrt{n-1}\, s_x}.$$

7. The formula for $s_{y \cdot x}$ is

$$s_{y \cdot x} = s_e \sqrt{1 + \frac{1}{n} + \frac{z_x^2}{n-1}} \cong s_e \sqrt{1 + \frac{1 + z_x^2}{n}}$$

where

$$z_x = \frac{x - \bar{x}}{s_x}.$$

Note: For large n, $s_{y \cdot x} \cong s_e$.

We should emphasize that $s_{y \cdot x}$ is the estimated standard deviation of y as a function of x and, as the x deviation from the mean value of x increases, $s_{y \cdot x}$ increases. The variable, z_x, measures the number of deviations x is from $\bar{x}$. Inspection of the above form shows clearly that, for any reasonable value of z_x and a large sample size, s_e will be a good approximation for $s_{y \cdot x}$.

8. The formula for $s_{\hat{\mu} \cdot x}$ is

$$s_{\hat{\mu} \cdot x} = s_e \sqrt{\frac{1}{n} + \frac{z_x^2}{n-1}} \cong s_e \sqrt{\frac{1 + z_x^2}{n}}.$$

Note: For $x = \bar{x}$, $z_x = 0$ and

$$s_{\hat{\mu} \cdot x} = \frac{s_e}{\sqrt{n}}.$$

You should recall that $s_{\hat{\mu} \cdot x}$ is the estimated standard deviation of the *mean value* of y as a function of x.

Refer to Table 10.5 for the basic calculations for the Casey's Department Store example. The necessary summations are at the bottom of each column. We will parallel the sequence of calculations listed above. Results will, of course, be the same as those in the original work earlier in the chapter. See Chapter 1 for comments on rounding and rounding errors.

1. From the work presented in Figure 10.4,

    $\Sigma y_i = 5,090$; $\Sigma x_i = 25.8$; $\Sigma y_i^2 = 2,214,100$;

    $\Sigma x_i^2 = 60.88$; and $\Sigma x_i y_i = 11,386$.

2. $$\bar{y} = \frac{5,090}{12} = 424.17$$

    $$\bar{x} = \frac{25.8}{12} = 2.150$$

    $$s_y = \sqrt{\frac{2,214,100 - 12(424.17)^2}{11}} = 70.75$$

    $$s_x = \sqrt{\frac{60.88 - 12(2.150)^2}{11}} = .7013$$

    and

    $$cov(x,y) = \frac{11,386 - 12(2.150)(424.17)}{11} = 40.22.$$

    At this point one should observe whether the calculated values of the univariate statistics are "reasonable." If not we will check our calculations before proceeding; major arithmetic errors will frequently be found with a quick survey of these results. For example, $\bar{y} = 424$ is certainly within the central cluster of the sales data; range of $y = 540 - 300 = 240$ and is about 3½ times s_y. (As a rough rule, the range will usually be 3 to 5 times the standard deviation. A standard deviation that is less than 2½ or more than 5½ times the range should be checked.) By similar comparison, do you think $\bar{x}$ and s_x are reasonable results? The sign of the covariance should be the same as that of the slope; this should be checked with the visual information gained from the scattergram and plot.

3. We next find the regression coefficients:

    $$b = \frac{cov(x,y)}{s_x^2} = \frac{40.22}{.4918} = 81.8$$

    and

    $$a = \bar{y} - b\bar{x} = 424.17 - (81.8)(2.15) = 248.3.$$

    The least squares linear regression function is: $\hat{y} = 248.3 + 81.8x$. This should now be plotted on the same graph as the scattergram—another check for the reasonableness of the results to this point.

4. Find r and r^2.

$$r = b\frac{s_x}{s_y} = 81.8\left(\frac{0.7013}{70.75}\right) = .811$$

and

$$r^2 = .657.$$

5. Find s_e.

$$s_e = s_y \sqrt{1 - r^2} \sqrt{\frac{n-1}{n-2}} = 70.75 \sqrt{.343} \sqrt{1.10}$$

$$s_e = 43.5$$

6. Find s_b.

$$s_b = \frac{s_e}{\sqrt{n-1}\, s_x} = \frac{43.5}{\sqrt{11}\,(.7013)} = 18.7.$$

7 and 8. Earlier in the chapter we found $s_{y \cdot x}$ and $s_{\hat{\mu} \cdot x}$ for $x = \bar{x} = 2.15$ and for $x = .50$. We will repeat the calculations for $x = .50$ but will choose $x = 2.50$ as the second example. We first find z_x and z_x^2 for both cases:

For $x = .50$,

$$z_x = \frac{.500 - 2.150}{.7013} = -2.35.$$

$$z_x^2 = 5.54.$$

For $x = 2.50$,

$$z_x = \frac{2.500 - 2.150}{.7013} = .50.$$

$$z_x^2 = .25.$$

(We will use the approximate formulas to illustrate that, even for a small sample of 12, the error is not significant.)

Find $s_{y \cdot x}$ for each case.

$$s_{y \cdot x} \cong s_e \sqrt{1 + \frac{1 + z_x^2}{n}}.$$

For $x = .50$,

$$s_{y \cdot x} \cong 43.5 \sqrt{1 + \frac{1 + 5.54}{12}} = 54.5.$$

For $x = 2.50$,

$$s_{y \cdot x} \cong 43.5 \sqrt{1 + \frac{1 + .25}{12}} = 45.7.$$

(For $x = .50$, $s_{y \cdot x}$ is 54.8, calculated correctly in the earlier discussion.)

Find $s_{\hat{\mu} \cdot x}$ for each case:

$$s_{\hat{\mu} \cdot x} = s_e \sqrt{\frac{1 + z_x^2}{n}}.$$

For $x = .50$,

$$s_{\hat{\mu} \cdot x} = 43.5 \sqrt{\frac{1 + 5.54}{12}} = 32.1.$$

For $x = 2.50$,

$$s_{\hat{\mu} \cdot x} = 43.5 \sqrt{\frac{1 - .25}{12}} = 14.1.$$

(For $x = .50$, the correct value of $s_{\hat{\mu} \cdot x}$ is 33.3. This is a fairly large relative error but, after all, n is only 12 and $x = .50$ is 2.35 standard deviations from the mean of x).

10.7 A Linear Regression by Computer

Frequent reference has been made to the fact that meaningful regression problems are arithmetically intractable without the computer tool. The multiple regression problem in Section 8 is discussed and analyzed in terms of the computer printout only. Before proceeding with this we would like to consider a computerized example and the associated printout of our simple two-variable problem. The data for our example are presented in the first two columns of Table 10.6. For the multiple regression example, the third data column is used along with the first two. We are dealing with some adjusted real-world data of the fictional firm of Butte-Anaconda Natural Gas Company, BANGCO. The dependent variable, y, represents daily sales in thousands of dollars during the winter months, and the independent variable, x, is mean daily temperature in degrees Fahrenheit. We asked the computer to determine the least squares linear function, $\hat{y} = a + bx$. Our thesis is, of course, that sales are an inverse function of temperature. The computer printout is reproduced in Table 10.7. The printout from most software linear regression programs is similar to that in Table 10.7.

We will temporarily skip a few lines to take a look at the equation, headed by "Variables in Equation." The least squares line of best fit is $\hat{y} = 646.2 - 8.36x$. The coefficients, a and b, are found under the heading "Coefficient," a in the row labeled "(Constant)" and b in that

Table 10.6 Data from BANGCO

Day	y Daily Sales ($1000's)	x_1 Mean Daily Temperature (Degrees Fahrenheit)	x_2 Mean Daily Wind Velocity (Miles/Hour)
1	350	33	6.4
2	346	36	14.3
3	354	37	21.8
4	402	32	14.4
5	455	25	10.2
6	482	20	6.6
7	497	16	4.3
8	442	24	8.4
9	517	16	8.4
10	513	18	9.3
11	514	18	11.1
12	461	25	13.4
13	403	29	6.1
14	399	30	7.5
15	385	31	8.2
16	364	34	6.8
17	378	35	14.7
18	367	34	6.7
19	354	33	5.5
20	370	32	4.1
21	367	34	4.5
22	341	37	6.5
23	344	34	1.2
24	378	31	5.3
25	399	28	2.8
26	426	26	6.2
27	427	24	0.7
28	431	25	5.5
29	415	26	2.5
30	412	25	1.1

labeled "x". Before proceeding with further analysis, we should interpret these coefficients. The y-intercept, $a = 646.16$, interpreted directly tells us that, if mean daily temperature were 0 degrees Fahrenheit, then sales would be estimated at $646,200. A mean daily temperature of 0 is some 2 standard deviations (note: $s_x = 6.40$) below the lowest temperature reading (x min is 16 degrees Fahrenheit). Because of this (and as the statistical consultants), we should warn BANGCO that $646,200 is not a reliable estimate of sales if mean daily temperature were to drop to 0. The slope coefficient, -8.36, is the estimated change in daily sales ($-8,360$) per degree change in mean daily temperature. For example, if the mean daily temperature were forecast to be 20 degrees tomorrow, estimated sales would be $\hat{y} - 646.2 - 8.36(20) = 479.0$ or $479,000. If the next day were to drop a degree to 19 degrees, the estimate of sales

Table 10.7 BANGCO Computer Printout
y = daily sales ($000); x = mean daily temperature in degrees Fahrenheit

NUMBER OF CASES   30
NUMBER OF VARIABLES 2

VARIABLE	MEAN	STD. DEVIATION
Y	409.933	54.508
X	28.267	6.397

COEFFICIENT OF CORRELATION   −.9654
STD. ERROR OF EST., S_E 14.4634

ANALYSIS OF VARIANCE

	DF	SUM OF SQUARES	MEAN SQUARE	F RATIO
REGRESSION	1	80306.562	80306.562	383.894
RESIDUAL	28	5857.298	209.189	

VARIABLES IN EQUATION

VARIABLE	COEFFICIENT	STD. ERROR	T
(CONSTANT)	646.15862		
X	−8.35703	.42653	−19.5931

would increase by $8,360 to $487,360; if it were to increase a degree to 21 degrees, the estimate would drop by $8,360 to $470,640.

With a sample of size 30 and a correlation coefficient of −.965 we are quite sure, even with our limited statistical experience, that the relationship is significant. The first check on this point would be to test the hypothesis, $H_o: B = 0$ against the hypothesis, $H_1: B \neq 0$; if the null hypothesis can be rejected, then accepting the alternate hypothesis is a verification of the original thesis, "natural gas sales are an inverse function of temperature." The standard error of b, s_b, is printed in the column headed "Std. Error" to the right of the coefficients. We see that $s_b = .427$. The test-t value is

$$t = \frac{b - B}{s_b} = \frac{-8.36 - 0}{0.427} = -19.6$$

and is printed in the next column. The slope coefficient is 19.6 standard deviations below the hypothesized value of $B = 0$. Clearly, the null hypothesis is rejected for any α-error we might choose; the result is significant at any level. With degrees of freedom of 28 and $\alpha = .01$, $t^* = 2.763$.

Let us find the 95 percent confidence interval for B. This would be of the form $b = ts_b = 8.36 \pm t(.427)$ where t is found from the Student-t table with $df = n - 2 = 28$ and a one-tailed area of .025. We find this

t-value to be 2.048. The confidence interval for B is: $-8.36 \pm (2.05)(.427) = -8.36 \pm .88$; we are 95 percent confident that $-9.24 \leq B \leq -7.48$.

Moving back to the top of the Step 1 printout, we see that the $r = -.9654$ and the "Std. Error of Est." is 14.46. The second number is the important standard error of the regression; $s_e = 14.46$.

Just below these 2 numbers we see "Analysis of Variance." We will say a few words about this during the discussion of multiple linear regression; Chapter 13 addresses this topic in detail. For the 2-variable equation this adds nothing to the analysis. Note at this point that the square root of this F-ratio, $\sqrt{383.9}$, equals the test-t value, 19.6.

In Figure 10.5 we have plotted the scattergram of the data and the linear function determined by the computer. We verified graphically that the point of means satisfies the function. We recommend this exercise as standard procedure for the 2-variable problem. Two conclusions are immediately obvious with just a cursory inspection of the

Figure 10.5 BANGCO Daily Sales as a Linear Function of Mean Daily Temperature

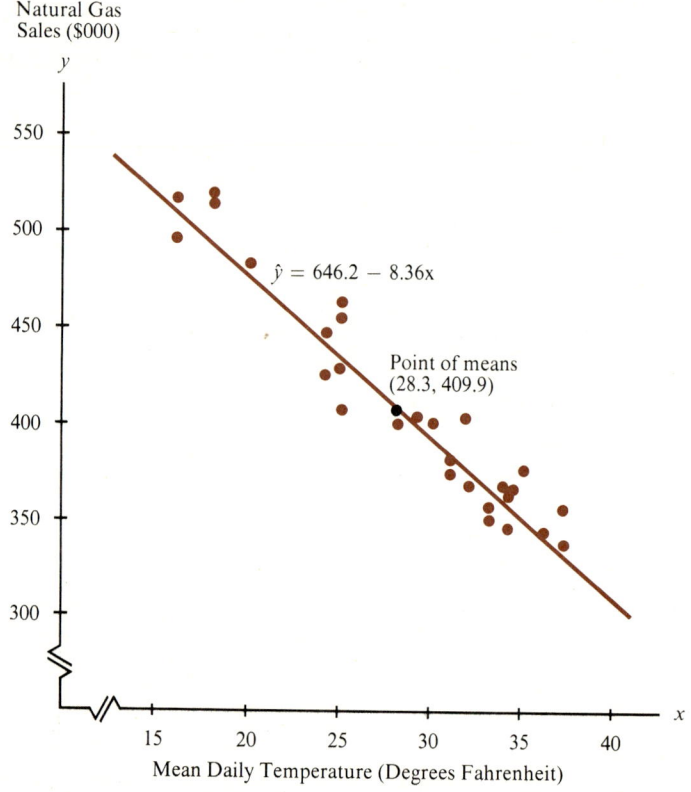

graph: (1) there are no major errors in our work with these 2 variables; (2) the linear model appears to be a very reasonable choice. If we really were making a statistical report to the nonstatisticians in management at BANGCO, the picture in Figure 10.5 would be invaluable in the discussion and explanation of our results.

 ## 10.8 Multiple Linear Regression

The assumptions for the classical linear regression model extend from the 2-variable problem to that with 3 or more variables. As with the 2-variable problem, y will be used to denote the dependent variable, and the 2 or more independent or explanatory variables will be x_1, $x_2, \ldots x_k$. Double subscripts are needed to indicate summations; e.g., Σx_{2i} will call for the sum of the n-values of the second independent variable, $x_{21} + x_{22} + \ldots + x_{2n}$. Much of the discussion that follows will address the 3-variable problem—1 dependent variable and 2 explanatory variables. Complications resulting from additional explanatory variables are primarily arithmetical in nature; we are going to assume that calculations for any meaningful multivariable regression problem will be done by a computer. It should be noted that increasing the number of variables in the regression problem increases the necessary arithmetic not linearly but roughly factorially. The total arithmetic "energy" required for a 4-variable problem is roughly 4 times that for a 3-variable problem, not a simple increase of 33⅓ percent. Recalling that 10! exceeds 1 million should increase our respect for the computer; once the data are in these machines, regression functions with a half-dozen variables and 100 data points will come spewing forth within seconds. (Going back several years, we know of an instance, not within the United States, where 6 graduate students worked the full summer with hand calculators on an 18-variable equation and never did succeed in getting an acceptable solution.)

Classical Assumptions for the Multivariable Model (Two Independent Variables)

Assumptions for the 3-variable Classical Regression Model are essentially the same as those for the 2-variable problem (see Section 3):

1. The mean value of y is a linear function of x_1 and x_2, where x_1 and x_2 are "fixed" variables; i.e., $\mu_y = A + B_1 X_1 + B_2 X_2$. We were able to look at the 2-dimensional analogue as a straight line in the 2-space; this function is a plane in Euclidean 3-space. If there were 4 or more variables, we could no longer visualize the geometry.

2. $y_i = A + B_1 X_{1i} + B_2 X_{2i} + \epsilon_i$. The dependent variable, y, is a random variable; it is a linear function of x_1 and x_2 plus a random error or disturbance term, ϵ_i.

3. The ϵ_i are normally distributed with a mean of 0 and a constant

variance, σ_ϵ^2, for all combinations of x_1 and x_2. (Again, this is the assumption of homoscedasticity.)

4. The random error terms are uncorrelated.

Discussion of the statistically determined equation, $\hat{y} = a + b_1x_1 + b_2x_2$, and the related statistical inference will be primarily in terms of the computer printout. Formulas for the equation coefficients and the important standard errors are not so complicated as they are arithmetically intractable.

Normal Equations for Three-Variable Regression

The extension of the normal equations from the 2-variable least squares determination to the 3-variable problem is interesting to note:

$$\Sigma y = na + b_1\Sigma x_1 + b_2\Sigma x_2$$
$$\Sigma x_1 y = a\Sigma x_1 + b_1\Sigma x_1^2 + b_2\Sigma x_1 x_2$$
$$\Sigma x_2 y = a\Sigma x_2 + b_1\Sigma x_1 x_2 + b_2\Sigma x_2^2.$$

The quite obvious extension pattern to be induced from the 2 × 2 set of equations (in black above) to the 3 × 3 set continues to the k-variable case. Dividing the first equation by n gives $\bar{y} = a + b_1\bar{x}_1 + b_2\bar{x}_2$ and, as with the 2-variable problem, we see that the least squares equation passes through the point of means, $(\bar{x}_1, \bar{x}_2, \bar{y})$, and that this is 1 of the 3 conditions imposed by the least squares criterion.

The data for our example, shown in Table 10.6, are hypothetical in the sense that they represent modest adjustments on some real data (the firm name is also fictional). Management of the Butte-Anaconda Natural Gas Company, BANGCO, was interested in how daily sales during the winter months were affected by temperature and wind velocity. Thirty days were selected from the three winter months, December, January, and February of 1978–79. Natural gas sales in thousands of dollars, mean daily temperature in degrees Fahrenheit, and the mean wind velocity were recorded for each of the sample days. The thesis would be that sales, y, are inversely related to temperature, x_1, and directly dependent upon wind velocity, x_2. After careful consideration, the classical linear model seemed to be acceptable to both management and the consultant statisticians.

The data were punched on computer cards and run on the computer using a multiple-regression, prepackaged computer program. There are a variety of multiple-regression software packages ("canned" programs), but the information on the computer printout, for most programs, is similar to that presented in Tables 10.8 and 10.9.

Several tables are printed before the equation printout; two of these are reproduced in Table 10.8. There we see a table of means and standard deviations for the variables y, x_1, and x_2. We need to keep in mind that y is sales, x_1 is temperature, and x_2 is wind velocity. Table

Table 10.8 Computer Printout of BANGCO Data: Means, Standard Deviations, and Correlation Coefficients

```
NUMBER OF CASES                        30
NUMBER OF ORIGINAL VARIABLES            3
NUMBER OF VARIABLES ADDED               0
TOTAL NUMBER OF VARIABLES               3
NUMBER OF SUBPROBLEMS                   1
THE VARIABLE FORMAT IS             (F3.0, F3.0, F5.1)
```

VARIABLE	MEAN	STD. DEVIATION
Y	409.93333	54.50842
X_1	28.26667	6.29687
X_2	7.48333	4.68306

CORRELATION MATRIX

VARIABLE NUMBER	Y	X_1	X_2
Y	1.000	−.965	.019
X_1		1.000	.197
X_2			1.000

10.8 also contains an upper triangular matrix with 1.000's on the main diagonal; this is the matrix of coefficients of linear correlation. Variables are labeled for both rows and columns, and the 1.000's on the main diagonal simply show that the correlation of a variable with itself is + 1.000. The other 2 numbers on the top row are the correlation coefficients of y with x_1 and x_2 respectively; i.e.,

$r_{y.x_1} = -.965$ and $r_{y.x_2} = .019$.

The last number in the matrix, the intersection of row 2 and column 3, is the correlation between x_1 and x_2, i.e.,

Table 10.9 Multiple Regression Computer Printout, BANGCO Data

```
MULTIPLE R          .9888
STD. ERROR OF EST.  8.4415
```

ANALYSIS OF VARIANCE

	DF	SUM OF SQUARES	MEAN SQUARE	F RATIO
REGRESSION	2	84239.870	42119.935	591.083
RESIDUAL	27	1923.989	71.259	

VARIABLES IN EQUATION

VARIABLE	COEFFICIENT	STD. ERROR	F TO REMOVE
CONSTANT	637.70359		
X_1	−8.72950	.25394	1181.7371
X_2	2.53678	.34145	55.1975

Multiple Linear Regression

$r_{x_1 x_2} = .197.$

The coefficient of correlation between any 2 variables is symmetrical; i.e., $r_{xy} = r_{yx}$; therefore, the bottom half of the matrix is left blank—there is no need for repetition. The correlation between sales and temperature, $-.965$, shows a very strong inverse linear relationship. This is not surprising. We expected sales to increase with wind velocity, but there appears to be very little if any relationship as shown by an $r = .019$. There also appears to be a mild positive correlation between the 2 independent variables. Even though the correlation coefficient,

$r_{x_1 x_2} = .197,$

is not significant, it indicates the possibility that winter wind velocity and temperatures move in the same direction.

In Table 10.9 we see the computer printout of the multivariable regression equation and related information. The functional coefficients are found below the part labeled "Variables in Equation." The constant term is a, the partial slope-coefficients for x_1 and x_2 are to the right of variables x_1 and x_2 respectively. From the computer printout we get

$\hat{y} = 637.7 - 8.73x_1 + 2.54x_2.$

The intercept, 637.7, is interpreted in much the same way as the intercept in the 2-variable equation. If both temperature and wind velocity were 0, then sales would be estimated at \$637,700. The slope coefficients are now "partial" coefficients. With wind velocity *fixed*, natural gas daily sales are estimated to decrease by \$8,730 per degree increase in mean daily temperature, and *fixing* temperature, daily sales are estimated to increase by \$2,540 per mile per hour increase in mean daily wind velocity. This notion of fixing the variables not under consideration is important and should be kept in mind when interpreting the meaning of the slope coefficients of a multivariable equation. You who have had calculus will recognize that the slope-coefficients are the partial derivatives of y with respect to each of the variables and that a partial derivative with respect to 1 variable is taken under the assumption that all other variables are kept constant.

The first and perhaps the most important question in any regression analysis is, Is the relationship significant? The answer is yes in the 2-variable problem if the hypothesis, $H_o: B = 0$, can be rejected. For the multivariable regression the answer is yes if at least 1 of the sample slope-coefficients is significantly different from 0. The F-ratio printed in the Analysis of Variance section of Table 10.9 tells us whether we are able to reject the hypothesis, $H_o: B_1 = B_2 \ldots = B_k = 0$. The F-ratio in our problem, 591.083, is large enough to reject the above hypothesis at any significance level. (See Chapter 13 for a detailed discussion of the F-ratio and analysis of variance.) This tells us that at least 1 of the sample slope-coefficents is significant; with this test, it could be the case that both are significant or it could be the case that 1 is significant and 1 is not. More to the point and giving us more detail are the individual t-tests on each of the slope-coefficients. Immediately to the right of the slope-coefficients on the printout are the standard errors of these coefficients. The hypotheses, $H_o: B_1 = 0$ and $H_o: B_2 = 0$, can each be very simply tested. For the first, $H_o: B_1 = 0$ ($H_a: B_1 \neq 0$), the test-t,

$$t = \frac{b_1 - B_1}{s_{b_1}} = \frac{-8.73 - 0}{.254} = -34.4.$$

For the second, $H_o: B_2 = 0$,

$$t = \frac{b_2 - B_2}{s_{b_2}} = \frac{2.54 - 0}{.341} = 7.45.$$

$H_a: B_2 \neq 0$.

Degrees of freedom in the k-variable equation are $n - k$; the critical t^* for 27 df and an α of .05 is $t^* = \pm 2.05$. Quite clearly, both slope-coefficients are significant; i.e., natural gas sales in our winter region are dependent upon both temperature and wind velocity changes.

The strength of the relationship is reflected by the magnitude of the multiple coefficient of determination R^2. At the top of the printout, Table 10.9, $R = .9888$; squaring this, $R^2 = .9777$. As with r^2 in the 2-variable problem, R^2 is the sample estimate of the proportion of variability in y explained by the variability in both x_1 and x_2. Only slightly over 2 percent of the variability in daily natural gas sales is left unexplained. The unexplained variance may be assumed to be random in nature.

The measure of the random error is the standard error of the regression s_e, and the printout shows this to be: $s_e = 8.4415$. For the 3-variable equation, s_e is defined as

$$s_e = \sqrt{\frac{\Sigma(y_i - \hat{y}_i)^2}{n - 3}},$$

the square root of the sum of the minimized residuals divided by the degrees of freedom, $n - 3$. In the neighborhood of $(\bar{x}_1, \bar{x}_2)$, the standard deviation of the individual y-values about the estimate, $\hat{y}$, given x_1 and x_2,

$s_{y \cdot x_1 x_2}$,

is roughly equal to s_e; and in this same neighborhood,

$s_{\hat{\mu} \cdot x_1 x_2}$ is approximately $\dfrac{s_e}{\sqrt{n}}$.

Consider the following example:

Example 10.8.1 (a) Estimate natural gas sales for a given day with mean daily temperature of 28 degrees and mean wind velocity of 8 miles per hour. (b) Estimate the average natural gas sales for all days with values of x_1 and x_2.

The point estimates for both cases are the same; $\hat{y} = \hat{\mu} = 637.7 - 8.73(28) + 2.54(8) = 413.6$. In both cases, sales would be estimated at $413,600.

A 95 percent confidence interval with 27 degrees of freedom requires $t = 2.05$. For the individual day, the 95 percent confidence interval is $413.6 \pm (2.05)(8.44) = 413.6 \pm 17.3$. For the average sales this interval is

Table 10.10 List of Residuals

CASE NUMBER	Y	$\hat{Y}$	RESIDUAL	X_1	X_2
1	350.0000	365.8655	−15.8655	33.0000	6.4000
2	346.0000	359.7176	−13.7176	36.0000	14.3000
3	354.0000	370.0140	−16.0140	37.0000	21.8000
4	402.0000	394.8893	7.1107	32.0000	14.4000
5	455.0000	445.3413	9.6587	25.0000	10.2000
6	482.0000	479.8564	2.1436	20.0000	6.6000
7	497.0000	508.9398	−11.9398	16.0000	4.3000
8	447.0000	449.5046	−2.5046	24.0000	8.4000
9	517.0000	519.3406	−2.3406	16.0000	8.4000
10	513.0000	504.1647	8.8353	18.0000	9.3000
11	514.0000	508.7309	5.2691	18.0000	11.1000
12	461.0000	453.4590	7.5410	25.0000	13.4000
13	403.0000	400.0225	2.9775	29.0000	6.1000
14	399.0000	394.8445	4.1555	30.0000	7.5000
15	385.0000	387.8907	−2.8907	31.0000	8.2000
16	364.0000	358.1507	5.8493	34.0000	6.8000
17	378.0000	369.4618	8.5382	35.0000	14.7000
18	367.0000	357.8971	9.1029	34.0000	6.7000
19	354.0000	363.5824	−9.5824	33.0000	5.5000
20	370.0000	368.7604	1.2396	32.0000	4.1000
21	367.0000	352.3161	14.6839	34.0000	4.5000
22	341.0000	331.2012	9.7988	37.0000	6.5000
23	344.0000	343.9448	.0552	34.0000	1.2000
24	378.0000	380.5341	−2.5341	31.0000	5.3000
25	399.0000	400.3806	−1.3806	28.0000	2.8000
26	426.0000	426.4647	−.4647	26.0000	6.2000
27	427.0000	429.9714	−2.9714	24.0000	.7000
28	431.0000	433.4184	−2.4184	25.0000	5.5000
29	415.0000	417.0786	−2.0786	26.0000	2.5000
30	412.0000	422.2566	−10.2566	25.0000	1.1000

FINISH CARD ENCOUNTERED
PROGRAM TERMINATED

$$413.6 \pm 2.05 \left(\frac{8.44}{30}\right) = 413.6 \pm 3.2.$$

Given a specific day with temperature and wind velocity measurements of 28 and 8 respectively, estimated sales are $413,600, and we are 95 percent confident that this is in error by no more than $17,300. For all days with the above values of x_1 and x_2, the estimated mean sales are $413,600, and we are 95 percent confident that this is in error by no more than $3,200.

As in the 2-variable problem, the standard errors $s_{y \cdot x_1 x_2}$ and $s_{\hat{\mu} \cdot x_1 x_2}$ are increasing divergent functions of the deviations of x_1 from $\bar{x}_1$ and x_2 from $\bar{x}_2$. Larger sample sizes will increase the region of validity for the approximations given above.

The final table of the printout, Table 10.10, shows the actual y data, y computed ($\hat{y}$), the residual ($e_i = y_i - \hat{y}_i$), and the actual values of the independent variables, x_1 and x_2. We can inspect the table to see if the data were entered properly. The individual residuals could be instructive and contribute to the analysis. For example, in Chapter 14, Nonparametric Statistics, there is a test that will check if the residuals could be a sample from a normal distribution. In this same chapter there is a runs test that can be used to check the randomness of the residuals as they vary above and below the regression line with increasing values of x. In a more subjective way, the more extreme values of the residuals could be inspected for possible recording errors or for other possible factors that might cause this large deviation from the regression function.

10A Appendix: The Analytical Geometry of a Straight Line

Analytical geometry is the branch of mathematics that deals with the geometry of functional relationships. The discussion here will address only the analytics of the straight line. In a Euclidean or plane 2-space, the rectangular (or Cartesian) coordinates system provides the orientation. Two number scales intersect at right angles; convention has the horizontal axis as the x-axis with x as the independent variable, and the vertical axis, the y-axis, with y as the dependent variable. The 2 scales intersect at 0 on each scale (the origin) with the positive x-direction to the right or east and the positive y-direction up or north. A pair of numbers, (x, y), the coordinates, locate a unique point in the space and conversely; the x-value is called the abscissa and the y-value the ordinate. The 4 quadrants, beginning with the quadrant where both x and y are positive, are numbered counterclockwise, I through IV. See Figure 10.6 where the above points are illustrated.

Figure 10.6 Rectangular Coordinate System

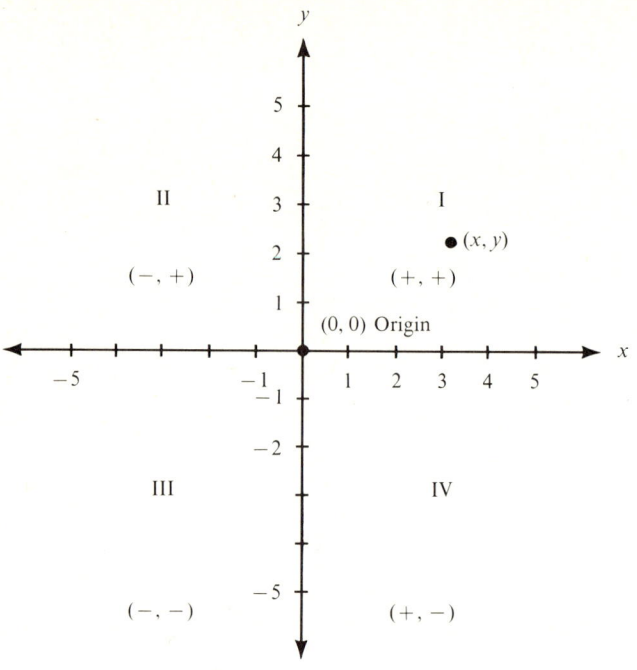

The concepts of distance and slope are important in our discussion of the straight line. Vertical and horizontal distances are treated as vectors; i.e., these distances have magnitude and direction. For example, the distance from the point $P_1 = (-2, 3)$ to the point $P_2 = (7, 3)$ is $7 - (-2) = +9$; whereas the distance from P_2 to P_1 is $-2 - (+7) = -9$. With y fixed, a horizontal distance from x_1 to x_2 is $x_2 - x_1$, or Δx. With x fixed, a vertical distance from y_1 to y_2 is $y_2 - y_1$. See Figure 10.7. For example, the distance from $P_1 = (-2, -8)$ to $P_2 = (-2, -1)$ is $(-1) - (-8) = +7$. An oblique distance (not necessary for our work) may be found by using the Pythagorean theorem: "In a right triangle the square on the hypotenuse is equal to the sum of the squares on the two legs." To illustrate this, let us find the distance from the point $P_1 = (-1, 3)$ to the point $P_2 = (11, -2)$. Complete the right triangle as in Figure 10.8. We find

$\Delta x = 11 - (-1) = 12$;
$\Delta y = -2 - (3) = -5$; and
$D^2 = \overline{\Delta y}^2 + \overline{\Delta x}^2$;

$D = \sqrt{\overline{\Delta y}^2 + \overline{\Delta} x^2}$ so,

$D = \sqrt{12^2 + 5^2} = \sqrt{144 + 25} = \sqrt{169} = 13.$

Figure 10.7 Horizontal and Vertical Distances

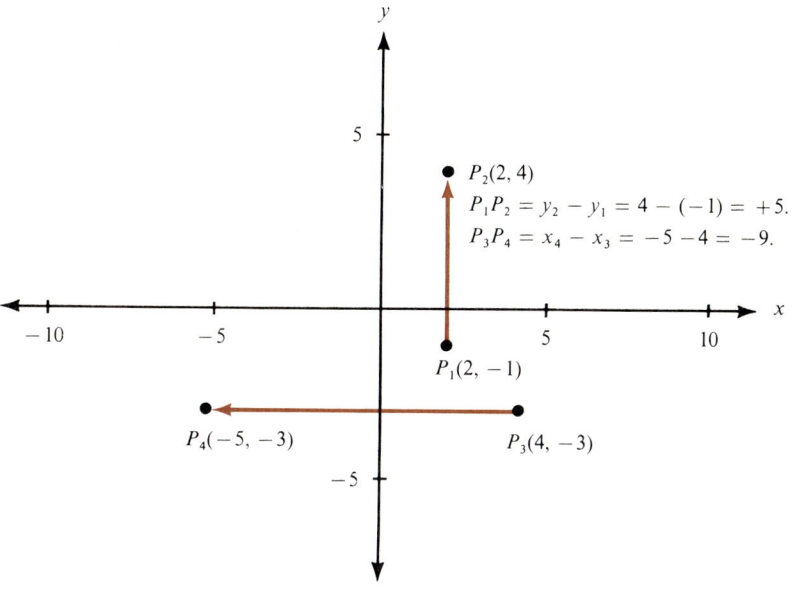

Oblique distances will be always positive.

The notion of slope is very important in our work with the straight line (the *linear function*). Slope is defined as the rate of change in y per unit change in x, or the ratio of the rise to the run. With the Δ-notation and using the letter *b* for slope,

Figure 10.8 Oblique Distance

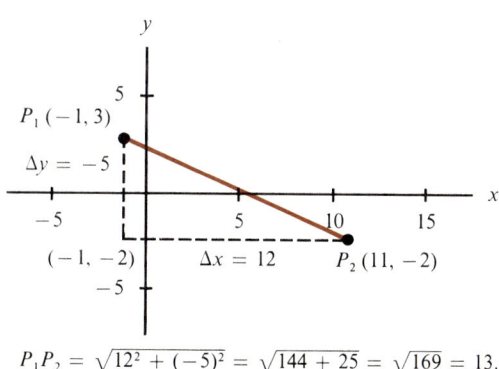

$P_1P_2 = \sqrt{12^2 + (-5)^2} = \sqrt{144 + 25} = \sqrt{169} = 13.$

Appendix to Chapter 10

$$b = \frac{\Delta y}{\Delta x} = \frac{y_2 - y_1}{x_2 - x_1}.$$

In the above example,

$$b = \frac{\Delta y}{\Delta x} = \frac{-5}{12} = -\frac{5}{12}.$$

For you who have studied trigonometry, the slope is the tangent of the angle of inclination of the line. With just a brush of calculus you will know that this is also the first derivative of the functional relation, $y = f(x)$ (yet to be determined in our discussion).

Possible values of b cover the full range of the number scale from $-\infty$ to $+\infty$. If y increases as x increases, b will be positive, and if y decreases with increasing values of x, b will be negative.

A horizontal line has a 0 slope, for no matter how large we take Δx, the numerator of

$$b = \frac{\Delta y}{\Delta x}$$

will be 0. The slope of a vertical line is undefined because the denominator of b can be nothing but 0. If a line with a positive slope is gradually moved to a vertical position, b approaches $+\infty$; if a line with a negative slope is similarly moved, b approaches $-\infty$. Technically incorrect but frequently used is the statement that "the slope of a vertical line is infinity." A 45 degree line has a slope of 1; steeper lines with positive slope have slopes greater than 1; and lines between the horizontal and 45 degrees have slopes between 0 and 1. A similar observation may be made for the negatively sloped lines.

We now want to show that the function, $y = a + bx$, where a and b are real numbers, is the general equation of a straight line, and then identify the geometrical meanings of a and b. When we say that this is the equation of a straight line, we mean that the set (or locus) of points that satisfy an equation of this type describes a straight line. In addition, every straight line in the 2-space may be written in this form with the exception of the vertical grid. Equations for vertical lines are of the form $x = k$, where k is a real number; for example, $x = 0$ is the equation of the y-axis, $x = 1$ is the equation of the vertical line 1 unit to the right of the y-axis, etc. Refer to Figure 10.9 as we show that $y = a + bx$ is a general form of the straight line. Consider a straight line through 2 fixed points, $P_1 = (x_1, y_1)$ and $P_2 = (x_2, y_2)$; then

$$b = \frac{y_2 - y_1}{x_2 - x_1}$$

by definition. Consider a third but variable point, that is, any point $P = (x, y)$ on this straight line. Because of similar triangles, the slope b will be

Figure 10.9 Slope

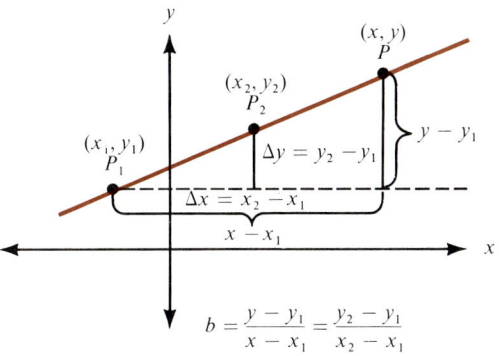

$$\frac{y - y_1}{x - x_1}$$

regardless of where the point (x, y) is located on the straight line. It follows that

$$\frac{y - y_1}{x - x_1} = \frac{y_2 - y_1}{x_2 - x_1};$$

multiplying both sides by $(x - x_1)$ we get

$$y - y_1 = \frac{y_2 - y_1}{x_2 - x_1}(x - x_1);$$

substituting b for

$$\frac{y_2 - y_1}{x_2 - x_1}$$

and multiplying through by b yields

$$y - y_1 = bx - bx_1;$$

and transposing y_1 we have

$$y = (y_1 - bx_1) + bx.$$

Inasmuch as y_1, b, and x_1 are ordinary real numbers, we let $a = y_1 - bx_1$ and have

$$y = a + bx.$$

In the derivation above, we have already shown that the coefficient of x, namely b, is the slope. The other coefficient, a, is the y-intercept—the value of y when x is 0. To illustrate: In the equation, $y = a + bx$, let x be 0; we then get $y = a + b(0) = a$.

Regardless of the form in which it is written, any equation with the x term(s) and/or y term(s) to the first degree only will be a straight line. There are a number of standard forms of the straight-line equation, but the *slope-intercept* form, $y = a + bx$, is completely adequate for our needs.

The analysis of an equation in this form is simple and efficient; for example, consider the equation $y = 2 + 3x$. The line goes through the y-axis at 2, or through the point (0, 2), and has a slope of 3 or, for every unit change in x, y changes 3 units in the same direction. This equation can be sketched rapidly (see Figure 10.10) and, with little effort, one can picture the function without a sketch. On Figure 10.10 we have also sketched 3 more equations: (1) $y = -3 + 0.4x$; (2) $y = 5 - x$; (3) $y = 2.8$. Do you agree with these graphs? Note that "y equals a constant" maps the horizontal grid; i.e., $b = 0$ so y is simply equal to a.

As already mentioned, two conditions determine a unique straight line; with more than two conditions, such as n sample points as we had in our statistical work earlier in the chapter, the line is overdefined. With only one condition, no unique line is determined, and we speak of a *family* of straight lines. In the synthesis or in the building of the equation of a straight line, for given conditions, the form, $y = a + bx$, is still adequate for all problems and actually the preferred form for statistical determination. Other forms are perhaps more convenient for synthesis under some conditions.

To clarify this point, let us consider an example with several parts.

Figure 10.10 Graphs of Equations

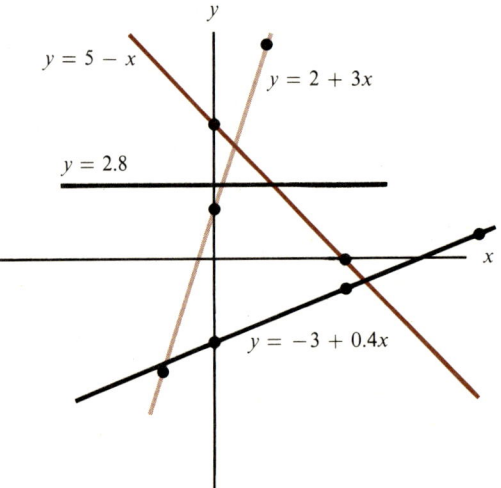

Example 10A.1 Find the equation of a straight line: (a) with slope of 2 passing through the point $(0, -5)$; (b) passing through the points $(-1, -3)$ and $(2, 8)$; (c) with an x-intercept of 3 and a y-intercept of 5.

In the first part we are given the y-intercept and the slope, so we simply need to substitute these in $y = a + bx$ to get $y = -5 + 2x$.

The second case isn't so easy. We first find the slope,

$$b = \frac{\Delta y}{\Delta x} = \frac{-3 - 8}{-1 - 2} = \frac{-11}{-3} = \frac{11}{3}.$$

We now know that our specific equation belongs to the family of lines with a slope of 11/3; substituting either pair of coordinates will allow us to solve for a. Choosing the first point,

$$-3 = a + \frac{11}{3}(-1),$$

and solving for a, we get

$$a = \frac{2}{3}.$$

The required equation is

$$y = \frac{2}{3} + \frac{11}{3}x.$$

The third case is similar to the second case above, but with the y-intercept given we need only to find the slope. We are given 2 points $(0, 5)$ and $(3, 0)$; so

$$b = \frac{\Delta y}{\Delta x} = \frac{5 - 0}{0 - 3} = -\frac{5}{3}$$

and $a = 5$. The required equation is

$$y = 5 - \frac{5}{3}x.$$

10B Appendix: Derivation of Least Squares Estimators

Given a set of paired data points, (x_i, y_i), we wish to derive the coefficients, a and b, of the linear function, $\hat{y} = a + bx$, such that $\Sigma(y_i - \hat{y}_i)^2$ is minimum. Recall that this is the least squares criterion.

Substituting for $\hat{y}_i$ the equivalent $a + bx_i$,

$$\Sigma(y_i - \hat{y}_i)^2 = \Sigma(y_i - a - bx_i)^2.$$

We want to find the minimum of this function of a and b; i.e., minimize $f(a,b) = \Sigma(y_i - a - bx_i)^2$.

Optimization methods of calculus call for setting both partial derivatives equal to 0. This results in 2 equations in the parameters, a and b; whether or not these conditions define a minimum is left to be tested, as we shall see later. The 2 partial derivatives are given below. (Remember that the derivative of a sum is the sum of the derivatives.)

$$\frac{\partial f}{\partial a} = 2\Sigma(y_i - a - bx_i)(-1)$$

$$\frac{\partial f}{\partial b} = 2\Sigma(y_i - a - bx_i)(-x_i).$$

Setting both of these equal to 0 and dividing through both by -2, we have

$$\Sigma(y_i - a - bx_i) = 0$$
$$\Sigma(y_i - a - bx_i)(x_i) = 0.$$

In the first equation, move the summation sign through and transpose the last two terms; in the second equation, multiply through by x_i and repeat above. We now have the 2 normal equations:

$$\Sigma y_i = na + b\Sigma x_i \quad \text{(Note: } \Sigma a = na\text{)}$$
$$\Sigma x_i y_i = a\Sigma x_i + b\Sigma x_i^2.$$

Do these conditions define a minimum? The simplest test, if conclusive, involves the second-order partial derivatives. We find

$$\frac{\partial^2 f}{\partial a^2} = 2n; \quad \frac{\partial^2 f}{\partial b^2} = 2\Sigma x^2{}_i; \quad \frac{\partial^2}{\partial a\, \partial b} = 2\Sigma x_i.$$

The test is conclusively minimum if both

$$\frac{\partial^2 f}{\partial a^2} \text{ and } \frac{\partial^2 f}{\partial b^2}$$

are positive and if

$$\left(\frac{\partial^2 f}{\partial a^2} \frac{\partial^2 f}{\partial b^2} - \left[\frac{\partial^2 f}{\partial a\, \partial b} \right]^2 \right)$$

is also positive.

We see that both second-order straight partials are always positive; $2n$ is positive because n is a positive integer, and $\Sigma x^2{}_i$ is nonnegative and 0 only if all $x_i = 0$, which would never be the case in our problem.
   Substituting the partials in the second condition, we get

$$\frac{\partial^2 f}{\partial a^2} \frac{\partial^2 f}{\partial b^2} - \left(\frac{\partial^2 f}{\partial a\, \partial b} \right)^2 = 4n\Sigma x^2{}_i - 4(\Sigma x_i)^2.$$

Factoring $4n$, we get $4n(\Sigma_i^2 - n\bar{x}^2)$, and we know that this equals (see Chapter 2) $4n[\Sigma(x_i - \bar{x})^2]$, clearly a positive number.

We have derived the normal equations and have shown them to define a minimum value for $\Sigma(y_i - \hat{y}_i)^2$.

10C Appendix: Notes and Comments on Nonlinear Regression Models

Many nonlinear regression models may be transformed to be consistent with the linear model and the parameters determined by the related formulas or through the linear model computer program.[7]

The exponential model, $\hat{y} = ab^x$, becomes linear in the following sense if we take the logarithm of both sides: $\log \hat{y} = \log a + x \log b$. Set $\hat{Y} = \log \hat{y}$, $c = \log a$ and $d = \log b$. We get $\hat{Y} = c + dx$; so, if we regress the log y against x we will get the logarithmic form of the exponential model. It is important to note that we are minimizing, by this approach, $\Sigma(\log y_i - \widehat{\log y_i})^2$, and not $\Sigma(y_i - \hat{y}_i)^2$. Methods for minimizing $\Sigma(y_i - \hat{y}_i)^2$ in the exponential model are iterative in nature and demand access to the computer.

The polynomial model, $\hat{y} = a + b_1 x + b_2 x^2 + b_3 x^3 + \cdots + b_k x^k$, using the least squares criterion, parallels the multivariable model directly if we let $x_1 = x$, $x_2 = x^2$, $x_3 = x^3$, etc. For example, the normal equations for the second degree function,

$$\hat{y} = a + b_1 x + b_2 x^2,$$

are

$$\Sigma y = na + b_1 \Sigma x + b_2 \Sigma x^2$$
$$\Sigma xy = a\Sigma x + b_1 \Sigma x^2 + b_2 \Sigma x^3$$
$$\Sigma x^2 y = a\Sigma x^2 + b_1 \Sigma x^3 + b_2 \Sigma x^4.$$

The general power function, $\hat{y} = ax^b$, as with the exponential model above, becomes linear (in one sense) if we take the logarithm of both sides:

$\log \hat{y} = \log a + b \log x$.

Let $\hat{Y} = \log \hat{y}$, $c = \log a$, and $X = \log x$. We then get $\hat{Y} = c + bX$ and simply regress log y against log x.

Problems

1. Given the following data:

   x 1 3 4 7 10
   y 30 22 15 12 5.

---

7. See Appendix 11A for additional detail on nonlinear models.

a. Plot a scattergram on arithmetic grid.
   b. Compute the first-degree least squares line of best fit, with y as a function of x.
   c. Plot the result of b on the same grid as a.
   d. Do you think you made a mistake? Why or why not?
   e. Find r and s_e.

2. Consider the following data:

x	y
−3	30
−1	18
2	25
5	10
7	11
12	2

   a. Calculate the least squares line of best fit.
   b. Calculate the coefficient of linear correlation.
   c. Plot a scattergram of the data and the equation in a on the same grid.

3. From a sample of 64 different-size production runs, the Dipsey Cola Corporation derived the following least squares cost function, where y is the total cost in dollars for a production run and x is the number of cans of cola produced in the run:

   $$\hat{y} = 300 + 0.024x.$$

   a. Interpret the two coefficients.
   b. If Dipsey Cola sells for $.15 per can, find the break-even size production run. (Show your work.)
   c. The maximum size production run is 5,000. What is the estimated profit for a run of this size?

4. For a sample of 100 from a large population, the x's summed to 120, the y's to 450, the x-squares to 244, the y-squares to 2,925, and the xy's to 340.
   a. Find $\hat{y} = a + bx$, the least squares line of best fit.
   b. Find r^2, s_e and s_b.
   c. Give the 95 percent confidence interval for B.
   d. Estimate the mean value of y for $x = 1$ and give the 95 percent confidence interval for this mean value.

5. The following regression equation is the result of analysis of a random sample interview of 150 firms (with 10 or more employees) from the retail sales sector of a large metropolitan area.

   $$\hat{y} = -25 + 52.5x,$$

   where x is the number of employees (full-time equivalent) and y is sales in thousands of dollars.

a. The average sales for the 150 firms was 1,700 thousands of dollars. What is the average number of employees in the firms sampled? What is the total number of employees?
b. If the average wage is $5,000, write a new equation with x as dollar wages and y as sales in dollars.
c. (Referring to the original variables) $s_x = 15$ and $s_y = 1,000$. Find r and $s_{e \cdot x}$.
d. Considering all firms with 50 employees, we can estimate the average sales to be _____, and we would be 80 percent confident that this is in error by no more than _____.
e. Find s_b. The Utah retail association claims that incremental sales per employee should increase by $60,000. Can you reject this claim at the .025 significance level?

6. The following consumption function and related results were determined from a sample of 100 households in a large metropolitan area. (y is consumer expenditures in dollars and x is household income in dollars.)

$\hat{y} = 1,800 + .75x$; $s_x = 4,800$; $s_y = 4,500$; and $\bar{y} = 10,200$.

a. Find the following:
   (i). the sample mean income;
   (ii). the coefficient of correlation;
   (iii). the standard error of the regression;
   (iv). the standard deviation of the slope parameter.
b. Also answer the following questions.
   (i). Estimated consumer expenditures exceed household incomes for incomes less than what amount?
   (ii). What proportion of the variability in y is explained by the variability in x?
   (iii). Would you reject H_o? $B = 0$. Why or why not?
   (iv). Several years ago a similar study resulted in a marginal propensity to consume of .85. Would you reject the hypothesis that B is at least .85? Why or why not?
   (v). The J.J. household has an income of $20,000. What is the best estimate of its consumer expenditures?
   (vi). The odds are 19:1 that this estimate is in error by no more than what amount?
   (vii). What is the estimated mean consumption of all households with $16,000 income?
   (viii). It is an even-money bet that this estimate is in error by no more than what amount?

7. Say as much as you can about each of the following; in some cases you will be able to give an actual number; in some, your answer will be more general.
   a. If $s_x = 6s_y$ and $r = -.80$ find b.
   b. Find r^2 if $s_y = 4s_e$ and n is very large.

c. If $r^2 = 1$, then r will be _____.
d. If $\bar{x} = \bar{y} = 1$, then $a + b =$ _____.
e. If $n = 100$ and $s_e = 10$, then $s_b =$ _____.
f. If $b = 0$, then $\hat{y} =$ _____.
g. If $\hat{y} = a + bx$ goes through points (1,8) and (10,0), find a and b.
h. What does an r-value of 1.24 indicate?

8. In connection with planning servicing facilities for machines that require attention, a company wants to study the relationship between the number of machines waiting for attention at a given time and the average time required by operators to service the machines. More specifically, the company wants to know whether there is a tendency for operators to work faster (and reduce the service time) when the number of machines waiting for service is large. Accordingly, the company randomly selects 8 records showing the number of machines in a line at the beginning of a given time period and the number of services completed by an operator during the period. The following are the data:

Machines in line, x 3 6 5 4 4 6 8 7
Number of completed services, y 3 2 3 5 3 6 6 4

a. Find least squares line of best fit.
b. Plot a scattergram of the data and line obtained in part a on the same grid.
c. Predict the average number of services an operator will complete during a period when there are 5 machines in line at the beginning of the period.

9. Find the regression equation of weight (y) on height (x). Find standard error of the regression and the coefficient of correlation.

Height (x)   68   72   71   74   70   67   73   69   66   75
Weight (y)   145   150   185   190   170   165   200   160   140   190

10. Consider the following results: $n = 6$, $\Sigma x = 18$, $\Sigma y = 30$, $\Sigma xy = 60$, $\Sigma x^2 = 64$, and $\Sigma y^2 = 240$.
    a. Compute the least squares line of best fit, with y as a function of x.
    b. Do the same for x as a function of y.
    c. What is the coefficient of linear correlation?
    d. Calculate s_e.

11. Consider the following least squares equation relating the Federal Reserve Bank interest rate in percent, x, to the commercial bank rate, y: $\hat{y} = 2.4 + .7x$. Given also, $\bar{x} = 4.3$, $s_x = 1.56$, $s_y = 1.40$ and $n = 1,800$.
    a. Plot the equation on arithmetic grid from $x = 2.00$ to $x = 6.50$. Locate and read from your graph the best estimates for com-

mercial bank rates for a federal rate of 5 percent and of 3 percent.
b. Compute r, the coefficient of linear correlation, from the above data.
c. Figure the average y. Figure the standard error of the estimate.

12. The following are results from a study made of 300 college males with weight, y, as a linear function of height, x. Given:

$\bar{y} = 171, \bar{x} = 70, s_y = 20, s_x = 2.5$, and $\overline{xy} = 12{,}000$. Find:

a. the least squares line of best fit;
b. the coefficient of linear correlation;
c. the standard error of the regression.
d. Choose your own α and test each of the hypotheses:
   (i). $B = 0$
   (ii). $B = 7.0$
e. Give the best estimate for average weight of men 6 feet tall. Also the 95 percent confidence limits for this average. Give also the 95 percent confidence limits on weight of an individual picked at random from those 6 feet tall.

13. Given the following calculations: $n = 20; \Sigma x = 92; \Sigma y = 150, \Sigma x^2 = 449; \Sigma y^2 = 1{,}190$; and $\Sigma xy = 716$. Find the following:

a. the least squares line of best fit;
b. s_e, the sample standard error of regression;
c. r, the sample coefficient of correlation.
d. Test the hypothesis (use an α of .05): $H_o: B = 0$.
e. Give the best point estimate and the 95 percent confidence interval estimate for both y and μ_y when $x = 2$.

14. With deliberate variation in the weekly price level of item A for a period of 18 weeks, Company JB came up with the following: A least squares regression equation, $\hat{y}$ in hundreds of dollars and x, price of item A, in dollars; $\hat{y} = 280 - 120x; s_y = 14$; and $s_x = .10$. Find or answer the following.

a. interpretation of the two coefficients;
b. the coefficient of linear correlation;
c. the standard error of the regression;
d. the 90 percent interval on sales, given a price of $3.50.

15. Given: $r = .60; s_x = 1.50; s_y = 2.00, \bar{x} = 10$; and $\bar{y} = 20$. Find y as a function of x.

16. The following is a least squares price-quantity equation relating Company A's sales in number of boomerangs, y, and a function of the price per unit, x (price in dollars). $\hat{y} = 4{,}060 - 900x$.

a. Plot the above equation on arithmetic grid, from $x = 0$ to $x = 5$.

b. From your graph, locate and read the quantity Company A can expect to sell if the price is set at $2.50 per boomerang.
c. From your equation, figure the expected sales for a price of $1.80.
d. Interpret the 2 coefficients in the above equation.

17. Consider the following regression function where y is consumer expenditures in dollars, x_1 is household income in dollars, x_2 is the number of members in the household, and x_3 equals 0 or 1 according as the head of the household is male or female:

$$\hat{y} = 600 + .60x_1 + 550x_2 - 400x_3.$$

a. Estimate consumer expenditures for a family of 4 with male head and a $12,000 annual income.
b. Compare the estimated surplus (for taxes and savings) between the family in part a and a single female living alone with an annual income of $7,000.
c. Interpret each of the coefficients.

18. (This is a long exercise involving several 2-variable parts as well as a multivariable problem.)

Table 10.11 shows data on recent home sales in a large city. The area of recorded sales was confined to several contiguous square miles so that location, being more or less fixed, would not affect selling price.

Table 10.11 Housing Study Data

House Number	Selling Price of House ($000) y	Finished Floor Space (sq. ft.) x_1	Number of Bedrooms x_2	Age of House (Years) x_3
1	52.50	1,880	4	29
2	67.00	2,128	4	4
3	59.90	2,149	5	35
4	98.50	3,375	6	12
5	45.90	1,300	3	29
6	41.95	1,375	3	12
7	86.90	3,250	6	25
8	57.90	1,400	2	15
9	34.50	1,250	4	33
10	23.50	900	2	45
11	59.50	2,500	3	10
12	72.00	2,310	5	8
13	40.75	1,170	3	25
14	44.00	1,615	4	30
15	77.50	2,800	5	11
16	49.95	2,370	2	9

a.
   (i). Plot the scattergram with selling price, y, a function of x_1, the finished floor space.
   (ii). Find the least squares regression line, $\hat{y} = a + bx$, and graph the line on the scattergram grid.
   (iii). Interpret the 2 coefficients, a and b. Does a have any real-world meaning?

b.
   (i). Find s_e, r, and s_b.
   (ii). Test the hypothesis H_o: $B = 0$. (What alternate hypothesis should you use?) Use $\alpha = .05$.
   (iii). Give the 90 percent confidence interval estimate of B, the universe slope coefficient.

c.
   (i). Mr. Jones's house has 2,000 square feet of finished floor space. Estimate the selling price of his house. Give the 90 percent confidence interval for this estimate.
   (ii). Estimate the mean selling price of homes with 2,000 square feet of finished floor space. Give the 90 percent confidence interval for this estimate.

d. Repeat parts a and b with y, the selling price of the house, as a function of number of bedrooms.

e. If a computer is available, run selling price of homes, y, as a function of the other three variables. Analyze the computer printout and discuss the real-world meaning of the results.

19. As a term paper and/or a major exercise, we suggest that members of the class, either individually or in teams of 2 to 4 students, gather their own data, run the data on a prepackaged program, and prepare a writeup of the results. The sample should be at least 30 with at least 3 variables, the dependent variable, y, and 2 independent x-variables. Your instructor should approve the project before you sample to get your data. Make a 2-variable computer run, $\hat{y} = a + bx$; chose the x-variable that you think will be the most important explanatory variable. Make a second computer run with y as a function of all your dependent variables. If this is not possible because of time and/or computer limitations, we have included at the end of the suggested writeup format several computer printouts. In lieu of gathering your own data, these could be utilized for the analysis and writeup. Following are our suggestions for the regression problem term paper.

   I. Introduction. *This should include:*
      A. A definition of each of your variables, the source of your data, and your sampling procedure.
      B. The name(s) of your partner(s), if any.
      C. A copy of the computer printout.

II. The Nontechnical Essay

Write a short report evaluating and interpreting your results. This should be directed toward a fictional person who is your supervisor or who has hired you to do this research. This person is very knowledgeable in the field of your research but has no background in statistics. This report could include such things as (1) an interpretation of your equations, particularly the slope-coefficients; (2) an explanation of the differences between the 2 variable relationships and the multivariable relationship; (3) your preconceptions in comparison with the actual results; (4) a criticism of your methodology; (5) suggestions for improvement for further investigation; (6) the nature of the general inferences that could be made and to what population; etc. (Specific examples usually make a report more interesting.)

III. The Technical Report

A. The 2-Variable Equation, $y = a + bx$.
 1. Plot a scattergram of the data and plot your equation on the same graph.
 2. Also plot 2 parallel lines with a vertical distance of $2s_e$ on either side of the regression line. (Note that these are not the 95 percent confidence bands; even so, most of your data points should fall in this region; at $x = \bar{x}$ this is very close to the 95 percent interval.)
 3. For some arbitrary x, say x_o, within 1 standard deviation of $\bar{x}$ (the mean of x), give both the point estimates and the 95 percent interval estimates for both $y(x_o)$, and $\mu(x_o)$.
 4. Give the 95 percent interval and also the point estimate for the slope parameter, B.
 5. Test the null hypothesis on B.

B. A second linear equation of the form $\hat{y} = a + bx$:
 From the printout of means, standard deviations, and correlations, derive the above equation for your second variable (or the second-most-important variable, if you have more than 2 independent variables). Also calculate s_e and s_b.

C. For the multivariable equation:
 1. Give the point and 95 percent interval estimates for each of the slope parameters.
 2. Test the null hypothesis on these B-values.
 3. Verify that the mean value of y is the best point estimate for $x_1 = \bar{x}_1$ and $x_2 = \bar{x}_2$, $x_3 = \bar{x}_3$, etc.
 4. Interpret R, the multiple coefficient of correlation.

20. Find the equation of the line passing through the points:

a. (3,2) and (4,7);
b. (3,3) and (−1,−5).

21. Write the equation of any line that is parallel to the line $y = 30.5 - 1.1x$. Write the equation of any line that is parallel to the x-axis.

22. a. Find an equation for the straight line whose slope is 2/3 and whose y-intercept is -3.
    b. Find the slope and y-intercept of the line whose equation is $3x - 5y = 20$.
    c. What is the equation of a line that is parallel to the line in (b) and that passes through the point $(2, -1)$?

23. Find the slope, the y-intercept, and the equation of the line passing through the points (5,4) and (2,8).

 **Printout 1**

Assume the following to be data gathered several years ago in a large metropolitan area in order to study factors influencing unskilled wage rates in the trades and services sectors. The variables are as follows:

- y—hourly wage rate
- x_1—employees' age less 17.00
- x_2—number of months worked for current employer
- x_3—$x_3 = 1$ if female
  $x_3 = 0$ if male.

Note that the last table, List of Residuals, also lists the actual data.

NUMBER OF CASES   40
NUMBER OF ORIGINAL VARIABLES   4

VARIABLE	MEAN	STD. DEVIATION
Y	2.89000	.38217
X_1	4.85000	4.35331
X_2	21.92500	13.66520
X_3	.60000	.49614

CORRELATION MATRIX

VARIABLE	Y	X_1	X_2	X_3
Y	1.000	.059	.447	−.434
X_1		1.000	.007	.197
X_2			1.000	−.197
X_3				1.000

Two-Variable Printout

R +.4474
S_E .3463

ANALYSIS OF VARIANCE

	DF	SUM OF SQUARES	MEAN SQUARE	F RATIO
REGRESSION	1	1.140	1.140	9.509
RESIDUAL	38	4.556	.120	

VARIABLES IN EQUATION

VARIABLE	COEFFICIENT	STD. ERROR	T
CONSTANT	2.61568		
X_2	.01251	.00406	3.08

Multivariable Printout

MULTIPLE R .5844
STD. ERROR OF EST. .3228

ANALYSIS OF VARIANCE

	DF	SUM OF SQUARES	MEAN SQUARE	F RATIO
REGRESSION	3	1.945	.648	6.224
RESIDUAL	36	3.751	.104	

VARIABLES IN EQUATION

VARIABLE	COEFFICIENT	STD. ERROR	T
CONSTANT	2.78545		
X_1	.01168	.01212	.9637
X_2	.01035	.00386	2.6813
X_3	−.29830	.10851	2.7491

List of Residuals

CASE NUMBER	Y	Ŷ	RESIDUAL	X_2	X_3	X_1
1	2.7500	3.0362	−.2862	35.0000	1.0000	16.0000
2	2.6500	2.8916	−.2416	8.0000	.0000	2.0000
3	2.9500	3.0038	−.0538	33.0000	1.0000	15.0000
4	2.8000	2.8653	−.0653	23.0000	1.0000	12.0000
5	2.7500	3.0362	−.2862	35.0000	1.0000	16.0000
6	2.6500	2.6140	.0360	10.0000	1.0000	2.0000
7	2.6500	2.8210	−.1710	30.0000	1.0000	2.0000
8	2.6500	2.5816	.0684	8.0000	1.0000	1.0000
9	2.6500	3.0275	−.3775	20.0000	.0000	3.0000
10	2.8000	2.7825	.0175	15.0000	1.0000	12.0000
11	2.8000	2.5572	.2428	.0000	1.0000	6.0000
12	2.7000	2.6878	.0122	16.0000	1.0000	3.0000
13	2.6500	2.8507	−.2007	34.0000	1.0000	1.0000
14	2.9000	2.9630	−.0630	7.0000	.0000	9.0000
15	2.8000	2.7061	.0939	11.0000	1.0000	9.0000
16	2.8000	3.0314	−.2314	17.0000	.0000	6.0000
17	2.6500	2.6968	−.0468	18.0000	1.0000	2.0000
18	2.8000	3.1763	−.3763	31.0000	.0000	6.0000

CASE NUMBER	Y	Ŷ	RESIDUAL	X_2	X_3	X_1
19	2.7500	2.9245	−.1745	40.0000	1.0000	2.0000
20	2.6500	2.7368	−.0868	23.0000	1.0000	1.0000
21	2.7500	2.5572	.1928	.0000	1.0000	6.0000
22	2.6500	3.0455	−.3955	24.0000	.0000	1.0000
23	2.9500	2.8186	.1314	23.0000	1.0000	8.0000
24	4.0000	3.2758	.7242	44.0000	.0000	3.0000
25	4.0000	3.2641	.7359	44.0000	.0000	2.0000
26	2.6500	2.6955	−.0455	19.0000	1.0000	1.0000
27	2.6500	2.6955	−.0455	19.0000	1.0000	1.0000
28	2.6500	2.8831	−.2331	36.0000	1.0000	2.0000
29	2.7500	2.5572	.1928	.0000	1.0000	6.0000
30	4.0000	3.2758	.7242	44.0000	.0000	3.0000
31	3.7500	3.0057	.7443	10.0000	.0000	10.0000
32	2.7500	2.5105	.2395	.0000	1.0000	2.0000
33	2.7500	3.2021	−.4521	38.0000	.0000	2.0000
34	3.0000	2.7485	.2515	23.0000	1.0000	2.0000
35	2.7500	2.9240	−.1740	10.0000	.0000	3.0000
36	2.8000	3.0365	−.2365	22.0000	.0000	2.0000
37	2.8000	3.0365	−.2365	22.0000	.0000	2.0000
38	3.4000	3.3703	.0297	52.0000	.0000	4.0000
39	3.0000	2.7836	.2164	23.0000	1.0000	5.0000
40	2.7500	2.9240	−.1740	10.0000	.0000	3.0000

Printout 2

The following printout results from a sample of evening clients in a local pub. The effort was to find out the dependence of the size of the tip to the waiter on the amount of billing and the number in the party.

The variables are as follows:
- ■ y—tip in dollars
- ■ x_1—bill in dollars
- ■ x_2—number of persons in the party.

NUMBER OF CASES   30
NUMBER OF ORIGINAL VARIABLES   3

VARIABLE	MEAN	STD. DEVIATION
Y	3.44000	2.06778
X_1	29.26900	17.93851
X_2	3.60000	2.11073

CORRELATION MATRIX

VARIABLE NUMBER	Y	X_1	X_2
Y	1.000	.919	.854
X_1		1.000	.969
X_2			1.000

Two-Variable Printout

R .9192
STD. ERROR OF EST. S_E .8289

ANALYSIS OF VARIANCE

	DF	SUM OF SQUARES	MEAN SQUARE	F RATIO
REGRESSION	1	104.758	104.758	152.471
RESIDUAL	28	19.238	.687	

VARIABLES IN EQUATION

VARIABLE	COEFFICIENT	STD. ERROR	T
CONSTANT	.33889		
X_1	.10595	.00858	12.3480

Multivariable Printout

R .9306
S_E .7842

ANALYSIS OF VARIANCE

	DF	SUM OF SQUARES	MEAN SQUARE	F RATIO
REGRESSION	2	107.391	53.695	87.308
RESIDUAL	27	16.605	.615	

VARIABLES IN EQUATION

VARIABLE	COEFFICIENT	STD. ERROR	T
CONSTANT	.49068		
X_1	.17162	.03276	5.2387
X_2	−.57605	.27842	2.0690

List of Residuals

CASE NUMBER	Y	Ŷ	RESIDUAL	X_1	X_2
1	3.8500	3.0506	.7994	20.3800	2.0000
2	2.0000	2.0817	−.0817	23.0400	4.0000
3	3.0000	3.0355	−.0355	21.9500	2.0000
4	2.5000	2.4801	.0199	18.5500	2.0000
5	4.0000	3.8820	.1180	36.6800	5.0000
6	1.9000	2.1469	−.2469	15.1000	2.0000
7	3.6100	4.6405	−1.0305	38.1400	4.0000
8	1.7500	1.7961	−.0461	14.1100	2.0000
9	1.0000	1.0982	−.0982	7.1600	1.0000
10	4.0000	2.6949	1.3051	26.2300	4.0000
11	2.0000	1.7066	.2934	14.3200	2.0000
12	.7500	1.4343	−.6843	7.0500	1.0000
13	1.5000	1.6723	−.1723	14.1100	2.0000
14	4.0000	3.2136	.7864	21.6300	2.0000
15	3.8500	3.7922	.0578	33.2000	4.0000
16	2.0000	2.1420	−.1420	16.4800	2.0000
17	6.5000	5.0206	1.4794	45.1700	6.0000
18	3.0000	4.2468	−1.2468	35.7300	4.0000

CASE NUMBER	Y	Ŷ	RESIDUAL	X_1	X_2
19	5.0000	6.1758	−1.1758	58.8600	8.0000
20	2.0000	2.9595	−.9595	24.9200	3.0000
21	2.5000	1.5012	.9988	12.8100	2.0000
22	7.6400	6.9142	.7258	64.7900	8.0000
23	6.2500	6.4683	−.2183	59.1300	7.0000
24	2.0000	3.4528	−1.4528	29.4600	4.0000
25	4.5000	4.1410	.3590	34.8300	4.0000
26	3.0000	2.1526	.8474	17.0500	2.0000
27	2.0000	1.9183	.0817	20.5200	3.0000
28	5.0000	5.4155	−.4155	49.2500	6.0000
29	2.1000	2.5422	−.4422	18.9300	2.0000
30	10.0000	9.4238	.5762	78.4900	8.0000

Printout 3

The data for this problem came from the 1970 United States census. Thirty cities with populations from 25,000 to 100,000 were selected, and data were recorded for the following variables:

- ■ y—total number of serious crimes, 1970, in thousands
- ■ x_1—population in thousands
- ■ x_2—percentage unemployed, 1970.

NUMBER OF CASES 30
NUMBER OF ORIGINAL VARIABLES 3

VARIABLE	MEAN	STD. DEVIATION
Y	1.65800	.90453
X_1	52.71700	17.39476
X_2	4.85667	1.57386

CORRELATION MATRIX

VARIABLE NUMBER	Y	X_1	X_2
Y	1.000	.628	.414
X_1		1.000	.185
X_2			1.000

Two-Variable Printout

R .6277
S_E .7166

ANALYSIS OF VARIANCE

	DF	SUM OF SQUARES	MEAN SQUARE	F RATIO
REGRESSION	1	9.349	9.349	18.205
RESIDUAL	28	14.378	.514	

VARIABLES IN EQUATION

VARIABLE	COEFFICIENT	STD. ERROR	T
CONSTANT	−.06270		
X_1	.03264	.00765	4.2667

Multivariable Printout

R .6970
S_E .6722

ANALYSIS OF VARIANCE

	DF	SUM OF SQUARES	MEAN SQUARE	F RATIO
REGRESSION	2	11.525	5.763	12.752
RESIDUAL	27	12.202	.452	

VARIABLES IN EQUATION

VARIABLE	COEFFICIENT	STD. ERROR	T
CONSTANT	−.76641		
X_1	.02967	.00730	4.0644
X_2	.17715	.08072	2.1946

List of Residuals

CASE NUMBER	Y	$\hat{Y}$	RESIDUAL	X_1	X_2
1	2.8920	3.0921	−.2001	86.9740	5.1000
2	1.2270	1.0951	.1319	36.4140	4.7000
3	2.0700	1.9529	.1171	57.2950	5.9000
4	4.4700	3.2996	1.1704	69.9470	10.3000
5	1.6780	1.0713	.6067	44.3420	3.1000
6	.5710	.9102	−.3392	32.6740	4.3000
7	2.8320	2.6788	.1532	85.9190	5.7000
8	1.4440	1.3143	.1297	40.4890	5.5000
9	.8700	1.1341	−.2641	37.5890	3.1000
10	2.2160	1.5948	.6212	57.0550	4.6000
11	1.0200	.7145	.3055	27.5550	4.1000
12	.3400	.9169	−.5769	33.1680	4.1000
13	1.5030	2.3976	−.8946	77.9980	6.2000
14	1.5510	1.2252	.3258	46.5460	4.1000
15	.9880	2.3948	−1.4068	70.8530	5.5000
16	2.0360	2.0369	−.0009	60.0400	3.1000
17	1.9610	1.6376	.3234	55.0010	4.5000
18	.8920	1.5180	−.6260	33.9080	6.8000
19	.8490	1.6268	−.7778	57.1240	4.4000
20	3.0950	2.1887	.9063	79.0250	4.1000
21	2.0060	1.0049	1.0011	39.3240	3.4000
22	.8290	1.5887	−.7597	56.1270	3.5000
23	1.9580	1.8994	.0586	74.9900	3.7000
24	.9940	1.0644	−.0704	36.1030	4.2000
25	2.0890	1.8499	.2391	59.8740	5.6000
26	1.3770	1.7529	−.3759	63.8840	4.0000
27	1.0390	1.2170	−.1780	30.5150	6.1000
28	1.5260	2.0044	−.4784	43.1180	8.2000
29	2.7630	1.4920	1.2710	46.8720	4.1000
30	.6540	1.0661	−.4121	40.7870	3.7000

CHAPTER 11

TIME SERIES

In this chapter we discuss, with illustrations, the important topic of time series analysis and explain the use of time series as tools for both historical analysis and forecasting.

Specific objectives are—

to discuss in a general way the components of a time series (Section 1).

to present in considerable detail methods of analyzing the secular trend component (Section 2, Appendix A, and Appendix B).

to present the methods of isolating the cycle and the irregular components of a historical time series (Section 3).

to present in great detail the methods of seasonal component analysis and the generation and use of seasonal indexes (Section 4 and Appendix B).

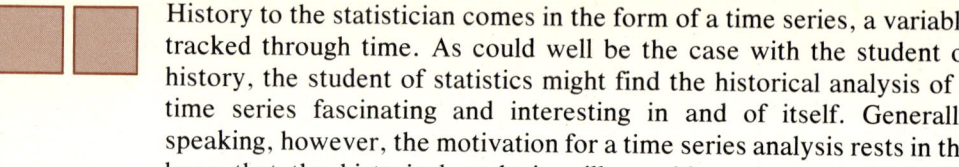

History to the statistician comes in the form of a time series, a variable tracked through time. As could well be the case with the student of history, the student of statistics might find the historical analysis of a time series fascinating and interesting in and of itself. Generally speaking, however, the motivation for a time series analysis rests in the hope that the historical analysis will provide an accurate basis for projection of the future.

Without the aid of formal statistical analysis, a good many decisions in business and government are made by simply extrapolating certain aspects of a time series. For example, leather goods sales in a large department store have, over the past 5 years, increased by an average of 12 percent. The manager of this department plans and orders for a 12 percent increase next year. (Perhaps he should consider what a 12 percent compounding would do for him by the year 2000.) As a second example, an automobile dealer notes that the number of units sold increased on the average by 200 units per year over the past 3 years, but he also makes note that population of the area is expected to continue its growth, the general level of business activity is strong, and no recession is predicted. He decides to increase his next year's order by 200 units. Even though a formal statistical model was not used, the automobile dealer was more sophisticated than the leather goods manager. Before simply extrapolating the trend, he gave consideration to several other variables which, from his experience, were determinants of automobile sales.

Time is the vehicle of all dynamics, but it is not a causal factor; i.e., change cannot take place without time, but we are unable to argue that time caused the change. As indicated in the above paragraph, changing sales are caused by changes in factors such as population, personal income, and consumer tastes. Clearly a word of warning must be issued: Extending a historical time pattern into the future, commonly called *naive* forecasting, whether in terms of a statistical model or not, is fraught with many dangers. Even a relatively short-run forecast of this sort can be disastrously wrong. Consider the case of the Bourbon Street merchants, hotel owners, and speakeasy operators. For decades they had experienced a 4-week-long Mardi Gras seasonal business explosion of magnitude 4 or 5 times their normal pattern. In addition, the Mardi Gras peak had been growing at the annual rate of 5 percent to 20 percent over the past 7 years. The 1979 business was projected accordingly; by year's end, 1978, inventories were either in or ordered, contracts for additional labor had been made, etc. With no warning and 10 days before the start of Mardi Gras, a citywide policemen's strike was called. In spite of all efforts, the strike was not settled before the big event. Major parades were cancelled, and others were moved to the suburbs. The effects of the strike were compounded by nationwide inclement weather and some local problems with several airlines. All of this could not have been predicted in time to keep the Bourbon Street businessmen from the serious consequences of gearing up for a business boom that didn't show. In the statistical language of time series, an unpredictable event such as the policemen's strike is called an *episodic* or an *accidental* event.

Consider another case which illustrates other problems with the use of time series forecasting. The United States ceased the military draft of 18 year olds in early 1972. Enrollments in institutions of higher education dropped dramatically during the 1972-73 academic year. As far as we know, nobody predicted this; many of the statisticians involved in the forecasting of enrollments were self-critical—they felt that, inasmuch as they knew about cessation of the draft, the cause-effect relationship should have been picked up. Higher education enrollments are usually projected on the basis of very accurate estimates of the college-age population coupled with a consideration of several other variables related to the propensity of this group to go to college. (For example, a propensity-related variable seems to be the unemployment rate; there is an inverse relationship between this and higher education enrollments.) In agreement with the self-critical statisticians and in retrospect, it seems quite apparent that many young men were registered in our colleges and universities in order to avoid being drafted. Even though it was well known that the draft would cease, the causal effect of this event upon higher education was missed completely by large numbers of competent statisticians and administrators.

Analysis of time series today has reached some quite sophisticated and highly theoretical levels. We will proceed with detail on the traditional and more descriptive methods with no more than a hint of what is being done in the modern vein. The notions we consider are, however, fundamental to any level of analysis.

11.1 The Components of a Time Series

Time Series Components

A time variable is traditionally viewed as the aggregation of four components:

1. *The trend (or secular trend).* This is the long-run movement of the time series, the basic building block from which the other components are measured.

2. *The cycle component.* This is, as the term suggests, the effect of the business cycle upon the time series.

3. *The seasonal component.* Over the period of a year, there may be a seasonally related effect upon the variable being measured through time. This seasonal component can be measured, then, only if the time readings on the variable are something less than

annual; i.e., they must be monthly, quarterly, or weekly, for example.

4. *Irregular variation.* Variations that cannot be attributed to trend, cycle, or season are called *irregulars*; in addition to the purely random variation of a statistical series, irregular variation includes that due to episodic or accidental events, events that are unpredictable. (For example, President Eisenhower had a heart attack while in office. The Dow-Jones average, and other measures of stock prices, dropped significantly the next day but returned to the normal level a few days thereafter.)

In retrospect, the four components can be more or less isolated; i.e., methods exist for a fairly detailed analysis of historical time series. From the standpoint of time series analysis and its use in forecasting, the statistician can do a great deal with the trend and the seasonal components. Cycles are typically variable in both magnitude and period and therefore do not lend themselves to the elegance of simple statistical analysis. In fact, it is frequently difficult to even recognize what part of a cycle is currently being experienced, and certainly statistical forecasting of cycles is hazardous.

Table 11.1 Average Annual State Government Employment in Utah
(Excluding Higher Education and Employment Security)
Employment in Thousands from 1963 to 1977
(Table includes calculations for least squares linear trend line.)

x' (year)	y (employment)	x	xy	x^2
1963	6.36	−7	−44.52	49
1964	6.74	−6	−40.44	36
1965	6.88	−5	−34.40	25
1966	7.16	−4	−28.64	16
1967	7.42	−3	−22.26	9
1968	7.77	−2	−15.54	4
1969	8.04	−1	−8.04	1
1970	8.39	0	0	0
1971	8.79	1	8.79	1
1972	9.43	2	18.86	4
1973	9.86	3	29.58	9
1974	10.01	4	40.04	16
1975	10.41	5	52.05	25
1976	10.49	6	62.94	36
1977	11.18	7	78.26	49
Total	128.93	0	96.38	280

SOURCE: *Utah Statistical Abstract*, Bureau of Economic and Business Research, University of Utah (Salt Lake City, January, 1979).

11.2 Trend Analysis (The Linear Model)

The nature of the secular trend is probably best determined by an inspection of the time series plotted as a scattergram. If an equation is desired to describe this long-run movement, we will borrow a page from the previous chapter and use ordinary least squares to determine the coefficients of the functional relationship we choose. Let us first give consideration to a time series that appears to be quite linear in trend. We present, in Table 11.1 and Figure 11.1 respectively, the data series and a scattergram plot. This is a series of the annual average number of employees (in thousands and not including employees in higher education) in state government in Utah for the 15 years from 1963 through 1977.

We want a trend line of the form, $\hat{y} = a + bx$, where y is the employment and x is time in years. It is always convenient to pick a new, but arbitrary, origin or 0-point on the time axis—even if a computer is being used. We would be exchanging 1- and 2-digit values of x for 4-digit values. If the arithmetic required by the least squares

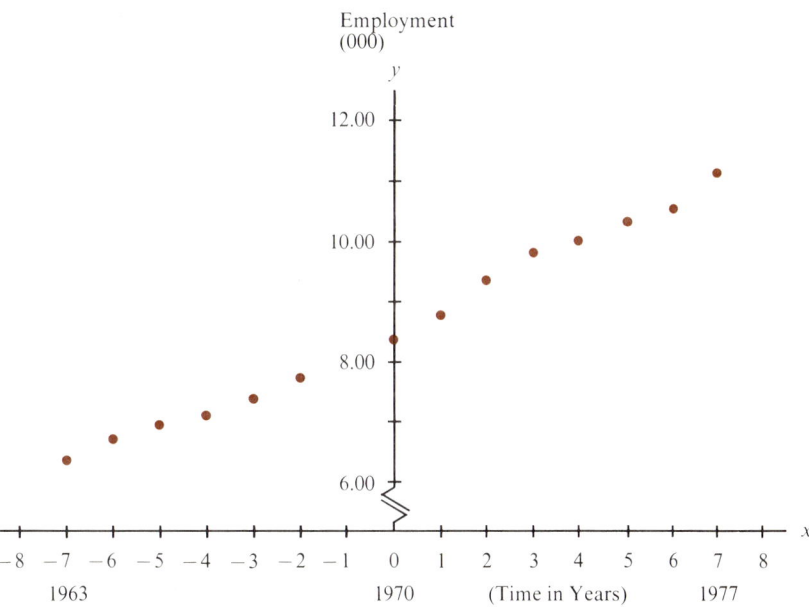

Figure 11.1 Scattergram, Utah State Government Employment*

SOURCE: *Utah Statistical Abstract*, Bureau of Economic and Business Research, University of Utah (Salt Lake City, January, 1979).

*Excludes higher educaton and employment security

method is being done without the aid of a computer, it is particularly efficient to place the new origin at the center of the data.

Formulas for Coefficients of Least Squares Linear Trend

In our example, if we map 1970 into the new 0, the sum of the new x's will be 0 as well as $\bar{x}$.[1] Because of this, the formulas we borrowed from the previous chapter for b and a in the least squares line of best fit are simplified: i.e.,

$$b = \frac{n\Sigma xy - \Sigma x \Sigma y}{n\Sigma x^2 - (\Sigma x)^2} = \frac{n\Sigma xy}{n\Sigma x} = \frac{\Sigma xy}{\Sigma x}$$

if $\Sigma x = 0$; and

$$a = \bar{y} - b\bar{x} = \bar{y}$$

if $\bar{x} = 0$.

For our example we refer to Table 11.1 where we have completed the additional columns, i.e., the xy and the x^2 columns, and the necessary summations for the least squares trend line. We find

$$b = \frac{96.38}{280} = .344 \text{ and } a = \frac{128.96}{15} = 8.60.$$

The least squares linear trend line is $\hat{y} = 8.60 + .344x$, where y is employment in thousands and x is in years with 1970 as the origin. (It is important in a trend function that we identify the time origin as well as the units of both x and y.) By interpreting the 2 coefficients, we can reach 2 conclusions. (1) The intercept of 8.60 estimates the 1970 trend value of 8,600 employees; i.e., when $x = 0$, $\hat{y} = 8.60 + .344(0) = 8.60$, the trend value at the origin, 1970. The difference between the actual value of 8,390 employees for 1970 and this trend value would be attributed to cycle and/or irregular effects.[2] (2) The slope value of .344 is the historical estimate of the average annual increase in thousands of employees—344 employees per year. Recall from Chapter 10 that the slope-coefficient is the incremental change in y per unit of x. For example, the 1971 trend value would simply be .344 added to the 1970 trend of 8.60; $\hat{y} = 8.60 + .344(1) = 8.94$.

We will discuss forecasting in some detail in a later section of this chapter, but let us note the ease with which we can make a naive forecast of employment through 1980:

1. In our example the number of years is 15, an odd number. When n is odd, there is a middle year where, if we place the new origin, the sum of the x-values will be 0. If n is an even number, there is no middle year. In this case we can (1) make n odd by either finding another year's datum or discarding the oldest datum point or (2) place the origin between the 2 middle years and let the new variable, x, be in half-years instead of years. This would lead to a set of positive and negative odd integers for values of x, but both Σx and $\bar{x}$ would be 0. For example, assume we had data for 1974 through 1979. The origin of the new variable, x, would be placed at the end of 1976 and the beginning of 1977; x would be in half-years and would take on only the odd integral values as shown below.

Years	1974	75	76	↓	77	78	79
x	−5	−3	−1	0	1	3	5

2. For additional comments, see Section 11.4.

1978: ŷ(8) = 8.60 + .344(8) = 11.35 (11,350 employees).
1979: ŷ(9) = 8.60 + .344(9) = 11.70 (11,700 employees).
1980: ŷ(10) = 8.60 + .344(10) = 12.04 (12,040 employees).

The plot of the trend line along with the scattergram in Figure 11.2 and the calculation of the deviations from trend in Table 11.2 show that the linear trend is an excellent historical description; the largest-magnitude deviation is 220 employees in 1969 with actual employment being about 2.7 percent below trend. In terms of materiality (borrowing a term from our friend, the auditor), we might well quit now with the observation that there appears to have been a mild employment cycle in our data with a possible episodic event in 1976, slowing down the normal growth; but this seems to have recovered in 1977.

11.3 Moving Averages, Cycles, and Irregulars

There are several additional notions, bearing on the historical analysis of time series, that we would like to consider. The technique of moving averages is used to smooth a time series; this becomes a method of isolating the cycle if the random and episodic effects are not too violent.

Figure 11.2 Linear Trend Line, Utah State Government Employment

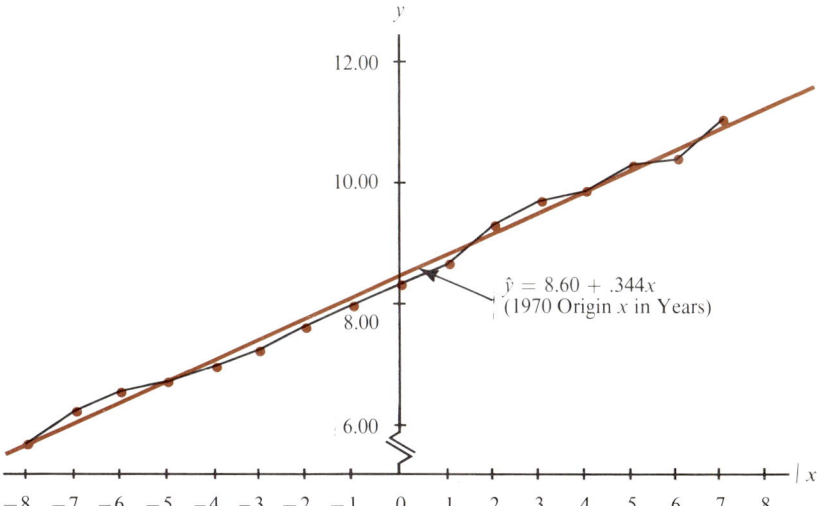

$\hat{y} = 8.60 + .344x$
(1970 Origin x in Years)

SOURCE: *Utah Statistical Abstract*, Bureau of Economic and Business Research, University of Utah (Salt Lake City, January, 1979).

Table 11.2 Deviations from Trend: Utah State Government Employment

Year	Employment (000) y	Trend $\hat{y}$	Deviations from Trend $y - \hat{y}$
1963	6.36	6.19	.17
1964	6.74	6.54	.20
1965	6.88	6.88	00
1966	7.16	7.22	−.06
1967	7.42	7.57	−.15
1968	7.77	7.91	−.14
1969	8.04	8.26	−.22
1970	8.39	8.60	−.21
1971	8.79	8.94	−.15
1972	9.43	9.29	.14
1973	9.86	9.63	.23
1974	10.01	9.98	.03
1975	10.41	10.32	.09
1976	10.49	10.66	−.17
1977	11.18	11.01	.17

SOURCE: *Utah Statistical Abstract,* Bureau of Economic and Business Research, University of Utah (Salt Lake City, January, 1979).

To illustrate what is meant by a *moving average* and the term *centered,* let us use the employment time series data from Table 11.2. The 1963 employment in thousands, 6.36, is y_1, y_2 equals 6.74 etc. A 2-year moving average, MA_1, would begin with

$$\frac{y_1 + y_2}{2} = \frac{6.36 + 6.74}{2} = 6.55 = MA_1;$$

the second moving average, MA_2, would be calculated by dropping y_1 and adding y_3,

$$MA_2 = \frac{y_2 + y_3}{2};$$

$$MA_3 = \frac{y_3 + y_4}{2};$$

etc. To illustrate further, a 5-year moving average (see Table 11.3) would begin with

$$MA_1 = \frac{y_1 + y_2 + y_3 + y_4 + y_5}{5}$$

$$= \frac{6.36 + 6.74 + 6.88 + 7.16 + 7.42}{5}$$

$$= \frac{34.56}{5} = 6.91.$$

Table 11.3 Moving Average and Cycle Measurements: Utah State Government Employment

Year	Employment (000) y	Five-Year Moving Total	Five-Year Moving Average MA	Cycle Differences $MA - \hat{y}$	Cycle Relatives $MA/\hat{y}$
1963	6.36	—	—	—	—
1964	6.74	—	—	—	—
1965	6.88	34.56	6.91	.03	1.004
1966	7.16	35.97	7.19	−.03	.996
1967	7.42	37.27	7.45	−.12	.984
1968	7.77	38.78	7.76	−.15	.981
1969	8.04	40.41	8.08	−.18	.978
1970	8.39	42.42	8.48	−.12	.986
1971	8.79	44.51	8.90	−.04	.996
1972	9.43	46.48	9.30	.01	1.001
1973	9.86	48.50	9.70	.07	1.007
1974	10.01	50.20	10.04	.06	1.006
1975	10.41	51.95	10.39	.07	1.007
1976	10.49	—	—	—	—
1977	11.18	—	—	—	—

SOURCE: *Utah Statistical Abstract,* Bureau of Economic and Business Research, University of Utah (Salt Lake City, January, 1979).

The second moving average is calculated by dropping y_1 and adding y_6:

$$MA_2 = \frac{y_2 + y_3 + y_4 + y_5 + y_6}{5}$$

$$= \frac{6.74 + \cdots + 7.77}{5} = \frac{35.97}{5} = 7.19.$$

You should verify several more; e.g.,

$$MA_3 = \frac{37.27}{5} = 7.45$$

$$MA_4 = \frac{38.78}{5} = 7.76 \quad \text{etc.}$$

Using an odd number of years will "center" the moving average whereas an even number of years will place the average "off center." Let us see what we mean by this. The 2-year average viewed in the time context can represent neither 1963 nor 1964; being an average employment for both years, it better represents the last 6 months of 1963 combined with the first 6 months of 1964. As such it is centered, if you please, at the end of 1963 and the beginning of 1964. Looking at the 5-year moving average, we see that there is a middle year, the third year of the 5-year sequence, and the 5-year moving average may be logically "centered" at this third year. For annual data, the length of the moving average is usually taken as either 5 or 7 years for the following 2 reasons: (1) An odd number is used because the moving average will be "centered." Using an even number of years would place the average in between 2 periods; (2) A relatively small number, k, is used because $(k - 1)/2$ years of data averages are lost on both ends of the series: $k = 3$ may be too small to accomplish the necessary smoothing.

The five-year moving totals, an intermediate step in the calculations, and the five-year moving averages are presented in Table 11.3. Note that there are two years on either end of the data where we were unable to calculate a moving average. Under the assumption that the moving average exhibits the combined cycle and trend, differences between the moving averages and the trend values estimate the cycle effects. If we were considering an *additive* model, these differences would be the cycle measures. For a *multiplicative* model, the ratio of the moving average to the trend, an index-type measure, gives the cycle-effect estimate and is called the *cycle relative*. The cycle differences and relatives are calculated and presented in Table 11.3 along with the moving averages.

The residuals, or the irregulars, are then estimated by the differences between the actual employment numbers and the moving averages (the combination of trend and cycle). The irregular differences are calculated in Table 11.4. The graphs in Figure 11.3 illustrate these deviations. The baseline, the x-axis, is the trend. Graph 1 shows the deviations of actual data from trend, and Graph 2 shows the moving average cycle about the trend. The difference between these two graphs would be the irregulars as calculated in the last column of Table

Table 11.4 Irregulars: Utah State Government Employment

Year	Employment (000) y	Five-Year Moving Average MA	Irregulars $y - MA$
1963	6.36	—	—
1964	6.74	—	—
1965	6.88	6.91	−.03
1966	7.16	7.19	−.03
1967	7.42	7.45	−.03
1968	7.77	7.76	.01
1969	8.04	8.08	−.04
1970	8.39	8.48	−.09
1971	8.79	8.90	−.11
1972	9.43	9.30	.13
1973	9.86	9.70	.16
1974	10.01	10.04	−.03
1975	10.41	10.39	.02
1976	10.49	—	—
1977	11.18	—	—

Time Series

Figure 11.3 Graph of Cycle and Deviations from Trend: Utah State Government Employment

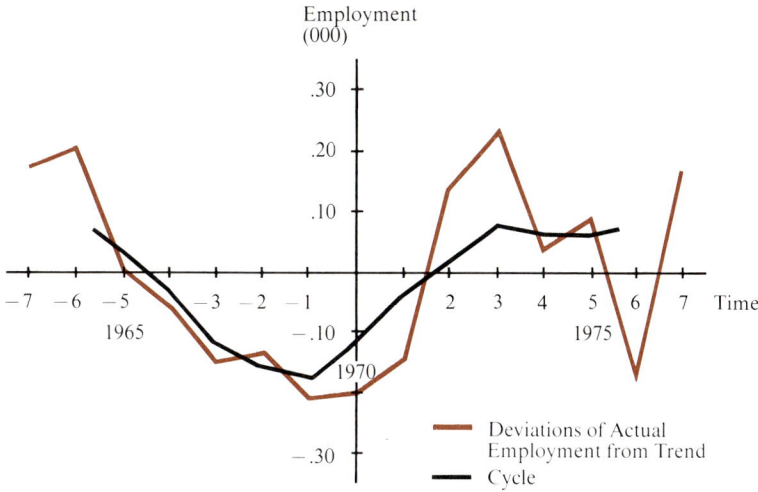

SOURCE: *Utah Statistical Abstract*, Bureau of Economic and Business Research, University of Utah (Salt Lake City, January, 1979).

11.4. Observe that the method of moving averages may well isolate the historical cycle effects, but it does not lend itself to long-term forecasting.

If it were important to understand and explain the cycle and the deviations about the cycle, some detail regarding the relevant political and economic history would be needed. What are some of the factors that might impact state government employment? We list a few: population growth, personal income change, availability of federal funds for road building and social services, political change and mix in the legislature and in the executive branch of state government, etc.

 ## 11.4 Measuring Seasonal Effects

If a relatively stable seasonal pattern persists in a time series, several methods exist that will isolate and measure it. We will first give consideration to a method that is standard and conceptually simple but does involve a great deal of arithmetic. (A second method for estimating seasonal effects—a multiple regression model—will be presented in Appendix B of this chapter.) The model we are going to discuss here is a multiplicative model; i.e., the values in the time series are viewed as the product of the four components, trend, cycle, seasonal, and irregular. In a model of this type, the trend is the building block and is measured in the same units as those of the time series; cycle, seasonal, and irregular effects must, then, be an index in decimal fraction form. This point will become clear as we proceed with our example. (See Column 3 of Table 11.5.) We are going to use quarterly gross earnings in thousands of dollars of the Utah State Liquor Monopoly for fiscal years 1964 through 1971. The data in Table 11.5 and the scattergram and line graph in Figure 11.4 clearly illustrate a reasonably stable quarterly pattern.

Table 11.5   Earnings from Utah State Liquor Monopoly

Year	Quarter	Time Variable x	Gross Earnings y	Centered Moving Average MA	Ratio y/MA
1963	3	1	1,493	—	—
	4	2	2,074	—	—
1964	1	3	1,445	1,652	.875
	2	4	1,565	1,667	.939
	3	5	1,557	1,677	.928
	4	6	2,128	1,698	1.253
1965	1	7	1,469	1,709	.860
	2	8	1,708	1,716	.995
	3	9	1,505	1,736	.867
	4	10	2,233	1,742	1,282
1966	1	11	1,524	1,764	.864
	2	12	1,701	1,797	.946
	3	13	1,694	1,817	.933
	4	14	2,307	1,835	1.257
1967	1	15	1,604	1,851	.867
	2	16	1,771	1,860	.952
	3	17	1,746	1,874	.932
	4	18	2,328	1,894	1.229
1968	1	19	1,694	1,916	.884
	2	20	1,842	1,957	.941
	3	21	1,854	1,996	.929
	4	22	2,547	2,014	1.265
1969	1	23	1,787	2,034	.879
	2	24	1,891	2,079	.909
	3	25	1,966	2,130	.923
	4	26	2,799	2,180	1.284
1970	1	27	1,940	2,230	.870
	2	28	2,141	2,266	.945
	3	29	2,113	2,303	.917
	4	30	2,944	2,344	1.256
1971	1	31	2,084	—	—
	2	32	2,327	—	—

SOURCE: *Annual Report,* Utah Liquor Control Commission, fiscal year 1963–64 through fiscal year 1970–71 (Salt Lake City, 1964–71).

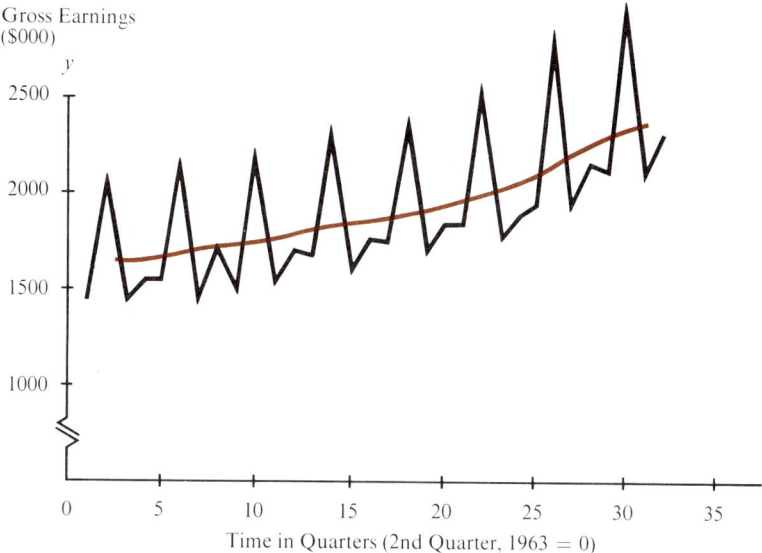

Figure 11.4 Utah State Gross Earnings from Liquor Monopoly Q_3, 1963, to Q_2, 1971

SOURCE: *Annual Report*, Utah Liquor Control Commission, fiscal year 1963-64 through fiscal year 1970-71 (Salt Lake City, 1964-71).

We now wish to find a set of seasonal indexes, 1 for each quarter, each of which will estimate the quarter's deviation from the combined trend and cycle. As mentioned above, we are introducing a multiplicative model, and the set of 4 seasonal indexes will be in decimal fraction form. (Multiplied by 100, they will give percentage measures.) Our effort is to estimate the "normal" seasonal impact on liquor earnings over and above the combined trend and cycle effects. Inspecting the data in Table 11.5, we see that the fourth calendar quarter is the peak earnings quarter each year. This fourth-quarter seasonal index will be something greater than 1.00; i.e., fourth-quarter earnings are always greater than 100 percent of the trend and cycle pattern. If there were no seasonal impact—i.e., if deviations about the trend and cycle were more or less random—then each of the 4 quarterly seasonal indexes would be 1.00; the sum of these would be 4.00. We will see in the following exercise that, even with definite seasonal patterns, the method used will quite automatically lead to an average of 1.00 for the 4 indexes—the sum will be close to 4.00. More often than not, the sum will be something slightly different from 4.00, and therefore the average will not be 1.00. In order to avoid gaining or losing something from series magnitude, we insist that the average seasonal index be 1.00 and make an artificial adjustment to accomplish this. (See Table 11.6 and

Measuring Seasonal Effects

Table 11.6 Quarterly Ratios by Quarter

	Quarter			
Year	1	2	3	4
1964	.875	.939	.928	1.253
1965	.860	.995	.867	1.282
1966	.864	.946	.933	1.257
1967	.867	.952	.932	1.229
1968	.884	.941	.929	1.265
1969	.879	.909	.923	1.284
1970	.870	.945	.917	1.256
Total	6.099	6.627	6.429	8.826
a. Average	.871	.947	.918	1.261
b. Adjusted Average	.871	.948	.919	1.262
c. Median	.870	.945	.928	1.257

(Note: The four medians sum to 4.00, so no adjustment is necessary.)

d. Weighted Average
   (Weighting last three
   years by 2 and first
   four years by 1)   .873   .942   .920   1.263
e. Adjusted   .873   .943   .920   1.264

---

related discussion.) The method will involve a series of comparisons, or ratios, of the actual quarterly gross earnings with a centered moving annual average. The series of centered moving averages are presented in Column 2 of Table 11.5 and are graphically shown in Figure 11.4 along with the original series. We can arrive at this series with either of 2 calculations.

1. Calculate a four-quarter moving average. This will not be centered because it falls at the beginning of one quarter and at the end of the previous quarter. A 2-quarter moving average of the 4-quarter averages will center on the specific quarters. Below are a few quarters from the data in Table 11.5 to illustrate this procedure.

Quarter	Time Variable x	Gross Earnings y	Four-Quarter Noncentered Moving Average	Two-Quarter Centered Moving Average
1963 Q_3	1	1,493		
Q_4	2	2,074		
1964 Q_1	3	1,445	1,644.25	
Q_2	4	1,565	1,660.25	1.652
Q_3	5	1,557	1,673.75	1,667
Q_4	6	2,128		

2. The second notion, which gives results identical to those of the

first method, is that of calculating a weighted five-quarter (naturally centered) moving average as follows:[3]

$$\bar{y}_3 = \frac{\frac{1}{2}y_1 + y_2 + y_3 + y_4 + \frac{1}{2}y_5}{4},$$

which is equal to

$$\bar{y}_3 = \frac{y_1 + 2(y_2 + y_3 + y_4) + y_5}{8}.$$

Inspection of procedure 1 shows that, in arriving at the first moving average of 1,652, the first- and fifth-quarter y-values were used only once, and the second, third, and fourth values were used twice, which is the exact nature of the second form of the five-quarter moving average.

For calculating the full series of moving averages, the second form above lends itself readily to a simple computer program and is easy enough if hand calculation is necessary. For the second moving average, drop 2 of the y-values, y_1 and one of the y_2's, and then add 2 y-values, another y_5 and y_6; the denominator remains fixed at 8. This process continues through the series.

We illustrate the calculations for the first two weighted five-quarter centered moving averages in Table 11.5:

$$MA_1 = \frac{1{,}493 + 2(2{,}074 + 1{,}445 + 1{,}565) + 1{,}557}{8}$$

$$MA_2 = \frac{2{,}074 + 2(1{,}445 + 1{,}565 + 1{,}557) + 2{,}128}{8}$$

$$= \frac{13{,}336}{8} = 1{,}667.$$

Note that MA_2 also equals the moving total from MA_1 less $y_1 + y_2$ plus $y_5 + y_6$ divided by 8:

---

3. The following shows that the four-quarter moving average followed by a two-quarter moving average is equivalent to the weighted five-quarter moving average: The first two four-quarter moving averages are

$$_4MA_1 = \frac{y_1 + y_2 + y_3 + y_4}{4}$$

and

$$_4MA_2 = \frac{y_2 + y_3 + y_4 + y_5}{4}.$$

The average of these two,

$$_2MA = \frac{\frac{y_1 + y_2 + y_3 + y_4}{4} + \frac{y_2 + y_3 + y_4 + y_5}{4}}{2}$$

$$= \frac{\frac{y_1 + 2y_2 + 2y_3 + 2y_4 + y_5}{4}}{2}$$

$$= \frac{y_1 + 2y_2 + 2y_3 + 2y_4 + y_5}{8},$$

which is identical to the second form of the five-year moving average given above.

$$MA_2 = \frac{13{,}218 - (1{,}493 + 2{,}074) + (1{,}557 + 2{,}128)}{8}$$

$$= \frac{13{,}336}{8} = 1{,}667.$$

Starting with 8 years or 32 quarters of data, y_i, in the liquor earnings example, we salvage 7 years or 28 quarters of moving averages, MA_i; i.e., we lose 2 quarters on each end of the data set. The last column of Table 11.5 presents a series of ratios, the quotient of y_i by $\bar{y}_i$. Each of these decimal fractions is, in one sense, an estimator of a quarterly index. This ratio gives the proportion of the annual average represented by the actual earnings, with the specific quarter as the center of the year. For example, the first moving average, 1,652, is centered at the first quarter of 1964, y_3. The actual y_3 is 1,445 and the ratio,

$$\frac{1{,}445}{1{,}652} = .875$$

The actual first quarter, 1964, earnings are only 87.5 percent of the average annual quarterly earnings with this specific quarter as the center of the year. There are 7 potentially different index estimators for each of the 4 quarters; this suggests that some sort of averaging might be appropriate in arriving at the final set of 4 quarterly index estimates. In Table 11.6 we have separated the ratios from Table 11.5 by quarters. If we are able to assume that, in addition to stability of the seasonal pattern, all irregular effects are strictly random in nature, the arithmetic mean of the 7 estimators for each quarter will yield a set of unbiased estimators for the 4 quarterly seasonal indexes. Row a in Table 11.6 gives this average, and Row b is the adjusted and final result of a simple arithmetic average of the sets of 7. (As mentioned above, the sum of the set of 4 seasonal indexes should be 4.000. Because of rounding errors, the original sum was 3.998. Adjusting each of the indexes on a proportional basis leads to adding 0.001 to each of the 2 largest indexes.) An inspection of the ratios in Table 11.6 shows some deviations that would appear to be too large to be random (e.g., quarters 2 and 3 of 1965). If we were familiar with the detailed economic history of Utah, we might well identify some episodic event in 1965 that would have caused this deviation from the normal seasonal pattern. Inasmuch as episodic events could well create a bias in the mean value estimators, consideration should be given to several alternatives. The choice is judgmental.

1. Use the *medians* from each of the quarterly sets. These are given in Row c of Table 11.6.

From the first through the fourth quarter respectively, the medians of the sets of ratios are .871, .948, .919, and 1.262; these add up to exactly 4.000. No adjustment would be required, and the medians would form our final set of seasonal indexes. We were lucky; there is no mathematical logic to ensure that the medians will sum to 4.000. An adjustment is usually required.

To illustrate again the idea of adjustment, let us assume that the medians were .880, .950, .920, and 1.280 respectively for the 4 quarters. This set of unadjusted seasonal indexes sums to 4.030. The adjusted sum should be 4.000; it is therefore necessary to subtract a total of .030 from the set of 4 indexes. This surplus of .030 will be subtracted proportionately from each of the unadjusted indexes, the proportion being the ratio of each of these to the total,

$$\frac{.880}{4.030}, \frac{.950}{4.030}, \frac{.920}{4.030}, \text{ and } \frac{1.280}{4.030}$$

respectively. These adjustment ratios are .218, .236, .228, and .318; as a fraction of .030, these give .007, .007, .007, and .009. Subtracting these, in our hypothetical example, from each of the unadjusted indexes, we have for quarters 1 through 4 respectively .873, .943, .913, and 1.271 for the final set of seasonal indexes that sum to exactly 4.00.

2. Discard the largest and the smallest number in each set, and average the middle 5 numbers.

3. Discard the extremes on the basis of judgment, and average the remaining numbers.

4. For each set of quarterly estimates, calculate the standard deviation, s. Discard numbers within each set that fall outside an arbitrary interval, $\bar{r} \pm ks$. (Use $k = 1.960$ or 1.645, for example.) The average ratios, $\bar{r}_i$, were calculated (Row a, Table 11.6). Calculate a new average after discarding these outliers.

5. A weighted average is frequently used; for example, weight the first 4 years by 1 and the last 3 years by 2, for a total weight of 10. (Other judgment weightings are, of course, possible.) The above weightings were prompted by the notion that the more recent experience has higher validity than the older experience. The indexes calculated using these weights are in Row d of Table 11.6, and the adjusted set is in Row e.

A comparison of results in Rows b, c, and e shows but modest differences in the final set of seasonal indexes. This indicates, as observed at the outset, that the seasonal pattern on liquor earnings in Utah was reasonably stable over the eight years. As a word of warning, however, one should always be aware that episodic events can have considerable effect on a quarterly time series (and perhaps even a greater relative effect on a monthly time series). A computer program dealing with either simple or weighted averages in the final calculation step could lead to considerable error in the set of seasonal index estimates.

The procedures for estimating a set of seasonal indexes for a monthly time series are similar to those for a quarterly series. The weighted moving average includes 13 months' data, the first and last month weighted by unity and the middle 11 months weighted by 2; the constant divisor would, of course, be 24. Twelve seasonal indexes would be generated, the average of which should be 1.00 and the sum 12.00.

The term *seasonally adjusted* is not unfamiliar to most of you. Unemployment percentages are usually given as seasonally adjusted numbers. For example, if both January and September unemployment percentages in the Northwest were reported as 5.8 percent, both seasonally adjusted, the actual percentage for September would be somewhat less than 5.8 and that for January would be considerably greater. The seasonal adjustment allows us to make valid comparisons on the unemployment problem throughout the year. Seasonally adjusting, or "deseasonalizing" a time series refers simply to the process of removing the seasonal effect. This is accomplished by dividing the time series's actual readings by the appropriate seasonal index.

To further illustrate this notion, consider again the liquor earnings example. We will use the seasonal indexes from Row e of Table 11.6—.873, .943, .920, and 1.264 for quarters 1 through 4 respectively, and deseasonalize the complete 32-quarter historical series. This series, yd_i, is shown along with the original series, y_i, in Table 11.7. The first yd is obtained by dividing 1,493 by the third quarter index, .920; yd_2 equals y_2, 2,074, divided by 1.264; etc. We observe that this new series is, for the most part, very close to the moving average series, also presented in Table 11.7. The centered moving average series reflects primarily the combination of trend and cycle. (This is the usual assumption and is realistic without episodic effects of large magnitude.) The differences between yd_i and y_i are then simply our estimates of the irregulars; i.e.,

$IR_i = yd_i - MA_i$

where

IR_i = the irregular effect for the ith period

yd_i = the deseasonalized value for the ith period

and

MA_i = the centered weighted moving average for the ith period.

Inasmuch as most of the irregulars will probably be random or accidental variations, we should have a good measurement of the magnitude of this component of our series. The series of irregulars ($yd_i - MA_i$) is shown in the last column of Table 11.7 and is graphically presented in Figure 11.5. A detailed historical analysis would hardly be appropriate for our purposes, but we make several observations. Most of the yd_i fall within ±30 of the corresponding moving average. Only 4

Table 11.7 (y) Gross Earnings from Utah State Liquor Monopoly

Year	Quarter	y_i	Deseasonalized Series yd_i	Centered Moving Average (MA_i)	Irregulars $yd_i - MA_i$
1963	3	1,493	1,623		
	4	2,074	1,641		
1964	1	1,445	1,655	1,652	3
	2	1,565	1,660	1,667	−7
	3	1,557	1,692	1,677	15
	4	2,128	1,684	1,698	−14
1965	1	1,469	1,683	1,709	−26
	2	1,708	1,811	1,716	95
	3	1,505	1,636	1,736	−100
	4	2,233	1,767	1,742	25
1966	1	1,524	1,746	1,764	−18
	2	1,701	1,803	1,797	6
	3	1,694	1,841	1,817	24
	4	2,307	1,825	1,835	−10
1967	1	1,604	1,837	1,851	−14
	2	1,771	1,878	1,860	18
	3	1,746	1,898	1,874	24
	4	2,328	1,842	1,894	−52
1968	1	1,694	1,940	1,916	24
	2	1,842	1,953	1,957	−4
	3	1,854	2,015	1,996	19
	4	2,547	2,015	2,014	1
1969	1	1,787	2,047	2,034	23
	2	1,891	2,005	2,079	−74
	3	1,966	2,137	2,130	7
	4	2,799	2,214	2,180	34
1970	1	1,940	2,222	2,230	−8
	2	2,141	2,270	2,266	4
	3	2,113	2,297	2,303	−6
	4	2,944	2,329	2,344	−15
1971	1	2,084	2,387		
	2	2,327	2,468		

SOURCE: *Annual Report*, Utah Liquor Control Commission, fiscal year 1963–64 through fiscal year 1970–71 (Salt Lake City, 1964–71).

of the irregulars, the second and third quarters of 1965, the fourth quarter of 1967, and the second quarter of 1969, appear to be extremes. These extremes could well be bureaucratic bookkeeping problems. The first 2 mentioned above are sequentially offsetting, and the last 2 are all but offset within 2 or 3 quarters.

Seasonal indexes are also very useful in forecasting a monthly or a quarterly time series. See Appendix C of this chapter for a discussion of this topic.

Measuring Seasonal Effects

Figure 11.5 Graph of Irregulars

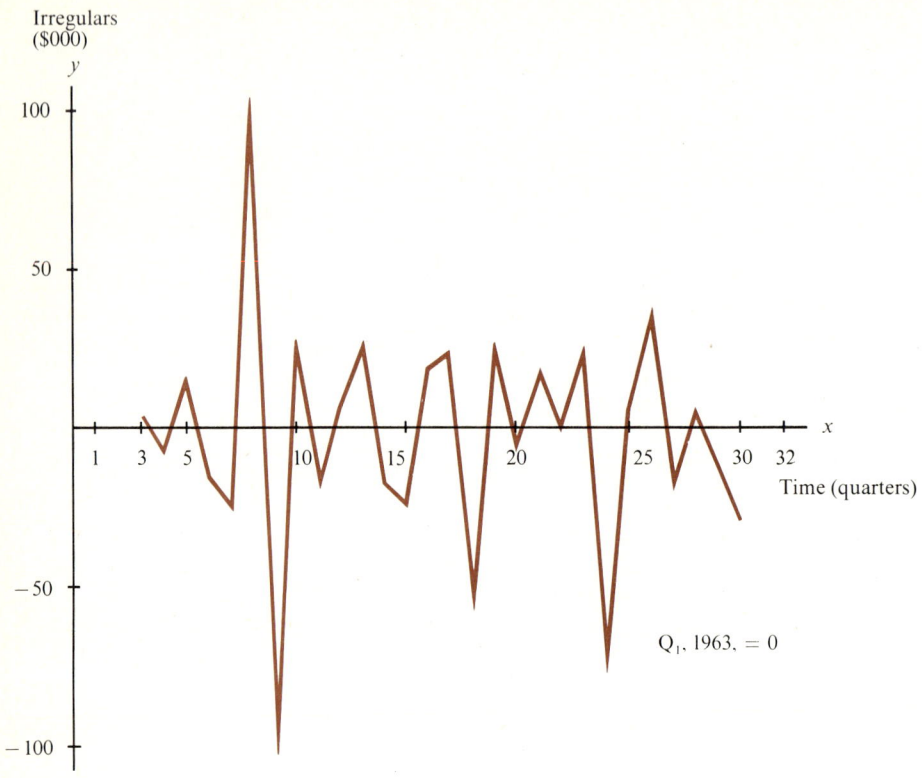

11A Appendix: Nonlinear Trend Functions

There are a number of functions that may be used to describe a nonlinear trend. We will give consideration to two, both of which are relatively simple in application and interpretation.

1. **The second degree polynomial function or the parabola,** $\hat{y} = a + bx + cx^2$.

   The coefficient, a, as with the linear trend, is the y-intercept or the trend value at the origin. The change in y is now dependent upon x^2 as well as x, and it is this dependence upon x^2 that adds curvature to the trend. The magnitude of the coefficient, c, tells us something about the rate of change and the nature of the slope. If c is relatively large in absolute value, the slope will change rapidly; a relatively small value of c indicates that the slope changes slowly. If c is positive, the trend will be concave upwards, and if c is negative it will be concave downwards (or convex).

There are two examples of this type of function sketched on a graph in Figure 11.6. The equations are written alongside the graph. The graph of this kind of function is called a parabola. It is symmetrical about a vertical axis which passes through the vertex—the maximum or minimum point on the graph. It is easy to prove that the abscissa, or x-value, at the vertex of a parabola is always

$$-\frac{b}{2a}.$$

For example, the abscissa of the vertex of one of our examples, $y = 30 - 10x + x^2$, is

$$x = -\frac{(-10)}{2(1)} = 5;$$

and when x is 5, y is also 5. The coordinates of the vertex are at (5, 5). Verify that for the second example, $y = 16 + 2x - .25x^2$, the coordinates of the vertex are (4, 20). Note that $c = 1$ in the first function and $c = -.25$ in the second; the parabola of the first example opens upward and has greater curvature than that of the second example, which opens downward.

The above discussion is to acquaint you with some of the mathematical properties of the second-degree function. More often than

Figure 11.6 Two Examples of Second-Degree Polynomial Functions

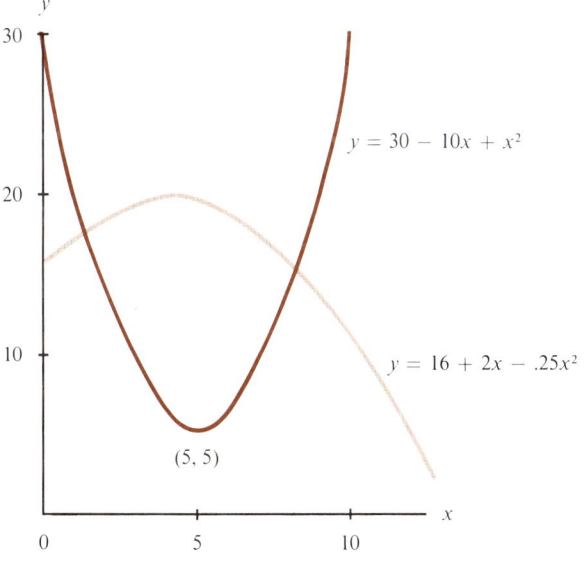

not, when a second-degree function is being used to estimate a trend, only a rising or a falling section of the parabola is actually used with no intention of forecasting or extending the use of the function beyond the vertex.

A least squares second-degree trend line was fit to the liquor earnings example discussed earlier in the chapter. The resulting equation was

$$\hat{y} = 1{,}653.8 + 2.53x + .66x^2,$$

with x in quarters, y in ($000), and the origin, i.e., $x = 0$, at the second quarter of 1963. The trend value at this 0 quarter is 1,653.8, and the slope or increase in earnings per quarter at the origin is 2.53. The coefficient of x^2, .66, tells us that the increase in earnings per quarter is increasing at an increasing rate. The abscissa of the vertex is

$$x = \frac{-2.53}{2(.66)} = -1.92,$$

or at roughly the fourth quarter of 1962. This is the bottom or the minimum value of the parabola, and if we were to use this parabola to estimate y before the fourth quarter of 1962, liquor earnings estimates would rise as we went back in time. Liquor earnings during the 1950s and early 1960s were lower than they were in 1962, and it would be a misuse of the trend function to pass beyond the vertex. This simply illustrates what was said earlier regarding the inclusion of the vertex in a parabolic trend—and, moreover, one of the hazards in the extension of a trend line much beyond the domain of the data.

We also found the least squares linear trend line for the liquor earnings, $y = 1{,}530.9 + 24.23x$. We knew from our earlier work (see Figure 11.4) that the trend was not linear. To illustrate the superior fit of the second-degree function, see Figure 11.7 where we have plotted the moving average scattergram along with both the linear and the second-degree least squares functions. We will address this again in Appendix C, "Forecasting."

2. **The exponential growth (or decay) function, $y = AB^x$ where B is a positive number.**

Because there is no simple set of formulas for a mathematical "fit" to the exponential function, this is usually discussed as a logarithmic model, *the semilog model*, (taking the log of both sides), $\log y = \log A + (\log B)x$. Inasmuch as we are taking the log of all y's, they must be positive; if there are negative y-values, we must add some constant to all y-values such that they will all be positive. The $\log y$ is now a linear function of x; e.g., letting $Y = \log y$, $a = \log A$ and $b = \log B$, we have $Y = a + bx$. We simply use the linear regression formulas to regress $\log y$ on x.

Figure 11.7 Gross Liquor Earnings: Linear Trend and Second-Degree Trend Compared with Moving Average

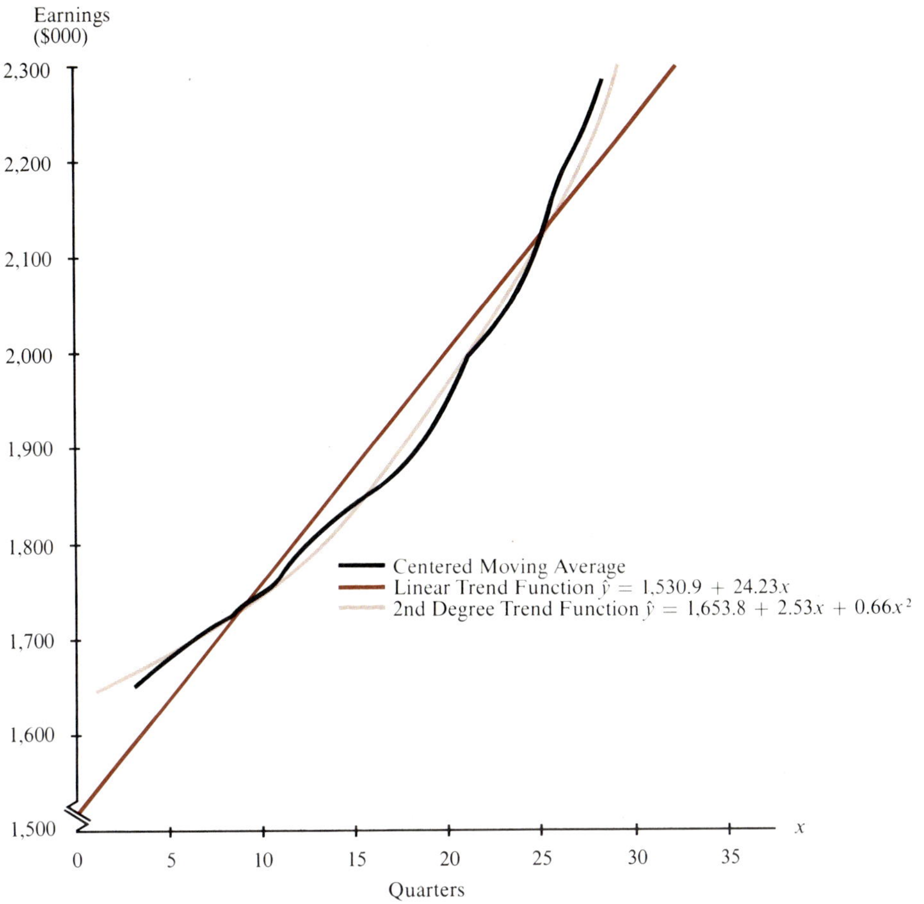

Most of you are somewhat familiar with an example of this function, the compound interest formula, $P_n = P_o(1 + r)^n$, where r is the rate of interest per period, n is the number of periods, P_o is the initial deposit, and P_n is the amount on deposit at the end of n periods. If $r > 0$, then P_n is a growth function; if $r < 0$, P_n decays through time; and if $r = 0$, we have the trivial case where all P_n equal P_o. In the general model, $\hat{y} = AB^x$, A is the y-intercept and B is 1 plus the rate of growth. If B is less than 1, then r is, of course, negative and $\hat{y}$ decays through time. See Figure 11.7 for graphs of the 2 examples of this type of function.

It is important to remember that the traditional least squares fit to this model is in terms of the semilog model; i.e., we are minimizing

$$\Sigma(\log y_i - \widehat{\log y_i})^2$$

and not

$$\Sigma(y_i - \hat{y}_i)^2.$$

The exponential form of the equation will not, in general, be the "best" fit to the original data; it is the least squares fit to the semilog model. For this reason it may not be a good fit to the actual data. There is no simple, mathematically elegant method to derive the exponential fit directly; computerized iterative techniques have been developed recently that handle this problem.

The semilog function was fitted to the liquor earnings example. The least squares semilog function was

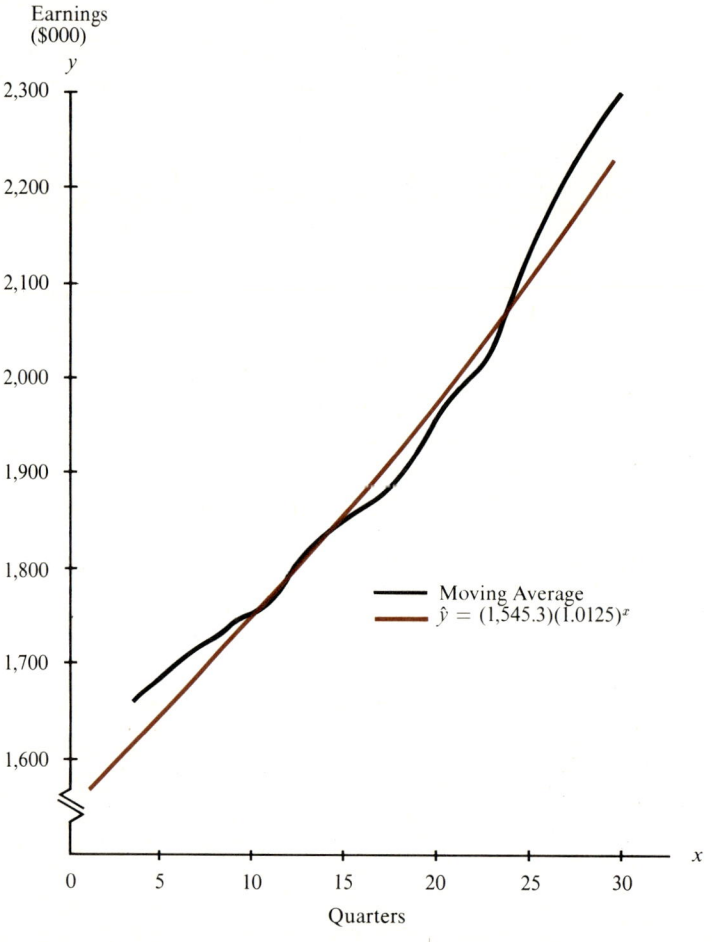

Figure 11.8 Gross Liquor Earnings: Exponential Trend and Moving Averages

$$\hat{Y} = 7.343 + .0124x,$$

with x in quarters, y in ($000) and $x = 0$ at the second quarter of 1963. The corresponding exponential form of the equation is

$$\hat{y} = (1,545)(1.0125)^x.$$

Interpreting this result: (1) The trend value at the origin is $1,545 thousands; and (2) the estimated quarterly average growth rate is 1.25 percent.

You will be disappointed, as we were, with the fit to the actual data. See Figure 11.8 where the exponential function is graphed with the moving average scattergram. Compare this with the second-degree fit in Figure 11.6.

A scattergram plot of the time series and/or the series of moving averages on semilog paper will give a visual test of the linearity of the semilog model. Semilog paper is scaled logarithmically on the y-axis with the usual arithmetic scale on the x-axis. If the data plot appears linear on this paper, then the exponential model will be a good fit. See Figure 11.9 for a plot of the moving averages of liquor earnings and the exponential fit to the data.

Figure 11.9 Gross Liquor Earnings: Exponential Trend and Moving Averages (Semi-Log Grid)

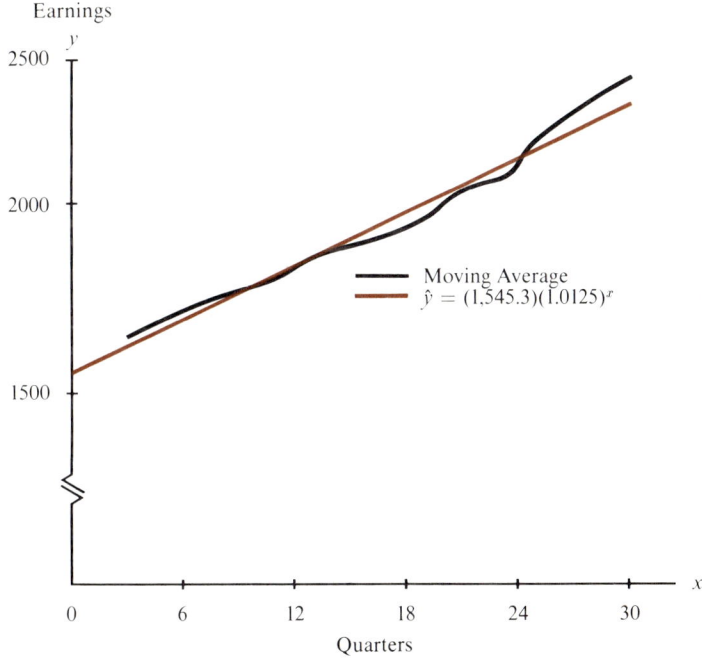

11B Appendix: A Multivariable Model That Includes Trend and Seasonal Patterns

With the use of dummy variables, we can formulate an interesting and efficient multivariate additive model that incorporates seasonal patterns along with the trend. For a linear trend and quarterly data, the least squares statistical result would be of the form

$$\hat{y} = a + b_1 x_1 + b_2 x_2 + b_3 x_3 + b_4 x_4,$$

where x_1 is the typical quarterly time variable and the other three x-variables are (1,0) dummy variables; $x_2 = 1$ if a specific quarter and 0 if not; and similarly for x_3 and x_4 for two other quarters. Dummy variables are attached to 3 of the 4 quarters, and we note that, if all 3 of these dummy variables are 0, we have identified the fourth quarter or what might be called the "base" quarter.

For a parabolic trend model, the equation would have the form

$$\hat{y} = a + b_1 x_1 + c x_1^2 + b_2 x_2 + b_3 x_3 + b_4 x_4,$$

with the variables defined as above.

To illustrate and to give us a chance to interpret this model, we fit the second-degree trend with seasonal dummy variables to the liquor earnings example with the following result:

$$\hat{y} = 1{,}487.3 + 2.50 x_1 = .67 x_1^2 - 97.8 x_2 + 52.0 x_3 + 654.0 x_4,$$

where y is earnings in ($000); x_1 is in quarters with second quarter, 1963, origin; x_2 equals 1 or 0; according as first quarter or not; and x_3 and x_4 are defined similarly for second and fourth quarters respectively. The first part of the equation,

$$\hat{y} = 1{,}487.3 + 2.50 x_1 + .67 x_1^2,$$

is the base quarter or third-quarter trend function; i.e., for x_1 equal to 1, 5, 9, etc., this equation would give actual direct estimates of third-quarter earnings. Inasmuch as this is a third-quarter trend only, seasonality is automatically incorporated in the estimate. The other 3 coefficients tell us that the first quarter is, on the average, $97.8 thousands below the third-quarter trend; the second quarter is estimated to be, on the average, $52.0 thousands above the third-quarter trend; and that of the fourth quarter, $654.0 thousands above.

The use of this model implies the assumption that the seasonal absolute deviations are constant through time, whereas using seasonal indexes derived earlier in the chapter implies that the relative seasonal deviations are constant through time.

Time Series

11C Appendix: Forecasting

Certainly the most obvious way to estimate the future is to extrapolate the past. As we, in our introductory course, view the so-called naive forecasting techniques, we can take comfort in the fact that, regardless of the degree of sophistication of a forecasting model, historical relationships are basic to the forecast. For quarterly (or monthly) data, two different approaches have been developed based upon statistical analyses of historical data—each of which lends itself almost automatically to forecasting.

1. Derive the historical trend function that appears to be the best fit to the actual data and/or to the moving average smoothed data. Estimate a set of seasonal indexes using methods described in Section 3 of this chapter. Forecast trend values by simply substituting the proper future values of the time variable, x, in the trend function. The final forecasts are obtained by seasonalizing the trend, i.e., by multiplying each trend forecast by the correct seasonal index.

2. Use the additive, multivariable, trend-seasonal model discussed in the previous section, and calculate the forecasts directly with the correct choice of the time variable and the seasonal dummy variables.

For an interesting comparison and in order to illustrate forecasting with these two models, let us assume that it is July, 1971, and we have completed all the analyses on our liquor earnings data. The governor has asked us to forecast the next two fiscal years' earnings. (A fiscal year lasts from July 1 to June 30; e.g., fiscal 1972 includes the third and fourth quarters of 1971 and the first and second quarters of 1972.) Tables 11.8 and 11.9 present the intermediate calculations and the forecasts for models 1 and 2 respectively.

Table 11.8 Liquor Earnings Forecast with Second-Degree Trend and Seasonal Indexes (forecast 1)

	x	Trend ($\hat{y}$)	Seasonal Index (I)	Forecast ($I\hat{y}$)
Fiscal '72	33	2,453	.920	2,257
	34	2,500	1.264	3,160
	35	2,548	.873	2,224
	36	2,597	.943	2,449
Fiscal '73	37	2,648	.920	2,436
	38	2,699	1.264	3,412
	39	2,753	.873	2,403
	40	2,807	.943	2,647

Table 11.9 Liquor Earnings Forecast with Multivariable Model (forecast 2)

	x	3rd Quarter Trend ($\hat{y}_3$)	Seasonal Dummy (d)	Forecast ($\hat{y}_3 + d$)
Fiscal '72	33	2,299	0	2,299
	34	2,347	654	3,001
	35	2,396	−98	2,298
	36	2,446	52	2,498
Fiscal '73	37	2,497	0	2,497
	38	2,550	654	3,204
	39	2,604	−98	2,506
	40	2,659	52	2,711

The second-degree trend function, $y = 1{,}653.8 + 2.53x + .66x^2$, was chosen for the trend forecasts; forecasts are needed for quarters 33 through 40. The weighted average set of seasonal indexes, row e from Table 11.6—.873, .943, .920, and 1.264 for the first through the fourth quarters respectively—was chosen for the seasonal adjustment. See Table 11.8 for this first forecast; Column 2 gives trend values, the seasonal indexes are listed in Column 3, and the products for the quarterly forecasts appear in Column 4, the last column of the table.

The multivariable model from the previous section of this chapter was used to calculate the forecasts in Table 11.9. The third-quarter trend values are in Column 2, and the seasonal adjustment part of the function is in Column 3. The final quarterly forecasts in Column 4 are the sums of the second and third columns.

Inasmuch as we have survived the forecast period, you may be interested in the quality of the forecasts. The actual earnings are com-

Table 11.10 Comparison of Forecasts with Actual Earnings

	x	Actual	Forecast 1	Forecast 2
Fiscal '72	33	2,310	2,257	2,299
	34	3,263	3,160	3,001
	35	2,301	2,224	2,298
	36	2,500	2,449	2,498
Annual Total		10,374	9,990	10,096
Fiscal '73	37	2,478	2,436	2,497
	38	3,390	3,412	3,204
	39	2,546	2,403	2,506
	40	2,609	2,647	2,711
Annual Total		11,023	10,898	10,918

Figure 11.10 Graphical Comparison of Liquor Earnings Forecasts with Actual Earnings

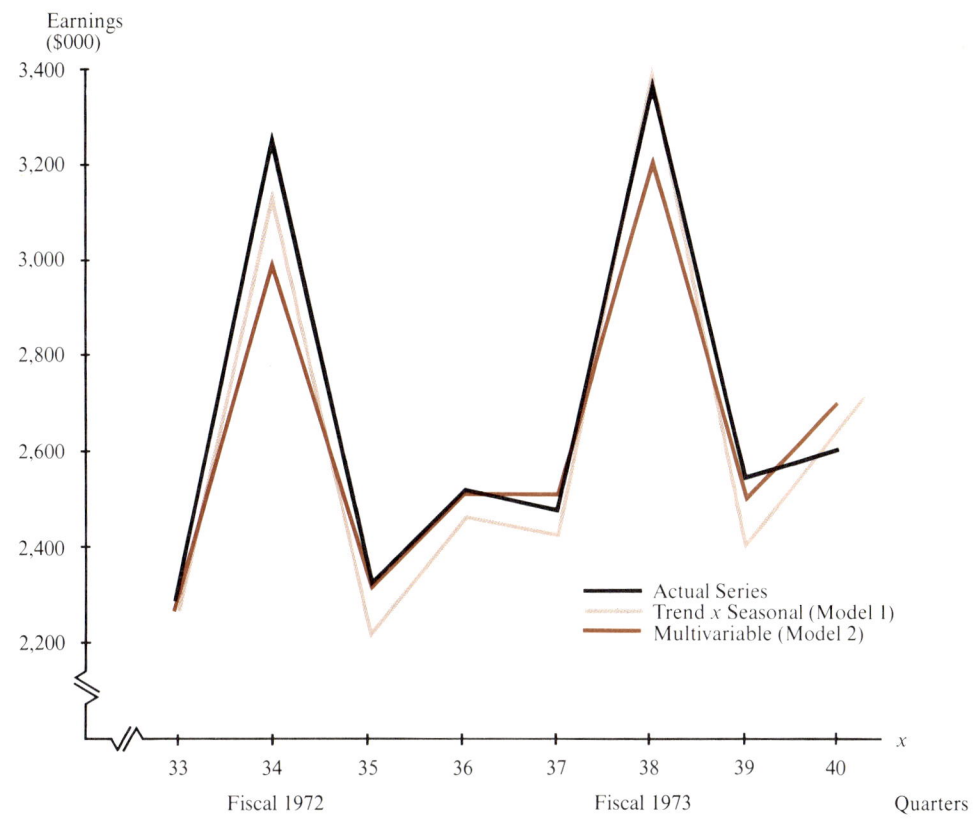

pared with the forecasts in Table 11.10, and a graphical comparison is made in Figure 11.10.

Model 1 appears to be doing a better job of tracking the seasonal variations; this would indicate that seasonal variations tend to be proportional to the level of earnings. Both models give a better forecast for fiscal 1973 than they do for fiscal 1972; the error for the fiscal year total in 1973 is less than 1 percent in both cases. The actual total for fiscal 1972 was higher than either of the forecasts by some 3 percent. Looking back to Figure 11.6 we discover a possible reason: All 4 quarters of fiscal 1971 were above trend, which we would attribute to cycle. We might have chosen, in July of 1971, to make an upward adjustment for cycle effects with the idea that this could well continue through another fiscal year.

Problems

(Note: The * means appendix material is required.)

1. The following data are on capital investment of a certain company.

Year	y (in millions of dollars)
1969	18.0
1970	19.2
1971	23.5
1972	22.1
1973	27.8
1974	34.7
1975	39.2

   a. Plot the data on arithmetic grid, draw a freehand line, and write its equation.
   b. Calculate the first-degree least squares trend line and plot this on the same grid as a.
   *c. Calculate the second-degree trend line.

2. $\hat{y} = 150 + 7x$ is the computed trend line for annual GNP of White-crusshhi in billions of buckniks, x in 5 years, and 1964 origin.
   a. Interpret the two coefficients.
   b. Forecast the 1980 GNP.
   c. Write the equation with the origin shifted to 1970.
   d. Identify the year in which GNP had a trend value of 0.

3. The profits per share of WAMMCO for the period from 1972 through 1978 are as follows:

Year	Profit per share
1972	$1.10
1973	1.16
1974	1.22
1975	1.28
1976	1.35
1977	1.42
1978	1.49

   a. Plot the scattergram on arithmetic grid.
   b. Figure a three-year moving average; plot this, and extrapolate to 1980.
   c. Calculate a first-degree least squares trend line. Estimate per-share profit for 1979 and 1980 using your linear trend, and compare with the judgment extrapolation in b.

4. The secular trend for sales for JDCP Company is accurately described by the equation $\hat{y} = 120{,}000 + 100x$, where x is in months, December, 1975, is the origin, and y is in dollars. The seasonal

indexes for department store sales in this area, beginning with January, are as follows: 100, 80, 90, 120, 115, 95, 75, 70, 90, 95, 120, and 150.
  a. Is the above set adjusted? Why or why not?
  b. Ignoring cyclical and random influences, forecast sales for (1) February, 1980; (2) May, 1982; (3) December, 1981; (4) the first half-year of 1980.

5. Given the following on unemployment for the state of nature: The least squares trend line for average annual unemployment is $\hat{y} = 13,000 + 720x$, 1972 origin and x in years. Seasonal indexes, January through November respectively, are 130, 141, 124, 104, 84, 95, 85, 88, 74, 75, and 90. Also note that the unemployment is not an aggregate number, but a continuous variable; therefore, the monthly trend line would be derived by merely dividing the annual slope coefficient by 12. Answer the following:
  a. Compute the seasonal index for December.
  b. Forecast the average annual unemployment for 1980.
  c. Assuming 0 for both cycle and accidental effects, forecast unemployment for February, July, September, and November, 1980.
  d. June, 1973, actual unemployment was 17,500. Assuming 0 accidental effect, what part of this unemployment would you attribute to the cycle? Does this indicate a period of recession or prosperity?

6. Consider the following data:

Year	Blitzkrieg Inc. Sales in Billions of $	Price Index (1967 = 100)
1969	3.46	110
1970	4.09	116
1971	4.34	121
1972	4.18	125
1973	4.45	133
1974	4.46	148
1975	4.79	161
1976	4.92	171
1977	5.48	182

  a. Develop a new series of sales in billions of 1967 real dollars; i.e., deflate the BI sales with the corresponding price index.
  b. Find the least squares line of best fit to the original data. Plot this and the scattergram on the same grid. Interpret the two coefficients.
  c. Repeat b above for the deflated series.
  d. Compare the results in b and c and comment.

Table 11.11 Northern Slobovia Nonagricultural Employment (y) Original Series 1973–79

Year	Jan	Feb	Mar	Apr	May	Jun	Jul	Aug	Sept	Oct	Nov	Dec	Average
1973	240,475	240,037	245,049	251,473	256,761	261,188	260,568	256,290	255,544	254,443	256,414	258,431	253,056
1974	247,937	252,744	255,338	263,506	266,325	269,835	267,777	269,567	273,232	269,030	267,181	266,556	264,086
1975	256,908	256,327	259,796	266,548	272,376	278,606	726,787	279,826	285,405	282,261	280,953	280,768	273,047
1976	270,046	270,035	274,544	283,077	288,039	292,352	292,447	295,520	297,805	295,466	291,293	292,851	287,009
1977	280,716	280,931	284,349	289,642	295,633	300,529	298,245	300,012	301,893	303,250	300,378	300,288	294,656
1978	285,344	284,692	286,694	291,734	296,637	300,466	291,980	294,263	299,903	301,201	299,158	297,756	294,152
1979	286,045	286,474	290,944	294,948	298,702	302,047	304,744	307,348	312,849	306,881	308,090	308,601	300,639

7. Presented in Table 11.11 are 7 years of monthly employment data from Northern Slobovia. In addition to the original series and the annual average employment, we present, in Table 11.12, a series of 13-month weighted moving averages and a series of ratios of the actual employment divided by the moving average.

   a. Develop a set of seasonal indexes from these data. (These are monthly data—you will have an index for each month, and

Table 11.12

Period	13-Month Weighted Moving Average	Ratio y/MA	Period	13-Month Weighted Moving Average	Ratio y/MA
July 73	253,367	1.0284	July 76	287,453	1.0174
Aug 73	254,207	1.0082	Aug 76	288,352	1.0249
Sep 73	255,166	1.0015	Sep 76	289,214	1.0297
Oct 73	256,096	.9935	Oct 76	289,897	1.0192
Nov 73	256,996	.9977	Nov 76	290,487	1.0049
Dec 73	257,754	1.0026	Dec 76	291,144	1.0059
Jan 74	258,415	.9595	Jan 77	291,726	.9623
Feb 74	259,269	.9748	Feb 77	292,155	.9616
Mar 74	260,559	.9800	Mar 77	292,512	.9721
Apr 74	261,904	1.0061	Apr 77	293,007	.9885
May 74	262,960	1.0128	May 77	293,683	1.0066
Jun 74	263,747	1.0231	Jun 77	294,346	1.0210
Jul 74	264,459	1.0125	Jul 77	294,848	1.0115
Aug 74	264,983	1.0173	Aug 77	295,198	1.0163
Sep 74	265,318	1.0298	Sep 77	295,452	1.0218
Oct 74	265,630	1.0128	Oct 77	295,637	1.0258
Nov 74	266,009	1.0044	Nov 77	295,766	1.0156
Dec 74	266,627	.9997	Dec 77	295,805	1.0152
Jan 75	267,367	.9609	Jan 78	295,542	.9655
Feb 75	268,170	.9558	Feb 78	295,041	.9649
Mar 75	296,105	.9654	Mar 78	294,719	.9728
Apr 75	270,163	.9866	Apr. 78	294,550	.9904
May 75	271,289	1.0040	May 78	294,414	1.0076
Jun 75	272,455	1.0226	Jun 78	294,258	1.0211
Jul 75	273,594	1.0117	Jul 78	294,182	.9925
Aug 75	274,713	1.0186	Aug 78	294,285	.9999
Sep 75	275,898	1.0345	Sep 78	294,536	1.0182
Oct 75	277,202	1.0183	Oct 78	294,847	1.0215
Nov 75	278,543	1.0087	Nov 78	295,067	1.0139
Dec 75	279,768	1.0036	Dec 78	295,129	1.0086
Jan 76	280,994	.9610	Jan 79	295,817	.9670
Feb 76	282,300	.9566	Feb 79	296,894	.9649
Mar 76	283,471	.9685	Mar 79	297,979	.9764
Apr 76	284,537	.9949	Apr 79	298,755	.9873
May 76	285,545	1.0087	May 79	299,364	.9978
Jun 76	286,505	1.0204	Jun 79	300,188	1.0062

Problems for Chapter 11

the sum of this set will be 12.000 or 1200.0, depending upon whether you use decimal fractions or percentages.)

b. Plot the series of actual data and the moving average series on the same grid.

c. Deseasonalize the actual series for 1978 and the first 6 months of 1979. Calculate the set of differences between these and the moving averages for the same period. In your opinion, do any of these differences appear to be other than random?

■ The following problems, at least in part, require material covered in the appendixes to this chapter.

8. With yearly data and 1953 as the origin, a least squares log trend line was fitted to the yearly sales of gasoline (in thousands of dollars) of a certain oil company. The following equation was obtained: $\log \hat{y} = 2.8754 + .0281x$.
   a. Write the equation in the form, $\hat{y} = ab^x$.
   b. Interpret the a and b in part a above.
   c. What year would you forecast sales to reach $1 million?
   d. What trend value do you compute for 1980?

9. Consider the following per-capita income series for the United States, 1961 through 1977:

Year	Per-Capita Income	Year	Per-Capita Income
1961	$2,270	1970	$3,970
1962	2,370	1971	4,200
1963	2,460	1972	4,490
1964	2,590	1973	4,980
1965	2,770	1974	5,430
1966	2,990	1975	5,860
1967	3,190	1976	6,403
1968	3,460	1977	7,020
1969	3,730		

   a. Fit a first-degree least squares trend line; use a 1960 origin.
   *b. Find the second-degree line.
   *c. Regress log y against time.
   *d. Plot each of the above 3 models with the data. In your judgment, which best describes the 17-year historical trend in per-capita income?
   *e. Forecast per-capita income for 1978 and 1979 with all 3 models, and compare with actual results.

10. In the table below are presented 6 years of quarterly unemployment numbers (in thousands of persons) for the hypothetical metropolitan area of Tirrenica.
    a. Fit a least squares linear trend line to the 24 data points. Let

Q_4 of 1976 be the origin. Interpret the coefficients. Plot the data and the function on the same grid.

b. Fit a least squares linear trend line to the 6 annual averages. Let 1976 be the origin. Interpret the coefficients. Compare this result with that in part a.

c. Calculate a 5-quarter weighted moving average (or a 4-quarter unweighted moving average followed by a 2-quarter unweighted moving average). Use this to derive a set of seasonal indexes. Plot the set of moving averages on the same grid with the data and the linear function from part a.

d. Using the results in a and c, forecast quarterly unemployment for 1980 and 1981.

*e. Fit a second-degree trend line to the 24 quarterly data points. Use second quarter of 1976 for the origin. Plot this function on a grid along with the scattergram.

*f. If a computer and a prepackaged multivariable regression program are available, determine the seasonal additive model with Q_2, 1976, as the origin and use (1, 0) dummy variables for Q_1, Q_3, and Q_4. Interpret the coefficients. Forecast 1980 and 1981 unemployment by quarters.

Unemployment in Thousands
Tirrenica Metropolitan Area

	Q_1	Q_2	Q_3	Q_4	Annual Average
1974	11.7	7.9	6.6	8.8	8.8
1975	13.4	11.4	7.5	8.4	10.2
1976	14.5	10.7	7.7	9.0	10.5
1977	16.8	12.5	9.5	11.8	12.6
1978	21.1	16.2	12.2	12.8	15.6
1979	21.3	15.0	13.6	13.8	15.9

11. In the table below are presented 6 years of quarterly employment data (in thousands of persons) for this same metropolitan area of Tirrenica.

a, b, c, d, e, and f. Repeat the corresponding exercises in Problem 10 above for the employment data below.

Employment in Thousands
Tirrenica Metropolitan Area

	Q_1	Q_2	Q_3	Q_4	Annual Average
1974	222.5	235.7	241.8	240.6	235.2
1975	231.0	241.9	248.1	245.2	241.6
1976	229.6	240.0	247.7	250.1	241.9
1977	242.2	256.8	257.3	256.4	253.2
1978	251.9	266.3	270.4	267.2	264.0
1979	257.2	272.4	281.3	281.2	273.0

12. This problem involves both the unemployment and the employment data in Problems 10 and 11 above.
    a. Calculate a new series of 24 numbers on the percentage unemployment rate for Tirrenica. (The unemployment rate for a given period equals the number of unemployed divided by the sum of the number unemployed and the number employed; this ratio is multiplied by 100 in order to get a percentage unemployment rate.)
    b. Derive a set of seasonal indexes for this new series. (See part c in Problems 9 and 10 for suggested method.) Interpret this set of indexes.
    c. The 1980 percentage unemployment rates were 8.4, 6.5, 4.1, and 6.2 for quarters 1, 2, 3, and 4 respectively. These were reported to the public as "seasonally adjusted" rates. What rates were reported to the public for the year 1980?

CHAPTER 12

INDEX NUMBERS

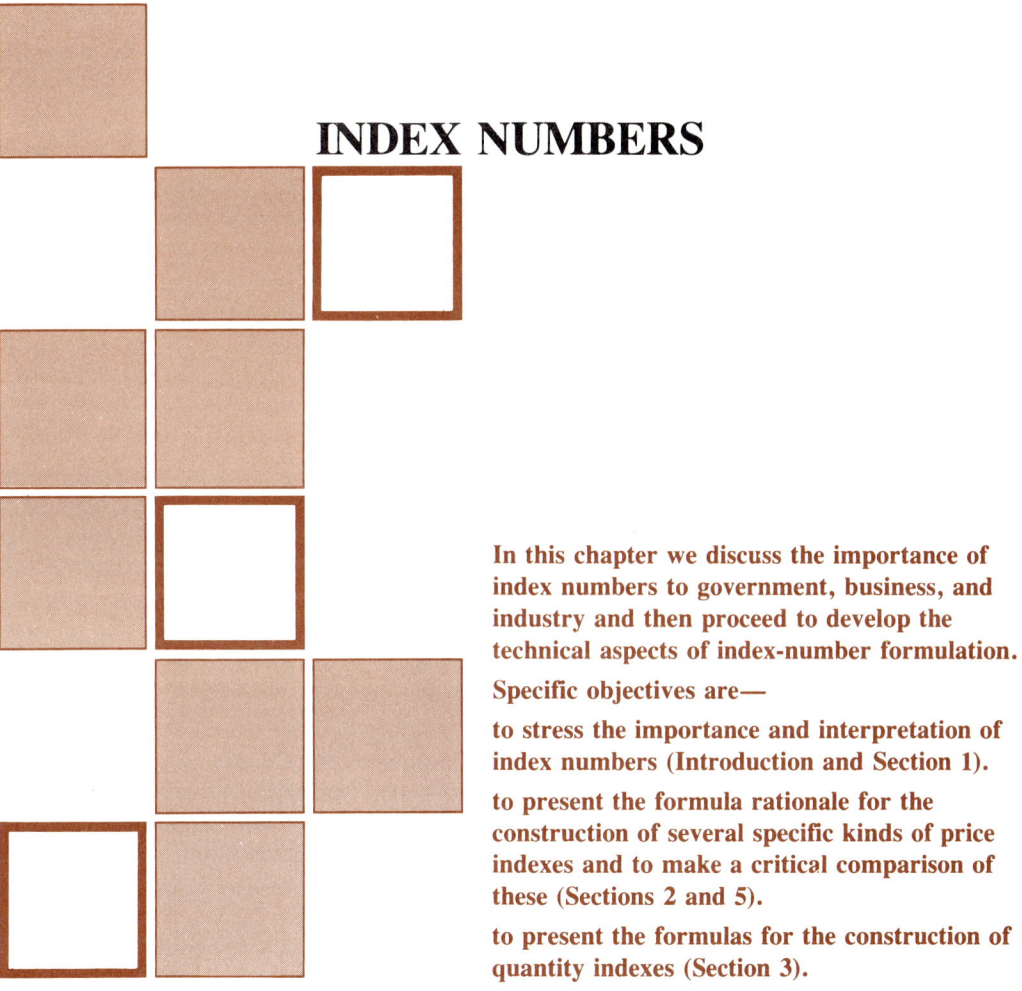

In this chapter we discuss the importance of index numbers to government, business, and industry and then proceed to develop the technical aspects of index-number formulation.

Specific objectives are—

to stress the importance and interpretation of index numbers (Introduction and Section 1).

to present the formula rationale for the construction of several specific kinds of price indexes and to make a critical comparison of these (Sections 2 and 5).

to present the formulas for the construction of quantity indexes (Section 3).

The nation's most widely watched measure of inflation by far is the Bureau of Labor Statistics' consumer price index. As inflation has embedded itself in American life, the CPI has become the most important economic statistic issued by the government. Escalator clauses tie the incomes of perhaps half of all Americans to movements in the CPI; among them are 8.5 million wage earners, 31 million Social Security recipients, 20 million people who receive food stamps, and 2.5 million retired military and federal employees. But the index has had two serious drawbacks: it is based on the spending patterns of only urban blue-collar and clerical employees, who now constitute less than 45 percent of the population, and it was compiled by pricing a "market basket" of goods and services compiled back in 1963.

This is the first paragraph of an article, "Gauging Prices—and Spending," from *Time* magazine, March 13, 1978.[1] It illustrates quite vividly how important an index number can be and then points out two of the several problems facing the statisticians who build index numbers. The rest of the article is reproduced in the extract below. We will occasionally refer to this article throughout the chapter.

Gauging Prices— and Spending New CPI shows more inflation than expected

No more. Last week the bureau began issuing not one but two new CPIs. The first new index, still focused on blue-collar workers and clerical employees, updates their spending habits through surveys of family budgets taken in 1972–73 and rigorously analyzed ever since. The second (CPI-U) reflects the new spending patterns not just of wage earners but of "all urban consumers," including, for example, retired people and self-employed professionals; it is supposed to reflect the way 80% of Americans spend their money.

Unfortunately, both of the new CPIs showed inflation speeding up still more rapidly than had been supposed. Even the old CPI showed prices rising in January at an annual rate of 8.7%, about double the pace in November and December. But according to both of the new indexes, the rate was 10%, which reaches the dreaded double-digit range. The increase was exaggerated by ice and snow that snarled rails and roads in January, leading to shortages that jacked up food prices. But wholesale prices have been rising rapidly enough in the past few months to threaten more jarring consumer-price jumps. Julius Shiskin, the savvy Labor Department statistician who updated the CPI, concedes that the January jump is "cause for concern."

Had the news been better, announcement of the new indexes might have been a triumphant occasion. Shiskin regards them as "the best indexes in the world." To compile the old CPI, the Labor Department's 360 price inspectors (all but a handful of whom are housewives) had been checking the prices of some products that hardly anyone buys any more:

1. Reprinted by permission from *TIME, The Weekly Newsmagazine*; Copyright Time Inc. 1978.

pedal pushers, garter belts, bobby pins. Such obsolete articles were thrown out of the 400-item market basket and many newer ones substituted. The BLS shoppers will now price, for example, joggers' warmup suits, pocket calculators, birth control pills and wine.

The biggest change has been in the weights assigned to different categories of spending. The old CPI assumed that the typical family spent 34.9% of its budget on home furnishings and housing (mortgage payments or rent); CPI-U gives this category a 43.9% weight. The jump reflects both inflation in home prices and the determination of many Americans to make the home a very comfortable castle. Food went down from 25.2% to 18.8%. One reason: as people have more money to spend, they spend proportionately less on food. Also, consumers are eating out more; about a third of the food expenditures figured into CPI-U are assumed to pay for restaurant or snack-bar meals *v.* one-fifth in the old index.

Americans now spend almost as much of their budget (18%) on transportation—air travel, payments on the family car—as they do on food. But the relative importance of clothing purchases dropped from nearly 11% in the old CPI to 5.8% in CPI-U. Some reasons: the average family in the survey is smaller, 2.9 people *v.* 3.2 in the early 1960s. The population is getting older, and thus buys fewer new clothes. Young people tend to wear casual clothes, which cost less than the party dresses and formal suits of yesteryear.

Though Shiskin clearly regards the new urban consumers' index as the most sensitive measure of price trends, he was unable to get rid of the old index, which includes only wage earners. Labor leaders forced him to keep publishing it, but in revised form. Their argument: figures comparable to those used in the past are needed to determine wage increases under cost of living escalator clauses. Only the performance of the two indexes will show whether they made a wise decision. If CPI-U goes up faster than the new wage-earner index, union members may in the future demand that their wage boosts be tied to the broader measurement.

The Consumer Price Index (CPI) may be the most familiar of all price index numbers, but there are many other important and very useful price indexes, e.g., the Wholesale Price Index (just recently changed in name to the Producers Price Index) and the Gross National Product Implicit Price Deflator (GNPIPD). Price indexes are also calculated and published for many of the components of these major index numbers. In addition to price indexes, quantity or production index numbers are important. You may be acquainted with the Index of Industrial Production, the Federal Reserve Board index of quantity output. Index numbers are an important input to the policy and decision making of business, industry, and government. For the most part, important indexes are constructed by one of several government agencies, and each represents a considerable investment.

12.1 Interpretation of Index Numbers

Before giving consideration to the construction of index numbers, we will examine some selected values of three indexes given in Table 12.1 to be sure we understand the meaning of an index number. As stated above, the CPI gauges price changes—or measures inflation. For every index series there is a base period; this may be and often is a one-year period, but the base period may also cover two or three years. The rationale behind the choice of a base period will be discussed later. The average price (or quantity in the case of a quantity index) over the base period is arbitrarily assigned the index of 100.

All three series in Table 12.1 have the same 1967 index of 100; this is because 1967 is the base period. How should the consumer interpret the CPI of 80.1 for 1953, 91.7 for 1963, or 170.5 for 1976? The term *market basket* (see the first paragraph of the *Time* magazine article) is often used to explain or discuss the Consumer Price Index. The market basket that would have cost $100 in 1967 came to only $80.10 in 1953, but in 1976 it would have been worth $170.50. Index numbers can also be interpreted as percentages; there has been an increase of 80.5 percent from 1967 to 1976 in the general price level of the American consumer market basket. If we move the decimal 2 places to the left—i.e., change the index from a percentage to a decimal fraction—and then take the reciprocal, we get another easily interpreted and commonly quoted number—the value of the dollar. For example, for 1953, 1/.801 = 1.248; for 1963, 1/.917 = 1.091; for 1967, 1/1.00 = 1.00; and for 1976, 1/1.705 = .587. The 1967 dollar was worth almost $1.25 in 1953, but by 1976 it was worth something less than $.60.

Interpretation of the Producers Price Index is similar to that of the CPI but of course applies to the wholesale market rather than the retail market. Comparison of the 2 price series is interesting. From 1953 to 1972, wholesale price inflation was not nearly so rapid (87.4 to 119.1, a 36 percent increase) as consumer price inflation (80.1 to 125.3, a 56 percent increase); but from 1972 to 1977 the picture was just reversed.

Table 12.1 Some Index Numbers for Selected Years

	Consumer Price Index (CPI)	Producers Price Index (PPI)	Index of Industrial Production
1953	80.1	87.4	
1963	91.7	93.8	74.7
1967	100.0	100.0	100.0
1972	125.3	119.1	113.5
1974	147.7	160.1	129.3
1975	161.2	174.9	117.8
1976	170.5	182.9	129.8
1977	181.5	195.0	138.5

During this 5-year period, wholesale prices jumped by 64 percent, while consumer prices were up by 45 percent.

The Federal Reserve Board Index of Industrial Production shows that the 1977 industrial production was 38.5 percent higher than it was in 1967. Quantity output isn't growing nearly so rapidly as prices are. It is interesting to note that industrial production actually dropped from 1974 to 1975, and the 1976 level was just slightly higher than the 1974 level of industrial production.

12.2 Construction of Index Numbers: Price Relatives, Laspeyres Price Index, Paasche Price Index, and Fisher's Ideal Index

The following hypothetical example will be used to illustrate the methods of index number construction.

Example 12.2.1 As experienced statisticians and index-number specialists, we were called to consult with the government of the developing country of Lower Slobovia. The government was concerned with inflation in general, but the tzar and the prime minister were particularly upset with what they considered skyrocketing food prices. So we decided to tackle the food price problem first. Everybody in Lower Slobovia works at least part-time for the federal government; so the historical data we needed were accurate and readily available. Right away we could see there were only 3 major food items: venison, walrus, and caviar. Other items, such as polar bear, arctic birds, tundra beetles, and reindeer cheese, were seasonal in consumption and minor in money volume. We decided to base our price index series on the 3 major food items and proceeded to gather the data from 1965 to date. Presented in Table 12.2 are 3 selected years of data. The year 1970 was selected as a base year. All prices are the average annual price in

Table 12.2 Price and Quantity Data for Major Food Items

	Venison		Walrus		Caviar	
	p	q	p	q	p	q
1965	3.20	25	2.50	40	8.00	3
1970	3.80	40	4.00	42	10.00	7
1975	4.00	80	6.00	38	12.50	10

p = price
q = quantity

buckniks (BN), and BN1.962 = $1.00. The quantities of venison and walrus are in millions of keelows where .454 keelows = 1 pound; the quantities of caviar are numbers of tins, .02 keelow by net weight. We are asked to develop some indexes that can be used to measure and compare price changes.

A simple price index for only 1 item is called a *price relative*. Such indexes are nonaggregate and may or may not reflect general price changes. We may, however, find them useful in the determination of our aggregate price index. A price relative is the quotient of the item's price during a given year divided by the price of the good during the base year. This quotient is usually multiplied by 100 to change the decimal fraction to a percentage.

Definition of Price Relative

Formally defined, the price relative,

$$R_{ik} = \frac{p_{ik}}{p_{io}}(100)$$

where

R_{ik} = the price relative
p_{ik} = the price of the ith good during the kth time period
p_{io} = the price of the ith good during the base time period.

The i-subscript will be used to identify the item or good and the k-subscript identifies the time period with respect to the base period; o is used to designate the base period. Table 12.3 has the price relatives for all 3 goods for all 3 years.

The price relative for venison in 1965 is calculated as follows:

$$R_{1,-5} = \frac{p_{1,-5}}{p_{10}}(100) = \frac{3.20}{3.80}(100) = 84.2.$$

To find the price relative for walrus ($i = 2$) in 1975 ($k = 5$), we use the above formula to find $R_{2,5}$:

Table 12.3 Price Relatives (R)

	Venison $i = 1$	Walrus $i = 2$	Caviar $i = 3$
1965 ($k = -5$)	84.2	62.5	80.0
1970 ($k = 0$)	100.0	100.0	100.0
1975 ($k = 5$)	105.3	150.0	125.0

$$R_{2,5} = \frac{p_{25}}{p_{20}}(100) = \frac{6.00}{4.00}(100) = 150.0$$

We must now give consideration to the construction of an aggregate price index. Inasmuch as we began our discussion with price relatives, it seems reasonable that we might be able to use these in the formulation of an aggregate index. A simple, unweighted arithmetic average of the price relatives should be immediately discarded as a possibility; this kind of aggregation would give the least important good in the market basket the same weight as the most important good. If we then ask, What is the measure of importance of an item in the market basket? our answer would probably be that it depends upon the proportion of the budget spent upon the good. We have, then, what seems to be a reasonable formulation for an aggregate price index for the kth period:

$$I_k = \frac{\sum_i R_{ik} w_i}{\sum_i w_i}$$

where

I_k = the aggregate price index for the kth period
R_{ik} = the commodity relative price index
w_i = the weight assigned to the ith commodity.

The weight, w_i, reflects the "importance" (or the dollar volume) of the ith item in the market basket.

The question arises as to what should be used to weight the price relatives. Should the weights be determined by spending patterns during the base period or during the current period, or in some other way? Let us begin by letting the w_i be the item-specific expenditures during the base period; i.e., $w_i = p_{io}q_{io}$. This latter expression represents the total amount spent on the ith good during the base period. Using the base period expenditures as the weights, and recalling that $R_{ik} = p_{ik}/p_{io}$, we have

$$I_k = \frac{\sum_i R_{ik}(p_{io}q_{io})}{\sum_i p_{io}q_{io}} = \frac{\sum_i \frac{p_{ik}}{p_{io}}(p_{io}q_{io})}{\sum_i p_{io}q_{io}}(100).$$

Definition of Laspeyres Price Index

In the numerator of the second form of our index above, we note that we can cancel the p_{io} in every term of the summation and get a somewhat simplified form:

$$_LI_k = \frac{\sum_i p_{ik}q_{io}}{\sum_i p_{io}q_{io}}(100)$$

where

$_LI_k$ = Laspeyres Price Index for the kth time period
p_{ik} = the price of the ith good in the kth period
p_{io} = the price of the ith good in the base period
q_{io} = the quantity of the ith good in the base period.

This well-known and much-used form is known as Laspeyres Price Index.

In the Laspeyres form we have perhaps lost track of the fact that we started with a weighted average of price relatives, but note what we do have. The denominator is the total money volume expenditures on the sample items during the base period; and the numerator is what this expenditure would have been in the kth year if quantities had remained (or had been) the same as those during the base year. Continuing with our example, the necessary calculations for Laspeyres Price Index are given in Table 12.4. If we were satisfied with the Laspeyres type of price index we would report to the Slobovian government that the 1965 food price level was just 74.1 percent of the 1970 level and the 1975 food market basket was 28.1 percent higher than that of 1970.

As an alternative to the Laspeyres index form, prices could be weighted by the current (or k^{th}) period quantities; i.e., the q_{io}'s could be replaced by the q_{ik}'s in the Laspeyres index formula.

Table 12.4 Laspeyres Price Index Calculations

	$p_{ik}q_{io}$			
	Venison $i=1$	Walrus $i=2$	Caviar $i=3$	$\sum_i p_{ik}q_{io}$
1965 ($k = -5$)	128	105	56	BN 289 million
1970 ($k = 0$)	152	168	70	BN 390 million
1975 ($k = 5$)	160	252	87.5	BN 499.5 million

The Laspeyres Price Indexes are:

for 1965,
$$_LI_{-5} = \frac{\sum_i p_{i-5}q_{io}}{\sum_i p_{io}q_{io}}(100) = \frac{289}{390}(100) = 74.1$$

for 1970,
$$_LI_0 = \frac{\sum_i p_{io}q_{io}}{\sum_i p_{io}q_{io}}(100) = \frac{390}{390}(100) = 100.0$$

and for 1975,
$$_LI_5 = \frac{\sum_i p_{i5}q_{io}}{\sum_i p_{io}q_{io}}(100) = \frac{499.5}{390}(100) = 128.1$$

Definition of Paasche Price Index

This is known as the Paasche Price Index; the formula is given below.

$$_pI_k = \frac{\sum_i p_{ik}q_{ik}}{\sum_i p_{io}q_{ik}}$$

where
$_pI_k$ = Paasche Price Index
p_{ik} = the price of the ith good in the kth period
p_{io} = the price of the ith good in the base period
q_{ik} = the quantity of the ith good in the kth period.

The Paasche Price Index is the actual money volume of the sample market basket during the index (k^{th}) year divided by what the money volume would have been if quantities sold during the base period had been equal to those sold during the kth period. A series of Paasche-type indexes requires more calculations than a series of the Laspeyres type. The denominator of Laspeyres index is the same for every year, whereas a new denominator must be calculated for each period of the Paasche index sequence. Table 12.5 has the calculations necessary for computing the Paasche indexes for Lower Slobovia.

Table 12.5 Paasche Price Index Calculations

	$p_{ik}q_{ik}$			
	Venison $i = 1$	Walrus $i = 2$	Caviar $i = 3$	$\sum p_{ik}q_{ik}$
1965 ($k = -5$)	80	100	24	BN 204 million
1970 ($k = 0$)	152	168	70	BN 290 million
1975 ($k = 5$)	320	228	125	BN 673 million

	$p_{io}q_{ik}$			
	Venison $i = 1$	Walrus $i = 2$	Caviar $i = 3$	$\sum p_{io}q_{ik}$
1965 ($k = -5$)	95	160	30	BN 285 million
1970 ($k = 0$)	152	168	70	BN 390 million
1975 ($k = 5$)	304	152	100	BN 556 million

The Paasche Price Indexes are:

for 1965,
$$_pI_{-5} = \frac{\sum_i p_{i-5}q_{i-5}}{\sum_i p_{io}q_{i-5}}(100) = \frac{204}{285}(100) = 71.6$$

for 1970,
$$_pI_o = \frac{\sum_i p_{io}q_{io}}{\sum_i p_{io}q_{io}}(100) = \frac{390}{390}(100) = 100.0$$

and for 1975,
$$_pI_5 = \frac{\sum_i p_{i5}q_{i5}}{\sum_i p_{io}q_{i5}}(100) = \frac{673}{556}(100) = 121.0$$

It seems quite clear that inflation has visited Lower Slobovia, but the differences between the Laspeyres and the Paasche indexes for 1965 and 1975 could well leave us in a quandary as to what we should report to the Slobovian government. Actually, the hypothetical data in this example were deliberately construed to point up a problem (perhaps unfairly—see later discussion regarding Laspeyres-type versus Paasche-type indexes). With the Laspeyres Price Index, one must assume that the *substitution effect* is 0; i.e., consumers *do not* shift from the relatively higher-priced goods to the relatively lower-priced goods when prices of goods within the sample market basket are changing at different rates. When substitution does occur, then the Laspeyres Price Index will overestimate the true aggregate price level change. On the other hand, if it can be argued that the shift in the current year's demand was caused by a temporary price aberration, then Paasche Price Index would tend to underestimate the price level change.

The early twentieth-century American economist Irving Fisher suggested a compromise known as Fisher's Ideal Index. (The discussion in Section 4 of this chapter explains why the term *ideal* is used.)

Definition of Fisher's Ideal Index

Fisher's Ideal Index is defined as the geometric mean of the Laspeyres and Paasche Indexes:

$$_FI_k = \sqrt{_LI_k \, _PI_k}$$

where

$_FI_k$ = Fisher's Ideal Price Index for the kth period
$_LI_k$ = Laspeyres Price Index for the kth period
$_PI_5$ = Paasche Price Index for the kth period.

As an example, Fisher's Ideal Price Index for 1975 in Lower Slobovia would be:

$$_FI_5 = \sqrt{_LI_5 \, _PI_5} = \sqrt{(128.1)(121.0)} = 124.5.$$

Table 12.6 compares the three price indexes discussed up to this point.

It is possible to devise other price indexes for making price com-

Table 12.6 Price Indexes

Year	Laspeyres	Paasche	Fisher
1965	74.1	71.6	72.8
1970	100.0	100.0	100.0
1975	128.1	121.0	124.5

parisons. For example, the weights in the basic index formula used at the beginning of this section could be varied to give indexes similar to, but different from, Laspeyres or Paasche. Or, one might consider various combinations of the Laspeyres and Paasche indexes, such as the arithmetic average of the two indexes. The three indexes discussed in this section—Laspeyres, Paasche, and Fisher's Ideal—are the ones most often referred to in basic discussions of price indexes. Before considering the relative merits of these indexes, let us proceed with a look at quantity index numbers.

12.3 Quantity Indexes: Quantity Relatives, Laspeyres and Paasche Quantity Indexes

The formulas for quantity indexes parallel those for price indexes, with the quantities weighted by prices for Laspeyres and Paasche indexes.

Definition of Quantity Relative

The quantity relative of the ith product in the kth time period is defined as

$$Q_{ik} = \left(\frac{q_{ik}}{q_{ok}}\right)(100)$$

where, as with the price relative, the i subscript identifies the good and k denotes the period.

Referring to the data in Table 12.2, we have for example,

$$Q_{1,5} = q_{15}/q_{05} = (80/40)(100) = 200.0;$$

i.e., the consumption of venison in 1975 is 200 percent (exactly double) of that in 1970. All quantity relatives for the Slobovian example are presented in Table 12.7. The student should check these.

Table 12.7 **Quantity Relatives**

	Venison $i = 1$	Walrus $i = 2$	Caviar $i = 3$
1965 ($k = -5$)	62.5	95.2	42.9
1970 ($k = 0$)	100.0	100.0	100.0
1975 ($k = 5$)	200.0	90.5	142.9

Definition of Laspeyres and Paasche Quantity Indexes

The formulas for Laspeyres Quantity Index and Paasche Quantity Index are given below:

$$_LG_k = \frac{\sum_i q_{ik} p_{io}}{\sum_i q_{io} p_{io}} (100)$$

and

$$_PG_k = \frac{\sum_i q_{ik} p_{ik}}{\sum_i q_{io} p_{ik}} (100)$$

where

$_LG_k$ = Laspeyres Quantity Index for the kth time period
$_PG_k$ = Paasche Quantity Index for the kth time period
q_{ik} = the quantity of the ith good in the kth period
q_{io} = the quantity of the ith good in the base period
p_{ik} = the price of the ith good in the kth period
p_{io} = the price of the ith good in the base period.

Note that the denominator of Laspeyres Quantity Index, the total money volume of the market basket during the base year, is the same as the denominator of Laspeyres Price Index. The quantities in both the numerator and denominator of Laspeyres Quantity Index are weighted by the base period prices. The current year's prices are used as weights in Paasche Quantity Index. The summations needed for the calculation of these two indexes for Lower Slobovia are already given in Tables 12.4 and 12.5. Table 12.8 has the final calculation for these two quantity indexes. Fisher's Ideal Quantity Index, $_FG_k$, is also presented. The latter is found in a manner analogous to calculations for the corresponding price index:

$$_FG_k = \sqrt{_LG_k \; _PG_k}.$$

Table 12.8 Quantity Index Comparison

Year	Laspeyres Quantity Index	Paasche Quantity Index	Fisher's Ideal Quantity Index
1965	$\frac{285}{390}(100) = 73.1$	$\frac{204}{289}(100) = 70.6$	$\sqrt{73.1(70.6)} = 71.8$
1970	$\frac{390}{390}(100) = 100.0$	$\frac{390}{390}(100) = 100.0$	$\sqrt{(100)(100)} = 100.0$
1975	$\frac{556}{390}(100) = 142.6$	$\frac{673}{499.5}(100) = 134.7$	$\sqrt{142.6(134.7)} = 138.6$

12.4 Laspeyres versus Paasche and Other Comments

Several observations regarding the construction of index numbers should be made at this point:

1. Judgment is exercised in the choice of the bill of goods included in the index survey.
2. Judgment is also exercised in the choice of weights and even in the general nature of the formula to be used.
3. Choice of the base period is also judgmental; this is not a major problem, but we must be aware of this base period when comparing two different series.
4. Because of 1 and 2, statistical inference is not possible; i.e., we cannot present a statistical error for an index number.

Most index number series are either a Laspeyres type, a Paasche type, or some combination thereof. The CPI has traditionally been a Laspeyres-type index with (for a recent past) 1967 as its base year—but with quantity weights coming from the period 1963! Read the *Time* magazine article again. Will the new CPIs still be the Laspeyres type?

The ongoing construction of a meaningful index sequence is, at best, a very expensive operation. In spite of the flaw—the substitution effect—in Laspeyres index, this basic type is frequently chosen in preference to the Paasche type. The construction of a Laspeyres series is significantly less expensive than that of any other type; once base year quantities are determined, only prices need be tracked through time. For a Paasche index, as well as for most other types of indexes, quantities as well as prices must be determined for each time period. Statisticians using the Laspeyres construction may do several things in order to minimize the substitution effect: (1) the base period can be changed frequently; i.e., it is possible to keep a reasonably current base period; (2) the base quantities should come from a "normal" period; i.e., consumer patterns should be natural and not artificially enforced as during war years, oil embargos, droughts, etc. If it is difficult to determine a normal year, the averages of several years' quantities may be used for base weights.

Aside from the cost factor, it would seem that a Paasche-type index sequence would be preferable to the Laspeyres type and Fisher's Ideal Index would be preferable to both. However, in our real world of index number construction, differences between the Laspeyres and Paasche indexes would not approach the magnitude of those in the Lower Slobovian example. Recall that there are some 400 goods and services in the CPI market basket. The relative demand for most of these items would change slowly through time; a significant change in the relative demand for just a few items would not have a controlling effect upon the Paasche index number magnitude. The Implicit Price Deflators published by the Bureau of Economic Analysis are Paasche-type series. The Personal Consumption Expenditures Implicit Price Deflator (PCEIPD) is roughly comparable to the Consumer Price Index (CPI). It is a Paasche-type index that includes rural consumption expenditures along with the urban consumption. The PCEIPD and the CPI are compared for selected years in Table 12.9. The former has a base year of 1972, and the first column in Table 12.9 shows the CPI series as published. The second column is an adjustment of the CPI that makes it comparable to the Implicit Price Deflator by giving it the same base period, 1972. The percentage index for 1972, 125.3, is changed to a decimal, 1.253, and divided into each of the CPI numbers. Thus the 1972 adjusted CPI is 100.0 and percentages are measured from a 1972 base instead of 1967.

Since 1972 the Consumer Price Index has exceeded the Implicit Price Deflator, and the relative difference as well as the absolute difference between the indicated rates of inflation is increasing. This difference could be explained by the substitution effect and/or by the proposition that inflation in the rural United States is not so rapid as that in urban America (the Consumer Price Index doesn't include rural areas). Because the CPI is figured on a 1963 market basket, the substitution effect, if any, would also tend to overestimate the 1967 price index. But we note that the PCEIPD figures in this year actually exceed the corresponding CPI figures. This cursory examination would lead us to believe that (1) inflation in rural America has probably been less than that in the urban areas, and (2) there is no clear evidence of substitution; if there was a substitution effect, it was not significant until after 1972.

Table 12.9 Comparison of Consumer Price Index with Personal Consumption Expenditures Implicit Price Deflator

	Consumer Price Index (CPI)	CPI Adjusted to 1972 Base Year	Personal Consumption Expenditures Implicit Price Deflator (PCEIPD)
1963	91.7	73.2	74.7
1967	100.0	79.8	81.3
1972	125.3	100.0	100.0
1974	147.7	117.9	116.9
1975	161.2	128.7	126.5
1976	170.5	136.1	133.2
1977	181.5	144.8	140.0
1978	195.3	155.9	150.0

As a final note in this chapter, it is of interest to observe why the term *ideal* is used in Fisher's Ideal Index. To this point we have said nothing about a possible index that simply measures ratios of total money expenditures. For example, in Lower Slobovia the total sample market basket expenditure in 1975 was 673 buckniks, and in 1970 this was 390. The ratio, 673/390, multiplied by 100 equals 172.6, which indicates that the combined change in both prices and quantities has increased 72.6 percent over the 5-year period. One criterion for an ideal index is that the product of the decimal fraction forms of the price index and the quantity index should equal the index of total expenditures. Recalling our previous calculations of Laspeyres and Paasche price and quantity indexes, we see that neither of these two types satisfy this criterion; for Laspeyres,

$$\frac{_LI_5}{100} \cdot \frac{_LG_5}{100} = \frac{(128.1)}{100} \frac{(142.6)}{100} = (1.281)(1.426) = 1.827,$$

and for Paasche,

$$\frac{_PI_5}{100} \cdot \frac{_PG_5}{100} = (1.210)(1.347) = 1.630.$$

The reader should verify that Laspeyres Price Index multiplied by the Paasche Quantity Index equals 1.726, as does the product of the Paasche Price Index and Laspeyres Quantity Index. Since the square root of the product of these 4 numbers is 1.726, it is clear that Fisher's Ideal Index is indeed ideal by this test. Thus, we have

$$\frac{_FI_5}{100} \cdot \frac{_FG_5}{100} = \left(\frac{138.6}{100}\right)\left(\frac{124.5}{100}\right) = (1.386)(1.245) = 1.726.$$

The general algebraic proof is quite obvious if you set up the formulas for the several indexes.

A second criterion would demand that, given a change in the base year, the decimal fraction form of the new index for the original base year be the reciprocal of the old index for the new base year. Again, we find the interesting asymmetry between Laspeyres and Paasche; i.e., the reciprocal of Laspeyres index would be the Paasche index and vice versa. The reader should verify this. See Problem 4 at the end of the chapter. If the above is true, it is again quite obvious that Fisher's Ideal Index satisfies this second criterion.

For your information and also for use in several of the chapter problems, we have included, in Table 12.10, the monthly CPI series for all urban consumers from 1967 through 1979. Refer to the *Time* magazine article in this chapter, and note that this is one of the two new series started early in 1978. The abbreviation for this index is CPI-U. The index series previous to 1978 is, of course, the old CPI series.

Table 12.10 U.S. Department of Labor—Bureau of Labor Statistics, Washington, D.C. 20212
Consumer Price Index for All Urban Consumers, U.S. City Average, All Items (1967 = 100)

Year	Jan.	Feb.	Mar.	Apr.	May	June	July	Aug.	Sep.	Oct.	Nov.	Dec.	Avg.
1967	98.6	98.7	98.9	99.1	99.4	99.7	100.2	100.5	100.7	101.0	101.3	101.6	100.0
1968	102.0	102.3	102.8	103.1	103.4	104.0	104.5	104.8	105.1	105.7	106.1	106.4	104.2
1969	106.7	107.1	108.0	108.7	109.0	109.7	110.2	110.7	111.2	111.6	112.2	112.9	109.8
1970	113.3	113.9	114.5	115.2	115.7	116.3	116.7	116.9	117.5	118.1	118.5	119.1	116.3
1971	119.2	119.4	119.8	120.2	120.8	121.5	121.8	122.1	122.2	122.4	122.6	123.1	121.3
1972	123.2	123.8	124.0	124.3	124.7	125.0	125.5	125.7	126.2	126.6	126.9	127.3	125.3
1973	127.7	128.6	129.8	130.7	131.5	132.4	132.7	135.1	135.5	136.6	137.6	138.5	133.1
1974	139.7	141.5	141.1	143.9	145.5	146.9	148.0	149.9	151.7	153.0	154.3	155.4	147.7
1975	156.1	157.2	157.8	158.6	159.3	160.6	162.3	162.8	163.6	164.6	165.6	166.3	161.2
1976	166.7	167.1	167.5	168.2	169.2	170.1	171.1	171.9	172.6	173.3	173.8	174.3	170.5
1977	175.3	177.1	178.2	179.6	180.6	181.8	182.6	183.3	184.0	184.5	185.4	186.1	181.5
1978	187.2	188.4	189.8	191.5	193.3	195.3	196.7	197.8	199.3	200.9	202.0	202.9	195.4
1979	204.7	207.1	209.1	211.5	214.1	216.6	218.9	221.1	223.4	225.4	227.5	229.9	217.4

Table 12.11 Percentage of Increase as of September 30 of Each Year Indicated

Year	Sep.
1967	2.7%
1968	4.4%
1969	5.8%
1970	5.7%
1971	4.0%
1972	3.3%
1973	7.4%
1974	12.0%
1975	7.8%
1976	5.5%
1977	6.6%
1978	8.3%
1979	12.1%

Problems

1. Consider the following data on price and production of the three basic softwoods in the United States.

	Price (per thousand bd. ft.)		Production (million bd. ft.)	
	1945 p_0	1950 p_1	1945 q_0	1950 q_1
Douglas Fir	$35	$80	540	800
Southern Pine	70	150	600	830
Western Pine	40	70	400	640

Using 1945 as the base year,
  a. compute Laspeyres Price Index for 1950;
  b. compute Paasche Price Index for 1950;
  c. compute Laspeyres Quantity Index for 1950;
  d. interpret each of the above three indexes.

2. During wage negotiations with the union in 1956, Semco Steel and the union presented the following two sets of indexes:

Semco's Year	1945 base	
	Wage Rates	Steel Prices
1945	100	110
1950	136	115
1955	140	125

The Union's Year	1939 base	
	Wage Rates	Steel Prices
1939	100	100
1945	120	150
1950	163	172
1955	168	188

Both parties used the same data and the same methods for computing the indexes. The union, of course, was arguing for a wage increase on the basis that steel prices had increased more than wages, and Semco was taking the opposing position.
  a. Explain the apparent contradiction between the above two sets of numbers and tell what additional information must be known in order to settle the issue of which of the sets should be used.
  b. Compute the two 1939 indexes so that Semco's table will be complete.
  c. Verify that the two tables are not contradictory.

3. Members of the Beach Bums' Local sent the following lists to LBJ to bear out their charge that the real necessities of life in their profession had been subject to a five-year inflationary spiral.

	1960		1965	
	p	q	p	q
Surfboards	$27	80	$42.00	100
Swimming Trunks	4	350	4.50	500
Gin	3	800	4.50	1000

Use 1960 as a base year and (1) calculate both Laspeyres and Paasche Price Indexes for 1965; (2) calculate Paasche Price Index for 1965; (3) decide whether the profession appears fairly attractive in spite of rising prices.

4. Use the Lower Slobovian data in Table 12.2 and with a 1975 base year, calculate the following six index numbers for 1970: Laspeyres, Paasche, and Fisher's Price and Quantity Indexes. Check the propositions in the last paragraph of this chapter.

5. Beef, pork, and lamb were used to establish a meat price index for Piute County, with 1967 as a base year. With the following data, calculate Laspeyres and Paasche Price Indexes for 1972.

	1967		1972	
	p	q	p	q
Beef	.75	200	1.00	240
Pork	.60	50	.70	100
Lamb	1.00	30	.90	50

6. Following are data on the three major items (food) in the market basket of Star Valley, Wyoming, consumers:

	1967		1976	
Item	Price	Quantity Consumed	Price	Quantity Consumed
Milk (qt.)	$.20	100	$.39	90
Bread (loaf)	.25	300	.38	350
Swiss cheese (lb.)	.60	50	1.25	40

a. Find price relatives for each of the three food items.
b. Find Laspeyres 1976 food price index using 1967 as the base year.
c. Find Paasche 1976 food price index using 1967 as the base year.
d. Assume Laspeyres consumption index to be a measure of population change. What is your estimate of the population change in Star Valley over the period?

7. The following is a time series of mean salaries in thousands of dollars of full professors at Nutmeg University from 1967 through 1979.

1967	1968	1969	1970	1971	1972	1973	1974	1975
14.7	15.1	16.3	17.4	17.7	18.3	19.5	21.8	23.5

1976	1977	1978	1979
24.9	26.7	28.6	31.5

    a. Deflate this series using the CPI annual averages from Table 12.10.

    b. Plot a line chart of the series of deflated salaries (salaries in 1967 "real" dollars) on a grid with the series of actual salaries ("current" dollars).

    c. Say a word or two about the change in real income and buying power of the "average" full professor at Nutmeg University. Are you in any way critical of this method of presenting and comparing real income through time?

8. Refer to the time series on gross earnings from the Utah State Liquor Monopoly in Table 11.5.

    a. Use the middle month CPI for each quarter from 1967 through 1970 and deflate the quarterly centered moving averages.[1]

    b. Plot the deflated moving average and the original series on the same grid. What happened to Utah's real income from its State Liquor Monopoly over this four-year period?

    c. The CPI component for alcoholic beverages has been significantly lower than the CPI. For example, the 1979 CPI for alcoholic beverages in the western United States is approximately 180. Assume that the ratio of this component of the CPI to the full CPI has remained fairly constant since 1967 and also assume that the percentage consumer mix of alcoholic beverages hasn't changed appreciably. Comment on the volume of consumption of alcoholic beverages from 1967 through 1970.

9. Consider the following time series on annual profits, in millions of dollars, by Super Oats Cereal, Inc.

1969	1970	1971	1972	1973	1974	1975	1976	1977
46.3	52.4	57.2	65.6	76.5	88.1	103.4	119.6	132.0

1978	1979
146.2	165.7

    a. Deflate this series using the annual CPI-U from Table 12.10.

    b. Comment regarding growth in real dollar profits. Prepare a chart for the board of directors meeting next week exhibiting growth in both current dollars and in real dollars.

---

1. One could use a three-month average, either arithmetic or geometric, or even a first and last two-month average. Quite obviously the quarter middle month CPI will usually be quite close to any of these averages. Which average would be technically correct?

10. The following data relate to production and prices of Mac A. Wroni and Sons, a large Midwest pasta manufacturer. Production is in thousands of pounds, and prices are the average annual wholesale price in dollars per pound.

	1971		1973		1977	
	p	q	p	q	p	q
Macaroni	.25	1,830	.30	2,060	.40	2,370
Lasagna	.50	240	.55	585	.60	1,090
Spaghetti	.30	666	.45	672	.65	681
Ravioli	.75	133	.95	222	.80	520

a. With a 1971 base, find Laspeyres pasta production index for 1973 and 1977; also find Paasche production indexes for these years with the same base year. Which in your opinion is the better measure of increased production?

b. With 1971 as a base year, find both Laspeyres and Paasche pasta price for 1973 and 1977. Compare and comment.

c. Deflate all prices using the CPI-U from Table 12.10. Calculate new pasta price indexes using the deflated prices. Comment on the real pasta price change over the six years. Criticize the method.

CHAPTER 13
ANALYSIS OF VARIANCE AND STATISTICAL INFERENCE RELATED TO VARIANCE

In this chapter we introduce an important statistical procedure known as *analysis of variance*. Analysis-of-variance models are many and varied, and some are quite sophisticated; we present several basic models.

Specific objectives are—

To first present two fundamental distributions that describe, under certain conditions, the distribution of the sample variance and the ratio of two sample variances (Section 1).

To present several analysis-of-variance models and illustrate the application of these models (Sections 2, 3, and 4).

To return to analysis of variance as it relates to the regression problem presented in Chapter 10 (Section 5).

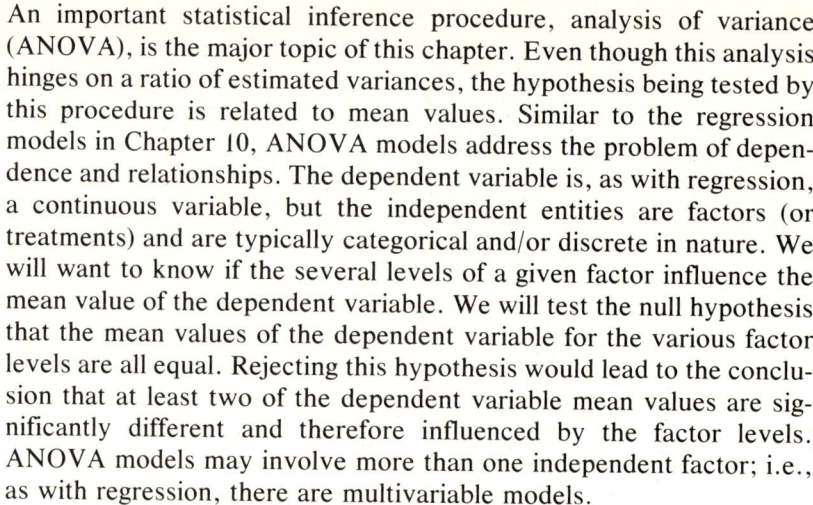

An important statistical inference procedure, analysis of variance (ANOVA), is the major topic of this chapter. Even though this analysis hinges on a ratio of estimated variances, the hypothesis being tested by this procedure is related to mean values. Similar to the regression models in Chapter 10, ANOVA models address the problem of dependence and relationships. The dependent variable is, as with regression, a continuous variable, but the independent entities are factors (or treatments) and are typically categorical and/or discrete in nature. We will want to know if the several levels of a given factor influence the mean value of the dependent variable. We will test the null hypothesis that the mean values of the dependent variable for the various factor levels are all equal. Rejecting this hypothesis would lead to the conclusion that at least two of the dependent variable mean values are significantly different and therefore influenced by the factor levels. ANOVA models may involve more than one independent factor; i.e., as with regression, there are multivariable models.

The test of equal means in analysis of variance is based upon a ratio of two components of the variance of the dependent variable. This is known as the Fisher Test or simply the F-test, and to implement this test we have included a table of values for the F-distribution, Table E, at the end of the textbook. The F-distribution and another new distribution, the chi-square, may both be used on statistical inference problems related directly to variance. Before we begin the ANOVA discussion, we will introduce both of these distributions relating to statistical inference. If time is a constraint and ANOVA procedures have a higher priority than inference related to variance, this discussion, in Section 1, may be bypassed. However, inasmuch as the Fisher distribution is basic to all ANOVA models, we suggest that you at least read Section 1 before proceeding to ANOVA.

13.1 The Chi-Square Distribution and F-Distribution

Our discussions on statistical inference to this point have been related to either means or proportions. Might it be necessary—and if so is it possible—to give a confidence interval estimate on a population standard deviation? To test a hypothesis regarding a population variance? The answer to the above and similar questions is a qualified yes.

Two distributions play key roles in statistical inference related to variances. The first distribution is introduced with the following theorem:

The Chi-Square Distribution

If the variable, x, is normally distributed, then the ratio,

$$\frac{(n-1)s^2}{\sigma^2},$$

has a chi-square, (χ^2), distribution with $n-1$ degrees of freedom.[1]

In Chapter 14 we will see another application of the chi-square distribution to a quite different kind of problem.

Like the Student-t distribution, the χ^2 distribution is actually a family of distributions. Unlike the symmetrical t-distribution, this new distribution is skewed to the right; the skewness decreases with increasing degrees of freedom. The χ^2 has a lower bound of 0; in the ratio in the theorem above, σ^2 is constant and finite, $n-1$ is positive ($n \geq 2$ for s^2 to exist), and s^2 is at least 0. The expected value of χ^2 is $n-1$. (We have stated before that s^2 is an unbiased estimator of σ^2; i.e., $E(s^2) = \sigma^2$. Therefore,

$$E\left(\frac{s^2}{\sigma^2}\right) = 1,$$

and it follows that $E(\chi^2) = n - 1$.)

If we are able to assume that the variable, x, is normally distributed (or approximately so), the above theorem may be used to place a confidence interval on σ^2 given a sample variance, s^2, or it may be used to test a hypothesis regarding the population variance. The following examples will illustrate these applications.

Example 13.1.1

Completion time in minutes for a certain task is assumed to be normally distributed. A random sample of 10 completion times resulted in a sample variance of $s^2 = 9.56$ min.^{2}. Find the 95 percent confidence interval for σ^2.

Tabular values of the chi-square function for selected degrees of freedom are given in Table D at the end of the textbook. The tabular values are given for $P(\chi^2 \geq c)$ where c is some fixed value. In the above problem for a sample of 10, the degrees of freedom are 9. With 9 degrees of freedom, values of χ^2, c-values, that satisfy the statement $P(x^2 \geq c) = k$ (for $k = .99, .975, .95, .05, .025$ and $.01$) are read from the row with $df = 9$. For example, $P(\chi^2 \geq 19.023) = .025$, and $P(\chi^2 \geq 2.700) = .975$. The second statement also implies $P(\chi^2 < 2.700) = .025$, inasmuch as the total area under the function is unity.

From this we are able to state:

---

[1]. Please note that the apparent exponent, 2, on the χ^2 notation is not an exponent but simply an integral—and perhaps unfortunate—part of the variable symbol.

$P(2.700 \leq \chi^2 \leq 19.023) = .95;$

from the theorem above,

$$\frac{(n-1)s^2}{\sigma^2} = \chi^2,$$

we can substitute the ratio to get

$$P\left(2.700 \leq \frac{9s^2}{\sigma^2} \leq 19.023\right) = .95.$$

Remember we have a sample s^2 of 9.56 in the original problem. Substituting this we get

$$\text{Conf: } \left\{2.700 \leq \frac{9(9.56)}{\sigma^2} \leq 19.023\right\} = .95.$$

Inverting each of the above fractions reverses the inequality signs:

$$\text{Conf: } \left\{\frac{1}{2.700} \geq \frac{\sigma^2}{9(9.56)} \geq \frac{1}{19.023}\right\} = .95.$$

Multiplying through by 9(9.56), or 86.04, gives

$$\text{Conf: } \left\{\frac{86.04}{2.700} \geq \sigma^2 \geq \frac{86.04}{19.023}\right\} = .95;$$

and we get

$$\text{Conf: } \{31.87 \geq \sigma^2 \geq 4.52\} = .95.$$

Also, we are 95 percent confident that σ is at least $\sqrt{4.52} = 2.13$ and at most $\sqrt{31.87} = 5.65$.

Example 13.1.2 Would we be able to reject the hypothesis, $H_o: \sigma^2 = 5.00$ against $H_a: \sigma^2 > 5.00$ with a sample, s^2, of 9.56? Use an $\alpha = .025$. For $df = 9$, $P(\chi^2 \geq 19.023) = .025$.

If the test value of χ^2 exceeds 19.023, we can reject the null hypothesis.

The test χ^2 value $= \frac{(n-1)s^2}{\sigma^2} = \frac{9(9.56)}{5} = 17.21.$

We cannot reject the null hypothesis. The sample result of $s^2 = 9.56$ could have come from a universe with $\sigma^2 = 5.00$. Note that if we had chosen $\alpha = .05$, the critical value of chi-square, χ^{2*}, is 16.919, and we could, by a whisker, reject H_o. This indicates that the sample result is somewhat borderline.

The second distribution dealing with variances is the Fisher distribution, or simply the F-distribution. This is introduced with the following theorem:

The F-Distribution   If independent samples of size n_1 and n_2 are selected from two normally distributed populations (or the same population) with $\sigma_1^2 = \sigma_2^2 = \sigma^2$, then the ratio,

$$\frac{s_1^2}{s_2^2},$$

has an F-distribution with $df_1 = n_1 - 1$ in the numerator and $df_2 = n_2 - 1$ in the denominator.

The F-distribution has a lower bound of 0 inasmuch as both s_1^2 and s^2 are nonnegative. As we will see later on in this chapter, one of the major applications of the F-ratio relates to the testing of a hypothesis on the equality of several means—the important applied area of analysis of variance. Before proceeding with our discussion of this subject, let us note briefly the direct application to testing the hypothesis of the equality of two variances, $H_o: \sigma_1^2 = \sigma_2^2$. If x_1 and x_2 are normally distributed, the F-ratio provides a very simple test for this hypothesis. Consider the following example:

Example 13.1.3   For some time Jones has been tracking the daily price of 2 different stocks and has found that, on the average, an investment in either would have yielded about the same return. To further investigate the stocks, he decides to use the variance as a measure of risk; he will invest in the stock with the lower risk. For the past 10 days, $s_1^2 = 388$, and for the past 12 days, $s_2^2 = 147$. Assuming variance is a good measure of risk, can he conclude that stock number 2 is the lower risk; or, in a formal way, can he reject the following null hypothesis; $H_o: \sigma_1^2 = \sigma_2^2$, and accept the alternative hypothesis, $H_a: \sigma_1^2 > \sigma_2^2$?

The test F-ratio, $F = 388/147 = 2.64$, is quickly calculated. We know that, if the null hypothesis is true, the expectation of the ratio is unity, and we need to know whether 2.64 could be a chance result or is a clear indicator that $\sigma_1^2 > \sigma_2^2$.

As you may already have speculated, the answer to this question will be found in the tabular presentation of selected F-probabilities in the back of the textbook. See Table E. We are somewhat limited in these tables because, as you will see, considerable space is required for simply one upper tail—e.g. $P(F \geq c) = k$. With different degrees of freedom in both numerator and denominator, the first F-table gives the constant value of c for $P(F \geq c) = .05$. This is illustrated graphically in Figure 13.1.

The second F-table gives values of F for an upper tail of .01. Returning to the problem, note that $df_1 = 9$ and $df_2 = 11$. If $\alpha = .05$, then

Figure 13.1 Sketch of F-Distribution and Area Given in Table E.

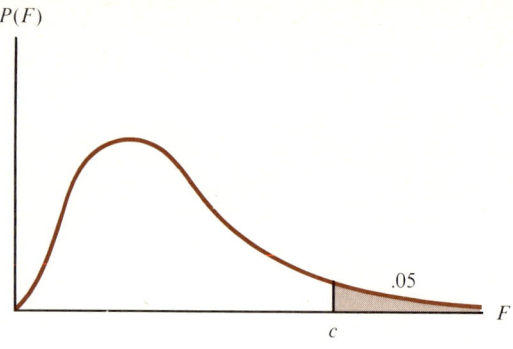

F^*, the critical value of F, is 2.90. With a test-F of 2.65, Jones is unable to reject the null hypothesis. He may still choose to invest in stock number 2, but his data do not clearly indicate that it is the lower-risk stock. If Jones had chosen $\alpha = .01$, note that the critical value of F, F^*, equals 4.63. Remember also that the above analysis is valid only if it can be assumed that daily stock prices of both stocks are normally distributed.

For you who prefer the formal four-step process of hypothesis testing discussed in Chapter 8, we present this process to solve the above example:

1. $H_o: \sigma_1^2 = \sigma_2^2$
   $H_a: \sigma_1^2 > \sigma_2^2$.
2. $\alpha = .05$, $F = s_1^2/s_2^2$ with $df_1 = n_1 - 1 = 9$ and $df_2 = n_2 - 1 = 11$.
3. Reject H_o if $F > 2.90$; i.e., $F^* = 2.90$.
4. $F = s_1^2/s_2^2 = 388/147 = 2.64$; therefore we cannot reject H_o.

13.2 One-Way Analysis of Variance

You will recall from Chapter 10 that the F-ratio and related analysis of variance are used to test the null hypothesis on the slope-coefficients of the regression function: $H_o: \beta_1 = \beta_2 = \ldots = \beta_k = 0$. As promised in Chapter 10, we will be looking at this problem in this chapter. The logic of analysis of variance is here introduced for a problem that is related (but not obviously so). The t-test is available for testing the equality of two means, $H_o: \mu_1 = \mu_2$; but we may be interested in testing the equality of several means, $H_o: \mu_1 = \mu_2 = \ldots = \mu_k$. Obviously we could use the t-test on all pairwise combinations of the above means. If there

Table 13.1 Three-Area Wage Rate Sample

	Area 1 Las Vegas	Area 2 The Bay Area	Area 3 Boise
	$6.35	$5.75	$4.80
	6.50	6.05	6.20
	7.10	5.50	5.40
	5.80	5.20	4.75
	6.25	5.00	4.85
$\bar{y}$	$6.40	$5.50	$5.20

$$\bar{\bar{y}} = \frac{6.40 + 5.50 + 5.20}{3} = 5.70.$$

were three populations, three pairwise tests would be needed; for four populations, six t-tests would be required; and, in general, for k populations, $C_2^k = k(k-1)/2$ pairwise tests would be necessary.[2] The analysis of variance on means from three or more populations is usually much more efficient.

We begin with the following hypothetical example, illustrative of a one-way analysis of variance (ANOVA) with equal sample sizes. Because of the relationship between analysis of variance and regression, we will use y (instead of x) for the basic variable.

Example 13.2.1 The beginning hourly wage rate for unskilled labor was taken from five contractors in each of three labor force areas, with the results shown in Table 13.1.

Mean wage rates, $\bar{y}_1, \bar{y}_2, \bar{y}_3$, were calculated to be $6.40, $5.50, and $5.20 respectively, and the grand mean, $\bar{\bar{y}}$, was calculated to be $5.70. The grand mean is the average of all 15 numbers and, with equal sample sizes in each of the areas, the grand mean is also the average of the 3 means.

We wish to test the hypothesis that mean beginning wage rates for unskilled labor in the three labor force areas are all the same; i.e., H_o: $\mu_1 = \mu_2 = \mu_3$. Note that rejection of this hypothesis leads us to conclude that there is at least one mean that is different from the rest; it is not necessarily so that all three means are different from each other. The above hypothesis is tested under two assumptions: (1) y is normally distributed in each of the areas; (2) σ^2, the variance of y, is the same in each area.

Under these assumptions, we are now ready to calculate two different estimates of σ^2. The *within-group* estimate of σ^2 is a pooled

---

2. Note: The stated α level is not preserved if this approach is taken. The actual α becomes increasingly larger as more tests are made. Hence this method is technically invalid.

variance similar to that introduced in Chapter 9. (The group in our example is the *area*. This is also frequently called *treatment*, a term passed on from some of the early ANOVA applications in agriculture.) The pooled variance is an average of the three area variances. Each of the three variances,

$$s_j^2 = \frac{\sum_{i=1}^{5}(y_{ij} - \bar{y}_j)^2}{4},$$

is calculated. For example,

$$s_1^2 = \frac{(6.35 - 6.40)^2 + (6.50 - 6.40)^2 + (7.10 - 6.40)^2 + (5.80 - 6.40)^2 + (6.25 - 6.40)^2}{4}$$

$$= .22125.$$

Similarly, $s_2^2 = .17625$, and $s_3^2 = .38125$. The pooled or within-group estimate of σ^2 is, then,

$$s_w^2 = \frac{s_1^2 + s_2^2 + s_3^2}{3} = .2596.$$

For the between-group estimate of σ^2 under the null hypothesis that $\mu_1 = \mu_2 = \mu_3$, we first calculate an estimate of $\sigma_{\bar{y}}^2$.

$$\hat{\sigma}_{\bar{y}}^2 = \frac{\sum_{i=1}^{3}(\bar{y}_j - \bar{y})^2}{2} = \frac{(6.40 - 5.70)^2 + (5.50 - 5.70)^2 + (5.20 - 5.70)^2}{2}$$

$$= .3900.$$

Recalling that $\sigma_{\bar{y}}^2 = \sigma^2/n$, it follows that $\sigma^2 = n\sigma_{\bar{y}}^2$, and we get our second estimate, the between-group estimate, of the population variance:

$$s_B^2 = n\hat{\sigma}_{\bar{y}}^2 = 5(.3900) = 1.9500.$$

If the null hypothesis is not true, then at least two of the population means are different. This tends to be reflected in the sample results. A large difference in sample means is in turn reflected as a large between-group estimator of the variance.

We have an efficient and simple answer to the problem; if the null hypothesis is true, the ratio,

$$\frac{s_B^2}{s_w^2},$$

has an F-distribution with degrees of freedom in the denominator of $k(n - 1)$ or 12, where k is the number of groups, and df in the numerator of $k - 1$ or 2. If this ratio exceeds the critical value of F, then quite clearly it is due to excessive differences between the means resulting in a relatively large s_B^2.

The critical F-value for $df_1 = 2$, and $df_2 = 12$ is $F_{.05}^* = 3.88$. The test-F is calculated to be

$$F = \frac{1.9500}{.2596} = 7.51.$$

The null hypothesis is rejected; we can conclude that at least one pair of the labor force areas differ in mean beginning wage rates for unskilled workers.

The Analysis of Variance Identity

Calculations for ANOVA are typically tedious and, as with regression analysis, there are a number of prepackaged ANOVA computer programs. The organization and presentation of the analysis is usually shown as illustrated in Table 13.2, and in the General Form of ANOVA Table, whether hand calculated or computerized. This presentation is based upon the well-known and quite famous analysis of variance identity:

$$\sum_j \sum_i (y_{ij} - \bar{y})^2 = \sum_j \sum_i (y_{ij} - \bar{y}_j)^2 + n \sum_j (\bar{y}_j - \bar{y})^2.$$

The total sum of the squares of all sample numbers about the grand mean is equal to the sum of the within-group sum of squares and the between-group sum of squares.

The first index of summation, i, runs from 1 through the sample size for each sample. In our example, this is from 1 to 5. The second index, j, is the group index from 1 to k—in our example $k = 3$. The total sum of squares, the term on the left, is simply the sum of the squared deviation of all nk (in our example $nk = 15$) sample numbers about the grand mean. The first term on the right side of the identity, the within-group sum of squares, calls for the sum of the squared deviations of the sample numbers about the specific group mean and then sum k (3 in our example) sets of squared deviations. The second term calls for a sum of

Table 13.2   ANOVA Table (Example)

Nature of Variation	Sum of Squares	Degrees of Freedom	Mean Sum of Squares	F
Between Groups	3.900	2	1.950	$\frac{1.950}{.2596} = 7.51$
Within Groups	3.115	12	.2596	
Total	7.015			

One-Way Analysis of Variance

the squared deviations of each group mean from the grand mean and this sum to be multiplied by n (5 in our example).

General Form of ANOVA Table

Nature of Variation	Sum of Squares	Degrees of Freedom		Mean Sum of Squares	F
Between Groups	$n\sum_j (\bar{y}_j - \bar{y})^2$	$k - 1$	$s_B^2 = \dfrac{n\sum (\bar{y}_j - \bar{y})}{k - 1}$		$\dfrac{s_B^2}{s_w^2}$
Within Groups	$\sum_j \sum_i (y_{ij} - \bar{y}_j)^2$	$n(k - 1)$	$s_w^2 = \sum_j \sum_i \dfrac{(y_{ij} - y_j)^2}{n(k - 1)}$		
Total	$\sum_j \sum_i (y_{ij} - \bar{y})^2$	$nk - 1$			

Following is a summary of this same ANOVA example in the four-step format introduced in Chapter 8.

1. $H_o: \mu_1 = \mu_2 = \mu_3$.
   $H_a: \mu_i \neq \mu_j$ for at least one i and one j.

2. $\alpha = .05$, $F = \dfrac{s_B^2}{s_W^2}$ with $df_1 = k - 1 = 2$ and $df_2 = k(n - 1) = 12$.

3. Reject H_o if $F > 3.88$; i.e., $F_{.05}^* = 3.88$.

4. $F = \dfrac{s_B^2}{s_W^2} = \dfrac{1.9500}{.2596} = 7.51$; therefore reject H_o.

We conclude that at least one pair of population means are not equal.
   An identity from Chapter 2,

$$\sum_i (y_i - \bar{y})^2 = \sum y_i^2 - n\bar{y}^2,$$

in conjunction with the ANOVA identity above, gives us some alternative calculation procedures. The total sum of squares may be calculated by first summing the squares of all numbers in the table and then subtracting from this the product of nk and the square of the grand mean; i.e.,

$$\sum_j \sum_i (y_{ij} - \bar{y})^2 = \sum_j \sum_i y_{ij}^2 - nk\bar{y}^2.$$

The between-group sum of squares, $n\sum(\bar{y}_j - \bar{y})^2$, is equal to $n\sum \bar{y}_j^2 - nk\bar{y}^2$. Square each of the means, sum and multiply by n, then subtract the same term as we did above to get the total sum of squares.
   Finally, the within-group sum of squares will be the difference between the total sum of squares and the between-group sum of squares.
   Quite obviously, there is no reason why the sample sizes in each

group need be the same. On the other hand, many ANOVA problems are the result of a designed experiment, and the statistician will probably use equal sample sizes in the design more often than not. Be that as it may, the formula adjustment for unequal sample sizes is minimal:

$$\sum_j \sum_i (y_{ij} - \bar{y})^2 = \sum_j \sum_i (y_{ij} - \bar{y}_j)^2 + \Sigma n_j(\bar{y}_j - \bar{y})^2.$$

The last term, the between-group sum of squares, is the major difference. Each of the squared deviations of the group mean from the grand mean is multiplied by the sample size before being summed. (Note that, if all n_j are equal, this can be factored from the summation to give us the equal-sample-size identity.) Degrees of freedom are as follows: for total sum of squares,

$$df = \sum_j n_j - 1 = n_1 + n_2 + \ldots + n_k - 1;$$

for within-group sum of squares,

$$df_2 = (n_1 - 1) + (n_2 - 1) \ldots + (n_k - 1) = n_1 + n_2 + \ldots + n_k - k$$
$$= \sum_j n_j - k;$$

and for between-group sum of squares,

$$df_1 = k - 1.$$

The following very simple example will be used to illustrate a one-way ANOVA with unequal sample sizes. Only two groups are considered, and, quite clearly, the t-test for the difference of two means would be more efficient than an ANOVA on the data. This example does illustrate, however, the adjustments in method and calculations necessary for unequal sample sizes, and, in addition, we use it to point out an interesting relationship between the t-test and the F-ratio with one degree of freedom in the numerator.

Example 13.2.2 Typist A was given three speed tests and typist B was given five tests, with the following results in words per minute. Is there a significant difference in mean speed of the two typists?

	A	B
	120	116
	112	100
	125	110
		101
		118
Column sums	357	545

Averages $\bar{y}_1 = 119$ $\bar{y}_2 = 109$

Grand Mean $\bar{y} = \dfrac{357 + 545}{8} = 112.75.$

Note first that the grand mean is the average of all eight numbers, which of course is also the weighted average, weighted by sample size, of the two group means.

We will outline the method and sequence of calculations and give the results in the ANOVA table, Table 13.3.

The total sum of squares,

$$\sum_j \sum_i (y_{ij} - \bar{\bar{y}})^2,$$

the sum of the squares deviations of all 8 numbers, was calculated first:

$$\sum_j \sum_i (y_{ij} - \bar{\bar{y}})^2 = (120 - 112.75)^2 + \ldots + (118 - 112.75)^2 = 549.50.$$

For the total sum of squares, $df = n_1 + n_2 - 1 = 5 + 3 - 1 = 7$.

Calculation of

$$\sum_j n_j (\bar{y}_j - \bar{\bar{y}})^2,$$

the between-group sum of squares, is as follows:

$$\sum n_j (\bar{y}_j - \bar{\bar{y}})^2 = 3(119 - 112.75)^2 + 5(109 - 112.75)^2$$
$$= 117.19 + 70.31 = 187.50$$
$$df = k - 1 = 2 - 1 = 1.$$

The within-group sum of squares is normally treated as a residual for both the sum of squares and df; i.e., $df_2 = 7 - 1 = 6$. The within-group sum of squares equals the total sum of squares (SS) minus the between-group SS, or 362.0.

Finally, the means of the SS, SS/df, are presented in Column 3 of Table 13.3. We should not forget that both of these are estimates of the variance of y, typing speed, under the null hypothesis that the mean speed of typist A is no different from that of typist B. The critical F-value with an $\alpha = .05$ is $F^* = 5.99$. The F-ratio or test value of $F = 3.11$. (See last column of Table 13.3.) We are unable to reject the null hypothesis; on the basis of this experiment, we cannot conclude that the two typists are different in their typing speeds.

Table 13.3 Analysis of Variance Table

Variation	ANOVA Sum of Squares	Degrees of Freedom	Mean Sum of Squares	F-Ratio
Between Groups	187.5	1	187.5	$F = \dfrac{187.5}{60.3}$
Within Groups	362.0	6	60.3	$= 3.11$
Total	549.5	7		

Inasmuch as we are dealing with only two groups, the t-test would perhaps be more appropriate and certainly more efficient. Let us apply the t-test to this problem. The null hypothesis is $H_o: \mu_1 = \mu_2$ against the alternative $H_a: \mu_1 \neq \mu_2$ with $df = n_1 + n_2 - 2 = 6$. The critical t for $\alpha = .05$ is $t^* = \pm 2.447$. The test-t,

$$t = \frac{119 - 109}{\sqrt{60.3}\sqrt{\frac{1}{3} + \frac{1}{5}}} = \frac{10}{5.67} = 1.76.$$

The pooled variance, $s^2 = 60.3$, was already calculated, and

$$s\left(\sqrt{\frac{1}{n_1} + \frac{1}{n_2}}\right) = \hat{\sigma}_{\bar{y}_1 - \bar{y}_2}.$$

Again, we are unable to reject the null hypothesis. Note that $\sqrt{F^*} = t^*$; i.e., $\sqrt{5.99} = 2.447$; and $\sqrt{\text{test-}F} = \text{test-}t$, i.e., $\sqrt{3.11} = 1.76$.

This is generally true: with degrees of freedom of 1 in the numerator of an F-distribution, the square root of F equals t and df equals to df in the denominator of the F-ratio.

13.3 Two-Way Analysis of Variance

In our first example, the groups were identified by the city—a categorical variable, or what we will chose to call a *factor*. The factor in the second example with unequal sample sizes was the typist. Quite obviously, we may want to consider the effects of more than one factor. As an example consider the following data.

Example 13.3.1 Each of three typists was given one speed test on each of four different typewriters. The results, in words per minute, are recorded in Table 13.4.

We can ask and provide an answer to both of the following questions.

1. Are there significant differences among or between the typists (factor 1) in typing speed?
2. Are there significant differences among or between the typewriters (factor 2) as these machines react to speed typing?

(A technical assumption should be mentioned at this point: We are assuming that there is no interaction between factors 1 and 2; i.e., translated to our problem, we assume that the typists do not respond

Table 13.4 Words per Minute

	Typist				
Typewriter	A	B	C	Σ	Average
I	120	116	109	345	115
II	112	100	97	309	103
III	125	110	101	336	112
IV	131	110	101	342	114
Σ	488	436	408	1,332	
Average	122	109	102		$\bar{\bar{y}} = 111$

psychologically, either positively or negatively, to the specific machines.)

With a two-factor or two-way analysis of variance, we need to be a bit more careful with subscripts on the notation. The variable, y_{ij}, refers to measurement in i^{th} row and j^{th} column. The row means will be denoted by $\bar{y}_{1.}, \bar{y}_{2.}, \bar{y}_{3.}$, and $\bar{y}_{4.}$ respectively, and column means by $\bar{y}_{.1}, \bar{y}_{.2}$, and $\bar{y}_{.3}$. In conjunction with the data presentation, Table 13.4, we have calculated the seven averages and the grand mean, $\bar{\bar{y}} = 111$.

In a formal way we will want to answer the two questions asked above by testing the hypotheses:

1. $H_o: \mu_{.1} = \mu_{.2} = \mu_{.3}$

and

2. $H_o: \mu_{1.} = \mu_{2.} = \mu_{3.} = \mu_{4.}$.

Accepting the first hypothesis would lead us to conclude that the typists are not different in their speed-typing ability; rejecting this hypothesis would say that at least one of the typists is faster than one of the others.

Accepting the second hypothesis would be equivalent to saying that no two machines have significant differences that affect typing speed. Rejecting the hypothesis would lead to the conclusion that at least two machines have differences that affect speed.

The Two-Way Analysis of Variance Identity

As with the one-way ANOVA, we have an important identity that is used for the required calculations:

Total sum of squares = sum of squares between columns + sum of squares between rows + the unexplained (or error) sum of squares.

(We will let k equal the number of columns and r the number of rows.)

Calculation of the total SS is the same as that for the one-way ANOVA, $\Sigma\Sigma(y_{ij} - \bar{\bar{y}})^2$:

between-column $SS = r\sum_j (\bar{y}_{.j} - \bar{\bar{y}})^2$;

between-row $SS = k\sum_i (\bar{y}_{i.} - \bar{\bar{y}})^2$.

In our example, $r = 4$ and $k = 3$.

These last two sums of squares are frequently called the "explained" sums of squares, since they are due to factors 1 and 2 respectively.

The unexplained sum of squares is found by subtracting the two explained sums of squares from the total sum of squares.

To illustrate, the total SS:

$$\sum_j\sum_i (y_{ij} - \bar{\bar{y}})^2 = (120 - 111)^2 + (112 - 111)^2 + \ldots + (101 - 111)^2$$
$$= 81 + 1 + 196 + 400 + 25 + 121 + 1 + 1 + 4 + 196$$
$$+ 100 + 100 = 1{,}226.$$

The between-column SS is that explained by factor 1:

$$r\sum_j (\bar{y}_{.j} - \bar{\bar{y}})^2 = 4[(122 - 111)^2 + (109 - 111)^2 + (102 - 111)^2] = 824.$$

The SS explained by factor 2 is

$$k\sum_i (\bar{y}_{i.} - \bar{\bar{y}})^2 = 3[(115 - 111)^2 + (103 - 111)^2 + (112 - 111)^2$$
$$+ (114 - 111)^2] = 270.$$

The unexplained or error SS would be equal to the total SS less the two explained SS, or $1{,}226 - (824 + 270) = 132$.

Refer to Table 13.5, the Analysis of Variance Table, for the pre-

Table 13.5 Two-way ANOVA Table

Variation	Sum of Squares	df	Mean Square	F-Ratios
Explained by Factor 1 (Typists)	824	$k - 1 = 2$	$MSS_1 = 412$	$\dfrac{MSS_1}{MSS_e} = 19.62$
Explained by Factor 2 (Machines)	270	$r - 1 = 3$	$MSS_2 = 90$	$\dfrac{MSS_2}{MSS_e} = 4.29$
Unexplained or Error	132	6	$MSS_e = 21$	
Total	1,226	$rk - 1 = 11$		

sentation of the above results along with the subsequent F-ratios—the key to whether or not we will be able to reject either one or both of the hypotheses.

The degrees of freedom associated with factors 1 and 2 are $(k - 1) = 2$, and $(r - 1) = 3$, respectively. For the total SS, $df_1 = rk - 1 = 11$, and the difference, $(rk - 1) - [(k - 1) - (r - 1)] = rk - r - k + 1 = 6$, gives the df for the error sum of squares. Each of three mean sums of squares is an estimate of the variance of y under the assumption that both hypotheses are true—i.e., all typists are equally speedy and all machines are equally adapted for speed typing. If the mean sum of squares for factor 1 is unusually large with respect to the mean square error, then all typists are not equally fast typists; similarly, if the mean sum of squares for factor 2 is large relative to the mean square error, then all typewriters are not equally good. These hypotheses are tested, as we have seen before, with the F-ratio. The critical F-values from Table E are: For factor 1, $df_1 = 2$ and $df_2 = 6$, $F_{.05}^* = 5.14$ and $F_{.01}^* = 10.92$; for factor 2, $df_1 = 3$ and $df_2 = 6$, $F_{.05}^* = 4.76$, and $F_{.01}^* = 9.78$. With the test-F on factor 1 equal to 19.62, we reject the hypothesis related to the typists for either $\alpha = .05$ or $\alpha = .01$. At least one of the typists is significantly faster than at least one of the others.

With the test-F factor 2 of 4.29, we are unable to reject the null hypothesis regarding the functioning of the typewriters; the test-F is smaller than the critical F-values for either significance level. As far as we know, a given typist could use any of the machines with no effect on typing speed.

13.4 Statistical Inference on the Pairwise Differences

Referring to the wage rate example in Section 1 of this chapter, recall that, as a result of the F-test, we concluded that the average beginning wage rate for unskilled labor was higher in at least one of the areas than it was in at least one of the other areas. It is now quite clear that the sample Las Vegas rate of $6.40 per hour is significantly higher than that for Boise of $5.20 per hour. Is the Vegas rate of $6.40 significantly higher than $5.50 in the Bay Area, and is $5.50 significantly higher than Boise's $5.20?

Perhaps the most efficient way to answer these and similar questions is to place a confidence interval on $\mu_1 - \mu_2$ and also on $\mu_2 - \mu_3$. The confidence interval for $\mu_1 - \mu_2$ will be of the form, $(\bar{y}_1 - \bar{y}_2) \pm \epsilon$, or $\$.90 \pm \epsilon$ where

$$\epsilon = t\hat{\sigma}_{\bar{y}_1 - \bar{y}_2}.$$

The appropriate t-value would have $n_1 + n_2 - 2 = 5 + 5 - 2 = 8$

degrees of freedom and would reflect our choice of confidence level. Assuming a 95 percent confidence level, $t = 2.306$.

Also,

$$\hat{\sigma}_{\bar{y}_1 - \bar{y}_2} = s_p \sqrt{\frac{1}{n_1} + \frac{1}{n_2}} = \hat{\sigma}_w \sqrt{\frac{1}{5} + \frac{1}{5}} = \sqrt{.2596} \cdot \sqrt{.40}$$
$$= \sqrt{.1038} = .322.$$

The pooled variance, s_p^2, would normally be calculated from the two sets of data. In the ANOVA model we will usually have several sets of data, in which case the pooled variance, the within-group estimate of σ^2, would be legitimate to use and would be considered a better estimate than that obtained from only two sets of data.

The confidence interval for $\mu_1 - \mu_2$ would be

$\$.90 \pm (2.306)(.322) = \$.90 \pm \$.74$;

or

Conf: $\{\$.16 \leq \mu_1 - \mu_2 \leq \$1.64\} = .95.$

Inasmuch as this interval does not include 0, we can conclude that the Vegas wage rate is significantly higher than that in the Bay Area—at the .05 2-tailed level or .025 1-tailed significance level.

The confidence interval for $\mu_2 - \mu_3$ is of the same form:

$(\bar{y}_1 - \bar{y}_3) \pm \epsilon = \$.30 \pm \epsilon.$

The ϵ for this statement is identical to that calculated above. So we have

$\$.30 \pm \$.74,$

or

Conf: $\{-\$.44 \leq \mu_2 - \mu_3 \leq \$1.04\} = .95.$

This interval includes 0, so quite clearly we are unable to conclude that the Bay Area wage rate exceeds that in Boise.

13.5 Analysis of Variance on the Regression Problem

Analysis of variance as applied to the regression function will be discussed in terms of the BANGCO data and the related computer printouts in Section 8 of Chapter 10, "Multiple Linear Regression." An almost complete ANOVA table for the two-variable equation is included in Table 10.6. This table is completed and presented in Table 13.6.

Table 13.6 ANOVA Table for Regression

Variance Due to	df	SS	Mean SS	F-ratio
Regression	1	80,306.6	80,306.6	$383.9 = \dfrac{80,306.6}{209.2}$
Residual	28	5,857.3	209.2	
Total	29	86,163.9		

The equation is of the form, $\hat{y} = a + bx$, and, in our problem, y is daily natural gas sales as a function of mean daily temperature, x.

The fundamental ANOVA identity for regression is really no different from the general identity given earlier: total sum of squares = explained SS + unexplained SS, or

$$\sum_i (y_i - \bar{y})^2 = \sum_i (\hat{y}_i - \bar{y})^2 + \sum_i (y_i - \hat{y}_i)^2.$$

The total sum of squares is, as we expected, the sum of squares of all data points around the mean, and for the two-variable function, $df = n - 1$. In our problem, $n = 30$; so $df = 29$ for total SS, and this total SS equals 86,163.9.

The residual (error or unexplained) sum of squares, $\Sigma(y_i - \hat{y}_i)^2$, is 5857.3 with $df = 28, n - 2$, for the two-variable problem. Note that $\Sigma(y_i - \hat{y}_i)^2$ is the function we minimized according to the least squares criterion, and the mean value, 209.2, is the square of the standard error of regression, s_e^2.

The explained sum of squares or the sum of squares due to the regression, $\Sigma(\hat{y}_i - \bar{y})^2 = 80,306.6$. This sum of squares has only 1 df.

In the regression problem, both the residual and the regression mean SS would be estimates of σ_y^2 if there were no linear relationship between y and x. This is equivalent to assuming that the population slope-coefficient, B, is 0.

This is precisely the hypothesis being tested, $H_o: B = 0$ against $H_a: B \neq 0$, for the two-variable relationship.

The critical F-value with $df_1 = 1$ and $df_2 = 28$ and $\alpha = .05$, $F^* = 4.20$. The test-F, the F-ratio from the ANOVA table, is 383.9. We clearly reject H_o and conclude that the linear functional relationship between sales and temperature is highly significant. Note the value of the t-test on the slope-coefficient, b:

$$t = \frac{b - B}{s_b} = \frac{-8.357}{.4265} = -19.59.$$

The square of this value, $(-19.59)^2$, equals the F-ratio, 383.9, again, because of the 1 df in the numerator.

Table 13.7 ANOVA Table

Variance	df	SS	MSS	F-ratio
Regression	2	84,239.9	42,119.9	$591.1 = \dfrac{42,119.9}{71.26}$
Residual	27	1,924.0	71.26	
Total	29	86,163.9		

Analysis of variance is something of an overkill in the two-variable problem, but it is quite useful in the three-or-more-variable problem.

ANOVA will always be testing the hypothesis that *all* of the slope-coefficients are 0, $H_o: B_1 = B_2 = \ldots B_k = 0$, against the alternative hypothesis that at least 1 of the slope-coefficients is not 0. Accepting H_o would be equivalent to declaring that we have shown no significant relationship between y and any 1 or more of the explanatory x-variables.

Table 13.7 is the complete ANOVA table from the three-variable equation printout displayed in Table 10.9.

The total sum of squares and the associated df remain at 86,163.9, and 29 respectively as the third variable is introduced into the regression model. With an additional variable, the residual sum of squares will usually decrease from the previous step; the degrees of freedom also decrease by 1 for each additional explanatory variable; so the mean square error may or may not decrease. Degrees of freedom for the regression sum of squares increase by 1 for each additional variable.

For our specific example, we would be testing $H_o: B_1 = B_2 = 0$. The F-ratio from Table 13.7 is 591.1, which, as we would have expected from the 2-variable result, is far in excess of $F^*_{.05} = 3.35$ with $df_1 = 2$ and $df_2 = 27$. We reject the null hypothesis, but all we know is that at least 1 of the sample slope-coefficients is significantly different from 0. Referring back to the t-tests on each of the 2 slope-coefficients in Chapter 10, we see that, in this example, both values of b are significant; i.e., we have enough evidence to show that daily sales of natural gas are a function of both temperature and wind velocity.

Problems

1. Test the following hypotheses on σ^2 with the given sample size, variance, and significance level.
   a. $H_o: \sigma^2 = 225$
      $H_a: \sigma^2 \neq 225$, with $\alpha = .05$, $n = 15$, and $s^2 = 83$.

b. $H_o: \sigma^2 = .1960$
$H_a: \sigma^2 > .1960$, with $\alpha = .02$, $n = 10$, and $s^2 = .6400$.
c. $H_o: \sigma^2 = 16$
$H_a: \sigma^2 < 16$, with $\alpha = .05$, $n = 4$, and $s^2 = 3.7$.

2. Place a 95 percent confidence on σ^2 given:
   a. $n = 20$ and $s^2 = 34.2$;
   b. $n = 12$ and $s^2 = 1,025$;
   c. $n = 6$ and $s^2 = 120$.

3. The Bear Foot Manufacturing Company has set up quality control on the variance of the voltage (measured in square volts) as well as the mean voltage of its standard nine-volt cell. Its control on variance is set up to test the hypothesis, $H_o: \sigma^2 = .04$, against the alternative, $H_a: \sigma^2 > .04$. If H_o is accepted, the variability of the battery voltages is assumed to be in control; if H_o is rejected, then the manufacturing process is stopped and all procedures are carefully checked. A sample of 15 nine-volt cells yielded a sample s^2 of .25. Do you reject H_o, and what action, if any, do you recommend? What assumption was necessary regarding the distribution of x, the individual battery voltage?

4. The manager of a large supermarket recorded daily sales in the meat department for the past 12 days. We copied his record (all in dollars): 987, 842, 730, 1,040, 1,293, 1,865, 786, 784, 729, 993, 1,165, 1,924.
   a. Find the mean and variance of daily meat sales.
   b. Place a 95 percent confidence interval on mean sales.
   c. Place a 95 percent interval on the variance of sales; on the standard deviation of sales.
   d. To solve Parts b and c, was it necessary to assume that daily sales are normally distributed? If so (and without a formal test), does a cursory inspection of the data make you a bit apprehensive about the results in these two problems?

5. Random samples of size 18 and 22 senior students respectively from West High and East High (both are very large high schools) were given a standard mathematics aptitude test. For the West High seniors, s^2 was 384, and for East High seniors, s^2 was 84. Test the hypothesis $H_o: \sigma_1^2 = \sigma_2^2$ against $H_a: \sigma_1^2 \neq \sigma_2^2$. Use $\alpha = .05$. Interpret this result so that the 2 high school principals, both nonstatisticians, would understand the implications of the conclusion you make.

6. Following is a tabular presentation of family incomes in a random sample of six families in each of three cities. Prepare an ANOVA table and test the hypothesis ($\alpha = .025$) that mean family incomes are equal among the three cities.

	Family Income ($000)	
Bliss	Mount Angel	Saint George
13.2	14.2	7.8
19.0	12.3	29.3
17.1	9.2	18.9
25.5	22.7	19.7
16.9	10.5	21.4
12.1	13.3	14.5

7. Three salespersons work in the notions department of the local department store. Total sales for five working days last week are recorded in the following table:

Day of the Week	Salesperson		
	A	B	C
M	210	194	168
T	145	161	140
W	192	182	172
Th	183	198	169
F	240	265	201

Use $\alpha = .05$

a. Perform a one-way ANOVA on sales by the salespersons.
b. Perform a two-way ANOVA with salesperson and day of the week as the two factors.

8. Consumer Test, Inc., tested three different brands of motor bikes for mileage (all bikes had the same size motor). They tested three brand A bikes, four brand B bikes, and six brand C bikes, with the following results in miles per gallon of gasoline:

	Brand of Bike		
Sample No.	A	B	C
1	85	92	82
2	79	86	93
3	88	97	97
4		99	90
5			81
6			75

Test the hypothesis that the population means of miles per gallon are equal for all three brands. Use $\alpha = .05$.

9. A variant of the Sig-Young Performance Perception Test was given to 3 employees in the mini-vacuum-tube section of PDQ Electronics, Inc. Six different tasks were described in detail, and the employees were asked to estimate the completion time for each of the 6 tasks. The next day, the employees actively per-

formed each task and were timed from start to completion for each. The differences to the nearest 1/2 minute between estimated time and actual time for the 3 employees and the 6 tasks are recorded below.

Task	Employee		
	A	B	C
I	3.0	−1.0	1.5
II	7.5	2.5	5.5
III	4.0	3.0	4.0
IV	−1.5	.0	−2.0
V	5.5	−2.0	6.0
VI	4.5	−3.0	4.5

Use $\alpha = .05$.

a. Perform a one-way ANOVA on the employees' time differences and interpret your results.
b. Perform a two-way ANOVA and interpret your results.
c. If you were assured that variances on actual completion times for each of the six tasks were all equal, would this help in any way in satisfying the ANOVA normalcy assumption for this problem?

CHAPTER 14

NONPARAMETRIC STATISTICS

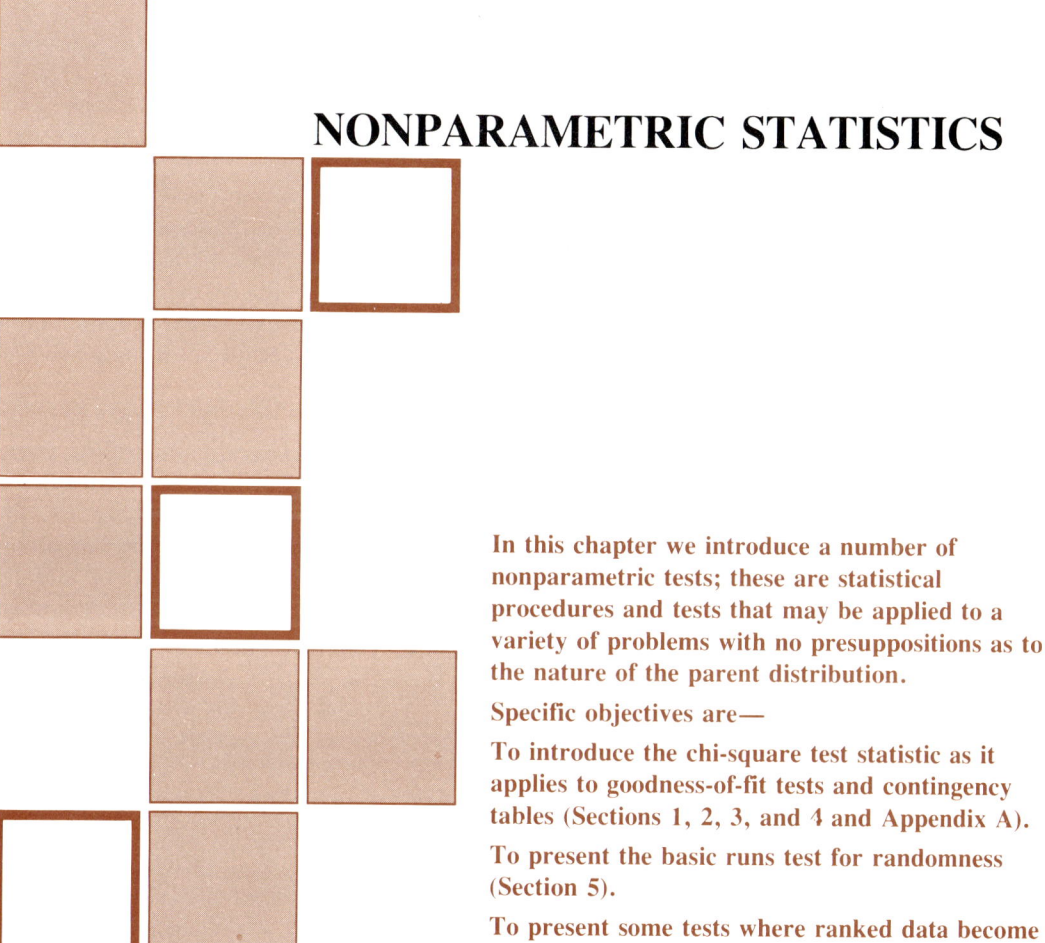

In this chapter we introduce a number of nonparametric tests; these are statistical procedures and tests that may be applied to a variety of problems with no presuppositions as to the nature of the parent distribution.

Specific objectives are—

To introduce the chi-square test statistic as it applies to goodness-of-fit tests and contingency tables (Sections 1, 2, 3, and 4 and Appendix A).

To present the basic runs test for randomness (Section 5).

To present some tests where ranked data become the consideration (Sections 6, 7, and 8).

The use of the t-statistic in Chapters 7, 8, and 9 and of the F-statistic in Chapters 10 and 13 requires the assumption that the populations sampled follow a normal distribution. The methods discussed in this chapter are traditionally referred to as nonparametric or distribution-free methods because there are no restrictions as to the nature of the population sampled. For this reason, the test statistics introduced in this chapter are quite different from those we have been studying. The chi-square (χ^2) statistic and its applications discussed in this chapter are not conceptually difficult and, because chi-square tests of this type are used frequently and in many disciplines, the use of this statistic is perhaps the most important of the topics covered here.

There are a great number of distribution-free methods and tests; we are presenting those that, in our judgment, are easily understood and have important applications in the fields of business and economics.

14.1 A General Chi-Square Test Statistic

Usually, the expression "perform a chi-square test" refers to a hypothesis test using a test statistic that has a chi-square or approximately chi-square distribution. There is a large variety of types of hypothesis-testing problems that can be handled by using a chi-square test statistic. A large group of these problems uses a test statistic that is essentially a direct application or special case of the chi-square test statistic given below.

The main characteristic of the group of hypothesis tests that use the chi-square test statistic given here is the classification of the population into three or more mutually exclusive categories. (If the population is classified into only two mutually exclusive categories, the hypothesis test for the population proportion discussed in Chapter 8 might be appropriate.) The description of the population of interest may directly classify the population into mutually exclusive categories. For example, the transistors being produced on an assembly line may be classified as grade A, grade B, or rejects. The population of interest is "all transistors being produced," and each of the transistors will fit into one and only one of the three categories—grade A, grade B, or reject. Often the initial description of the population doesn't directly specify categories, but categories can be defined so that a hypothesis test can be performed using the chi-square test. For example, the members of our population of interest may be described by their annual income. We may arbitrarily define a specific number of mutually exclusive and exhaustive income groups covering the incomes of all members of the population. Doing so in essence classifies each member as belonging to one and only one of the income groups or categories.

Information given in the statement of the null hypothesis fixes the respective proportions of the population belonging to each of the categories. For a given sample size, the expected number of sample observations in each category is calculated assuming the null hypothesis is true. The number of sample items actually observed in each category is then compared with the expected number of observations and, on the basis of this comparison, one decides whether the null hypothesis should be rejected.

The General Chi-Square Test Statistic

The general form of the chi-square test statistic for the goodness-of-fit tests and tests of independence is as follows:

$$\chi^2 = \sum_{i=1}^{k} \frac{(o_i - e_i)^2}{e_i} \text{ with } df = k - d - 1$$

where

k = the number of categories
e_i = the expected number of sample observations in the ith category
o_i = the actual number of sample observations observed in the ith category
d = the number of parameters that have to be estimated from the sample data.

The expression,

$$\sum_{i=1}^{k} \frac{(o_i - e_i)^2}{e_i},$$

is approximately distributed as a chi-square distribution with $k - d - 1$ degrees of freedom if n is large and if $e_i > 5$ for all i—i.e., the expected number must be greater than 5 for all k categories.

The chi-square distribution is a 1-parameter distribution. That parameter is traditionally referred to as the degrees of freedom. (You will recall that the parameter of the t-distribution is also referred to as the degrees of freedom.) The symbol χ^2 is used for the chi-square variable; note that the apparent exponent, 2, is not really an exponent (χ by itself is not used) but an integral part of the variable notation. As indicated in Figure 14.1, the chi-square distribution is skewed to the right, and the chi-square random variable never assumes negative values. The mean of the chi-square distribution is equal to df, the degrees of freedom, and the standard deviation of the chi-square distribution is equal to $\sqrt{2(df)}$. As the degrees of freedom increase, the chi-square distribution approaches a normal distribution. If $df > 30$ (note this is

Figure 14.1 The Chi-Square Distribution

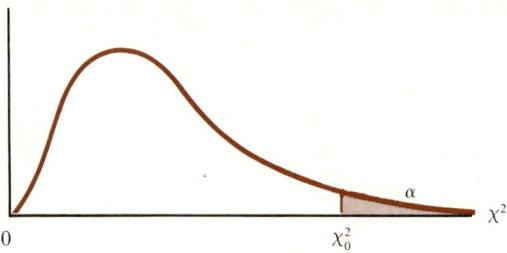

degrees of freedom, *not* sample size), the normal distribution provides a good approximation of the chi-square probabilities.

For hypothesis-testing purposes we will use the values of the chi-square distribution provided in Table D. In this table, the column headings are probabilities, and the degrees of freedom are given in the left-hand column with each row representing different degrees of freedom. The values in the body of the table are the values of χ^2 such that the $P(\chi^2 > \chi_o^2) = \alpha$ (see Figure 14.1); i.e., the probability, α, is in the upper tail. When the chi-square test statistic given above is used, the null hypothesis is rejected if the value of χ^2 calculated from the sample results is greater than χ_o^2. This is explained in more detail in the example worked below.

 ## 14.2 Goodness-of-Fit Test— The Multinomial Population

A multinomial population is one in which all elements are classified in one of three or more mutually exclusive categories. The assembly line transistor example referred to above illustrates this type of population; each transistor is classified as either grade A, grade B, or reject. If we let π_i be the proportion of all transistors that are in category i, then

$$\sum_{i=1}^{k} \pi_i = \sum_{i=1}^{3} \pi_i = \pi_1 + \pi_2 + \pi_3 = 1.$$

The sum of the proportions must equal 1. This follows from the requirements that each item in the population must be assigned to a category and the categories must be mutually exclusive—i.e., an item cannot be assigned to two or more categories. For hypothesis tests dealing with this type of population, the values of the population proportions are implicit in the hypothesis.

Example 14.2.1 If everything is functioning properly on the assembly line, then 50 percent of the transistors produced should be grade A, 40 percent should be grade B, and 10 percent should be rejects. The transistors in a random sample of size 100 have been classified as follows:

Grade A 39
Grade B 46
Rejects 15.

Test the hypothesis that the assembly line is operating properly. Use a significance level of .10.

The null hypothesis for this problem is that the assembly line is operating properly; this is expressed in terms of the proportions that should prevail for each of the respective categories if the assembly line is operating properly. The complete solution to this problem is given below and then discussed in detail.

1. $H_o: \pi_1 = .5, \pi_2 = .4, \pi_3 = .1$.
   $H_a: H_o$ is not true.
2. $\alpha = .10$.

$$\chi^2 = \sum_{i=1}^{k} \frac{(o_i - e_i)^2}{e_i}$$

with $df = k - 1 = 3 - 1 = 2$.

3. Reject H_o if and only if $\chi_c^2 > 4.605$.
4.

π_i	$e_i = n\pi_i$	o_i
.5	50	39
.4	40	46
.1	10	15
	100	100

$$\chi_c^2 = \frac{(39-50)^2}{50} + \frac{(46-40)^2}{40} + \frac{(15-10)^2}{10} = 5.82.$$

Reject H_o.

The assembly line is not operating properly.

The alternate hypothesis is simply stated as "H_o is not true" because there is more than one way in which H_o might be false—i.e., all the proportions could be wrong, π_1 and π_2 could be wrong, π_1 and π_3 could be wrong, or π_2 and π_3 could be wrong. Clearly, for a larger number of categories, there is a larger number of ways in which the null hypothesis can be wrong.

The degrees of freedom for this problem are $k - 1$ because it was not necessary to calculate any parameters using the sample data in order to obtain the e_i's (the expected number of observations for each category). The expected values were calculated using the sample size and the hypothesized proportions stated in the null hypothesis. (The degrees of freedom for the general chi-square test statistic are $k - d - 1$ where d is the number of parameters estimated from the sample data. For this case, $d = 0$.)

The fact that the decision rule should be to reject only for large calculated values of the test statistic can be ascertained by observing what happens to the test statistic when the null hypothesis is false. The expected values, e_i's, are calculated assuming the null hypothesis is true. If the null hypothesis is true, then we expect the observed number in each category, o_i, to be close to the expected number, e_i. If o_i is close to e_i, then the numerator of each term of the test statistic, $(o_i - e_i)^2$, will be relatively small, as will the calculated value of χ^2; but if the null hypothesis is false, then we expect the o_i's to be quite different from the e_i's, and $(o_i - e_i)^2$ will be a relatively large positive number (positive because the difference is squared). Thus, the calculated value of the chi-square test statistic tends to get large when the null hypothesis is false.

The degrees of freedom for this hypothesis test are $k - 1 = 3 - 1 = 2$, inasmuch as there are three categories into which the population is classified. We want to reject for only large values of the chi-square statistic; therefore, the entire value of the significance level, $\alpha = .10$, will be in the upper tail of the distribution. The chi-square table, Table D, is set up with the probability in the upper tail; so the critical value for the decision rule is read under the column headed ".10" and along the row for 2 degrees of freedom. This value is 4.605, as indicated in the decision rule above. See Figure 14.2.

The expression for finding e_i given as a column heading in Step 4 is a general formula for finding the expected sample number in each category. The formula has intuitive appeal and is conceptually the same as for the binomial expression. If we know that 10 percent of the

Figure 14.2 Determining the Value of χ^2 for $df = 2$ and $\alpha = .10$

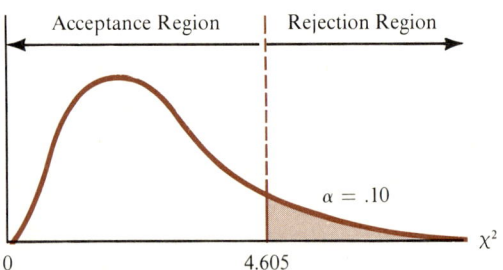

apples in an orchard are bad and we randomly select 100 apples, then we expect, on the average, 10 percent of the 100 we selected to be bad. This logic is reflected in the formula for the expected number: $e_i = n\pi_i$ where n is the sample size and π_i is the hypothesized proportion of the population classified as belonging to the ith category.

A special case of this type of problem is one in which it is hypothesized that the proportions are equal for all the categories. As is readily apparent from the above illustration and the following example, the expected frequencies in each category are all equal.

Example 14.2.2 A book publisher would like to know if customer preference is the same for 4 different book cover designs. The title, author, and price are the same for each of the designs. A random sample of 113 customers was observed and the following results were obtained.

Design	Number of Customers Selecting the Design
1	30
2	36
3	25
4	22

Perform the chi-square goodness-of-fit test with a significance level of .05.

Recall that the respective category proportions must sum to 1. There are 4 categories; so the hypothesized proportion for each category is 1/4. The degrees of freedom for this problem are $k - 1 = 4 - 1 = 3$. The solution to this problem is given below.

1. $H_o: \pi_1 = \pi_2 = \pi_3 = \pi_4 = 1/4$.
   $H_a: H_o$ is not true.

2. $\alpha = .05$

$$\chi^2 = \sum_{i=1}^{k} \frac{(o_i - e_i)^2}{e_i}$$

   with $df = k - 1 = 3$.

3. Reject H_o if and only if $\chi_c^2 > 7.815$.

4.

π_i	$e_i = n\pi_i$	o_i
.25	28.25	30
.25	28.25	36
.25	28.25	25
.25	28.25	22
	113.00	113

Goodness-of-Fit Test—The Multinomial Population

$$\chi_c^2 = \frac{(30-28.25)^2}{28.25} + \frac{(36-28.25)^2}{28.25} + \frac{(25-28.25)^2}{28.25} + \frac{(22-28.25)^2}{28.25}$$

$$= 3.99.$$

Do *not* reject H_o.

The publisher is unable to reject H_o; as far as marketing the book is concerned, there is no indication that any specific book design is any better or any worse than the others.

14.3 Contingency Table Tests— Tests of Independence

The chi-square statistic can be used to test hypotheses concerning the independence of two-way classifications within a population. For example, it might be of interest to know whether a person's feeling regarding the president's economic policy is independent of his or her type of employment; or, if preference for different types of television programs is independent of the educational level of the viewer. When this type of hypothesis test is performed, the two-way display of the sample data is traditionally referred to as a contingency table, and this type of test of independence is frequently identified as a contingency table test.

Example 14.3.1

We are interested in determining whether voters' attitudes regarding tax proposition number 2 are independent of the income group to which they belong. We have arbitrarily defined 4 mutually exclusive and exhaustive income intervals and 3 attitudinal levels. Every voter will fall into 1 of the 12 categories displayed in Table 14.1. The results from a random sample of 320 voters are reported in the table. Perform a chi-square test of independence at a significance level of .05.

Table 14.1 Sample Data for Example 14.3.1

Attitude	Income Groups				Totals	
	1	2	3	4		
For	33	38	35	25	131	r_1
Against	37	36	20	15	108	r_2
Indifferent	32	14	21	14	81	r_3
Totals	102	88	76	54	320	
	c_1	c_2	c_3	c_4		

Nonparametric Statistics

The data in Table 14.1 were obtained by determining how a voter felt about tax proposition number 2 and to which income group the voter belonged. For example, in the sample of 320 voters there were 20 voters belonging to the third income group who were against tax proposition number 2. The cross-classification of voters in the population, and thus in the sample, results in 12 mutually exclusive categories into which each member of the population can be classified. The subscripted r's and c's in Table 14.1 are strictly notational devices that will prove to be useful when the calculation of the expected number of sample observations is explained below.

The null hypothesis for a contingency table test is that the 2 broad classifications are independent. For this problem we are interested in determining whether a voter's opinion about tax proposition number 2 is independent of the income group to which he or she belongs. The null hypothesis can be formally stated as follows:

H_o: A voter's attitude towards tax proposition number 2 is independent of his or her income.

This hypothesis does not specify the proportions for the 12 cross-classification categories; however, the assumption of independence is used in calculating the category proportions.

You will recall from Chapter 3 the theorem that, if two events—A and B—are independent, then the probability that both events will occur simultaneously is equal to the product of the probabilities of the independent events: $P(AB) = P(A)P(B)$. If the null hypothesis of independence is assumed to be true, then the proportion of members of the population who fall in the first category—for tax proposition number 2 and in income group number 1—is equal to the proportion of the population for tax proposition number 2 *times* the proportion of the population in income group number 1. We do not know what these population proportions are, but they can be estimated using the sample data from Table 14.1.

The total of the observations in the first row of the table is 131. This is the number of people in the sample of 320 people who were for tax proposition number 2. The sample proportion of voters for tax proposition number 2, 131/320, is our estimate of the proportion of voters in the population who are for tax proposition number 2. If we let the subscripted r's (see Table 14.1) be the respective row totals, then we can write,

The estimated proportion of voters in the population for tax proposition number 2 = r_1/n.

The letter n, as usual, is defined as the total number of sample observations.

From the data in Table 14.1 we observe that 102 of the 320 voters sampled are in income group number 1. If we let the subscripted c's be the respective column totals, then we can write,

The estimated proportion of voters in the population belonging to income group number $1 = c_1/n$.

If the null hypothesis of independence between voters' opinions on the issue and their income group is assumed to be true, then the proportion of voters in income group number 1 for tax proposition number 2 is equal to the product of the 2 probabilities estimated above:

The estimated proportion of voters in the population for tax proposition number 2 belonging to income group number 1

$$= \left(\frac{r_1}{n}\right)\left(\frac{c_1}{n}\right) = \left(\frac{131}{320}\right)\left(\frac{102}{320}\right) = .1305.$$

Multiplying this proportion by the total number of voters in the sample results in the expected number of sample observations in the category defined by the intersection of the first row and first column in Table 14.1 *if* the null hypothesis is true. Note that, while r_1/n and c_1/n are estimated proportions, the proportion for the first category is calculated from these 2 estimates assuming independence. The expected number of observations in the first category is $(.1305)(320) = 41.76$.

Let e_{ab} be the expected number in the category represented by the intersection of row a and column b in the contingency table in Table 14.1. Then we can write as a general formula for finding the expected number of sample observations for each of the categories in a contingency table as follows:

$$e_{ab} = \left(\frac{r_a}{n}\right)\left(\frac{c_b}{n}\right)n = \frac{(r_a)(c_b)}{n}.$$

The expected number of observations in the category represented by the intersection of the second row and third column can be calculated using this formula:

$$e_{23} = \frac{(r_2)(c_3)}{n} = \frac{(108)(76)}{320} = 25.65.$$

Table 14.2, part a, has the calculations for each of the 12 categories of the contingency table from Table 14.1, and part b has the end results of the calculations. Part b is a table of expected number of sample observations for each of the 12 categories when the null hypothesis is assumed to be true.

The calculated value of the chi-square test statistic for this problem is obtained by comparing the observed number for the 12 categories given in Table 14.1 with the expected number for the same categories given in Table 14.2, part b:

Table 14.2 a. The calculation of the expected number in each category.

Attitude	Income Group			
	1	2	3	4
For	$\dfrac{(131)(102)}{320}$	$\dfrac{(131)(88)}{320}$	$\dfrac{(131)(76)}{320}$	$\dfrac{(131)(54)}{320}$
Against	$\dfrac{(108)(102)}{320}$	$\dfrac{(108)(88)}{320}$	$\dfrac{(108)(76)}{320}$	$\dfrac{(108)(54)}{320}$
Indifferent	$\dfrac{(81)(102)}{320}$	$\dfrac{(81)(88)}{320}$	$\dfrac{(81)(76)}{320}$	$\dfrac{(81)(54)}{320}$

b. The expected number of sample observations in each cell assuming the null hypothesis is true.

Attitude	Income Group			
	1	2	3	4
For	41.76	36.03	31.11	22.11
Against	34.43	29.70	25.65	18.23
Indifferent	25.82	22.28	19.24	13.67

$$\chi_c^2 = \sum_{i=1}^{12} \frac{(o_i - e_i)^2}{e_i} = \frac{(33 - 41.76)^2}{41.76} + \frac{(38 - 36.03)^2}{36.03}$$

$$= \frac{(35 - 31.11)^2}{31.11} + \frac{(25 - 22.11)^2}{22.11} + \frac{(37 - 34.43)^2}{34.43}$$

$$+ \frac{(36 - 29.70)^2}{29.70} + \frac{(20 - 25.65)^2}{25.65} + \frac{(15 - 18.23)^2}{18.23}$$

$$+ \frac{(32 - 25.82)^2}{25.82} + \frac{(14 - 22.28)^2}{22.28} + \frac{(21 - 19.24)^2}{19.24}$$

$$+ \frac{(14 - 13.67)^2}{13.67} = 10.88.$$

The degrees of freedom for the test statistic when a contingency table test is being performed is equal to $(r - 1)(c - 1)$ where r is the number of rows and c the number of columns in the contingency table. Thus, for this example, the degrees of freedom are $(3 - 1)(4 - 1) = 6$. The derivation of this useful $(r - 1)(c - 1)$ rule for degrees of freedom from the general rule given for the general chi-square test statistic is discussed below. The complete solution to this contingency table hypothesis test problem is as follows:

1. H_o: Voter preference for tax proposition number 2 is independent of the income group to which the voter belongs.
   H_a: H_o is not true.

2. $\alpha = .05$

   $$\chi^2 = \sum_{i=1}^{k} \frac{(o_i - e_1)^2}{e_i}$$

   with $df = (r - 1)(c - 1) = 6$.

3. Reject H_o if and only if $\chi_c^2 > 12.592$.

4. $\chi_c^2 = \dfrac{(33 - 41.76)^2}{41.76} + \dfrac{(38 - 36.03)^2}{36.03} + \cdots$

   $= \dfrac{(21 - 19.24)^2}{19.24} + \dfrac{(14 - 13.67)^2}{13.67} = 10.88.$

   Do *not* reject H_o.

On the basis of these sample results we are unable to conclude that voter attitude regarding the tax proposition is dependent on income. The critical value of the chi-square statistic used for the decision rule in Step 3 was found from Table D under the column headed .05 along the row for 6 degrees of freedom.

The degrees of freedom for the general chi-square test statistic in Section 14.1 was given as $k - d - 1$ where k is the number of classes and d the number of parameters that need to be estimated in order to calculate the e_i's. For the above problem there were 12 categories. The proportions of the voter population belonging to each of the 4 income groups had to be estimated from the sample data for 3 of the groups. The proportion for the fourth group can be calculated from the other 3 because the sum of the proportions must sum up to 1. Similarly, it was necessary to estimate 2 of the 3 proportions for the breakdown of the population into the 3 groups, For, Against, and Indifferent. Thus, we had to estimate $3 + 2 = 5$ proportions or parameters from the sample data. Once the row and column proportions were estimated, then the intersection proportions could be calculated, not estimated, and the expected number of observations for each category could then be determined. The degrees of freedom for this problem can be calculated from the general formula as $k - d - 1 = 12 - 5 - 1 = 6$. This is the same value as was obtained from the formula $(r - 1)(c - 1)$.

In general, there are rc categories for any contingency problem. The number of estimated parameters from the sample data is $(r - 1) + (c - 1)$. Substituting these amounts in the general rule results in $k - d - 1 = rc - [(r - 1) + (c - 1)] - 1 = rc - r - d + 1$. The reader can verify that $(r - 1)(c - 1) = rc - r - d + 1$.

14.4 Goodness-of-Fit Test— The Normal Distribution

The general chi-square test statistic discussed in Section 13.1 can be used to test hypotheses concerning the distribution of a population from which sample data have been randomly selected. For a population with a continuous distribution, it is necessary to arbitrarily define a finite number of mutually exclusive categories into which the population values can be classified. This is accomplished by defining intervals as categories.

Example 14.4.1   Test the hypothesis that the random sample data in Table 14.3 were drawn from a normally distributed population with mean, μ, equal to 50 and standard deviation, σ, equal to 10. Use $\alpha = .05$ for the significance level.

The chi-square statistic can be used as the test statistic for this problem if we can classify the population into mutually exclusive categories. This is accomplished, as indicated in Figure 14.3, by arbitrarily defining intervals of 5 units length symmetrical about the hypothesized mean. There are 2 open-ended intervals and 8 closed intervals. This effectively classifies the population into the 10 categories indicated by the first column in Table 14.4. The plus sign is used at the beginning of the intervals to assure that they are mutually exclusive; i.e., 40.0 falls in the 35+ to 40 interval but not in the 40+ to 45 interval.

In the second column of Table 14.4, the sample observations in Table 14.3 are tallied into each of the respective intervals. For example, 4 of the 80 sample values are less than 30: 26.7, 23.2, 27.1, and 29.8.

The third column of Table 14.4 has the respective proportions for each of the intervals. This is the proportion of the items in the popula-

Table 14.3   Random Sample of 80 Values

40.3	35.1	44.8	30.9	53.8	37.5	43.5	36.2
56.8	62.7	23.2	64.6	41.8	63.9	49.1	51.2
70.6	46.3	56.1	36.0	52.5	36.9	32.3	59.4
55.4	42.3	47.4	53.9	39.3	50.4	59.5	44.1
47.8	62.3	38.6	46.1	61.8	29.8	45.2	70.1
33.7	51.4	49.2	41.2	41.7	61.4	58.1	48.2
44.3	35.6	54.7	55.2	39.8	49.8	44.2	34.0
26.7	74.7	58.9	31.2	46.8	60.3	37.1	66.2
50.3	57.3	60.2	57.6	68.9	52.3	64.7	51.6
38.7	42.5	27.1	40.9	43.4	36.8	48.7	45.6

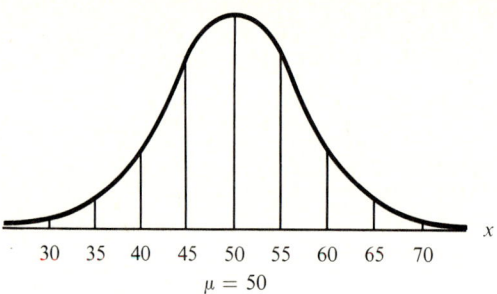

Figure 14.3 Hypothesized Normal Distribution for Example 14.4.1

tion that will belong to that interval if the null hypothesis is true—i.e., if the population has a normal distribution with $\mu = 50$ and $\sigma = 10$. These proportions are calculated using the method for finding probabilities of a normal distribution discussed in Chapter 5. The proportion for the fourth interval, π_4, is found as follows:

$$\pi_4 = P(40 < x < 45) = P\left(\frac{40 - 50}{10} < z < \frac{45 - 50}{10}\right)$$

$$= P(-1.0 < z < -.5) = .3413 - .1915 = .1498.$$

You will recall that this is the area under the curve over the respective intervals in Figure 14.3. Once the proportions consistent with the null hypothesis have been calculated, then the expected number of sample items in each interval is calculated as in the previous problems by

Table 14.4 Calculation of the e_i's for Example 14.4.1

Interval	o_i	π_i	$e_i = n\pi_i$
30 or less	4	.0228	1.824
30+ to 35	5	.0440	3.520
35+ to 40	12	.0919	7.352
40+ to 45	13	.1498	11.984
45+ to 50	12	.1915	15.320
50+ to 55	10	.1915	15.320
55+ to 60	10	.1498	11.984
60+ to 65	9	.0919	7.352
65+ to 70	2	.0440	3.520
more than 70	3	.0228	1.824
	80	1.0000	80.000

multiplying these proportions by the sample size. The e_i's are given in the fourth column of Table 14.4.

With the calculations in Table 14.4 completed, we would be ready to finish with the hypothesis testing except for one problem. You will recall that one of the conditions for using the test statistic in Section 14.1 was that all of the e_i's should be greater than 5. This is not the case for the first, second, ninth, and tenth categories in Table 14.4. The first and tenth categories have expected numbers of 1.824, and the second and ninth categories have the expected values of 3.520. Before we can continue with the hypothesis test, it will be necessary to redefine the categories so that the expected number, e_i, is greater than 5 for all categories. We can accomplish this by combining the first 2 categories into a single category and similarly combining the ninth and tenth categories. We now have 8 mutually exclusive categories with $e_i > 5$ for all categories (see Table 14.5). Note also that the observations of the 2 categories are added together for the combined category.

With the data in Table 14.5, we are now ready to perform the hypothesis test.

1. H_o: The sample data come from a normal population with $\mu = 50$ and $\sigma = 10$.
   H_a: H_o is not true.

2. $\alpha = .05$

   $$\chi^2 = \sum_{i=1}^{k} \frac{(o_i - e_i)^2}{e_i}$$

   with $df = k - 1 = 7$.

3. Reject H_o if and only if $\chi_c^2 > 14.067$.

Table 14.5 Data from Table 14.4 with Categories Redefined to Assure $e_i > 5$ for All Categories

Interval	o_i	π_i	$e_i = n\pi_i$
35 or less	9	.0668	5.344
35+ to 40	12	.0919	7.352
40+ to 45	13	.1498	11.984
45+ to 50	12	.1915	15.320
50+ to 55	10	.1915	15.320
55+ to 60	10	.1498	11.984
60+ to 65	9	.0919	7.352
more than 65	5	.0668	5.344
	80	1.0000	80.000

4. $\chi_c^2 = \dfrac{(9-5.344)^2}{5.344} + \dfrac{(12-7.352)^2}{7.352} + \dfrac{(13-11.984)^2}{11.984}$

$+ \dfrac{(12-15.320)^2}{15.320} + \dfrac{(10-15.320)^2}{15.320} + \dfrac{(10-11.984)^2}{11.984}$

$+ \dfrac{(9-7.352)^2}{7.352} + \dfrac{(5-5.344)^2}{5.344} = 8.81.$

Do *not* reject H_o.

The sample data do not indicate that we should reject the hypothesis that the population has a normal distribution. The problem with goodness-of-fit tests is that we demonstrate our conjecture by not rejecting the null hypothesis, and it is difficult to calculate the probability of a Type II error for this type of test.

The degrees of freedom for this problem are $k - 1 = 8 - 1 = 7$ since we have 8 categories and the e_i's could be calculated without using the sample data to estimate parameters. We were able to do this because values of the population mean and standard deviation were included in the null hypothesis. Had the null hypothesis been simply that the population was normal, then $\bar{x}$ and s would have been calculated from the sample data. These sample values would then have been used as estimates of μ and σ in order that the π_i's could be calculated. This would make the degrees of freedom equal to $k - d - 1 = 8 - 2 - 1 = 5$ because 2 parameters, μ and σ, had to be estimated from the sample data before the e_i's could be calculated.

14.5 Runs Tests for Randomness

Sample data are frequently taken from a continuous population such as that from a production line. For example, the sample may be the next 75 items produced by a machine. It may be necessary to test a hypothesis regarding the average breaking strength of the parts produced by this machine. If there is some concern about the randomness of the sample, then 1 of the 2 runs tests discussed in this section can be used to test for randomness. To use this type of test, it is necessary to record the sample data according to the order in which they were taken. If there are patterns in the ordered array of sample values, then the randomness of the sample data might be questioned.

Another application of the runs tests for randomness has to do with so-called random number generators. Simulation or Monte Carlo techniques are used more and more in business applications. The use of these techniques depends very much upon the ability of the user to obtain a set of random numbers. Random numbers may be obtained from an external source, or they may be generated as part of a computer program designed to solve business problems using simulation methods. In either case, it may be of interest to know whether or not the set of numbers used as random numbers is really random.

The first runs test discussed in this section is concerned with the sign of the difference between consecutive sample values. This test is sometimes referred to as the sign test for randomness.

Example 14.5.1 The following set of 2-digit numbers is a column of numbers from a table that purports to be a set of random digits. Test the hypothesis that this sample of numbers, selected by going down a column of this table, is a random sample of 2-digit numbers. (Use $\alpha = .05$.)

84, 69, 41, 03, 54, 48, 20, 29, 58, 85, 96, 58, 45, 62, 37, 37, 48, 66, 33, 72, 84, 98, 86, 63, 60, 72, 35, 15, 36, 30.

The first step in using this type of runs test is to replace this set of data with a string of pluses and minuses determined by comparing consecutive sample values. This is done by assigning a plus or minus sign to each successive pair of numbers according to whether or not the latter observation in each pair is greater than or less than the first observation in the pair. For example, the first pair of numbers in this sample is 84 and 69. The second number, 69, is less than 84; so we replace this pair with a minus sign. The second pair of numbers in the sequence is 69 and 41. For this pair, the second number is less than the first; so this pair is also replaced by a minus sign. Continuing with this procedure results in the following string of plus and minus signs. (The sign for ties is assigned randomly. For this example the plus sign for the pair 37 and 37 was determined by flipping a coin.)

− − − + − − + + + + − − + − + +
+ − + + + − − − + − − + −

The next step is to count the number of runs in the resulting string of plus and minus signs. A run is defined as a string of pluses or minuses uninterrupted by the other sign. Thus, the first 3 minus signs are a run; the first plus sign is a run; the 2 minus signs following the plus sign constitute a run, etc. There are 15 runs in this string of plus and minus signs. Let R be the number of runs. If we assume the null hypothesis is true—i.e., assume the sample is random—then the expected number of runs, μ_R, and the standard deviation of the number of runs, σ_R, can be calculated from the following formulas:

$$\mu_R = \frac{2n - 1}{3},$$

which is approximately $\frac{2}{3}$ the sample size, n, and

$$\sigma_R = \sqrt{\frac{16n - 29}{90}}.$$

If the null hypothesis is true, then the distribution of the number of runs, R, is approximately normal for large sample sizes.

Test Statistic for the Sign Test

For $n \geq 20$ the z-statistic can be used as a test statistic:

$$z = \frac{R - \mu_R}{\sigma_R} = \frac{R - \frac{2n-1}{3}}{\sqrt{\frac{16n-29}{90}}}.$$

Note that this is consistent with the z-statistic used in applications covered in previous chapters; i.e., the right-hand side is equal to the sample statistic, in this case R, minus its expected value divided by its standard deviation.

If there are too many runs, this would suggest an artificial pattern in the data—a tendency for large numbers to be followed by small numbers and vice versa—inconsistent with the null hypothesis. On the other hand, if there are too few runs, this would suggest a possible trend and/or long cycle in the data, which would again be inconsistent with the null hypothesis of randomness. Thus, the decision rule is to reject for both large and small calculated values of z. The complete solution to this problem is given below.

1. H_o: The sample data are random.
   H_a: H_o is not true.

2. $\alpha = .05$

$$z = \frac{R - \frac{2n-1}{3}}{\sqrt{\frac{16n-29}{90}}}.$$

3. Reject H_o if and only if $z_c > 1.96$ or $z_c < -1.96$.

4. $z_c = \dfrac{15 - \frac{2(30)-1}{3}}{\sqrt{\frac{16(30)-29}{90}}} = -2.08.$

Reject H_o.

We conclude that the sequence of numbers is not a random sequence.

The second type of runs test considers runs of the sample values above and below the median instead of runs in the signs of consecutive differences. The following example illustrates this test.

Example 14.5.2 A sample of 28 light bulbs has been selected from an assembly line at 28 different times, equally spaced throughout the day. You are concerned that the work habits of the workers may vary during the day and thus affect the randomness of the sample. Test the hypothesis that the sample is random. Use a significance level of .10. The length of life of the light bulbs in the sample is given below in the order in which the sample items were drawn.

866, 958, 891, 923, 841, 862, 916, 905, 950, 878, 853, 849, 828, 883, 917, 839, 821, 941, 768, 832, 796, 869, 857, 759, 819, 803, 788, 780.

The median for this set of numbers is 855. If we replace all numbers above the median with the letter A and all numbers below the median with the letter B we get the following sequence:

A A A B A A A A A B B B A A B B A B B B A A B B B B B.

If the sample size is odd, then 1 of the sample values will be equal to the median. In this case, the median can be deleted (reducing the sample size by 1) or else A or B can be randomly assigned to the median. The number of runs is determined the same way as it is for the sign test example. For this string of letters there are 10 runs.

Test Statistic for Runs Above and Below Median Test

If the null hypothesis is true and $n \geq 25$, then the z-statistic can be used in this runs test with

$$\mu_R = \frac{n+2}{2}$$

and

$$\sigma_R = \sqrt{\frac{n(n-2)}{4(n-1)}}.$$

$$z = \frac{R - \mu_R}{\sigma_R} = \frac{R - \frac{n+2}{2}}{\sqrt{\frac{n(n-2)}{4(n-1)}}}$$

where

R = the number of runs above and below the median
n = sample size.

The argument for a 2-tail test is the same for this test as that for the sign test: too many runs or too few runs would suggest that the sample is not random. The complete solution to this problem is given as follows:

Runs Tests for Randomness

1. H_o: The sample is a random sample.
   H_a: H_o is not true.
2. $\alpha = .10$

$$z = \frac{R - \frac{n+2}{2}}{\sqrt{\frac{n(n-2)}{4(n-2)}}}.$$

3. Reject H_o if and only if $z_c < -1.65$ or $z_c > 1.65$.

4. $z_c = \dfrac{10 - \frac{28+2}{2}}{\sqrt{\frac{28(28-2)}{4(28-1)}}} = -1.93$.

Reject H_o.

We conclude that the sample is not random; selecting a light bulb as it comes off the assembly line at equally spaced intervals during the day does not result in a random sample. The number of runs is on the low side, suggesting a possible cycle in operation.

14.6 The Wilcoxon Test

The Wilcoxon Test is used to test the hypothesis that two populations are identically distributed.

Example 14.6.1

A bank wants to determine if there is a difference in the time it takes to serve customers at the city branches and the suburban branches. The results from independent random samples of the times, in minutes, required to serve customers at city banks and at suburban banks are given below.

City Customers		Suburban Customers	
3.6	3.5	3.2	8.2
5.8	7.1	4.5	5.4
2.1	4.6	2.8	3.1
1.2	8.4	5.6	7.2
4.7		1.9	6.8
		7.7	2.2

On the basis of these sample data, can we conclude there is a difference in service times for the two types of customers? (Use $\alpha = .02$.) Service times tend to follow an exponential distribution, which means the normality assumption would not be valid for this problem.

Nonparametric Statistics

At first glance this problem appears to be one that we could solve by using the t-statistic for testing the equality of 2 population means as was done in Chapter 9. The t-statistic would be an appropriate statistic except for the fact that the populations from which the samples were taken are not normally distributed. If both samples were large (greater than 30), then the z-statistic for testing the equality of population means could be employed. The Wilcoxon Test is a nonparametric test and thus is not dependent upon the distribution of the populations being sampled. It can be used when both sample sizes are greater than 8 and the samples are independent.

The Wilcoxon Test is also called the *rank sum test* because the sample statistic used in the test is obtained by summing ranks. The sample values of the 2 samples are combined and ranked from the lowest to the highest (see Table 14.6). The sample statistic of interest is the sum of the ranks for 1 of the samples. The sum of the ranks for the 21 combined sample values is equal to

$$21 \left(\frac{21 + 1}{2} \right) = 231.$$

Table 14.6   Combined Sample

C = City S = Suburban	Ranks	City Ranks	Suburban Ranks
1.2 C	1	1	
1.9 S	2		2
2.1 C	3	3	
2.2 S	4		4
2.8 S	5		5
3.1 S	6		6
3.2 S	7		7
3.5 C	8	8	
3.6 C	9	9	
4.5 S	10		10
4.6 C	11	11	
4.7 C	12	12	
5.4 S	13		13
5.6 S	14		14
5.8 C	15	15	
6.8 S	16		16
7.1 C	17	17	
7.2 S	18		18
7.7 S	19		19
8.2 S	20		20
8.4 C	21	21	
	231	97	134

The sum of the ranks is simply equal to the sum of the integers from 1 to 21. (You may recall from your algebra course that the sum of the first n positive integers is equal to

$$\frac{n(n+1)}{2}.)$$

If we let n_1 and n_2 equal the respective sample sizes, then substituting $n_1 + n_2$ for n in the latter formula gives us

$$\frac{(n_1 + n_2)(n_1 + n_2 + 1)}{2},$$

a formula for the sum of the ranks expressed as a function of n_1 and n_2.

If the two populations are the same, then we would expect the sum of the ranks in the first sample to be proportional to the size of the first sample. Multiplying the above expression for the sum of the ranks by the proportion of the first sample size to the size of the combined sample,

$$\frac{n_1}{n_1 + n_2},$$

gives us the expected sum of the ranks for the first sample:

$$E(R_1) = \mu_{R_1} = \frac{n_1(n_1 + n_2 + 1)}{2}$$

where

$R_1 = $ the sum of the ranks for values in the first sample
$n_1 = $ sample size of first sample
$n_2 = $ sample size of second sample.

A similar, but more complex, analysis yields the following formula for the standard deviation of R_1:

$$\sigma_{R_1} = \sqrt{\frac{n_1 n_2 (n_1 + n_2 + 1)}{12}}.$$

Test Statistic for the Wilcoxon Test

If both sample sizes are greater than 8, then the normal distribution is a good approximation of the distribution of the sum of the ranks. This observation provides us with a test statistic:

$$z = \frac{R_1 - \mu_{R_1}}{\sigma_{R_1}} = \frac{R_1 - \dfrac{n_1(n_1 + n_2 + 1)}{2}}{\sqrt{\dfrac{n_1 n_2 (n_1 + n_2 + 1)}{12}}}.$$

If the sum of the ranks for the first population tends to be either much smaller or much larger than the expected sum of the ranks, we would conclude that the population distributions are different; thus, we reject the null hypothesis for both large positive and large negative values of z. The complete solution to the hypothesis-testing problem is given below. Note that we are letting the sample of times from the city branches be the first sample; i.e., from Table 14.6 we have $R_1 = 97$. Also, n_1 is equal to 9, and n_2 is 12.

1. H_o: The distribution of service time per city branch customer is the same as the distribution of service time per suburban branch customer.
   H_a: H_o is not true.

2. $\alpha = .02$

$$z = \frac{R_1 - \frac{n_1(n_1 + n_2 + 1)}{2}}{\sqrt{\frac{n_1 n_2(n_1 + n_2 + 1)}{12}}}.$$

3. Reject H_o if and only if $z_c > 2.33$ or $z_c < -2.33$.

4. $z_c = \dfrac{97 - \frac{9(9 + 12 + 1)}{2}}{\sqrt{\frac{9(12)(9 + 12 + 1)}{12}}} = -.142.$

Do *not* reject H_o.

The sample data do not indicate that there is a difference in the service time for city branch and suburban branch customers.

Situations can arise where a one-tail test would be appropriate when using the Wilcoxon Test. We might be arguing that the values of one population tend to be more (or less) than the values in the other population. A one-tail test is appropriate for the following problem.

Example 14.6.2 You believe that family income has deteriorated from 1965 to 1977 in your metropolitan area. You have available a random sample of family incomes taken in 1965. These sample values are inflated to 1977 dollars. A random sample of family incomes taken in 1977 is presented below with the 1965 sample (in 1977 dollars). (Amounts are in thousands of dollars.)

1965		1977	
16.1	9.9	8.1	12.5
8.2	13.9	9.8	7.9
13.4	8.2	15.2	10.3
12.4	15.4	11.1	6.5
10.5	16.6	6.8	
14.6			

Test the hypothesis that the incomes are the same against the alternative you have proposed. Use $\alpha = .05$. Note that income distributions are typically highly skewed. The normality assumption on income distributions would not be appropriate.

Here again is an example where the sample sizes are too small to use the Central Limit Theorem (and thus the z-statistic), and because the populations are not normally distributed, the t-test for the difference in two population means would not be appropriate. The Wilcoxon Test makes no assumptions about the population distributions, and it is applicable to this type of problem.

In Table 14.7 the data from the two samples have been combined and ranked from the lowest to the highest. The sum of the ranks of the values in each of the two samples is also given in Table 14.7. Let R_1 be the sum of the ranks for the 1965 sample. The alternate hypothesis for this problem is that family income has deteriorated from 1965 to 1971 in this particular area. If the alternate hypothesis is true, then we would expect the 1965 sample values to have a more-than-proportionate number of higher ranks; i.e., the sample values in 1965 would tend to be higher than the sample values in 1977. If the ranks tend to be proportionately higher, then the sum of the ranks would tend to be higher

Table 14.7 Combined Sample

$A = 1965$ $B = 1977$	Ranks	1965 Ranks	1977 Ranks
6.5 B	1		1
6.8 B	2		2
7.9 B	3		3
8.1 B	4		4
8.2 A	5.5	5.5	
8.2 A	5.5	5.5	
9.8 B	7		7
9.9 A	8	8	
10.3 B	9		9
10.5 A	10	10	
11.1 B	11		11
12.4 A	12	12	
12.5 B	13		13
13.4 A	14	14	
13.9 A	15	15	
14.6 A	16	16	
15.2 B	17		17
15.4 A	18	18	
16.1 A	19	19	
16.6 A	20	20	
	210	143	67

than the expected sum under the null hypothesis. We would reject the null hypothesis for a large value of z_c. The complete solution to this problem is given below.

1. H_o: The distribution of family income for 1977 is the same as for 1965.
   H_a: Family income in 1977 is lower than it was in 1965.

2. $\alpha = .05$.

$$z = \frac{R_1 - \frac{n_1(n_1 + n_2 + 1)}{2}}{\sqrt{\frac{n_1 n_2 (n_1 + n_2 + 1)}{12}}}.$$

3. Reject H_o if and only if $z_c > 1.65$.

4. $z_c = \dfrac{143 - \frac{11(11 + 9 + 1)}{2}}{\sqrt{\frac{11(9)(11 + 9 + 1)}{12}}} = 2.09$.

Reject H_o.

We reject the null hypothesis and conclude that real family income was higher in 1965 than it was in 1977 for families living in the metropolitan area under study.

In Table 14.7, 2 of the sample values were the same. When ranking data, if ties occur, the average of the ranks that the same value would receive is assigned to each of the values. The 2 values 8.2 would be assigned the fifth and sixth ranks. The average of the 2 ranks, 5.5, is assigned to the 2 respective 8.2 values in the second and third columns of Table 14.7.

You might be inclined to ask, Why not use the Wilcoxon Test even if the populations are normally distributed? The Wilcoxon Test is certainly applicable when the two populations are normal; however, if the two populations are normal, the t-test can be used, and it is a more powerful test and, therefore, preferable. A test is said to be more powerful if it performs better with respect to a Type II error. That is, for a given significance level, the probability of a Type II error is smaller if the t-test is employed than if the Wilcoxon Test were used for the same problem.

14.7 The Kruskal-Wallis Test

The Kruskal-Wallis Test can be used to compare three or more populations in a manner similar to that in which the Wilcoxon Test was used to compare two populations. This test, like the Wilcoxon Test, makes

Example 14.7.1

It is desired to determine whether family income is the same for three regions of the city. Random samples were taken from each of the regions. The results are reported below in thousands of dollars.

Region 1	Region 2	Region 3
11.6	8.1	10.8
16.8	15.2	9.6
12.7	11.1	7.4
17.3	12.5	12.9
14.2	13.6	11.8
		13.2

Test the hypothesis that family income is the same for all three regions. Use a significance level of .05.

If the populations were normally distributed, this problem could be worked using the analysis of variance method discussed in Chapter 13. However, family income generally has a skewed distribution, which means it can't be normally distributed. The Kruskal-Wallis Test can be used for this problem because it is not dependent on any assumptions about the probability distribution of the populations from which the samples were taken.

The test statistic for the Kruskal-Wallis Test is a function of the ranks of the sample values obtained by ranking the combined sample values, as was done for the Wilcoxon Test. The combined sample values are ranked from lowest to highest as in Table 14.8, and then the sums of the ranks of the values for each of the respective samples are obtained.

Test Statistic for the Kruskal-Wallis Test

If all of the sample sizes are 5 or more, then the following test statistic has an approximately chi-square distribution.

$$\chi^2 = \frac{12}{n(n+1)} \sum_{j=1}^{m} \frac{R_j^2}{n_j} - 3(n+1)$$

with
$df = m - 1$

where

m = the number of populations

R_j = the sum of the ranks for the jth population sample

n_j = the number of observations in the sample from the jth population

Table 14.8 Combined Sample

A = Region 1 B = Region 2 C = Region 3	Ranks	Region 1 Ranks	Region 2 Ranks	Region 3 Ranks
7.4 C	1			1
8.1 B	2		2	
9.6 C	3			3
10.8 C	4			4
11.1 B	5		5	
11.6 A	6	6		
11.8 C	7			7
12.5 B	8		8	
12.7 A	9	9		
12.9 C	10			10
13.2 C	11			11
13.6 B	12		12	
14.2 A	13	13		
15.2 B	14		14	
16.8 A	15	15		
17.3 A	16	16		
	136	59	41	36

$n = \sum_{j=1}^{m} n_j$; i.e., n is the combined number of sample observations.

If the null hypothesis that the population distributions are the same is *not* true, then the chi-square test statistic tends to get large. Thus, the null hypothesis is rejected for large values of the test statistic. There are 3 populations being compared; so the degrees of freedom for this example are $3 - 1 = 2$. The complete solution to this problem is given below.

1. H_o: The distribution of family incomes is the same for the 3 regions.
   H_a: H_o is not true.

2. $\alpha = .05$

$$\chi^2 = \frac{12}{n(n+1)} \sum_{j=1}^{m} \frac{R_j^2}{n_j} - 3(n+1) \text{ with } df = m - 1 = 2$$

3. Reject H_o if and only if $\chi_c^2 > 5.991$.

4. $\chi_c^2 = \frac{12}{16(16+1)} \left[\frac{(59)^2}{5} + \frac{(41)^2}{5} + \frac{(36)^2}{6} \right] - 3(16+1)$

   $= .0441(1248.4) - 51 = 4.05.$

   Do *not* reject H_o.

The Kruskal-Wallis Test

We are unable to conclude that there are differences in family incomes among the 3 regions of the city.

If the populations were normally distributed, the analysis of variance model would be preferred to the Kruskal-Wallis Test because of the lower probability of Type II error for a given significance level. The ranking procedure required for the Kruskal-Wallis Test is tedious. This might be a pragmatic reason for preferring the analysis of variance model.

14.8 Spearman's Rank-Correlation Coefficient

The rank-correlation coefficient measures the correlation between two variables based on the ranks of the paired observations.

Example 14.8.1   A nationwide distributing firm is interested in knowing whether or not sales of one of its products varies as per-capita income changes. A sample of per-capita income and annual per-capita sales of the product in 20 randomly selected areas produced the data in the first 2 columns of Table 14.9. Calculate Spearman's Rank-Correlation Coefficient for these sample data.

The ranks of the values in column 1 from the lowest to the highest appear in the third column of Table 14.9. In the case of ties, the average of the ranks is assigned to each of the tied numbers. (The number 4.5 occurs at the tenth and eleventh ranks; so both values of 4.5 receive the rank of $(10 + 11)/2$ or 10.5.) Similarly, the fourth column of Table 14.9 gives the ranks of the values in column 2.

Spearman's Rank-Correlation Coefficient   Spearman's Rank-Correlation Coefficient, r_s, is calculated from the paired ranks by the following formula:

$$r_s = 1 - \frac{6\Sigma d_i^2}{n(n^2 - 1)}$$

where
  d_i = the difference between the paired ranks of the ith observation
  n = the number of paired observations in the sample.

The rank-correlation coefficient can assume values from -1 to $+1$; i.e., $-1 \leq r_s \leq 1$. If the ranks are identical for each pair, then the sample values have perfect positive rank correlation, and $r_s = 1$. If the ranks for the 2 sets of values are exactly reversed, then $r_s = -1$, which

Table 14.9 Calculation of Σd_i^2 for Example 14.8.1

Paired Observations		Paired Ranks			
Per-Capita Income (1000's)	Per-Capita Sales	Per-Capita Income	Per-Capita Sales	d_i	d_i^2
4.5	24.1	10.5	8	2.5	6.25
5.1	32.8	17	20	3	9
3.4	19.7	1	4	3	9
4.9	31.2	15	18	3	9
3.9	20.6	5.5	5	.5	.25
3.8	25.2	4	12	8	64
4.3	22.3	8	7	1	1
4.7	26.1	13.5	14	.5	.25
5.5	18.2	19	2	17	289
5.0	27.3	16	15	1	1
3.9	24.2	5.5	9	3.5	12.25
5.6	32.3	20	19	1	1
4.5	25.7	10.5	13	2.5	6.25
5.3	28.7	18	17	1	1
4.7	24.5	13.5	11	2.5	6.25
4.0	18.6	7	3	4	16
4.6	24.4	12	10	2	4
4.4	27.9	9	16	7	49
3.7	17.8	3	1	2	4
3.6	21.0	2	6	4	16
					504.5

represents perfect negative correlation. If $r_s = 0$, then there is no correlation between the 2 sets of ranks.

Column 5 in Table 14.9 has the absolute value of the differences between the paired ranks. The absolute value of the difference is sufficient inasmuch as the difference is squared before it is summed. Column 6 has the squared differences and the column sum, Σd_i^2. Using this information, Spearman's Rank-Correlation Coefficient for these sample data can readily be calculated:

$$r_s = 1 - \frac{6\Sigma d_i^2}{n(n^2 - 1)} = 1 - \frac{6(504.5)}{20(20^2 - 1)} = .62.$$

Test Statistic for Spearman's Rank-Correlation Coefficient

If the number of sample paired observations is greater than or equal to 20, $n \geq 20$, then the following function of the sample rank correlation has an approximate t-distribution with $n - 2$ degrees of freedom if the population rank-correlation coefficient is equal to 0.

$$t = \frac{r_s \sqrt{n - 2}}{\sqrt{1 - r_s^2}} \text{ with } df = n - 2.$$

Spearman's Rank-Correlation Coefficient

This provides us with a test statistic for testing the hypothesis that the population rank correlation coefficient, ρ_s, is 0.

Example 14.8.2 Test the hypothesis that the sample of paired observations in Table 14.9 came from a population with 0 rank correlation. Use $\alpha = .05$.

1. $H_o: \rho_s = 0$.
   $H_a: \rho_s \neq 0$.

2. $\alpha = .05$

$$t = \frac{r_s\sqrt{n-2}}{\sqrt{1-r_s^2}}$$

$df = n - 2$.

3. Reject H_o if and only if $t_c > 2.101$ or $t_c < -2.101$.

4. $t_c = \dfrac{.62\sqrt{20-2}}{\sqrt{1-(.62)^2}} = 3.35$.

Reject H_o.

We conclude that sales do tend to be higher in the areas of higher per-capita income.

If the original variables, say x and y, are each the set of positive integers 1 through n, then Spearman's Rank-Correlation Coefficient is identical to the coefficient of linear correlation introduced in Chapter 10. In the above example, 14.8.1, it was necessary to rank the original data before calculating the rank-correlation coefficient. Examples arise where the original data are possible only in ranked form. The rank-correlation coefficient would be applicable for these situations. Consider the following simple example.

Example 14.8.3 The two department managers were asked to rank the eight secretaries from the secretarial pool according to overall performance. The president of the company received the report shown in Table 14.10. Can he conclude that the two managers used somewhat the same criteria in their evaluations?

In addition to the data, Table 14.10 also includes calculations of Σd_i^2 and r_s. Inasmuch as $n < 20$, the test of the null hypothesis, $H_o: \rho_s = 0$, is not valid. Purely as an exercise, test H_o according to the procedure outlined in Example 14.8.1. You will find that you will be unable to reject the null hypothesis. In any case, it appears that criteria for evaluation of overall performance are not altogether consistent between the 2 managers. Inspection of the data shows that the greatest contribution to Σd_i^2 lies in the number 2 rank. Interchange ranks 2 and 5 for manager 2 and calculate a new r_s. Might the president's conclusion

Table 14.10 Data for Example 14.8.3

Secretary	Paired Ranks		d_i	d_i^2
	Manager 1	Manager 2		
A	2	5	3	9
B	8	7	1	1
C	1	1	0	0
D	7	8	1	1
E	5	4	1	1
F	3	3	0	0
G	4	6	2	4
H	6	2	4	16
				$\Sigma d_i^2 = 32$

$$r_s = 1 - \frac{6(32)}{8(63)} = .62.$$

have been quite different had the data been presented with this adjustment?

As a final note, if the original data are measured as in Example 14.8.1, the coefficient of rank correlation is usually a good estimate of the coefficient of linear correlation—if there are no extreme values in either x or y. In fact, the significance test on r_s is borrowed from and identical to that on r, the linear corrleation coefficient. The rank-correlation coefficient provides us with an accurate calculation method for estimating the coefficient of linear correlation.

14A Appendix: Goodness-of-Fit Test— The Poisson Distribution

The Poisson random variable and its probability distribution were the subject of Appendix 4D at the end of Chapter 4. It was noted that the Poisson Distribution has an important application in queuing theory or waiting-line models. The chi-square statistic can be used to test the hypothesis that the random variable of interest follows a Poisson Distribution.

Example 14A.1 A bank has been assigning tellers according to calculations based on the assumption that the number of customers arriving at the bank follows a Poisson Distribution with an average arrival rate of 2 customers per minute. In the past, the assignment of tellers using the calculated results of the model proved to be effective and efficient; but the current results are very unsatisfactory. It is decided to randomly sample 100 minutes during the week. The sample results are shown in Table 14.11.

Table 14.11 Sample Data for Example 14A.1

Number of Customers Arriving per Minute	Sample Frequency
0	2
1	11
2	24
3	20
4	18
5	10
6	8
7	4
8	2
9	0
10	1
	100

Test the hypothesis that the number of customers arriving per minute is a Poisson Distribution with an average of 2 customers per minute. Use $\alpha = .05$.

Before the hypothesis test can be performed, it is necessary to calculate the expected number, e_i, for each category and determine the maximum number of categories with $e_i > 5$. The e_i's are calculated assuming that the null hypothesis is true: the random variable has a Poisson Distribution with a mean, λ, equal to 2.0. The calculations are carried out in Table 14.12. In the third column of this table, there are a number of e_i's less than 5. The last 6 categories have been combined

Table 14.12 Calculation of the e_i's for Example 14A.1

x Number of Customers Arriving	π_i Probabilities Calculated from Table F for $\lambda = 2.0$	e_i $n\pi_i$	o_i Sample Frequency	x, e_i, and o_i for Hypothesis Test		
				x	e_i	o_i
0	.1353	13.53	2	0	13.53	2
1	.2707	27.07	11	1	27.07	11
2	.2707	27.07	24	2	27.07	24
3	.1804	18.04	20	3	18.04	20
4	.0902	9.02	18	4	9.02	18
5	.0361	3.61	10	5 or more	5.27	25
6	.0121	1.21	8		100.00	100
7	.0034	.34	4			
8	.0009	.09	2			
9 or more	.0002	.02	1			
	1.0000	100.00	100			

Nonparametric Statistics

into 1 category, "5 or more," in the last 3 columns of Table 14.12. All of the e_i's are now greater than 5, and these data can be used to perform the desired hypothesis test. Note that we now have 6 categories. The degrees of freedom for this hypothesis test are $k - 1 = 6 - 1 = 5$.

1. H_o: The number of bank customers arriving per minute is a Poisson Distribution with $\lambda = 2.0$.
   H_a: H_o is not true.
2. $\alpha = .05$

$$\chi^2 = \sum_{i=1}^{k} \frac{(o_i - e_i)^2}{e_i} \text{ with } df = k - 1.$$

3. Reject H_o if and only if $\chi_c^2 > 11.070$.

4. $$\chi_c^2 = \frac{(2 - 13.53)^2}{13.53} + \frac{(11 - 27.07)^2}{27.07} + \frac{(24 - 27.07)^2}{27.07}$$
$$+ \frac{(20 - 18.04)^2}{18.04} + \frac{(18 - 9.02)^2}{9.02} + \frac{(25 - 5.27)^2}{5.27} = 102.73$$

Reject H_o.

The null hypothesis is rejected, and we conclude that customer arrivals do not follow a Poisson Distribution with $\lambda = 2.0$. Assuming a Poisson Distribution with $\lambda = 2.0$ could very well be the cause of the poor performance of the waiting-line model currently in use by the bank. The null hypothesis tested was a strong hypothesis in that, not only was a Poisson Distribution hypothesized, but the value of the parameter was also specified. It could well be that the bank's customers arrive according to a Poisson Distribution with another mean.

Example 14A.2 Use the sample data given in Example 14A.1 to test the hypothesis that the number of arrivals per minute of bank customers follows a Poisson Distribution.

Before the probabilities necessary for determining the expected number in each category can be calculated, a value for the mean, λ, of the Poisson Distribution must be available. The value of λ is estimated from the sample data by finding the sample mean number of arrivals. The sample mean is the weighted average of the number of customers arriving with the sample frequencies used as weights. Using the data given in Example 14A.1, we have

$$\bar{x}_w = \frac{\Sigma w_i x_i}{\Sigma w_i} = \frac{2(0) + 11(1) + 24(2) + \cdots + 2(8) + 0(9) + 1(10)}{2 + 11 + 24 + \cdots + 2 + 0 + 1}$$

$$= \frac{343}{100} = 3.43 \cong 3.4.$$

The expected values for each category are calculated in Table 14.13 for $\lambda = 3.4$.

The last 3 columns of Table 14.13 have the data to be used for the hypothesis test. Note that the categories had to be adjusted at the beginning of the distribution as well as at the end in order to assure that all of the e_i's are greater than 5. The degrees of freedom are calculated somewhat differently for this example than for the previous example. You will recall from Section 14.1 that the degrees of freedom for the chi-square distribution are $k - d - 1$ where k is the number of categories and d is the number of parameters estimated from the sample data. In the previous example, λ was given in the hypothesis, but in this example it was necessary to estimate λ from the sample data before the calculations in Table 14.13 could be made. From Table 14.12 we see that for this example there are 7 categories; thus, the degrees of freedom are $k - d - 1 = 7 - 1 - 1 = 5$.

1. H_o: The number of bank customers arriving per minute is a Poisson Distribution.
    H_a: H_o is not true.

2. $\alpha = .05$

    $$\chi^2 = \sum_{i=1}^{k} \frac{(o_i - e_i)^2}{e_i} \text{ with } df = k - d - 1.$$

3. Reject H_o if and only if $\chi_c^2 > 11.070$.

Table 14.13 Calculation of the e_i's for Example 14A.2

x Number of Customers Arriving	π_i Probabilities Calculated from Table F for $\lambda = 3.4$	e_i $n\pi_i$	o_i Sample Frequency	x, e_i, and o_i for Hypothesis Test		
				x	e_i	o_i
0	.0334	3.34	2			
1	.1135	11.35	11	1 or less	14.69	13
2	.1929	19.29	24	2	19.29	24
3	.2186	21.86	20	3	21.86	20
4	.1858	18.58	18	4	18.58	18
5	.1264	12.64	10	5	12.64	10
6	.0716	7.16	8	6	7.16	8
7	.0348	3.48	4	7 or more	5.78	7
8	.0148	1.48	2		100.00	100
9	.0056	.56	0			
10 or more	.0026	.26	1			
	1.0000	100.00	100			

4. $\chi_c^2 = \dfrac{(13 - 14.69)^2}{14.69} + \dfrac{(24 - 19.29)^2}{19.29} + \dfrac{(20 - 21.86)^2}{21.86}$

$+ \dfrac{(18 - 18.58)^2}{18.58} + \dfrac{(10 - 12.64)^2}{12.64} + \dfrac{(8 - 7.16)^2}{7.16}$

$+ \dfrac{(7 - 5.78)^2}{5.78} = 2.43.$

Do *not* reject H_o.

The sample results are consistent with the hypothesis that the Poisson Distribution is appropriate for the waiting-line model used by the bank. It does appear, however, that more satisfactory results would occur if the bank used a value of λ different from 2.0 in its application of the Poisson Distribution.

Problems

1. The production manager of a manufacturing firm has instituted a new quality control system with more emphasis on process control. The old system permitted too high a proportion of costly defects. He feels that the new system should change the distribution of defects with regard to their cost to the company. Under the old system, 10 percent of the defects cost over $1,000 each, 25 percent were between $500 and $1,000 in cost to the firm, and the cost of 45 percent of the defects was between $250 and $500. The remainder of the defects resulted in a loss to the firm of less than $250. A random sample of defects taken after the new system was instituted produced the following results:

Cost of Defect	Number of Defects
Over $1,000	4
$500 to $1,000	22
$250 to $500	61
less than $250	48

Is the distribution of defects, as measured by cost, for the new quality control system different than for the old system? Test at the 1 percent significance level.

2. An insurance firm determined at a meeting of its executive officers that it should be directing its sales efforts in such a way that 30 percent of its life policies are written for blue collar workers, 50 percent for white collar workers, and 20 percent for self-employed individuals (either blue collar or white collar types of businesses).

A random sample of 160 customers showed that 58 were blue collar workers and 38 were self-employed. Do these sample results indicate that the firm's life insurance customers are consistent with the desired occupation proportions? Test at the 5 percent significance level.

3. The marketing manager for a large manufacturing and retailing firm is interested in determining whether there is a relationship between the amount people are willing to pay for a given product and the type of packaging used for the product. Twelve different price-package combinations were offered on a specific day at 12 randomly selected retail outlets having similar characteristics. The number of units sold for each of the price-package combinations is given below.

Price	Package Design			
	Elegant	Fancy	Standard	Plain
$7.95	19	21	23	16
9.95	23	17	9	5
13.95	18	11	5	3

Is there a relationship between price and customer preference for the 4 package designs? Test at the 10 percent significance level.

4. The vice-president of finance for a large firm wants to determine whether there is a difference between the 2 internal auditors with regard to the types of errors they are able to detect in the customer billing statements. A random sample of 100 faulty billings was selected and classified according to type of mistake and the auditor detecting the mistake. The results are given below.

Auditor	Type of Error Detected			
	Addition	Prices Wrong	Wrong Items	Combination and Misc.
IEB	9	13	21	9
JBS	6	8	24	10

Is the type of mistake detected the same for the 2 auditors? (Use $\alpha = .01$.)

5. A die is tossed 120 times. The number of spots coming up was observed and recorded. The results are given below.

Number of Spots	1	2	3	4	5	6
Times Occurred	16	24	23	15	17	25

Can it be concluded at the 1 percent significance level that the die is not fair?

6. Components were manufactured from 4 different kinds of materials. Randomly selected components of the 4 types of materials were then subjected to extreme temperature changes. The components were then inspected and classified according to their condition after they were removed from the testing facility. The following results were obtained.

	Material I	Material II	Material III	Material IV
Broke Completely	27	41	40	17
Showed Defects	38	35	37	36
Remained Intact	35	24	23	47

Are the materials the same with respect to the proportion of components classified into the 3 categories? Test at the 1 percent significance level.

7. The outside diameters of 30 metal disks were measured at the rate of 1 per minute as they came off the stamping machine. The measurements are given below in the order they were selected. Perform a hypothesis test to determine if the sizes of the disks are random.

6.019, 5.994, 6.001, 6.008, 5.995, 5.983, 6.011, 5.996, 6.000, 6.007, 5.987, 6.016, 5.996, 5.989, 6.007, 5.995, 6.002, 5.996, 6.012, 6.011, 6.000, 5.997, 6.002, 6.001, 6.003, 6.006, 5.998, 5.995, 6.012, 6.003.

Use a 5 percent significance level.

8. A large manufacturer wants to determine whether the magnitude of billing errors made by the billing clerks is random throughout the day. The following error magnitudes were recorded in the order they occurred during a specific workday:

$17.84, 31.49, 15.85, 18.00, 21.64, 40.72, 20.79, 30.06, 24.55, 33.42, 28.59, 28.19, 21.44, 48.76, 1.52, 18.59, 5.53, 16.46, 10.32, 18.58, 20.96, 27.45, 12.15, 6.41, 32.65, 6.74, 16.58, 15.66, 0.21, 31.77, 31.06, 31.74, 24.20, 31.36, 8.94, 17.26, 0.10, 0.34, 19.67, 5.94, 16.12, 38.21.

Can the manufacturer conclude at the 1 percent significance level that magnitude of the billing error occurs randomly throughout the day?

9. The following digits are from a random number table in the order they appear in the table.

2, 5, 8, 8, 6, 0, 4, 3, 0, 3, 8, 5, 5, 6, 5, 6, 6, 8, 3, 9, 7, 8, 0, 3, 2, 1, 5, 8, 8, 1, 4, 7, 1, 5, 5, 7, 7, 1, 1, 6, 9, 0, 9, 0, 8, 6, 9, 9, 2, 4, 3, 9, 2, 3, 1, 0, 8, 3, 6, 2, 2, 1, 4, 6, 5, 9, 4, 0, 9, 0, 8, 7, 8, 9, 1, 3, 3, 7, 1, 9, 8, 8, 9, 1, 5, 7, 7, 5, 4, 8, 3, 9, 3, 4, 6, 5, 0, 7, 1, 1.

Use a runs test to test the hypothesis that this is a random set of numbers. Test at the 10 percent significance level.

10. The quality control engineer has determined that there is a higher-than-average percentage of defects of a particular component for those components produced between the end of the lunch period and the beginning of the afternoon break. The quality engineer wants to determine whether the percentage of defectives is the same during this time period for all 5 operators. The following results were obtained for a randomly selected day (the operators were not aware of the Q.C. manager's special interest in that day's output).

	Operator				
	1	2	3	4	5
Good	143	121	127	132	123
Defective	8	2	15	5	6
Total Components	151	123	142	137	129

Can the quality control manager conclude that the percent defectives is the same for the 5 operators? Test at the 5 percent significance level. (Note that the hypothesis "the percent of defects is the same for the 5 operators" is the same as the hypothesis "the classification of a component as good or defective is independent of the operator producing the component." This latter hypothesis suggests the use of the contingency table methodology.)

11. The design department has proposed 3 different package designs for the company's product. The marketing manager believes the first design would be twice as preferred as the second design and that the second design would be 3 times as preferred as the third design. Market test results show that 111 preferred the first design, 62 preferred the second, and 40 preferred the third design. Are these results consistent with the marketing manager's appraisal of the relative preference for the 3 designs? Test at the 5 percent significance level.

12. The finance vice-president in Problem 4 would like to know whether the billing clerks differ with regard to the types of mistakes made on customer billing statements. Two hundred faulty statements were randomly selected and classified according to the clerk preparing the statement and the type of error made. The following tabulations resulted.

	Type of Mistake			
Billing Clerk Number	Addition	Wrong Price	Wrong Items	Combination and Misc.
1	4	18	12	14
2	12	7	16	9
3	6	9	18	24
4	7	9	28	7

Are the clerks the same with regard to the types of billing errors they make? Test at the 5 percent significance level.

13. Refer to Problem 9. Set up a category for each of the 10 digits, and use the chi-square statistic to test the hypothesis that the set of numbers is random. (Use $\alpha = .10$)

14. A banker wanted to determine whether there is a difference in the number of checks written per month per account in 2 branches of the bank located in different parts of the city. He felt that the economic differences between the 2 parts of the city might be sufficiently great to affect the level of customer usage of checking accounts. Twelve checking account customers were selected randomly from the first branch, and 15 from the second branch. The following checks per month were recorded for the customers in the sample.

East Branch		West Branch	
23	32	17	12
10	25	9	21
27	17	13	13
18	21	10	6
19	35	27	7
8	23	16	18
		19	11
		7	

Can the banker conclude at the 10 percent significance level that there is a difference in checking account usage between the customers of the 2 branches?

15. The manager of a large retail men's store wants to determine whether the 4 tailors employed by the company work at the same rate of speed. Six randomly selected suit adjustments given to the first tailor at random times throughout the week resulted in the following completion times: 9.8, 14.2, 7.6, 5.4, 12.3, and 15.1 minutes. Similarly, 5 randomly selected suit adjustments were given to the second tailor, 8 to the third, and 7 to the fourth tailor. The completion times were as follows: Second tailor: 8.3, 17.6, 12.8, 14.2, 13.7. Third tailor: 5.8, 12.7, 14.7, 9.3, 12.6, 8.7, 9.6, 12.3.

Fourth tailor: 12.5, 15.8, 16.4, 17.3, 9.6, 19.4, 13.6. Is the distribution of completion times the same for all 4 tailors? Test at the 10 percent significance level.

16. A large distributor of dairy products wants to determine whether there is a difference in the amount of milk consumed per week for 3 different age groups. The following sample results were obtained from independent random samples of each of the respective age groups.

Age in Years	21-30	31-45	46-65
Quarts per Week	3.75	2.25	1.00
	1.50	1.75	2.50
	2.00	6.50	1.75
	8.25	5.75	3.00
	4.25	6.25	1.50
	3.50	7.25	
		4.25	

Is there a difference in milk consumption patterns for the 3 age groups? Test at the 5 percent significance level.

17. The production manager wanted to determine whether there is a difference in the time used to make the final inspection of electric motors at the 5 different inspection stations required to handle the large volume of motors produced by the company. The final inspection consists of observing whether the motor runs and listening for unusual sounds. The inspection stations were observed at random times during the day, and the time to inspect a motor was recorded. The following results were obtained.

Station 1	Station 2	Station 3	Station 4	Station 5
6.7	2.7	3.1	2.1	6.7
5.4	3.1	8.3	5.4	4.3
3.6	3.3	4.5	2.3	10.2
7.8	6.8	4.2	2.9	4.7
2.8	3.5	5.6	3.4	5.1
3.3	4.1	7.1		3.6
2.9				9.8
3.6				

Is the time required to inspect a motor the same for the 5 inspection stations? Test at the 5 percent significance level.

18. A die has 3 sides painted green, 2 sides painted yellow, and the remaining side painted blue. The die is tossed 150 times, and the color of the top side is noted and recorded.

Color	Sample Frequency
Green	65
Yellow	43
Blue	42

Can it be concluded at the 1 percent significance level that the die is not fair?

19. A large retailer is considering installing key duplicating machines in her stores. The machines require the sales clerks' time to operate them. She is considering 2 different makes of key duplicating machines, and she would like to determine if the machines are the same with respect to the time it takes to duplicate keys. Table 14.14 shows the results obtained when independent random samples of various types of keys were duplicated. On the basis of these sample data, can the retailer conclude at the 1 percent significance level that there is a difference between the 2 machines in the time required to duplicate keys?

Table 14.14 Machine Times for Problem 19

Time in Seconds	
Machine 1	Machine 2
5.6	7.2
11.3	3.1
4.1	2.7
3.4	4.1
15.8	3.8
9.2	2.6
5.4	5.6
3.4	2.9
	2.7
	3.2

20. The speeds of 400 westbound cars were recorded as they passed a checkpoint on State Highway 88 in Arizona. The mean of the sample of 400 cars was calculated before grouping the data; $\bar{x} = 52.5$ miles per hour and $s = 6.6$ miles per hour. The data were then grouped as follows:

Speed (miles per hour)	Number of Cars
40 or less	12
40+ to 45	26
45+ to 50	80
50+ to 55	130
55+ to 60	104
over 60	48

Test the hypothesis that the speeds of westbound cars passing the checkpoint follow a normal distribution. (Use $\alpha = .05$.)

21. Did the sample data in Problem 7 at the end of Chapter 2 come from a normally distributed population? Test at the 5 percent significance level.

22. The following values were taken from a table of alleged random deviates; i.e., they come from a standard normal population ($\mu = 0$, $\sigma = 1$). Test the hypothesis that the following data represent a sample taken from a standard normal population. (Use $\alpha = .10$.)

−.216, 1.149, −.615, −.20061, .16423, −.552, −1.849, −.142, −1.448, −.355, −.968, 2.072, .079, 1.007, .456, −.142, 1.342, .859, .819, .144, 2.876, .097, .746, −.786, −1.359, .766, .931, 1.821, −.388, 1.637, −1.407, .421, 1.174, −.034, −2.263, −1.991, 1.107, −.274, −1.106, 1.178, −2.306, −.434, −.342, −1.327, 1.266.

23. Consider the data shown in Table 14.15 on net profit and capital investment for 11 large U.S. corporations. Calculate the rank-correlation coefficient.

Table 14.15 Profit and Investment Data for Problem 23

Corporation	Net Profit (millions of $)	Capital Investment (millions of $)
A	85	460
B	81	510
C	38	22
D	35	28
E	30	18
F	28	19
G	27	17
H	25	21
I	24	20
J	21	11
K	16	13

24. The math entrance scores and grades in statistics of 21 students are shown in Table 14.16. Find the rank-correlation coefficient, and test the hypothesis that the rank-correlation coefficient for all students taking statistics is equal to 0. Test at the 1 percent significance level.

25. A certain drug is claimed to be effective in curing colds. Two hundred people were randomly selected for the experiment. Half were given the drug, and half were given sugar pills. Given the following table of results, test the hypothesis that the drug is no better than sugar pills for curing colds. (Use $\alpha = .01$.)

	Improved	Became Sicker	Remained the Same
Cold Drug	64	12	24
Sugar Pills	46	13	41

26. On the basis of the following data from a sample of 700 randomly selected salespersons, can it be concluded at the 1 percent significance level that sales ability is independent of sense of humor?

Volume of Sales	Sense of Humor		
	Nil	Average	Above Average
Low	60	50	40
Average	110	180	90
High	40	60	70

Table 14.16 Entrance Scores and Statistics Grades for Problem 24

Math Entrance Score	Statistics Grade
108	90
139	100
90	92
100	82
120	98
80	50
103	70
95	80
125	94
110	84
104	71
111	83
126	91
96	79
105	74
82	53
121	96
101	83
92	85
132	97
107	88

27. From the data in Problem 9 at the end of Chapter 2, can it be concluded that the time to complete a 3-game series in the U.C. Bowling League is normally distributed? Test at the 1 percent significance level.

28. Refer to the data in Problem 4 at the end of Chapter 2.
    a. Find the rank-correlation coefficient between "Ending May Balance" and "Amount Paid During Month."

b. Find the rank-correlation coefficient between "Ending May Balance" and "Amount Charged During Month."
c. Find the rank-correlation coefficient for "Amount Paid During Month" and "Amount Charged During Month."
d. Comment on your results.

29. A trade association wants to determine whether the relative preference for 4 different brands of soap was the same for 3 small cities of interest. The following sample results were obtained. Test at the 5 percent significance level.

	Preferred Brand			
City	W	X	Y	Z
A	21	29	11	9
B	43	34	19	8
C	28	26	38	16

30. The marketing vice-president of a large national retailing firm instructed the managers of its stores to use their own judgment in setting prices on the standard model color TV sold by the chain. He selected randomly 13 stores with about the same size of market and recorded the price for the color television set and the number of sets sold in the month of November. Table 14.17 shows the results. Calculate the rank-correlation coefficient for these sample data.

31. Find the coefficient of linear correlation for the rank data in Example 14.8.3. (This should be identical to r_s calculated in the example.)

Table 14.17 Price and Sales Data for Problem 30

Price	Number of Sets Sold
$390	18
450	15
385	21
395	25
500	9
415	20
425	16
375	24
465	12
362	31
435	13
495	11
440	14

32. From the formula

$$r_s = 1 - \frac{6\Sigma d_i^2}{n(n^2 - 1)},$$

it is obvious that, if the ranks are identical for each pair, each $d_i = 0$, $\Sigma d_i^2 = 0$, and $r_s = 1$. Show that, for the following example where one of the rank sequences is reversed, $r_s = -1$.

Rank I   9 8 7 6 5 4 3 2 1
Rank II   1 2 3 4 5 6 7 8 9

■ The remaining problems require material from the appendix.

33. A company observed that, in a 100-day period, there were 12, 25, 28, 16, 10, 5, 3, and 1 days on which there were, respectively, 0, 1, 2, 3, 4, 5, 6, and 7 record changes for a certain computer data file. Is it reasonable to suppose, at the .05 level of significance, that the distribution of the daily number of required changes may be satisfactorily described by a Poisson Distribution with a mean equal to 2?

34. The company auditor would like to determine whether billing errors per day are random or occur with some type of pattern. He decides that, if they are random, they should follow a Poisson Distribution. A sample of 197 days produced the results shown in Table 14.18. Does the number of billing errors per day follow a Poisson Distribution? Test at the 1 percent significance level.

35. A quality control engineer has taken 50 samples of size 200 from a production process. Before setting up her quality test criteria, she

Table 14.18 Billing Error Data for Problem 34

Number of Errors per Day	Sample Number of Days
0	6
1	14
2	25
3	43
4	36
5	30
6	21
7	9
8	8
9	2
10	2
11	0
12	1
	Total 197

would like to determine whether the number of defects per sample follows a Poisson Distribution. The sample results are recorded below.

Number of Defects	Sample Frequency
0	10
1	24
2	10
3	4
4	1
5	1

Perform the hypothesis test for the Q.C. engineer at the 1 percent significance level.

36. The owner of a small specialty-food family restaurant in a small town in Kansas was told by his CPA that he should consider using queuing theory to determine the amount of part-time help he needs during the evening hours. His son, who is currently enrolled in a business program, is concerned about the fact that the queuing theory model suggested by the CPA depends on the assumption that the number of customers coming to the restaurant each evening follows a Poisson Distribution. He has observed that they primarily come in groups and not independently of each other, as assumed for the Poisson Distribution. A sample of 103 evenings was selected by the owner's son. He recorded the results shown in Table 14.19. Does the number of customers per evening follow a Poisson Distribution? Test at the 5 percent significance level.

Table 14.19 Customer Data for Problem 36

Number of Customers	Sample Number of Evenings
5	12
6	13
7	18
8	6
9	5
10	4
11	7
12	3
13	6
14	12
15	17

Table 14.20 Sample Data for Problem 37

Defectives in Sample	Number of Samples
0	1
1	1
2	3
3	6
4	6
5	10
6	11
7	8
8	3
9	1

37. Fifty samples of size 100 were drawn from large lots of electrodes. The manufacturer claims that the fraction of defective electrodes is .05. The sample results obtained are shown in Table 14.20.

   a. Use the Poisson table to compute approximate theoretical frequencies.
   b. Test the manufacturer's claim using the chi-square goodness-of-fit test. (Use $\alpha = .05$.)

CHAPTER 15

DECISION THEORY

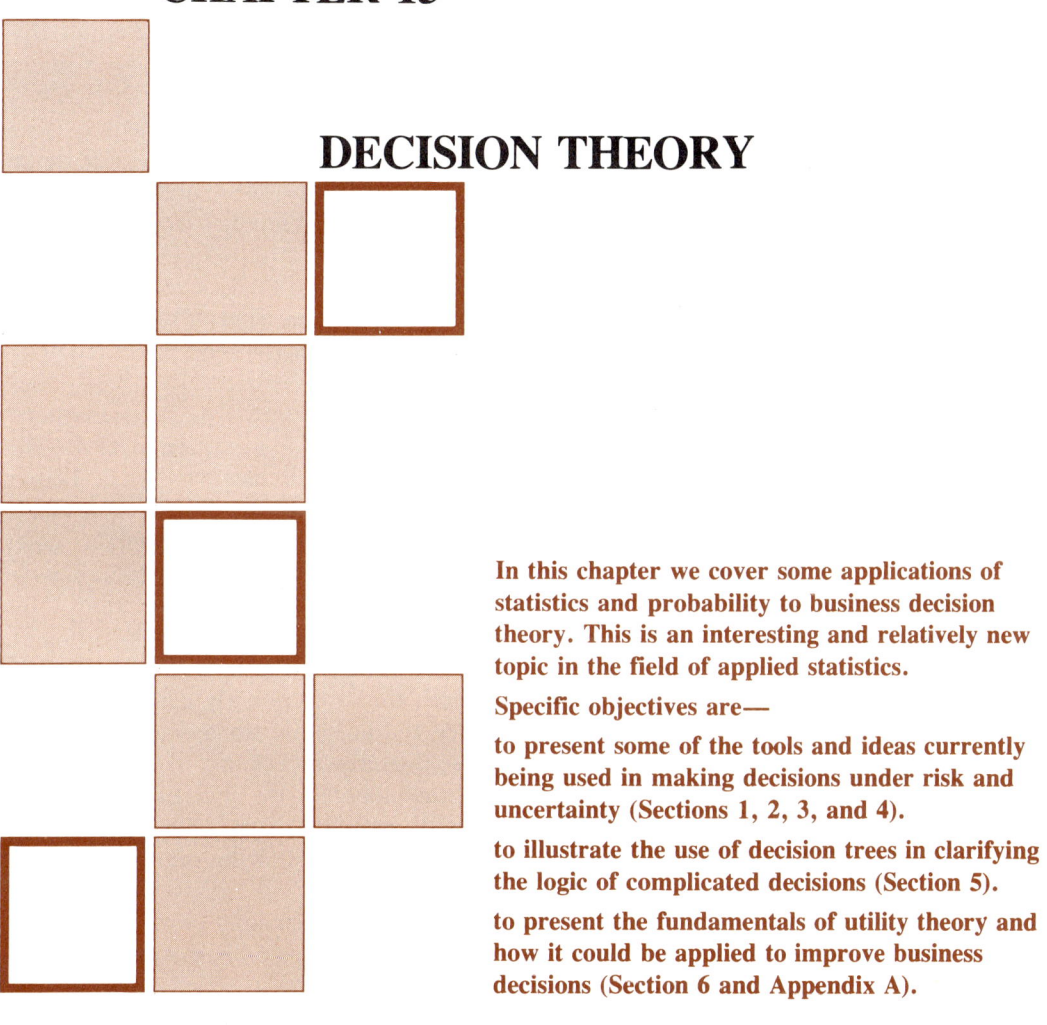

In this chapter we cover some applications of statistics and probability to business decision theory. This is an interesting and relatively new topic in the field of applied statistics.

Specific objectives are—

to present some of the tools and ideas currently being used in making decisions under risk and uncertainty (Sections 1, 2, 3, and 4).

to illustrate the use of decision trees in clarifying the logic of complicated decisions (Section 5).

to present the fundamentals of utility theory and how it could be applied to improve business decisions (Section 6 and Appendix A).

The growth in the application of quantitative methods to business and economics since World War II is no less spectacular than the technological mushrooming we have witnessed over the same period. The terms *model, mathematical model, statistical model,* etc.—heretofore the sole property of the hard sciences—are common terms in the business world today. Fear, awe, and distrust of mathematical methods are rapidly disappearing, and "seat-of-the-pants" management is being replaced with a successful combination of quantitative methods and highly trained management experts. It should be noted that the catalytic agent in this grand expansion was and is the computer—both the hardware and the software.

Much of the theoretical basis for the statistical work we have covered in the textbook to this point was developed before World War II, but applied statistics as a tool for business decisions is one of the important postwar developments in quantitative methods. We would like, in this chapter, to introduce another and not unrelated notion that is even more recent in its development and acceptance as a management tool.

In most quantitative applications, an optimization objective is assumed—we think for the most part rightly so. In the class of deterministic models, we make decisions *under certainty*; and the objective is to maximize profits, maximize payoff, minimize costs, minimize regret, or pursue some similar optimization criterion. This we refer to as *decision under certainty*, because all we really need is sufficient quantitative skills to solve the problem. This kind of model assumes that the world is fixed and the correct strategy is inherent in the problem.

Perhaps more realistic in our world of business are decisions *under uncertainty* or *under risk*. In decisions of this kind, the facts are still very important. The possible actions or strategies need to be known and listed; for each possible strategy over which the decision maker has control, there is another list of possible outcomes (events or states of nature) over which he or she has no control. If the probabilities of the events are known, the decision maker faces a *risk* situation; if the probabilities are not known, we speak of decision *under uncertainty*. Our major emphasis here will be on the case where the probabilities are not known but can be estimated; this is frequently called *subjective uncertainty*, but for convenience and to follow along with much of the literature in the field, we will use the term uncertainty in our discussion of these problems. Later in the chapter, we will discuss several techniques used to solve the true problem of decision under uncertainty—i.e., the case where the probabilities of the outcomes are not known nor can they be estimated.

The criterion or objective for decision making in the case where event probabilities are known or can be estimated is still one of optimization; but, for example, we cannot maximize profits—we can only maximize *expected* profits. We will be optimizing an expectation. The problem is further complicated by the decision maker's attitude toward risk; the notion of *utility* will be introduced in order to address this aspect. Assuming it is possible to measure a person's utility for the several event payoffs, the objective would then be to maximize expected utility.

Let us illustrate some of the above points with a simple example.

Example 15.0.1   For some unknown reason, you are offered the chance to play 1 of the following games:

1. An honest coin is tossed; if heads, you win $100; if tails, you lose $50.
2. An honest die is tossed; if a 6 shows, you win $600; if a 1 shows, you lose $300; if any other number shows, the payoff is 0.
3. If you can spell your last name correctly, you win $10; if not, you lose $10.

You may play only 1 of the 3 games. Which is your choice if: (1) the game you choose is played but once, or (2) you may play the game twice daily for the next year?

We want to help you make your decision. We will first calculate the expected payoff for each of the possible decisions. Let x be the payoff. Recall that $E(x) = \Sigma x p(x)$.

For game 1: $E(x) = \frac{1}{2}(100) + \frac{1}{2}(-50) = \$25$. (With an honest coin, $p(H) = p(T) = \frac{1}{2}$.)

For game 2: $E(x) = 1/6(600) + 1/6(-300) + 2/3(0) = \$50$. (With an honest die, each $p_i = 1/6$.)

For game 3, we will assume that the probability of a college student's misspelling his or her name is, for all practical purposes, 0: $E(x) = 1(10) + 0(-10) = \$10$.

Do these expected payoffs help you with your decision? Regardless of your current liquidity and your attitude toward risk, in the case where the game can be repeated a large number of times the correct choice is that with the largest monetary expectation—i.e., game 2. Over the long run, you can expect an average of $50 per game. (If you are nervous about the decision to play game 2 in the case of repeated plays, go see your most conservative local banker; he will be more than happy to back you for some fraction of the payoff.)

If the game may be played but once, your decision is very personal. For some of us, the possibility of losing $300, or even $50, would lead to a choice of game 1, with lower risk and lower payoff, or game 3 with an absolutely certain payoff of $10. That big crap table of the

Decision Theory

business world will, more often than not, allow a game to be played only one time—albeit the games to be played by a specific decision maker may be quite similar from one to the next. Which game you choose in the above example depends upon *your* utility—your attitude toward risk over the range of −$300 to +$600. It would be perfectly rational for 3 different people to make 3 different choices. The problem of utility will be discussed in Section 6 of this chapter. We want to first consider several aspects and complications of our problem and address the cases where maximizing expected payoff, minimizing expected costs, or minimizing expected opportunity loss seems to be a reasonable criterion.

Consider the following example:

Example 15.0.2

Cathy's Catering Service has been given considerable latitude (and a monopoly) in providing the lunch for the AFL-CIO Labor Day Celebration in Platinum City, Nevada. The cafeteria-style lunch may be served in the large labor union hall, or it may be set up outside on the green lawns of the city park. If the weather is nice, the outside picnic will yield considerable profits; if rain or snow falls, it would be better to hold the lunch inside. Cathy has worked up the following profit table.

	Profits	
States of Nature	Outside Picnic A_1	Inside Lunch A_2
N_1 (nice weather)	$800	+$200
N_2 (rain or snow)	−$100	$400

The two locations require different types and quantities of food, and Cathy must choose one a week in advance of the big celebration in order to get delivery on the food. The best information available from the weather bureau is that there is a 20 percent chance of precipitation on Labor Day. She wants to proceed to make her decision based upon this probability and assures us that the monetary risks involved do not make her nervous even though she doesn't relish the thought of losing $100. She wants our advice.

Calculating the conditional expected profits we have

$E(x/A_1) = .80(800) + (.20)(-100) = \$620$
and
$E(x/A_2) = .80(200) + (.20)(400) = \$240.$

The expected profit for the outside picnic is considerably greater than that for the inside lunch; we recommend that she plan for the outside picnic.

15.1 The Expected Monetary Value Criterion

In order to introduce other aspects of the notion of optimizing an expectation as the basis for decision making, and to delve more deeply into the subject, we consider the following example of a common type of inventory problem.

Example 15.1.1 The owner of Suzi's Submarine Sandwich Stand receives only 1 early morning delivery each day from the submarine sandwich supplier. Suzi prides herself on having fresh sandwiches for her customers; thus, any sandwiches left over at the end of the day are discarded. Demand is not the same every day. On the basis of past experience, Suzi has identified 4 demand levels, 1 of which occurs on any given working day, and the respective probabilities of the several demand levels.

Demand	Probability
25	.10
35	.30
45	.40
50	.20

The selling price for a submarine sandwich is $1.15, and the supplier charges Suzi $.60 each. Assume that, if customer demand is not satisfied for a given day, there is no loss of goodwill—i.e., demand for the next day is unaffected. The sale of other items is not affected by the number of submarine sandwiches sold. How many submarine sandwiches should Suzi order each day?

There are 4 alternative order quantities Suzi will consider. She will order either 25, 35, 45, or 50 sandwiches. For each alternative she selects, the gross profit (or contribution[1]) will vary according to the demand for that day. Suzi has no control over the daily demand. One of the 4 events—"demand is 25," "demand is 35," "demand is 45," or "demand is 50"—will occur. Such events are frequently referred to as *states of nature* in order to emphasize that the decision maker has no control over them. In order to use the expected monetary criterion, it is necessary that the events be mutually exclusive and exhaustive. This is clearly the case for this example; more than 1 of the 4 demand amounts cannot occur simultaneously on a given day, and no other demand is possible. Also, the probabilities assigned to the 4 events are all positive and sum to 1.

---

1. The term *contribution* is used to designate the dollar sales net of variable costs that contribute the fixed costs and net profit.

A payoff table for this problem is given in Table 15.1. Each column represents a possible choice or an alternative action for Suzi. Each row represents 1 of the 4 events that will occur after Suzi makes her choice. Reading down a column in the body of the table gives the conditional dollar contribution for the column alternative; these are conditional upon the event or outcome identified by that particular row. For example, the third column and second row has the monetary amount $13.25; this would be the payoff or the contribution if Suzi were to order 45 sandwiches on a given day (column 3) and the resulting demand were 35 (row 2).

The $13.25 value in the table is found as follows:

Total revenue = 35($1.15) = $40.25
Total cost = 45($.60) = $27.00
Total contribution to profit (and overhead) = $TR - TC$ = $40.25 - $27.00 = $13.25.

Revenue was received only for the 35 sandwiches sold; however, all 45 sandwiches ordered had to be paid for because the 10 sandwiches left over were discarded.

The dollar payoffs for the first alternative, "order 25," are identical because demand will always be at least 25; the full order of 25 sandwiches will be sold, and it is not possible to sell more than were ordered. For this action and for every demand, we have payoff = 25($1.15 - $.60) = $13.75. As another example, if 50 sandwiches are ordered and only 25 demanded, a loss is incurred; i.e., the payoff is negative: 25($1.15) - 50($.60) = $28.75 - $30.00 = -$1.25.

We will assume that you are able to calculate the balance of the payoffs shown in Table 15.1 by simply subtracting total costs from total revenues. This problem is a special case of a more general inventory problem. The following observations may help as we proceed with the analysis. Let x be demand (x = 25, 35, 45, or 50); let Q be order quantity (Q also equals 25, 35, 45, or 50), and let y be the payoff or contribution.

If $x \geq Q$, then all the sandwiches ordered are sold, there is no surplus, and $y = (1.15 - .60)Q = .55Q$. The per-unit payoff of $.55 is

Table 15.1 Payoff Table

		Alternatives			
Probabilities	Events	Order 25	Order 35	Order 45	Order 50
.10	Demand is 25	$13.75	$ 7.75	$ 1.75	-$ 1.25
.30	Demand is 35	13.75	19.25	13.25	10.25
.40	Demand is 45	13.75	19.25	24.75	21.75
.20	Demand is 50	13.75	19.25	24.75	27.50

realized for every sandwich ordered. This single equation can be used to calculate payoffs on the main diagonal of the payoff matrix and all those below the main diagonal.

If $x < Q$, there will be some unsold sandwiches. A payoff of $.55 will be realized for all those demanded, but Suzi will lose $.60 on each sandwich not sold; $y = .55x - .60(Q - x)$. This equation can be used to complete the payoffs above the main diagonal.

Except for the first alternative, "order 25," each alternative has associated with it more than 1 possible payoff. The expected monetary value for each of these alternatives is calculated by observing that the selection of an alternative is equivalent to the selecting of a random variable. For example, if the fourth alternative, "order 50," is selected, then Suzi in essence has selected a discrete random variable with 4 outcomes, each having a non-0 probability. There is a 10 percent chance she will lose $1.25; the probability she will make $10.25 is .30; there is a 40 percent chance she will make $21.75; and there is a probability of .20 that she will make $27.50. This is summarized below using y for the payoff and, consistent with our notation in Chapter 4, $p(y)$ for the respective probabilities.

y	$p(y)$
−1.25	.10
10.25	.30
21.75	.40
27.50	.20

Recalling the definition of expected value from Chapter 4, we have

$E(y) = \Sigma y p(y) = -1.25(.10) + 10.25(.30) + 21.75(.40) + 27.50(.20) = 17.15$.

In general the expected monetary value, EMV, of each alternative is calculated by multiplying each outcome of the alternative by the probability of that outcome and summing the products. The calculations are presented in Table 15.2.

Table 15.2 Calculation of Expected Monetary Value

First alternative—"order 25":

$EMV = 13.75(.1) + 13.75(.3) + 13.75(.4) + 13.75(.2) = 13.75$.

Second alternative—"order 35":

$EMV = 7.75(.1) + 19.25(.3) + 19.25(.4) + 19.25(.2) = 18.10$.

Third alternative—"order 45":

$EMV = 1.75(.1) + 13.25(.3) + 24.75(.4) + 24.75(.2) = 19.00$.

Fourth alternative—"order 50":

$EMV = -1.25(.1) + 10.25(.3) + 21.75(.4) + 27.50(.2) = 17.15$.

The optimal alternative using the expected monetary criterion is that alternative with the maximum expected value. From Table 15.2 we see that the third alternative has the largest *EMV*, $19.00. Suzi should order 45 submarine sandwiches.

Note that, if Suzi orders 45 submarine sandwiches for a given day, she will never earn $19.00 from the sandwiches, even though the expected value is $19.00. She will earn either $1.75, $13.25, or $24.75. But if she orders 45 sandwiches per day for a long time period, her average earnings per day will be close to $19.00. If she selects 1 of the other 3 alternatives she will, at least in the long run, make less money.

 ## 15.2 The Expected Value of Perfect Information

An interesting question to consider is, How much would perfect information be worth to Suzi? By "perfect information" we mean a perfect forecast of demand for the following day. If Suzi knew ahead of time what demand was going to be, she could order the number of submarine sandwiches that would be the most profitable for her, given the demand for that day. If someone approached Suzi and offered to forecast demand perfectly (assume he or she has inside information on what the company cafeterias are serving the next day and can forecast Suzi's demand given this information), how much would Suzi be willing to pay for this information? This, of course, depends upon how much Suzi can improve expected profits using this information.

It is important to note that the forecaster is not offering to control demand. He is only offering to forecast demand. He is not offering to guarantee Suzi a demand of 50 sandwiches every day. For any given day there is still just a 20 percent chance that demand will be 50 sandwiches, but if demand is going to be 50 sandwiches Suzi can know this ahead of time by using the forecaster's services. Each day she will be able to order the optimal quantity. Checking our intuition with the payoffs in Table 15.1, we see that maximum payoffs are achieved by ordering the exact quantity that will be demanded.

From Table 15.3 we observe that, if Suzi chooses to use the forecaster, then she has chosen a random variable with values given in the last column of Table 15.3 and corresponding probabilities given in the first column. Thus, for any given day there is a 10 percent chance that demand will be 25 (which will be duly forecasted) and Suzi will make $13.75; a 30 percent chance demand will be 35 and Suzi will make $19.25; and so on. The expected payoff or profit given perfect information is simply the expected value of this random variable.

The expected monetary value with perfect information
$= 13.75(.10) + 19.25(.30) + 24.75(.40) + 27.50(.20) = \$22.55.$

Table 15.3 Payoffs with Perfect Information

Probabilities	Events	Forecasted Amount	Amount Ordered	Profit
.10	Demand is 25	25	25	$13.75
.30	Demand is 35	35	35	19.25
.40	Demand is 45	45	45	24.75
.20	Demand is 50	50	50	27.50

If Suzi chooses to use the forecaster, then in the long run she will average $22.55 per day profit from her sales of submarine sandwiches.

To determine the value of the forecast information, we compare (1) the expected profit with perfect information to (2) the expected profit with no information. We determined above that the maximum expected value without other information on demand was $19.00; i.e., if Suzi orders 45 submarine sandwiches per day, her average profit per day will be $19.00. If she uses the forecaster's services, her average profit per day from sandwiches will be $22.55. The perfect information has increased the expected payoff by $3.55; thus, the expected value of perfect information (*EVPI*) is equal to $3.55. Suzi would be willing to pay up to $3.55 for the perfect information, but no more.

Expected Value of Perfect Information

In general we have,

the expected value of perfect information (*EVPI*) = (the expected monetary value with perfect information) − (the expected monetary value with no information).

15.3 The Expected Opportunity Loss Criterion

Another approach to the above problem is to select the alternative with the minimum expected opportunity loss. The alternative selected will be identical to the one selected by using the maximum expected value criterion. The concept of opportunity cost or opportunity loss is important for decision making. Opportunity loss is the difference between the payoff that could have been realized if the best alternative had been selected and the realized payoff. In Table 15.4 we illustrate the method for calculating opportunity losses from the payoff table given in Table 15.1.

For each state of nature, there is a "best" alternative. The best alternative is the one with the highest payoff. If an alternative with a lower payoff is selected, then there is an opportunity loss that arises because the decision maker failed to make the best choice possible for that state of nature. Recognize that, without perfect information, there will be opportunity losses; only with perfect information or through hindsight can the opportunity loss always be 0. If demand is 25 and Suzi orders 35 sandwiches, she will incur an opportunity loss of $13.75 − $7.75 = $6.00; if she had known that demand would be 25 sandwiches, she could have had a payoff of $13.75. By ordering 35 sandwiches instead of 25, Suzi incurs an opportunity loss of $6.00

In Table 15.4, part b, we present the calculations needed to obtain the opportunity loss table in part c. You will more easily follow the calculations if you keep in mind that they are made along a row rather than down a column. For example, the second row represents a demand of 35 sandwiches. If this event should occur, the best alternative is to order 35 sandwiches because $19.25 is the largest payoff in this row. All other payoffs in the second row are compared to $19.25. The differences between $19.25 and the other payoffs in the third row represent the opportunity losses for the other alternatives when demand is 35 sandwiches. Perhaps better than the row-column rule (there is no

---

Table 15.4 Opportunity Loss Table and Related Calculations
a. Payoff Table from Table 15.1

		Alternatives			
Probabilities	Events	Order 25	Order 35	Order 45	Order 50
.10	Demand is 25	$13.75	$ 7.75	$ 1.75	−$1.25
.30	Demand is 35	13.75	19.25	13.25	10.25
.40	Demand is 45	13.75	19.25	24.75	21.75
.20	Demand is 50	13.75	19.25	24.75	27.50

b. Calculation of Opportunity Losses

	Alternatives			
Events	Order 25	Order 35	Order 45	Order 50
Demand is 25	13.75 − 13.75 = 0	13.75 − 7.75 = 6.00	13.75 − 1.75 = 12.00	13.75 − (−1.25) = 15.00
Demand is 35	19.25 − 13.75 = 5.50	19.25 − 19.25 = 0	19.25 − 13.25 = 6.00	19.25 − 10.25 = 9.00
Demand is 45	24.75 − 13.75 = 11.00	24.75 − 19.25 = 5.50	24.75 − 24.75 = 0	24.75 − 21.75 = 3.00
Demand is 50	27.50 − 13.75 = 13.75	27.50 − 19.25 = 8.25	27.50 − 24.75 = 2.75	27.50 − 27.50 = 0

c. Opportunity Loss Table

		Alternatives			
Probabilities	Events	Order 25	Order 35	Order 45	Order 50
.10	Demand is 25	$ 0	$6.00	$12.00	$15.00
.30	Demand is 35	5.50	0	6.00	9.00
.40	Demand is 45	11.00	5.50	0	3.00
.20	Demand is 50	13.75	8.25	2.75	0

reason for the row-column to identify with the event-action) would be to remember the definition of opportunity loss. What is the best act for a given state of nature? Any other action leads to a smaller payoff for this specific state of nature and therefore an opportunity loss.

As with the payoff table, two formulas can be written for the opportunity loss (OL) function. If $x \geq Q$, then $OL = .55(x - Q)$; i.e., if demand exceeds the order quantity, the opportunity loss will be simply the product of per-unit contribution and the number of additional units that could have been sold. This is the equation for opportunity losses on and below the main diagonal of Table 15.4, part c.

If $x < Q$, then $OL = .60(Q - x)$; i.e., the number of units of overage are multiplied by the $.60 per-unit cost.

Note that numbers in the opportunity loss table are always positive. If at any point you have a negative number in the OL table, you have made an error. The term *loss* negates all numbers in the table, and zero is the very best position in an opportunity loss table.

The expected opportunity loss (EOL) for each alternative is calculated in the same manner as that for expected payoff. The opportunity loss for each alternative is multiplied by the probability that the loss will be incurred, and these products are summed to obtain the expected opportunity loss. As explained earlier, this is really the expected value of a discrete random variable. The calculation of the expected opportunity loss for each alternative is given in Table 15.5.

The best alternative is that with the minimum expected opportunity loss. From Table 15.5 we see that the third alternative—"order 45 sandwiches"—has the minimum EOL. This is also the alternative with the maximum expected monetary value (see Table 15.2); thus, the 2 approaches, maximize EMV and minimize EOL, lead to the same conclusion: Suzi's best alternative is to order 45 submarine sandwiches each day.

It is interesting to note that the minimum expected opportunity loss, $3.55, is exactly what was calculated to be the expected value of

---

Table 15.5   Calculation of Expected Opportunity Loss

First alternative—"order 25":

$0(.1) + 5.5(.3) + 11.00(.4) + 13.75(.2) = 8.80$.

Second alternative—"order 35":

$6(.1) + 0(.3) + 5.5(.4) + 8.25(.2) = 4.45$.

Third alternative—"order 45":

$12(.1) + 6(.3) + 0(.4) + 2.75(.2) = 3.55$.

Fourth alternative—"order 50":

$15(.1) + 9(.3) + 3(.4) + 0(.2) = 5.40$.

perfect information. The minimum expected opportunity loss—frequently called the *cost of uncertainty*—will always equal the expected value of perfect information. This is what we would intuitively expect; with perfect information, the future demand is known, and Suzi can select the best alternative for that particular demand. This means that the opportunity loss is always 0; thus, perfect information guarantees an expected opportunity loss of 0. The best possible action with no information is "order 45" with an $EOL = \$3.55$.

An Identity Relating *EMV*, *EOL*, and Expected Payoff Given Perfect Information

For any action, be it the optimal act or not, we have the following important identity:

$EML + EOL$ is constant and equal to the expected payoff given perfect information.

This relationship makes it quite obvious that maximizing the *EMV* will lead to the same action as minimizing the *EOL*. Another important general proposition for inventory problems of this type: the *EMV* is unimodal. If we find an order quantity with an *EMV* that exceeds the *EMV* for order quantities on either side, we have found the optimal *EMV*.

There is a somewhat subtle notion that leads to an efficient method for calculating the expected profit given perfect information. Earlier in the textbook we had the theorem that, if $y = ax$, then $E(y) = aE(x)$. In our problem, the payoff, y, under perfect information is simply $.55x$, where x is demand. Calculating expected demand (see Table 15.1), we have: $E(x) = .1(25) + .3(35) + .4(45) + .2(50) = 41.0$. Multiplying 41.0 by $.55 results in $22.55 for the expected profit given perfect information.

15.4 Decision Criteria when Probabilities Are Not Known

As mentioned earlier, there are decision criteria that can be used under true uncertainty—i.e., in those cases where the possible acts are known and the possible events are known, but the event probabilities neither are known nor can be estimated. In Table 15.6 we present a payoff table for a decision problem with four alternatives and three possible events or states of nature. A list of probabilities is also given to the right, and the *EMV*'s have been calculated and are presented at the bottom of the table. We list the probabilities and the *EMV*'s in order that we might compare these results with those obtained by applying the several criteria for decision under uncertainty.

One decision criterion that could be used to select the "best" alternative is the *maximax rule*: Select the set of maximums for each

Table 15.6 Payoff Table (thousands of dollars)

Probabilities	Events	Alternatives			
		A_1	A_2	A_3	A_4
.05	E_1	1	50	0	−10
.50	E_2	1	−100	20	25
.45	E_3	1	−50	30	−10
	EMV	1	−70	23.5	7.5

act and choose the act with the maximum of this set of maximums. This is equivalent to selecting the act associated with the largest payoff in the table.

From Table 15.6 we observe the following:

Alternative	Maximum Possible Payoff
A_1	1
A_2	50
A_3	30
A_4	25

Using the maximax rule, we would select the second alternative, A_2, as the "best" alternative; i.e., 50 is the maximum of the set of maximums. An obvious criticism of the maximax rule is that it ignores the other possible outcomes for the alternative selected. It is an essentially optimistic rule—perhaps too optimistic for survival if it were actually applied in the business world. With the selection of A_2 there is a possibility of incurring relatively large losses. As mentioned above, if the probabilities were known, neither the maximax nor any of the following rules for decision under uncertainty would be invoked. If the unknown probabilities were as we have given, note how bad the maximax decision could be. For the chosen act, the probability of incurring a large loss is very high, .95, and there is only a 5 percent chance of a positive payoff. This is a decision rule that might be used by the individual who is a risk taker.

In contrast to the optimistic outlook implied by the maximax rule, the second decision criterion we will discuss, the *maximin rule*, is essentially conservative. The maximin rule is as follows: Find the minimum payoff for each alternative; select the alternative associated with the maximum of the set of minimum payoffs. It is a no-risk rule. This same notion applied to a cost table or an OL table would be the *minimax rule*: Select the minimum of the set of maximums. In general, both rules can be summarized as, Select the best from the set of worsts. Presented below is the set of minimum payoffs from Table 15.6.

Alternative	Minimum Possible Payoff
A_1	1
A_2	-100
A_3	0
A_4	-10

The maximum of the set of minimums is 1; thus, according to the maximin rule, the best alternative is A_1. If the decision maker chooses A_1, he knows that the payoff will never be less than $1,000.

Another decision rule that has been suggested is the *most-likely-event rule*. This rule is as follows: Assume that the event with the highest probability will occur, and select the alternative that has the highest conditional payoff. From Table 15.6 we see that the most likely event is the second event, E_2. If this event were to occur, then the respective payoffs for the 4 alternatives would be 1, -100, 20, and 25. Alternative 4, with a payoff of $25,000, would be the "best" alternative under this criterion. This rule, like the maximax and the minimax rules, would be a poor substitute for the decision rule based upon the optimization of an expectation. We don't recommend any of the above decision rules, but perhaps the best of the lot is the maximin criterion, which is, at least, survival oriented.

15.5 Decision Trees

The decision tree provides a technique for addressing decision problems in which the act-event sequences are involved and somewhat complex. The Suzi's Submarine Sandwich Stand problem, treated in a previous section, can be handled with the decision tree method, but for a simple problem of this type the payoff table method is adequate and possibly more instructive. However, many decision problems involve a complex sequence of decisions and events, with the final outcomes dependent upon previous decisions and events. The decision tree technique provides a method for outlining the logic of the time-related decisions along with a method for arriving at the best course of action. In the decision tree problem, we will assume that the criterion for the best course of action is still that of maximizing expected payoff or minimizing expected opportunity loss.

Example 15.5.1 Allied Surplus has sold Christmas trees during the holiday season for the past 10 years. Three days before the deadline for placing its Christmas tree order, Allied learns that it has a chance to obtain exclusive rights to a nationally advertised toy line that currently is not being marketed in the area. Allied made an effort to obtain the exclusive rights earlier in the year, but the rights were awarded to the ABC store.

Now an agent for the toy company informs Allied Surplus that ABC Company has changed its mind regarding the line and he feels that there is a 70 percent chance that Allied Surplus can obtain the exclusive rights if it is still interested. The toy company will make a decision on exclusive rights within a week.

Allied must decide whether to order Christmas trees or try to sell toys. It is unable to do both. If trees aren't ordered within 3 days, that option will close. Allied already has considerable information regarding the toy market because it seriously considered that option earlier in the year. If it gets the exclusive toy line and sets a high price, observations from toy sales in other similar areas, indicate that Allied will have a 40 percent chance to make a profit of $4,000 and a 60 percent chance to make $3,000. If a low price is set, then there is a 70 percent chance to make $3,000 and a 30 percent chance to make $5,000. If Allied decides to sell toys and fails to get the exclusive line, it can buy a line of toys that are already being sold by a number of stores in the area. Company leaders feel that, if they obtain this common line of toys, they can make a profit of $600. If Allied chooses to sell Christmas trees, then it has to decide how many trees to order, and a single order must be placed within 3 days. Based upon past experience, there appears to be a 30 percent chance that demand will be sufficient to sell 500 trees, a 30 percent chance of selling 1,000 trees, and a 40 percent chance that the demand will be 2,000 trees. All trees sell for $4.00 each; purchasing and handling costs are $2.00 per tree. Those trees not sold during the season will be burned. For a seasonal sideline like Christmas trees, Allied feels the goodwill cost of not being able to meet customer demand is 0.

What course of action should Allied Surplus take with respect to selling Christmas trees or toys?

After having read through the above example, most would agree that it would be desirable to have a method for outlining a decision problem of this type. The decision tree approach to a problem isn't difficult to master; however, it is necessary to get clearly in mind the respective meanings of the two basic symbols used in drawing a decision tree, as well as the distinction between them.

The square symbol is used to identify a decision point.

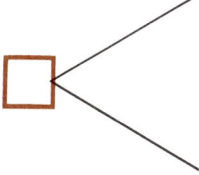

The rays to the right of the square represent alternatives open to the decision maker. A decision point represents a stage in the decision process at which the decision maker has the ability to make a choice. In contrast, the event point, the circle symbol, represents a stage in the decision process at which the decision maker has no control over the outcome.

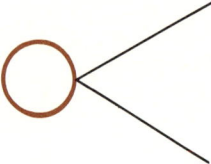

The rays to the right of the circle represent possible events that could occur. Event probabilities are placed on the rays. The decision maker has to accept the outcome that nature or the environment "deals" him.

In Figure 15.1 we present the decision tree that outlines the sequential logic of the Allied Surplus decision problem. The first decision point, on the left, represents the major decision that must be made—to sell Christmas trees or to sell toys. This prime decision is to be based upon the monetary expectation criterion. In order to calculate this for the "sell Christmas trees" branch, a decision must be made on the order quantity of Christmas trees, as illustrated by the second decision point on this upper branch. Each of the three rays from this decision point is in turn drawn with three radiating demands or events (let x equal demand):

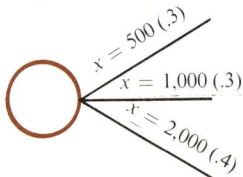

We have chosen to keep event rays to a minimum by using only those that are relevant to the order quantity decision. For example, if 500 trees are ordered, the probability that demand is at least 500 is 1.0; all trees will be sold and only one ray is necessary:

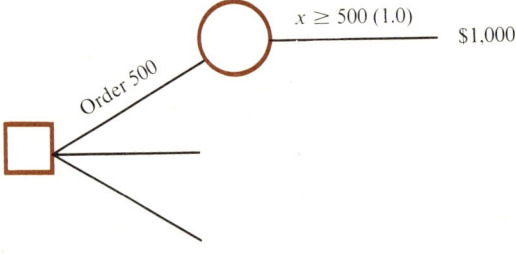

This part of the problem is another example of the inventory problem

Decision Theory

Figure 15.1 Decision Tree

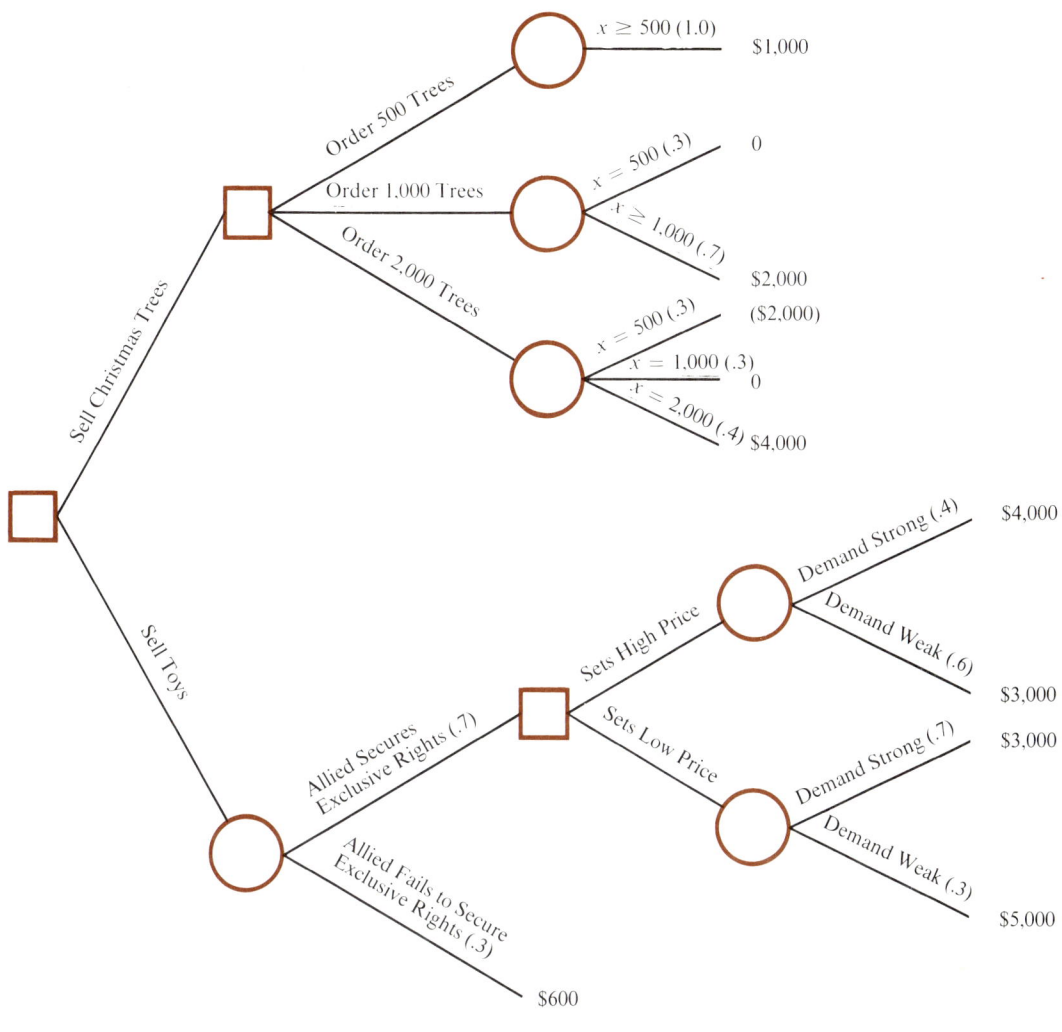

we discussed in Section 1 earlier. See Table 15.7 for the table of payoffs, and verify that these conditional payoffs are correct.

It should be noted that the events represented by the rays to the right of an event point must be mutually exclusive and exhaustive and the sum of the probabilities assigned to the events of a particular event point must be unity. The reader should observe that this is the case for all of the event points in Figure 15.1. (The number in parentheses is the probability that the event represented by the ray will occur.)

If Allied Surplus chooses to sell toys, it may or may not get exclusive rights to the nationally advertised toy line. This is represented by

Table 15.7 Payoff Table if Allied Decides to Sell Christmas Trees

		Alternatives		
Probabilities	Events	Order 500	Order 1,000	Order 2,000
.3	$x = 500$	$1,000	$ 0	−$2,000
.3	$x = 1,000$	1,000	2,000	0
.4	$x = 2,000$	1,000	2,000	4,000
	$E(x) = \Sigma xp(x)$	$1,000	$1,400	$1,000

the first event point at the bottom of Figure 15.1. If exclusive rights are granted, then Allied will have to decide on the pricing policy to pursue. Each policy has an uncertain outcome. A careful study of Figure 15.1 should convince the reader that the diagram does outline the decision problem given in Example 15.5.1.

The dollar amounts at the endpoints of the decision tree are the gains or losses (in parentheses) that are attained if the choices are made and events occur that lead to the particular endpoint. For example, if Allied decides to sell toys, if it receives exclusive rights, if it sets a low price, and if demand is strong, the payoff will be $5,000.

Finding the best action from the decision tree requires a "backward" solution. We begin at the outermost branches, calculate the expectations (and frequently expectations of expectations), and move to the leftmost decision point, where the act with greatest expected monetary payoff is chosen. Refer to Figure 15.2 as the specifics of the process are discussed. Near the right of the "sell Christmas trees" branch, the 3 expected payoffs for the possible order quantities (see Figure 15.1 again) are placed near the circle event points. The maximum expected payoff is $1,400 for the order quantity of 1,000 trees. If it decided to sell Christmas trees, Allied would order 1,000 trees, and so the $1,400 is placed at the square decision point; the other two possible rays (and actions) are slashed (#), indicating the alternatives not chosen.

Working backward on the "sell toys" lower branch, we see that the expectation for a high price policy is $3,400 and that for a low price policy is $3,600. The "high price" ray is slashed, and the expected payoff for the low price policy, $3,600, is placed at the rightmost decision square. We now have an expectation of an event ray. This $3,600 expectation along with the $600 conditional payoff yields a final $2,700 expectation for the "sell toys" decision branch. Comparing this with the expected payoff on the "sell Christmas trees" branch, only $1,400, we slash the latter ray. The correct decision is to sell toys and, if the exclusive line is secured, go with the low price policy, for a maximum expected monetary payoff of $2,700.

It might be helpful to the reader to discuss some additional details

Figure 15.2 Backward Solution to Decision Tree

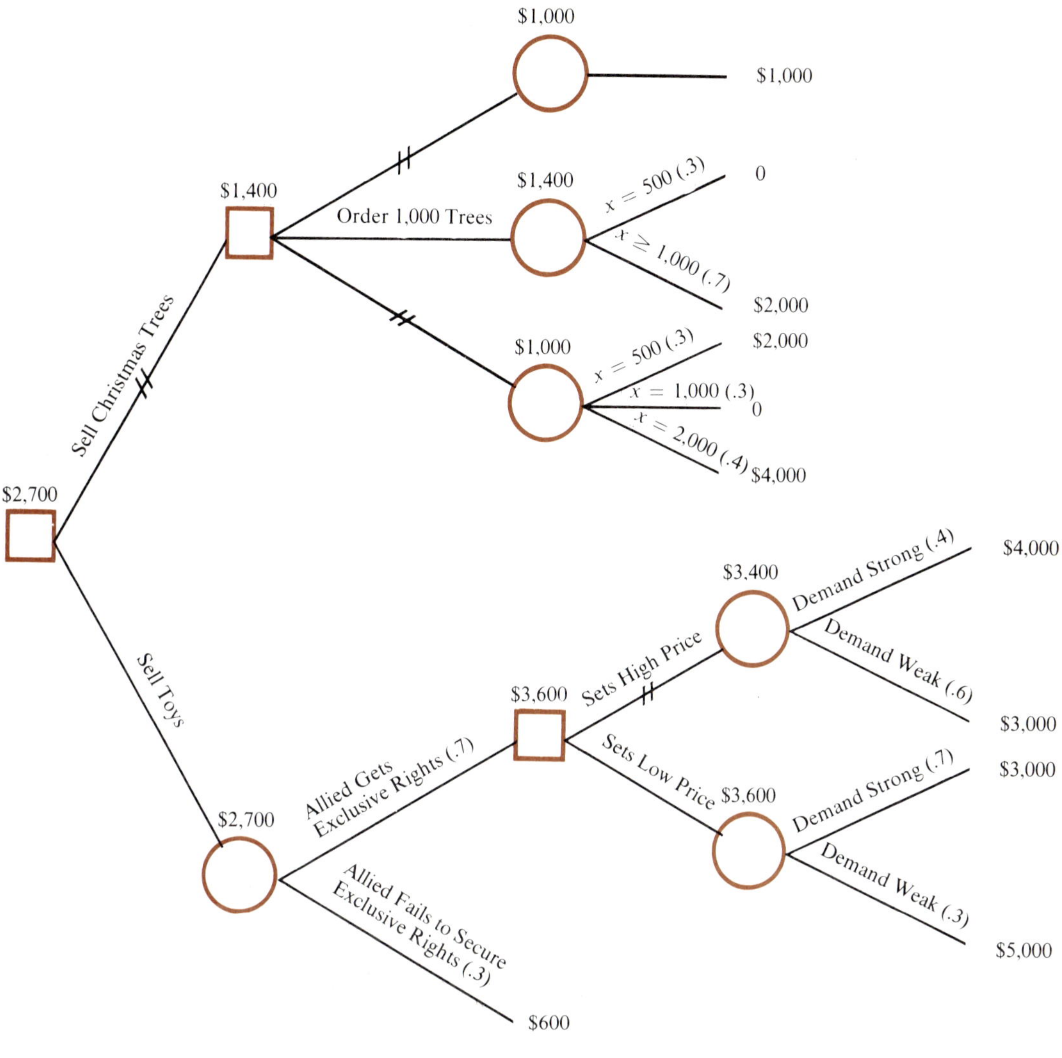

of evaluating expectations in the decision tree in Figure 15.2. The event point following the alternative "sets low price" is repeated here.

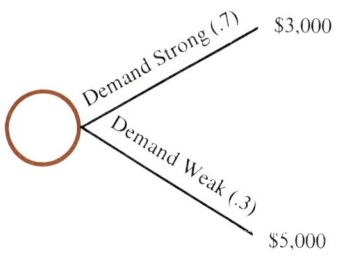

Decision Trees

This event point represents the two possible outcomes if a low-price strategy is followed. There is a .7 probability of making $3,000 and a .3 probability of making $5,000. The expected monetary value is (.7)($3,000) + (.3)($5,000) = $3,600. If Allied chooses to follow a high-price strategy, the expected monetary value would be (.4)($4,000) + (.6)($3,000) = $3,400. The better choice is to follow a low-price strategy with EMV = $3,600. This is indicated by writing $3,600 over the decision point and crossing out the ray representing "sets high price."

Continuing backward to the event point near the lower left of the diagram, we have the situation illustrated below.

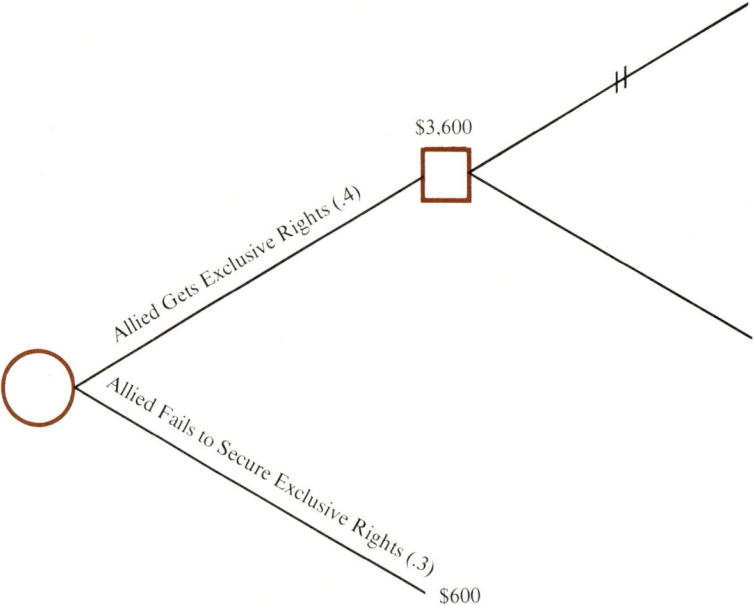

This represents the possible outcomes, along with their probabilities, that Allied will face if it chooses to sell toys. The expected monetary value for this event point is (.7)($3,600) + (.3)($600) = $2,700. This is the EMV for the alternative "sell toys."

15.6 Insurance, Gambling, and Utility Theory

The criterion used throughout this chapter for making the "best" choice when the outcomes are uncertain is that of selecting the alternative with the highest expected monetary value. This criterion will usually result in the "correct" decision for most of the problems facing the business person if there is relatively little variability among the monetary payoffs in the set. If relatively extreme payoffs are involved, then the expected monetary value may prove to be an unsatisfactory decision criterion. The following illustrates what is meant by "relatively extreme payoffs."

Consider an individual who has just purchased a new house for $50,000. An insurance agent approaches the new home buyer and offers him a fire insurance policy for a $200 annual premium. Assume that the probability that the house will burn down during any given year is 1/1,000. Should the new owner buy the insurance policy? (Assume that the premium offered by all other companies in the area is at least $200.) Table 15.8 has the payoff table and the expected monetary values for the 2 alternatives. The entries under the "buy insurance" alternative need some explanation. If insurance is purchased and the house doesn't burn down, then the homeowner has given up $200; i.e., his wealth is reduced by $200. If he or she buys insurance and the house burns down, the homeowner's net wealth reduction is still only $200 because the insurance policy pays for the house (but not the premium). If insurance is not purchased and there is no fire, the homeowner's wealth position remains unchanged; however, if the house burns down, his or her wealth is reduced by $50,000.

According to the expected monetary value criterion, the best alternative is to *not* buy insurance. The expected value of this alternative is −$50, which is greater than the −$200 expected value when fire insurance is purchased. Yet most homeowners would choose to buy the insurance even if their banker holding the mortgage didn't require it. Why would our homeowner buy the insurance? Because the loss of $50,000 is more painful to him or her than 250 times the pain suffered from giving up the $200 insurance premium (250 × 200 = 50,000).

The homeowner insurance company "game" is similar to the game between the player and the "house" on an Atlantic City crap table. The insurance company, like the house, is facing a large number of players. Also, the payoffs and the related probabilities are almost as accurately known for the insurance company as they are on the crap table. Both the insurance company and the gambling house calculate expectations adequate to cover costs and desired profits. The game

Table 15.8 Payoff Table for Insurance Problem

Probabilities	Events	Alternatives	
		Buy Insurance	Don't Buy Insurance
.999	No Fire	−$200	$ 0
.001	Fire Burns House Down	−$200	−50,000
Expected Monetary Value		−$200	−$50

being played, from the insurance company and house side, is a *repeated* game; as mentioned in the introduction to this chapter, expectations for a game that can be repeated a large number of times will actually be realized. For all practical purposes, the large insurance company is not playing a risk game. On the other hand, the homeowner and the individual player are facing a negative monetary expectation in a game where the number of repeats is limited. The player on the crap table is hoping to "get lucky" and win over a short-run series of plays. Most players are aware that the odds are against them. In a reverse way, the homeowner is guarding against a disastrous dollar loss and, at the same time, hoping *not* to "get lucky" and win $50,000 from the insurance company.

A similar situation arises when there is a possibility of receiving a large sum of money for a small investment. This is the case for many gambling situations—in particular a lottery with a high payoff. A nongambling *(risk averse)* individual might buy a $2 lottery ticket if the payoff were $1 million even if he or she was aware that a million tickets were being sold. The expected payoff to this individual would be negative,

$$EMV = -\$2(.999999) + \$999,998(.000001) = -\$1,$$

and yet many individuals will take this type of risk or bet. Why? Because the possibility of receiving $1 million causes an excitement and joy that exceeds the small pain of the loss of $2 by much more than the relative dollar amounts involved.

When it comes to buying house insurance, the individual is acting in a risk averse manner. The homeowner doesn't want to risk a large loss even though the cost of risk avoidance exceeds the expected return. The same individual could well act in a *risk preferring* manner and buy a lottery ticket even though the expected return is negative. This suggests that something other than expected monetary value is needed to describe the decision process when relatively extreme losses and/or gains are possible. Decision theorists have suggested the use of utility theory. A utility function is developed for the individual decision maker which reflects that person's relative feeling for different dollar amounts when uncertainty is involved. Friedman and Savage in their classic article on the topic[2] have suggested that most individuals in the everyday world of business are basically conservative and the general appearance of the utility curve in Figure 15.3, i.e., concavity to the money axis—would describe this conservatism.

In the discussion and examples that follow, we assume that it is possible to construct a cardinal utility function for an individual and, where we use actual numbers for utilities, that such a function has been derived. We will see, as this problem is discussed in Appendix 15A,

---

2. M. Friedman and I. I. Savage, "The Utility Analysis of Choices Involving Risk," *Journal of Political Economy,* August, 1948.

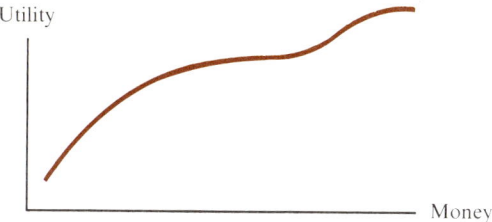

Figure 15.3 Utility Function

that it is not an easy task to derive a utility function. Before proceeding, several additional observations need to be made: (1) The scale on a person's utility curve is arbitrary; i.e., the magnitude of the utility has meaning only in comparison with another utility from the same person's utility curve. (2) Comparison of specific utility numbers between two different functions is meaningless. (3) It is axiomatic that the utility for a payoff, say x_1, is at least as large as the utility for a second payoff, x_2, if $x_1 > x_2$; that is, it is assumed that an individual is irrational if utility decreases for an increase in the payoff.

Once a utility function has been determined for the decision maker, the criterion for selecting the "best" alternative becomes, Select the alternative with the highest expected utility. The expected utility is found by replacing the dollar values with their respective utilities. For example, the person considering buying the lottery ticket in the above example has a choice of buying the ticket or not buying the ticket. If that person's utility function is such that the utility of $0 is 10, the utility of $-$2 is 8, and the utility of $999,998 dollars is 4 million, then the expected utility of buying a lottery ticket is approximately 12. The expected utility[3] of the lottery, 12, is compared with the utility of $0—10—the so-called do-nothing utility. Inasmuch as the expected utility of buying the lottery ticket is greater than the utility of not buying the lottery ticket, the person under discussion prefers the lottery, and it has a positive "cash equivalent" for him or her. The calculation of the expected utility of the lottery ticket is given below.

Utility of $0 = 10
Utility of $-$2 = 8
Utility of $999,998 = 4,000,000.

The lottery offers a .999999 probability of losing $2 and a .000001 probability of winning $1,000,000, which would be a net of $999,998. This is equivalent to a probability of .999999 that a utility of 10 will be obtained and a probability of .000001 that a utility of 4,000,000 will be

---

3. The expected utility of a gamble or a business decision is simply referred to as the "utility" of the lottery or decision.

Insurance, Gambling, and Utility Theory

obtained. The utility of purchasing the lottery ticket is calculated as follows:

Utility of lottery = expected utility = 8(.999999) + 4,000,000(.000001) = 12.0.

Utility theory could be applied to the decision tree problem discussed in the previous section.

Another example, less complicated than the Allied example, will illustrate the use of a utility function.

Example 15.6.1 The president and owner of a small construction firm went to China for a month's vacation. Before leaving, he told the general manager to consider one additional contract if the opportunity arose and the cash equivalent was at least $500. The president left his utility function with the general manager:

$$U(x) = 1 - \left(\frac{x - 10}{20}\right)^2$$

where

x = payoff in *thousands* of dollars
U = utility.

The president stressed that this function was good only for payoffs between −$5,000 and $10,000. If the dollar risks involved in a contract were outside these limits, the contract shouldn't be considered. The day after the president left, two possible contracts became available. Analyzing these on the basis of weather forecasts (true states of nature) the manager arrived at the payoffs shown in Table 15.9. Estimated probabilities for N_1, N_2, and N_3, respectively, were .3, .4, and .3. Note that the *EMV* for contract A is $1,500 and that for contract B is $1,240.

Utilities for the dollar amounts are presented in Table 15.10. The (expected) utilities for each contract are

$U_A = .3(1.0000) + .4(.75000) + .3(.4375) = .73125$
$U_B = .3(.8775) + .4(.7978) + .3(.7399) = .8422$.

From the above utilities we can make 2 observations: (1) contract B is

Table 15.9 Payoff Table

States of Nature	Contract A	Contract B
N_1	$10,000	$3,000
N_2	0	$1,000
N_3	−$5,000	−$200

Table 15.10 Utility Table

	Contract A	Contract B
N_1	1.0000	.8775
N_2	.7500	.7975
N_3	.4375	.7399

preferred to contract A; (2) the cash equivalent of contract B is positive because its utility, .80422, exceeds the do-nothing ($0) utility of .75. (Note that contract A has a negative cash equivalent, even with a higher monetary expectation than B. This results from the higher variability of the payoffs in A.)

The general manager must also know if the cash equivalent (CE) of this utility exceeds $500. This is answered by finding the inverse utility function. Solving for x as a function of U we get: $CE = x = 10 - 20\sqrt{1 - U}$ where in this form, x is the cash equivalent in thousands of dollars and U is the utility (expected utility). The CE is approximately $1,100. The general manager should accept contract B; this is the same view the owner would have taken had he been in town (and also assuming that he didn't lose a bundle on his overnight stop in Vegas).

Using expected utility instead of expected monetary value as a basis for decision making is theoretically appealing. There are, however, some problems involved when trying to apply utility theory to practical problems, not the least of which is determining the utility function for the decision maker. Utility is different for each individual and is further complicated by the fact that the function may change over time. Determining the utility function for a corporation is even more difficult. Appendix 15A illustrates how one might go about trying to determine a utility function.

Fortunately, many of the decision problems a corporation is confronted with involve dollar amounts over a range such that the use of expected monetary value as a criterion gives essentially the same results as would have been obtained if expected utility had been used as a criterion. For these cases, the decision maker is spared the onerous task of trying to determine the utility function for the corporation. The decision maker must, of course, make a judgment as to whether or not the dollar amounts involved are such that expected monetary value is an appropriate criterion. If the variability of the dollar amounts involved is relatively small, then the expected monetary value will generally be an adequate measure for comparing alternatives. When some of the alternatives have possibilities of very large losses and/or gains, expected monetary value may not be a satisfactory criterion for selecting among alternatives.

15A Appendix: Deriving a Utility Function

For purposes of this discussion, we will define a lottery as a combination of 2 mutually exclusive events 1 of which must occur; i.e., the sum of the probabilities assigned to the events must equal 1.[4] A point on the utility curve of an individual is determined by offering the person a lottery and asking what amount of money would seem as valuable as the lottery. For example, the individual may be presented with a 50 percent chance of losing $5,000 and a 50 percent chance of receiving $50,000. The expected monetary value of this lottery is ½(−5,000) + ½(50,000) = $22,500. Given the choice, if the individual is risk averse, he or she would indicate that $22,500 with certainty would definitely be preferable to the indicated lottery. Thus, the utility of $22,500 for that particular individual is greater than the utility of a lottery with a ½ probability of losing $5,000 and a ½ probability of gaining $50,000.

The next step is to have the individual indicate an amount of money that he or she would consider equivalent to the lottery; i.e., the person would have an equal preference for the specified amount with certainty and the lottery amounts with their respective probabilities. Assume the stated amount is $15,000. This means that any amount greater than $15,000 is preferred to the lottery and the lottery is preferred to any amount less than $15,000. Thus, the utility of $15,000 with certainty is equal to the expected utility of the lottery. The utility of the lottery is expressed as an index number. This in turn can be determined by assigning utility index numbers to the dollar amounts, −$5,000 and $50,000, substituting the index numbers for the respective amounts, and applying the probabilities of the lottery to calculate the expected utility of the lottery. The only information lacking at this point is the utilities of −$5,000 and $50,000.

For use in decision making, the utility function must be unique up to a linear transformation. What this mathematical jargon means is that we can arbitrarily assign utility values to dollar amounts, but then the other utilities must be determined using the method described in the previous paragraph. Returning to the example in the above paragraph, we can arbitrarily assign utility values to −$5,000 and $50,000. Let the utility of −$5,000 be 10 and the utility of $50,000 be 100. We write $U(-\$5,000) = 10$ and $U(\$50,000) = 100$. Utilities for any other dollar amounts must be calculated. In the above paragraph we determined that the utility for the lottery was the same as the utility of $15,000. The index number for expected utility of the lottery can be calculated using the utility values assigned to −$5,000 and $50,000:

---

4. The pioneering work in utility functions is the book by John Von Neuman and Oskar Morgenstern, *Theory of Games and Economic Behavior*, John Wiley and Sons, Inc., New York, 1944.

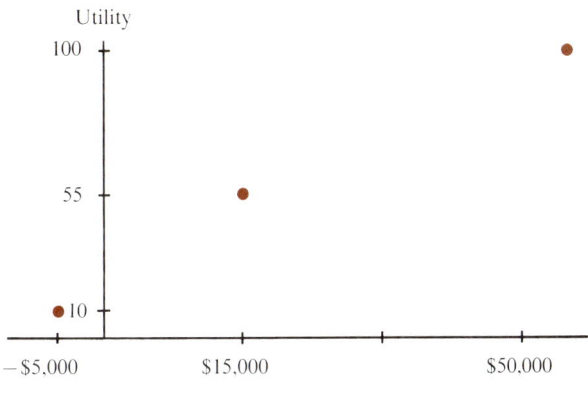

Figure 15.4 Three Points on the Utility Curve

$$U(-\$5,000) = 10$$
$$U(\$50,000) = 100$$
$$\text{Expected Utility} = \tfrac{1}{2}(10) + \tfrac{1}{2}(100) = 55.$$

The utility of $15,000 is equal to 55, the expected utility of the lottery.

We now have 3 points on the utility curve (see Figure 15.4)—those for −$5,000, $15,000, and $50,000. There are a number of ways to obtain additional points on the curve. One approach would be to assign different probabilities to the same dollar amounts (in our example, −$5,000 and $50,000) and ask the individual to indicate the sum of money that he or she considers to have equal utility with this lottery. Another approach would be to select a sum of money and ask the individual to indicate the probabilities that would be required to make the lottery as attractive as the sum of money. A third approach would be to fix the lottery probabilities (for example, ½ and ½), change the payoffs, and search, as above, for the certainty cash equivalent to the new lottery. Given that a great deal of subjectivity is involved in any of the above approaches, it is our opinion that this third approach is more easily and accurately comprehended by the average person.

We will use the third method to find additional points on the utility curve in our example. Using the third method and referring to Figure 15.4, we observe again that we have utility values for 3 dollar amounts. We can form a lottery using any 2 of these amounts and use this lottery to determine the utility of a fourth dollar amount. For example we could ask the individual for whom we are deriving the utility curve to indicate the amount of money that seems equivalent to a lottery with a 50 percent chance of obtaining $15,000 and a 50 percent chance of obtaining $50,000. If the individual decision maker indicates no preference between this lottery and $30,000, then the utility of $30,000 is the same as the expected utility of the lottery. The expected utility is calculated as follows:

Lottery:
½ probability of $15,000
½ probability of $50,000.

Lottery in terms of utility:
½ probability of 55
½ probability of 100.

Expected utility of the lottery:
½(55) + ½(100) = 77.5.

Thus, the utility of $30,000 is 77.5. Similarly, we obtain another point on the curve when an equal preference for $8,000 and a lottery with ½ probability of $30,000 and ½ probability of −$5,000 has been expressed. The expected utility of the lottery is ½(10) + ½(77.5) or 43.75, which is also the utility of $8,000.

Figure 15.5 shows the five points on the utility curve thus far determined. A smooth curve sketched through these points would represent the individual's utility function. The above process could be used to obtain additional points. Our experience indicates that plotting more than five or six points will frequently show what appears to be an irrationality; i.e., the axiom of a nondecreasing function will be broken. This is probably due to the subjective impreciseness of the method rather than any irrationality on the part of the subject. If the plot of the points is concave downward (or concave upward) throughout, a second-degree function could be fitted to the points (see Chapter 10). As a description of the utility, this mathematical function may be more accurate than the freehand curve fit and, clearly, the utility and cash-equivalent values determined from a formula are exact given the formula.

Figure 15.5 Five Points on the Utility Curve

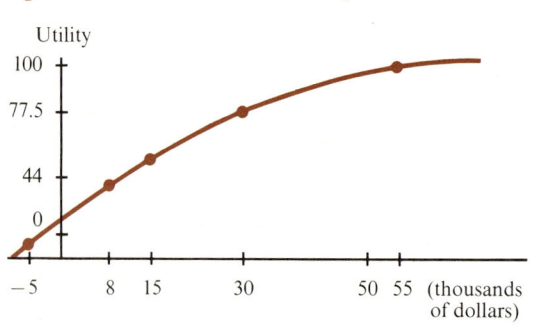

Problems

1. Consider the following payoff table where the A's are possible acts and the N's are states of nature.

	A_1	A_2	A_3
N_1	8	4	-2
N_2	1	2	5
N_3	-3	-1	2
N_4	-2	1	-1

   Show reasoning and/or work for each of the following:
   a. If the probabilities of the states of nature were .2, .4, .3, and .1 for N_1 through N_4 respectively, maximizing expected payoff would lead to act _____ with an expected payoff of _____.
   b. What is the *EVPI*?
   c. Work out the opportunity loss table.
   d. Which act is selected if the minimax (maximin) principle is applied to the payoff table?
   e. Which act is selected if the minimax (maximin) principle is applied to the opportunity loss table?

2. A retail-store manager is trying to decide how many boxes of Kalmack's most expensive Christmas cards to stock. These cost $8.00 per box and retail for $15.00. Boxes not sold during the season are closed out at $4.00 a box; no reorder is possible. From past experience, the manager has estimated the following demand function:

x	5	6	7	8	9	10	11
$p(x)$	.07	.08	.15	.20	.20	.15	.15

   a. What is the optimal order quantity?
   b. What is maximum expected profit?
   c. What is *EVPI*?
   d. What is expected profit given perfect information?

3. Consider the conditional value table below.

		Our Bid Price/Unit			
Probability	State of Nature	$1.00	$1.20	$1.30	$1.40
(.2)	We are sole bidder	40	60	80	100
(.3)	Jones bids too	60	80	60	40
(.5)	Both Jones and Smith bid	80	60	40	20

   (Conditional values are in thousands of dollars)

a. Find the *EMV* of each bid price, and select the bid price with the highest *EMV*.
b. Find the minimax values of each bid price, and select the best bid price according to this criterion.
c. Find the conditional opportunity loss table and the *EOL* of each bid price.
d. What is the maximum amount you would be willing to spend to find out about Jones's and Smith's intentions?

4. Tom's Toy Manufacturing, Inc., is heated by coal during the winter. The management is faced with the following problem: Coal ordered in September costs $15 per ton. (This price includes holding and interest costs to the firm.) If ordered during the winter the price is $19 per ton. Coal left over in the spring is sold back at $8 per ton (the coal storage space is needed for expanded summer operations). Winter may be mild, normal, or very cold; coal requirements are 20, 30, and 50 tons respectively.
    a. Set up Tom's problem in decision tree form.
    b. The historical probabilities are .2, .5, and .3 respectively for mild, normal, and cold winters. What is Tom's best strategy?
    c. If an old Indian friend were willing and able to predict with certainty the nature of the coming winter (based upon the September signs), what's the most Tom should be willing to pay him for this annual service?
    d. Use a cost table and an opportunity loss table to determine the best strategy.
    e. If Tom's cash position were very tenuous in September, 1971 (he didn't even want to put out the fee for the Indian), and he decided to order coal as he needed it, what would be the expected cost of his irrationality?

5. JB is currently under contract with the U.S. Post Office, trucking mail from Bliss to Iona. He is clearing $18,000 per year. He is considering bidding on the local poultry association contract, which involves trucking eggs from Bliss to the railhead at Shippey Hill and feed from the railhead to Bliss. His post office job is secure, but the annual agreement for this job must be signed before any decision will be made on the poultry contract. There is uncertainty with the poultry association because it is likely that there will be at least one other bidder. He figures there is a 75 percent chance of his getting the poultry contract. Even if he gets the contract, there is additional uncertainty regarding egg farming. If egg prices stay up and/or feed prices don't go too high, he will clear $30,000 per year; if things don't go well, his profit could be only $15,000. Things in general look pretty good, and he decides there is a .90 probability of good times in eggs. If he doesn't get the poultry contract his only alternative is to take local furniture

moving and coal hauling jobs. He has done this before and knows he can clear only $10,000 per year at this.
   a. Draw a decision tree of JB's problem.
   b. If he is willing to use expectation as a decision criterion, what should he do and why?
   c. What is his *EVPI*? Explain to JB what *EVPI* is.

6. Joe's Job Shop has signed a contract to produce 10 large filter unit castings for a chemical plant being built nearby. The variable production cost is $300 per unit; setup cost or fixed cost per production run is $1,200. If there are not enough nondefective units on the first run, additional units will be made 1 at a time (every unit produced individually is good) at a cost of $1,000 each. Excess production, either defective or nondefective, is a total loss; i.e., there is no salvage value. From long experience at this sort of work, Joe figures he is faced with the following probabilities:

Q	10	11	12	13
$P(x \geq 10) =$	.6	.90	.95	1.00
$P(x = 9) =$	.3	.08	.05	0
$P(x = 8) =$	.1	.02	0	0

(Q is the number of units on the initial production run, and x is the number of good or nondefective units.)
   a. Sketch the decision diagram.
   b. Using the costs as the criterion, calculate terminal costs for each act-event sequence.
   c. Using expected costs as the basis for decision (i.e., Joe's preference function is linear over the set of possible costs), which Q value should Joe choose? Show work.

7. Harry Cooley is considering buying a lot of 1,000 electric motors which he uses in the assembly of his Model HC-009 hair dryer. These are war surplus and have been to Vietnam and back. They were manufactured by his current supplier and if handled properly would be of the same quality, just 5 percent defective. If they were not handled properly on the overseas trip, salt spray and moisture would have made 50 percent defective. The vendor is noncommital on this point, but from past experience HC figures there is only about a .30 chance that the motors have been handled properly. He can buy the lot for $5,000, which is just half his current cost for new motors. Defective motors from the war surplus lot are worthless because his present supplier will not, of course, replace them with good ones, which he does do on the newly manufactured lots. The cost of handling defective motors on the assembly line is $3 each. HC needs 1,000 good motors right away. If he should buy the war surplus lot, he could fill in any shortage with an order from his supplier.

a. Should HC buy the war surplus lot? Show work to support your answer. What is the *EVPI*?
b. Assume the vendor will allow HC to test 3 motors if HC will pay him $50. (Other testing costs are negligible.) HC decides to do so and to make the purchase if there are no defectives; if one or more defectives are in the sample of 3, he will not buy. What is the unconditional expected risk of this decision? Did he make a good decision? Why?

8. A particular item that must be ordered daily has a retail unit selling price of $8.00. The item costs $2.00 per unit and the variable unit cost of selling the item over the counter is $.50. Items left over at the end of the day have no salvage value; in fact, the costs of disposal are $1.00 per unit. Let d be the demand variable and Q the order quantity.
   a. Write equations for both profit (contribution) and opportunity loss as functions of d and conditional upon Q.
   b. If the probability distribution for demand is as follows,

d	8	9	10	11
$P(d)$	.2	.4	.3	.3

   find the expected demand.
   c. Find expected loss for $Q = 10$; $Q = 9$.
   d. If I assure you that either 9 or 10 is the best order quantity, which is best and why? What is the cost of uncertainty?
   e. Calculate the expected profit for the best order quantity.

9. Frieda's Frisbee, Inc., is considering building a new plant in Mexico City. Frieda estimates that total contribution to profit and construction costs would be $1 million, $2 million, or $4 million depending upon the success of her advertising campaign. The prior probabilities are estimated at .2, .3, and .5 respectively for the above 3 states of nature. She can build a large plant for $1.8 million, which will handle all possible demands. She can build a small plant for $1.3 million; this will handle the 2 smaller demands, and if she is successful in creating the larger demand, she may double the size of the small plant for another $1.3 million cost.
   a. Using expected payoff as the decision criterion, what should Frieda do? What is her *EVPI*?
   b. Frieda was not too sure that she should use expectation as a decision criterion. Frieda's utility function is given below:

   $U(x) = 5x - x^2 \ (-1 \leq x \leq 3.0)$
   (x is millions of dollars.)

   What are her utilities for the possible acts? (Note that "do nothing" is a possible act.) What decision would you recommend? Find the cash equivalents for these acts. Classify Frieda with regard to her attitude toward risk.

10. Three linear segments provide good approximations of utilities to be used for making decisions in which the cash amounts involved range from −$5,000 to $10,000. Following are equations of the utility function:

   1. $U = .50 + .10x$ for $-5 \leq x < 0$.
   2. $U = .50 + .06x$ for $0 \leq x < 5$.
   3. $U = .60 + .04x$ for $5 \leq x \leq 10$.

   (x is in thousands of dollars.)

   ABC Corporation is faced with a decision on the following 2 possible acts with the given payoffs.

		payoffs	
N	$p(N)$	A_1	A_2
N_1	.2	−5,000	1,000
N_2	.5	2,000	2,000
N_3	.3	10,000	3,000

   a. Calculate utilities and cash equivalent for each action.
   b. Which action should ABC take?
   c. Assume that XYZ Corp. has linear utility over this complete range; i.e., EMV may be used as the decision criterion. Which action would XYZ select?

11. The management of Apex Corporation feels that there is a fair chance that the demand for a certain item may double next year. They are currently manufacturing and selling 100,000 units per year. The present cost is $.80 per unit and the selling price is $1.20 per unit. Additional equipment needed to double output would cost $30,000, but production costs would drop to $.70 per unit. The probabilities of N_1 (sales 100,000) and N_2 (sales 200,000) are estimated to be .4 and .6 respectively. Assume that only these 2 states of nature are possible and that, even if Apex tools up to meet the higher demand, it can still adjust production to the lower demand so that there will be no surplus.
   a. What would you recommend to Apex and why?
   b. If management felt uneasy about your decision in a and further investigation revealed a utility function, $U(x) = \sqrt{x + 40}$, where x is in thousands of dollars, what would you now recommend? Show your work.

■ The remaining problems require material covered in the appendix.

12. Mrs. Smith's utility for $20,000 is 10 and for $0 is −10. Mrs. Smith is indifferent between a lottery involving $20,000 and $0 and a certain cash payment of $15,000. Her utility for $15,000 is 6.

- a. What probabilities would you assign to the two lottery payoffs? Show work.
- b. Comparing the cash equivalent of Mrs. Smith's and the expectation, we would classify Mrs. Smith as risk _____ over the range from 0 to $20,000.

13. In a lottery, Pete has a .6 chance of winning $200 and a .4 chance of losing $50. He is willing to sell his ticket for $40; i.e., he is indifferent between the lottery and a certain $40.
    - a. If we arbitrarily assign a utility of 10 to $200 and a utility of 5 to a -50, what is his utility for $40?
    - b. Pete states that if the probabilities were changed from .6 and .4 to .5 and .5 on the original lottery, his cash equivalent would be 0. Find his "do-nothing" utility.
    - c. Plot these 4 points on an arithmetic grid and draw a smooth curve through them. Does Pete appear to be rational?

CHAPTER 16

SAMPLING

In this chapter we
look beyond simple random sampling to several
other kinds of designs and procedures.

Specific objectives are—

to present the fundamentals of stratified random
sampling with illustrative applications (Sections
1, 2, and 3).

to present the concepts of systematic sampling
and cluster sampling (Sections 5 and 6).

to discuss the problems related to implementing
randomness in applied sampling (Section 4).

In our previous discussions of sampling and sampling errors, we have considered the sample to be a simple random sample. There are some important and interesting alternative sampling procedures. Any one or all of the following motivate our consideration of some of these alternatives.

1. Some procedures yield more precise results than a simple random sample without increasing sample size; or, as a tradeoff, sample size can be somewhat reduced without decreasing precision.
2. It may be physically impossible to implement simple random sampling.
3. The cost of implementing simple random sampling may be prohibitive.

As we discuss alternative sampling procedures, it should be kept in mind that most of the applications to business and economics are related to statistical estimation rather than hypothesis testing. The term *survey sampling* is frequently used and specifically implies estimation.

16.1 Stratified Random Sampling

If in the population of interest there are strata[1] that are more homogeneous than the population itself and if it is possible to draw random samples within these strata, we will find that it pays to do so. This is called *stratified sampling*, and the payoff is in terms of increased precision of a statistical estimate for a given sample size.

Before proceeding with an example illustrating stratified random sampling and the resulting increase in precision, let us review the notation and the formulas associated with simple random sampling.

For the population,

1. The mean is μ;
2. the standard deviation is σ;
3. the variance is σ^2; in this chapter we find it convenient to use $V(x)$ or simply V for variance;
4. N is the population size.

For the sample,

1. the mean is $\bar{x}$;
2. the standard deviation is s;
3. the variance is s^2, or for convenience we may use $v(x)$ or v;
4. n is the sample size.

---

[1]. The strata are a set of subpopulations that, taken together, comprise the population. They are a partitioning; i.e., the strata are collectively exhaustive and mutually exclusive.

For a simple random sample, recall that—

1. $E(\bar{x}) = E(x)$ or $\mu_{\bar{x}} = \mu$;

2. $\sigma_{\bar{x}} = \dfrac{\sigma}{\sqrt{n}} \sqrt{1 - \dfrac{n}{N}}.$

Note: the *fpc* factor,

$$\sqrt{1 - \dfrac{n}{N}}$$

should be

$$\sqrt{\dfrac{N - n}{N - 1}};$$

dropping the 1 in the denominator gives

$$\dfrac{N - n}{N} = 1 - \dfrac{n}{N} - 1 - f$$

where f is the fraction of the population sampled. This can create significant errors in textbook examples where populations are small but is rarely significant in real problems.

3. If N is large relative to n,

$$\sigma_{\bar{x}} \cong \dfrac{\sigma}{\sqrt{n}}.$$

(If N is infinite, the equals holds.)

Additional Notation for Stratified Sampling

In the discussion of stratified sampling, we adopt the following additional notation.

1. The number of strata will run from 1 through L.
2. Stratum sizes will be denoted by N_h, where h goes from 1 to L; and we note that

$$\sum_{h=1}^{L} N_h = N.$$

3. We will frequently discuss stratum weights, W_h, where

$$W_h = \dfrac{N_h}{N},$$

the proportion of the population in the given stratum; note the

$$\sum_{h=1}^{L} W_h = 1.$$

4. Stratum means will be denoted by μ_h; also

$$\mu = \sum_{h=1}^{L} W_h \mu_h;$$

i.e., the population mean is the weighted average of the stratum means.

5. Stratum standard deviations are σ_h, and variances are σ_h^2 or $V_h(x)$. In general $V(x)$ is not the weighted average of the stratum variances. This relationship is discussed below.

Some hypothetical parameters from a manufacturing firm with 7,200 employees are presented in the first 3 columns of Table 16.1 below. In order to effectively illustrate the power of stratification, we are taking as given all relevant strata and population parameters. In the normal sample survey problem, our objective would be to estimate unknown parameters. The variable, x, in the following example is institutional age, the length of time an employee has been with the firm. (In Table 16.1 below we present the basic data in Columns 1, 2, and 3. Columns following the first 3 are calculations necessary to the discussion that follows.) The 7,200 employees are classified into 1 of 3 strata, Administrative, Clerical, or Blue Collar. The number of employees, the mean institutional ages, the standard deviations, and the variances for each stratum are given respectively in Columns 1, 2, 3, and 4. The stratum weights in the next column are simply each of the stratum sizes divided by 7,200; e.g.,

$$W_1 = \frac{900}{7,200} = .125.$$

In Column 6, the mean institutional age of 10.2 years is calculated as the weighted average of stratum means.

As pointed out earlier, the population variance is, in general, not the weighted average of the stratum variances. In order to calculate this variance, we use the following identity, which is a variation of the basic analysis of variance formula used in Chapter 13:

Table 16.1 Employee-Related Parameters in Hypothetical Firm

Personnel Stratum	1 N_h	2 μ_h	3 σ_h	4 $V_h^{(x)}$	5 W_h	6 $W_h \mu_h$	7 $W_h \sigma_h^2$	8 $W_h(\mu_h - \mu)^2$	9 $W_h \sigma_h$
Administrative	900	10.4	4.0	16.00	.125	1.3	2.00	.005	.50
Clerical	1,800	3.6	2.2	4.84	.250	.9	1.21	10.890	.55
Blue Collar	4,500	12.8	5.6	31.36	.625	8.0	19.85	4.225	3.50
Total	7,200				1.000	10.2	23.06	15.120	4.55

$$V(x) = \sum W_h V_h(x) + \sum W_h(\mu_h - \mu)^2.$$

For the example above, the weighted average of the stratum variances is 23.06 (Column 7 in Table 16.1), and the weighted average of the squared differences in means is 15.12 (Column 8). The sum of these, 38.18, is the population variance. Note that this is larger than any 1 of the 3 stratum variances, which is due to the fact that the stratum means are different. If the stratum means were all the same, the population variance would simply be the weighted average of the stratum variances.

We wish to sample the population of employees and will use stratified random sampling. The objective will be to estimate μ, the mean institutional age of all employees. If we define the stratified sample mean, $\bar{x}_{st}$, as

$$\bar{x}_{st} = \sum W_h \bar{x}_h,$$

this mean will be an unbiased estimate of μ; i.e., $E(\bar{x}_{st}) = \mu$.

Variance of Sample Mean, Stratified Sampling

The variance of the stratified sample mean is, in general,

$$V(\bar{x}_{st}) = \sum W_h^2 V(\bar{x}_h) = \sum W_h^2 \frac{\sigma_h^2}{n_h} (1 - f_h).$$

If the N_h are large with respect to the n_h, then

$$V(\bar{x}_{st}) \cong \sum W_h^2 \frac{\sigma_h^2}{n_h}.$$

How should we allocate the sample size among the strata? Using a fairly advanced technique from calculus, we have an optimal allocation formula, one that minimizes $V(\bar{x}_{st})$ for a given sample size. As a second method, the sample size could be allocated to the stratum in proportion to the population weights. This simple method ensures that $V(\bar{x}_{st})$ will be no larger than the variance of $\bar{x}$ from a simple random sample, $V(\bar{x}_{ran})$, and frequently proportional allocation gives a precision that is very close to the optimal allocation.

Variance of the Sample Mean, Proportionate Sampling

Considering proportional allocation first,

$$\frac{n_h}{n} = \frac{N_h}{N} = W_h \text{ or } n_h = W_h n.$$

Substituting $W_h n$ for n_h in the formula for $V(\bar{x}_{st})$, we get

$$V(\bar{x}_{prop}) = \frac{1-f}{n} \sum W_h \sigma_h^2$$

$\left(\text{note that all } f_h = f = \frac{n}{N} \right).$

For relatively large N,

$$V(\bar{x}_{prop}) \cong \frac{1}{n}\sum W_h \sigma_h^2.$$

Returning to the example above and choosing a somewhat arbitrary sample size of 224, let us find the variance of the sample mean with stratified proportional allocation and compare it with that resulting from a simple random sample. With an approximate 3 percent sample in each stratum, we will ignore the *fpc*.

$$V(\bar{x}_{prop}) = \frac{1}{n}\sum W_h V_h = \frac{1}{224}(23.06)$$
$$= .1029$$

(the 23.06 is the sum from Column 7 of Table 16.1); and

$$\sigma_{\bar{x}\,prop} = \sqrt{V(\bar{x}_{prop})} = .32 \text{ year}.$$

For simple random sampling,

$$V(\bar{x}_{ran}) = \frac{\sigma^2}{n} = \frac{38.18}{224} = .1704$$

and

$$\sigma_{\bar{x}\,ran} = .41 \text{ year}.$$

Proportional allocation for the given sample size of 224—i.e., $n_1 = 28$, $n_2 = 56$, and $n_3 = 140$—decreases the variance of the sample mean by 40 percent and the standard error by 22 percent.

The reduction in the variance of the estimated mean can be measured in terms of the squared differences between the stratum means and the population mean. Beginning with the ANOVA identity mentioned earlier,

$$V(x) \cong \sum W_h V_h(x) + \sum W_h(\mu_h - \mu)^2,$$

dividing through by n and transposing the first term on the right, we get

$$V(\bar{x}_{ran}) - V(\bar{x}_{prop}) = \frac{1}{n}\sum W_h(\mu_h - \mu)^2.$$

The gain in precision, the term on the left, is a function of the differences between strata means and the population mean. If, in a real-world problem, we are aware that there are significant differences in the mean values of the variable of interest within the several strata, then proportional stratified sampling will result in a significant payoff.

Variance of Sample Mean, Optimal Allocation

Assuming that unit costs of sampling among the strata are no different, optimal allocation in stratified sampling may be found from the formula

$$n_h = n \frac{W_h \sigma_h}{\sum W_h \sigma_h},$$

and the minimum variance is

$$V(\bar{x}_{min}) = \frac{1}{n} \left(\sum W_h \sigma_h \right)^2.$$

The stratum sample size is determined by the weighted standard deviation; i.e., it varies directly with the relative size of the stratum and the variability within the stratum. For our example, refer to column 9 of Table 16.1 for the calculation of $W_h \sigma_h$; the sample size of 224 would be optionally allocated among the strata as follows:

$$n_1 = \frac{.50}{4.55} = 25$$

$$n_2 = \frac{.55}{4.55} = 27$$

$$n_3 = \frac{3.50}{4.55} = 172.$$

The minimum variance,

$$V(\bar{x}_{min}) = \frac{1}{224} (4.55)^2 = .0924,$$

$$\sigma_{\bar{x}min} = .30 \text{ year}.$$

This is a modest gain in precision over the standard deviation of .32 year obtained from proportional allocation. It can be shown that the reduction in variance from stratified proportional allocation to the optimal allocation is given by

$$V_{prop} - V_{min} = \frac{1}{n} \sum W_h (\sigma_h - \mu_{\sigma h})^2.$$

(Note:

$$\mu_{\sigma h} = \sum W_h \sigma_h;$$

i.e., the mean value of the strata standard deviations is the weighted average of those standard deviations.)

The gain from proportional to minimal allocation is then a function of the weighted average of the squared differences between the stratum standard deviations and their weighted average. If the standard deviations are highly variable among the strata, the optimal allocations will lead to significant gain in the precision of the estimate relative to proportional allocation.

The above example was used to illustrate the possible gains due to stratified sampling. All relevant parameters were given and, as mentioned, this would not be the case in a real-world problem. We consider the following example, hypothetical but closely related to an actual problem.

Example 16.1.1   In a class action suit against Toby's Tall Shops and Toby himself, Toby was sued for usury. Throughout the full year of 1978, he had been charging 2 percent per month on unpaid balances of his charge accounts, and the legal limit was 1.5 percent. Toby was considering an out-of-court settlement but, in order to establish his bargaining position, he needed to know an estimate of the overcharges. Toby had 4 retail outlets in 4 different cities in Idaho. Time pressure and the high cost of a population review led Toby to decide upon sampling the accounts in order to estimate the overcharges. In 1978 there were 36,000, 24,000, 12,000 and 8,000 charge accounts respectively in the 4 cities. All were subject to finance charges. Using a sample size of 400 and stratified sampling, estimate the annual mean overcharge per account and the total 1978 overcharges.

Let the variable, x, be the 1978 annual dollar total per account subject to finance charge. The annual overcharge estimate, y, would be .5 percent of this annual total—i.e., 2 percent of the annual total less the legal 1.5 percent of this amount. Very little was known about the parameters in the population, except that the bookkeepers who did the billing were confident that the mean balances in the 2 larger cities were considerably higher than those in the smaller cities. Stratification by city seemed both natural and reasonable and, because of the above information, it could result in an estimate more precise than one obtained by simple random sampling. In order to make an optimal allocation, it would be necessary to have estimates of the stratum standard deviations. This would, in this example, require a stratified pilot sample. The pilot sample stratum standard deviations would then be used as estimates of the subpopulation standard deviations for approximate optimal allocation. In our example we will proceed with the proportionate allocation. In Table 16.2, the first 3 columns give N_h, W_h and n_h,

Table 16.2   Parameters for Toby's Tall Shops with Calculations for Stratified Sampling

City	1 N_h	2 W_h	3 n_h	4 $\bar{x}_h$	5 $W_h\bar{x}_h$	6 s_h	7 s_h^2	8 $W_h s_h^2$	9 $(\bar{x}_h - \bar{x})^2$	10 $W_h(\bar{x}_h - \bar{x})^2$
A	36,000	.45	180	1,200	540	800	640,000	288,000	52,900	23,800
B	24,000	.30	120	2,200	660	800	640,000	192,000	592,900	177,900
C	12,000	.15	60	1,000	150	500	250,000	37,500	184,900	27,700
D	8,000	.10	40	800	80	400	160,000	16,000	396,900	39,700
	80,000	1.00	400		$\bar{x} = 1,430$			533,500		269,100

Sampling

the proportional allocation of $n = 400$. In the other columns of Table 16.2 we give the sample results, $\bar{x}_h$ in Column 4 and s_h in Column 6, along with other necessary calculations to be discussed as we analyze the results.

The point estimate of μ_x is $\Sigma W_h \bar{x}_h$ in column 5, $\hat{\mu}_x = \bar{x} = \$1,430$. For Toby's purpose, $\hat{\mu}_y = \bar{y} = .005\bar{x} = \$7.15$, the estimated mean overcharge for the year 1978. This multiplied by 80,000, the total number of accounts, is the estimated total overcharge; $(\$7.15)(80,000) = \$572,000$. This is the estimate of the number Toby wanted to know.

Toby will also want to know how precise this is, and we want to discuss the estimation of precision when universe parameters are not known.

As with most formulas used in statistical estimation (see Chapter 7), we simply replace population parameters with the corresponding sample statistic in order to get an estimate of the required standard error.

Ignoring the fpc, we saw earlier that

$$V(\bar{x}_{prop}) = \frac{1}{n} \Sigma W_h \sigma_h^2;$$

when the σ_h^2 are not known,

$$V(\bar{x}_{prop}) \cong v(x_{prop}) = \frac{1}{n} \Sigma W_h s_h^2.$$

In the above example (see Column 8),

$$v(\bar{x}_{prop}) = \frac{1}{400}(533,500) = 1,333.75.$$

For the population standard error of the mean,

$$\hat{\sigma}_{\bar{x}\,prop} = \sqrt{v(\bar{x})} = \sqrt{\$1,333.75} = \$36.52;$$

recall that $\bar{y}$, the mean overcharge is .5 percent of $\bar{x}$; so $\sigma_y = .005(36.52) = \$.1826$.

Because the sample sizes are quite large, we can feel comfortable in using the normal curve in presenting Toby with a confidence interval. For 95 percent confidence, recall that $z = \pm 1.96$.

The 95 percent confidence interval for μ_y, the mean annual overcharge, is

$$7.15 \pm 1.96(.1826) = 7.15 \pm .36.$$

In a summary statement to Toby, we would say: "The best estimate of the mean annual overcharge is $7.15. We are 95 percent confident that this is at least $6.79 and no more than $7.51." These would be multiplied by 80,000 for the 95 percent interval on total overcharges; T: Conf($543,000 \leq T \leq 601,000) = .95$.

Before leaving Toby and his problem, we want to observe the gain in precision achieved by stratified proportionate sampling over that from a simple random sample of 400. Earlier we noted that the reduction in variance was given by

$$V(\bar{x}_{ran}) - V(\bar{x}_{prop}) = \frac{1}{n}\sum W_h(\mu_h - \mu)^2.$$

This reduction is estimated by

$$\frac{1}{n}\sum W_h(\bar{x}_h - \bar{x})^2.$$

For our problem, the estimated reduction in the variance of $\bar{x}$ would be

$$\frac{1}{400}(269,100) = 672.76.$$

(See Column 10 of Table 16.2.)
Recall that $v(\bar{x}_{prop})$ was 1,333.75; so $v(\bar{x}_{ran})$ would be 2,006.5; and $\hat{\sigma}_{\bar{x}ran} = \$44.80$ compared to $\$36.52$ for $\hat{\sigma}_{\bar{x}prop}$.

16.2 Estimation of Proportions with Stratified Sampling

The formulas for estimating the proportion of a given attribute and the related standard error through stratified sampling parallel those for estimating the mean and its standard error. We need the following additional notation.

1. π = population proportion of an attribute.
2. $\theta = 1 - \pi$.
3. p = sample proportion.
4. $q = 1 - p$.

For simple random sampling, recall that—

1. $E(p) = \pi$;
2. $\sigma_p = \sqrt{\dfrac{\pi\theta}{n}}\sqrt{1-f}$.

To illustrate stratified sampling for the estimation of proportions, return to our hypothetical manufacturing firm with 7,200 employees. The proportion of females will be the attribute of interest, and, as with our first example, population parameters will be given. Columns 1 and 2 of Table 16.3 give N_h and W_h identical with those in Table 16.1; in column 3 we give the proportion of females in each of the strata. The other columns in Table 16.3 show calculations necessary to the discussion that follows.

Sampling

Table 16.3 Attribute Stratified Sampling

	1 N_h	2 W_h	3 π_h	4 $W_h\pi_h$	5 $\pi_h\theta_h$	6 $W_h\pi_h\theta_h$	7 $\sqrt{\pi_h\theta_h}$	8 $W_h\sqrt{\pi_h\theta_h}$
Administrative	900	.125	.50	.0625	.25	.03125	.50	.0625
Clerical	1,800	.250	.80	.2000	.16	.04000	.40	.1000
Blue Collar	4,500	.625	.10	.0625	.09	.05625	.30	.1875
	7,200	1.000		= .3250		.12750		.3500

We first calculate π, the proportion of females in the population. This is the weighted average of the stratum proportions:

$$\pi = \sum W_h\pi_h = .3250$$

(see Column 4).

The general formula for the variance of the sample proportion for stratified sampling is

$$V(p_{st}) = \sum W_h^2 V(p_h) = \sum W_h^2 \frac{\pi_h\theta_h}{n_h}(1 - f_h).$$

If N_h are large with respect to n_h, we drop the *fpc*, and

$$V(p_{st}) \cong \sum W_h^2 \frac{\pi_h\theta_h}{n_h}.$$

If the allocation among the strata is proportionate,

$$V(p_{prop}) = \frac{1-f}{n}\sum W_h\pi_h\theta_h;$$

or, for relatively large N_h,

$$V(p_{prop}) \cong \frac{1}{n}\sum W_h\pi_h\theta_h.$$

With a sample size of 224 and using stratified proportionate sampling in the above example, i.e., $n_1 = 28$, $n_2 = 56$, and $n_3 = 140$,

$$V(p_{prop}) = \frac{1}{224}(.12750)$$

(see Column 6, Table 16.3)

$$= .0005691;$$

and

$$\sigma_{p\,prop} = \sqrt{.0005691} = .0239.$$

Note that

$$\sigma_{p\,ran} = \sqrt{\frac{(.325)(.675)}{224}} = .0313.$$

The standard error of the proportion for proportionate sampling is 24 percent smaller than that obtained from a simple random sample.

In the case of attribute sampling, optimal allocation is achieved by the following:

$$n_h = n \frac{W_h \sqrt{\pi_h \theta_h}}{\sum W_h \sqrt{\pi_h \theta_h}};$$

and the minimum standard error is

$$V(p_{min}) = \frac{1}{n}\left(\sum W_h \sqrt{\pi_h \theta_h}\right)^2.$$

For the above problem, optimal allocation of the sample of 224 would be (see Column 8 of Table 16.3):

$$n_1 = 224 \frac{.0625}{.3500} = 40$$

$$n_2 = 224 \frac{.1000}{.3500} = 64$$

and

$$n_3 = 224 \frac{.1875}{.3500} = 120.$$

Calculating the minimum variance of the estimate, we get

$$V(p_{min}) = \frac{1}{224}(.3500)^2 = .0005468;$$

and

$$\sigma_{p\,min} = \sqrt{.005468} = .0234.$$

As with the previous example on institutional age, there is little gain from proportional allocation to the optimal allocation.

 ## 16.3 Sample Size Estimation with Stratified Sampling

The formulas and methods for sample size estimation with stratified sampling parallel those developed and discussed in Chapter 7 for simple random sampling. As with simple random sampling, the researcher or statistician must state the desired precision; i.e., the maximum allowable error and the confidence level of this error.

Let ϵ be the maximum error and z be the standard normal curve variable associated with the confidence level. Then $\epsilon = z\sigma_{\bar{x}}$, and from this may be derived the desired standard error of the mean,

$$\sigma_{\bar{x}} = \frac{\epsilon}{z}.$$

With simple random sampling we also need to know or have an estimate of the population standard deviation; with stratified sampling we need this knowledge or estimate for each stratum. Strata standard deviations may come from a previous similar study; they may be estimated from a pilot sample; an educated guess may serve as a starting point; or some combination of the above could lead to reasonably accurate strata standard deviations.

Sample Size Estimation in Stratified Sampling

If the population size is large relative to the sample size, the following formulas apply for estimating the population mean.

1. For proportional allocation,

$$n = \frac{\Sigma W_h \sigma_h^2}{V(\bar{x})}$$

where $V(\bar{x}) = \sigma_{\bar{x}}^2$, the desired variance of the sample mean.

2. For optimal allocation

$$n = \frac{(\Sigma W_h \sigma_h)^2}{V(\bar{x})}.$$

If the sample size, n, derived above is more than 5 percent of the population size, N, then it should be adjusted downward as follows.

1. For proportional allocation, use the sample size derived above for proportional sampling as an initial estimate, n_o. The corrected sample size would be

$$n = \frac{n_o}{1 + \frac{n_o}{N}}.$$

2. For optimal allocation,

$$n = \frac{n'_o}{1 + \frac{n_o}{N}}$$

where n'_o is the initial sample size from the optimal allocation formula above. Note that the denominator, the *fpc* for sample size estimation, is the same for both corrections; i.e., the initial sample size, n_o, is that derived for proportional allocation.

To illustrate sample size estimation, let us use some of the data given earlier for an hypothetical manufacturing firm. See Table 16.4. Recall that the variable, x, is institutional age.

Assume we wish to estimate institutional age within a half year at the 95 percent confidence level. What sample size is needed? First, the desired standard error of the mean is

$$\sigma_{\bar{x}} = \frac{\epsilon}{z} = \frac{.50}{1.96} = .255$$

and $V(\bar{x}) = (.255)^2 = .065$.

For proportional allocation,

$$n = \frac{\Sigma W_h \sigma_h^2}{V(\bar{x})} = \frac{23.06}{.065} = 355.$$

(Note: From Table 16.4,

$$\Sigma W_h \sigma_h^2 = 23.06.)$$

A sample of 355 allocated proportionately—i.e., $n_1 = 44$, $n_2 = 89$, and $n_3 = 222$—would yield the desired precision.

A sample of 355 is just short of a 5 percent sample; we will, however, use the *fpc* formula here for illustrative purposes. With $n_o = 355$,

$$n = \frac{n_o}{1 + \frac{n_o}{N}} = \frac{355}{1 + \frac{355}{7,200}} = 339.$$

The corrected sample size, 339, will yield the exact desired precision, and 355 would do just slightly better.

If we wanted optimal allocation,

$$n = \frac{(\Sigma W_h \sigma_h)^2}{V(\bar{x})} = \frac{(.455)^2}{.065} = 319,$$

and the strata sample sizes would be

$$n_1 = 319 \left(\frac{.50}{4.55}\right) = 35$$

Table 16.4 Calculations for Sample Size Estimation

Stratum	N_h	W_h	σ_h	$W_h \sigma_h^2$	$W_h \sigma_h$
Administrative	900	.125	4.0	2.00	.50
Clerical	1,800	.250	2.2	1.21	.55
Blue Collar	4,500	.625	5.6	19.85	3.50
	7,200			23.06	4.55

$$n_2 = 319 \left(\frac{.55}{4.55}\right) = 39$$

$$n_3 = 319 \left(\frac{3.50}{4.55}\right) = 245.$$

For the optimal allocation, a sample size of 319 would yield the same precision as the proportional allocation of 355.

To complete the example, we correct the optimal sample size using the *fpc*:

$$n = \frac{n'_o}{1 + \frac{n_o}{N}} = \frac{319}{1 + \frac{355}{7,200}} = 304.$$

In the final analysis, $n = 304$ optimally allocated, i.e.,

$$n_1 = 304 \left(\frac{.50}{4.55}\right) = 33$$

$$n_2 = 304 \left(\frac{.55}{4.55}\right) = 37$$

and

$$n_3 = 304 \left(\frac{3.50}{4.55}\right) = 234,$$

and this is the smallest sample size that will yield the desired precision on mean institutional age.

If we are sampling to estimate π, the proportion of some attribute, and N is large relative to n, we have—

1. For proportional allocation:

$$n = \frac{\Sigma W_h \pi_h \theta_h}{V(p)};$$

2. For optimal allocation:

$$n = \frac{(\Sigma W_h \sqrt{\pi_h \theta_h})^2}{V(p)}.$$

The *fpc* sample size corrections for relatively small populations are identical to those given above relating to the mean.

16.4 Implementing Randomness in Sampling

We have defined the term *random sample*, and in all inference discussions to this point it has been implicit if not always explicit that the sample has been randomly drawn. It is frequently possible to implement random sampling in practice, but many times true randomness cannot be achieved. We want to discuss some methods of approximating randomness in real-world situations, but before doing this let us consider several alternatives for implementing true randomness.

If it is possible to identify each of the elements of a finite population with the set of positive integers, 1, 2, 3, . . . , N, then a random sample can be drawn from the population using a table of random digits. (Equivalent to this would be a random digit generator on the computer.) Random digits are generated from the set of 10 digits, 0, 1, 2, 3, 4, 5, 6, 7, 8, 9, with each of these having a .1 chance of being drawn at each step in the generation process. We have included a small table of random digits; see Table 16.5 on opposite page. This table was computer generated, but the most impressive collection of random digits is in a book by Rand Corporation, *A Million Random Digits with 100,000 Normal Deviates,* published in 1955 by the Free Press. To illustrate the use of this table, assume we wish to draw a random sample of size 20 from a population of 850 elements. Assume also that we are able to identify each of the population elements with the integers 001, 002, 003, . . ., 850. The table of random digits will be used to draw 20 different 3-digit numbers between 1 and 850 inclusive. If 000 or a number exceeding 850 is drawn, it will be discarded; a duplicate number will also be discarded. Inasmuch as these are random digits, we can start any place in the table and proceed by row or column. We will arbitrarily begin in the sixth row and twenty-third column with the number 723 (in color on the table) and take 3-digit numbers down the column. Our sample would be as follows:

723, 239, 694, 523, 598, 742, 329, 446, 266, 573, 213, 317, 157, 545, 312, 361, 806, 540, 768, 092.

Note that we drew 5 numbers exceeding 850 and 1 duplicate before we were able to get 20 valid elements from the population. Arranging the numbers from smallest to largest, the following would be our sample of 20 from the population of 850:

092, 157, 213, 239, 266, 312, 317, 329, 361, 446, 523, 540, 545, 573, 598, 694, 723, 742, 768, 806.

Another method for achieving randomness is to number chips or uniform pieces of paper, mix them in a bag or flower pot, and draw the sample from the mixed bag. This is feasible for a small population, but for large populations it would be too cumbersome and/or too costly.

Many industrial processes are considered to be random in nature; i.e., output sequences may be realistically assumed to approximate independent trials. In this case, the process output may be sampled in sequence or in some systematic manner and the sample will be random.

There are other creative ways to approximate randomness in the

Table 16.5 Table of Random Digits

79430	69992	68479	62421	97959	91499	38631	67422
16332	89246	26940	44670	35089	80336	49172	17691
57327	81336	85157	34679	62235	44104	89232	79655
15026	81330	11100	32439	58537	12550	02844	48274
65406	42237	16505	37920	08709	63606	40387	60623
44177	05798	26457	51171	08723	61196	80240	39323
99321	10836	95270	72173	56239	15474	44910	04595
86347	51770	67897	00926	44915	94557	33663	34823
19102	24358	97344	37420	41976	42481	86430	76559
68588	36879	73208	81675	15694	23523	31379	43438
32533	24826	75246	17767	14523	04493	98086	52494
04805	16232	64051	05431	94598	00549	33185	97654
68953	00406	26898	99634	81949	35963	80951	15307
02529	49140	45427	40200	73742	59808	79752	08391
99970	32537	01390	67348	49329	46058	18633	95236
74717	45240	87379	17674	90446	32179	74029	00597
10805	41941	20117	35635	45266	69234	54178	61406
77602	96382	01758	99817	28573	19565	11664	41430
32135	71961	19476	26803	16213	45155	48324	14938
45753	98145	36168	20505	78317	94864	69074	31994
84532	02438	76520	13618	23157	10097	85017	86952
82789	05403	64894	69041	05545	37542	16719	44109
27672	86529	19645	82186	14871	08422	65842	22115
20684	43204	09376	39187	38976	99019	76875	94324
39160	93137	80157	41453	97312	12807	93640	41548
70086	29004	34072	71265	11742	66065	99478	18226
26486	42753	45571	47353	43361	31060	65119	99436
21592	21828	02051	48233	93806	85269	70322	32584
41278	67954	05325	11697	49540	63573	58133	61777
81255	38537	03529	31133	36768	73796	44655	60452
13487	61303	14905	98662	07092	98520	02295	44673
54881	48493	39808	35587	43310	11805	85035	48897
86935	65710	06288	28021	61570	83452	01197	23350
19078	44369	86507	40646	31352	88685	97907	48625
96905	74294	87517	28797	57048	99594	63268	46359
59684	84369	17468	67411	09243	65481	52841	56092
71399	13018	17727	10916	07959	80124	53722	21225
51908	02400	77402	62171	93732	74350	11434	26958
99292	07289	66252	21177	72721	69916	62375	66995
20923	31374	14225	87929	61020	09893	28337	62841

sample for specific situations. We mention one to illustrate a point. The last 4 digits of Social Security numbers are random, and they may be used in sampling an adult population. This method becomes particularly efficient when sampling the employees of a firm or the students of an institution where the lists of names, Social Security numbers, and other data are on file in the computer. For example, assume we wish to draw a sample of 500 from a firm's population of 5,000 employees. Simply pick 1 random digit, and have the computer print all names whose Social Security numbers end with that digit. Desired sample sizes will not be precisely realized using this method, but they will be close. The expected sample size will be 500. Drawing a 10 percent sample is a very simple proposition, but note the ease with which we could draw approximately 200 (or for that matter any approximate sample size). Draw 4 random digits for the units place of the Social Security number and 1 random digit for the tens place; e.g., assume 0, 1, 4, and 7 were drawn for the units place and 5 was drawn for the tens place; then all Social Security numbers ending in 50, 51, 54, or 57 would make up the sample. The sample size would be approximately 4 percent of 5,000, or approximately 200.

16.5 Systematic Sampling

A formal method that is frequently used to approximate randomness is systematic sampling. If one has access to a listing of the population, either alphabetical or otherwise, or to a card file or folder file on the population, systematic sampling will usually save both time and money. Systematic sampling is implemented as follows: Divide the population size by the desired sample size to get a quotient, k, and a remainder, g, where g is less than n. In equation form,

$$\frac{N}{n} = k + \frac{g}{n}.$$

If we take a random start from 1 to k in the file or listing, choose that element, and then take every k^{th} element throughout, we will have a sample size of n or $n + 1$. Note that there are k possible samples (k is called the *skip interval*), and all will be of size n only if $g = 0$ and $nk = N$; otherwise some of the samples will be of size $n + 1$.

To illustrate systematic sampling, consider a population of 850 student loan files. The bank has these filed alphabetically in manila folders. Bank management would like to draw a sample of 50 to get a preliminary estimate of delinquency rate and dollar amount of delinquency. Dividing 850 by 50 gives 17 with a 0 remainder; there are 17 possible systematic samples of size 50. We choose a random 2-digit number between 01 and 17 inclusive, start with this account, and take every seventeenth account thereafter. If the sample were size 20 instead of 50, then 850 divided by 20 equals 42 with a 10 remainder. A random start between 01 and 42 inclusive and choosing each forty-second file thereafter would lead to 10 possible samples of size 21 and 32 of the correct size, 20.

If it can be argued that the population listing is random regarding the variable being studied (this seems to be the case for most alphabetical listings), then the expected values of variance of the sample mean as well as that of the sample proportion are equal to the corresponding variances for simple random sampling. In most situations where systematic sampling is suggested by the nature of the population identification, systematic sampling is a reasonable substitute for random sampling. The standard error of the sample mean as estimated from the sample data is generally as reliable as that obtained from a random sample.

A modification of systematic sampling would be to take 2 or more replications. In order to illustrate what is meant by replication, assume that we wanted to sample 100 of the 850 student loan files. We could take 2 systematic samples, 2 replicates, of size 50; or 4 samples of size 25 could be drawn. If we decided to take 2 samples of size 50, there would simply be 2 different random starts between 1 and 17 inclusive, both with skip intervals of 17. The replicate sample means as well as the sample variances could be tested for significant differences. (See Chapters 9 and 13.) If either the means or the variances were significantly different, we might well want to question the notion of systematic sampling in this particular population. If more than 2 systematic replicates were taken, an analysis of variance test could be made on the sample means. It is interesting to note that several auditing and estimating procedures of the United States Internal Revenue Service and other federal agencies specify replicated systematic sampling.

16.6 Cluster Sampling

The use of cluster sampling as an alternative to random sampling and the subsequent study of the associated sampling errors were motivated by (1) the natural clustering of certain populations being sampled and (2) the typically lower per-sample unit cost of cluster sampling when compared to random sampling. For example, labor force studies are frequently done by sampling households. A household may include 0, 1, 2, or more members of the labor force; the household is, therefore, a variable-sized cluster of basic population elements. Acceptance inspection sampling is frequently done by clusters. A large lot of items coming into an assembly plant from another manufacturer is frequently packaged in relatively small bundles. If, for example, there were 5 items per package and the predesigned sample size were 100, the quality control group would very likely open 20 packages and inspect all 5 items in each package. This is an example of a fixed-size cluster sample.

For a fixed sampling expenditure, it is nearly always possible to get a larger sample size using cluster sampling as opposed to random sampling; or, alternatively, for a fixed sample number, the cost of the cluster sample will be lower than that of a random sample. As students of statistics, we are concerned with the effect of cluster sampling on sampling errors. It is possible for the gain in sample size resulting from cluster sampling to be more than offset by a loss in the precision of the estimate. Before proceeding with the discussion of statistical errors related to cluster sampling, we introduce the following notations and definitions.

1. M = cluster size if constant.
2. M_i = size of ith cluster if cluster size is variable, and $\overline{M}$ = average cluster size.
3. k = number of clusters; so $kM = N$ or $k\overline{M} = N$, depending upon constant or variable cluster size.
4. $\mu_i = i^{\text{th}}$ cluster mean.
5. C = number of clusters in sample; i.e., $CM = n$ or $C\overline{M} = n$.
6. σ_b^2 = variance between clusters assuming constant cluster size:

$$\sigma_b^2 = M \left[\frac{\sum_{i=1}^{k} (\mu_i - \mu)^2}{k - 1} \right]$$

7. σ_w^2 = within cluster variance,

$$\sigma_w^2 = M \frac{1}{k} \sum_{i=1}^{k} \left[\frac{\sum_{j=1}^{M} (x_{ij} - \mu)^2}{M - 1} \right]$$

From the now familiar ANOVA identity we have the approximate relationship:

$$\sigma^2 \cong \frac{\sigma_b^2 + (M - 1)\sigma_w^2}{M}.$$

We also want to define the intracluster correlation coefficient, ρ:

$$\rho = \frac{E(x_{ij} - \mu)(x_{ik} - \mu)}{E(x_{ij} - \mu)^2}.$$

It can be shown that the variance of the sample mean given cluster sampling is approximated by

$$V(\bar{x}_{c1}) \cong \frac{\sigma^2}{n} [1 + (M - 1)\rho] \cong \frac{\sigma_b^2}{n}.$$

(For this relationship we are assuming a fixed cluster size and a large population relative to the sample size.)

Inspection of the above relationships leads to several observations regarding cluster sampling and the variance of the sample mean. Most of these observations are also intuitively satisfying.

1. If the elements within clusters are independent, i.e., $\rho = 0$, then

$$V(\bar{x}_{c1}) = \frac{\sigma^2}{n},$$

which is identical to $V(\bar{x}_{ran})$.

2. If within each cluster all values of x are equal—i.e., if all $x_{ij} = x_{ik}$—then $\rho = 1$ and

$$V(\bar{x}_{c1}) = M\frac{\sigma^2}{n} = \frac{\sigma^2}{c}.$$

In this case, as we would expect, the effective sample size is the number of clusters, c; i.e.,

$$\frac{\sigma^2}{c}$$

is the variance of the random sample mean with a sample of size c.

3. In general, if ρ is positive, the effective sample size will be something less than n but no smaller than c.

4. If ρ is negative, and keeping in mind that M is a positive integer greater than 1, then $[1 + (M - 1)\rho]$ will be less than unity, and $V(\bar{x}_{c1})$ will be less than $V(\bar{x}_{ran})$. This is an obvious conclusion given the above formulas but, because it is difficult to construe a set of population clusters with a negative intracluster correlation coefficient, this conclusion is not intuitively obvious.

In many applications of cluster sampling, the coefficient of intracluster correlation tends to the positive direction; i.e., the values of the variable within the cluster tend to be more homogeneous than those in the population. This increases the value of σ_b^2, and $V(\bar{x}_{c1})$ will exceed $V(\bar{x}_{ran})$.

In other applications it seems reasonable to argue that the elements within the clusters are independent—i.e., $\rho = 0$ and $\sigma_b^2 = \sigma_w^2 = \sigma^2$—and $V(\bar{x}_{c1})$ will equal $V(\bar{x}_{ran})$. It is rare in applied work that a statistician will argue for a gain in precision using cluster sampling; i.e., for the most part the best of all worlds in the use of cluster sampling would be for ρ to be 0 and the precision of cluster sampling to be the same as that for random sampling.

We should note further that, if cluster sizes are unequal, M is replaced by $\overline{M}$ in the formulas. Also, sample relationships parallel the population formulas. For example:

$$v(\bar{x}_{c1}) \cong \frac{s^2}{n}[1 + (M - 1)\hat{\rho}] \cong \frac{s_b^2}{n}.$$

Consider the following example to illustrate some of the concepts of cluster sampling.

Example 16.6.1   A company engaged in the assembly of small, high-speed fans receives shipments of small rods in lots of 5,000. The rods are packed 10 to a box. The sampling plan calls for a sample of 120 with the diameter, x, as the important variable. Twelve boxes of 10 were selected at random. Measurements were taken, and data were kept for cluster sampling with $M = 10$. Find

$$\hat{\sigma}_{\bar{x}cl}$$

and compare with

$$\hat{\sigma}_{\bar{x}ran}.$$

The following sample statistics were calculated:

$\bar{x} = 2.035$ millimeters (mm); $s_b^2 = .00250$ $s_w^2 = .00080$.

Using the above, we calculate the following:

1. $\hat{\sigma}^2 = s^2 = \dfrac{s_b^2 + (M - 1) s_w^2}{M} = \dfrac{.00250 + 9(.00080)}{10}$

   $= .00097$.

2. $\rho \cong \dfrac{s_b^2 - s^2}{(M - 1)s^2} = \dfrac{.00250 - .00097}{9(.00097)}$

   $= .18$.

3. $v(\bar{x}_{cl}) \cong \dfrac{s_b^2}{n} = \dfrac{.00250}{120} = .0000208$

   $\hat{\sigma}_{\bar{x}cl} \cong \sqrt{.0000208} = .0046$ mm.

Noting that the sample coefficient of intracluster correlation is positive, we can conclude that the sample of 120 did not yield the precision that the same size random sample would have given. This loss in precision can be estimated using the estimated variance, $\hat{\sigma}^2$:

$$V(\bar{x}_{ran}) \cong \dfrac{\hat{\sigma}^2}{n} = \dfrac{.00097}{120} = .0000081$$

and

$$\hat{\sigma}\bar{x}_{ran} \cong \sqrt{.0000081} = .0028 \text{ mm}.$$

We would estimate that a random sample of 44 would yield the same precision as the cluster sample of 120. This follows if we set the estimated variance of the random sample mean,

$$\frac{\hat{\sigma}^2}{n} = \frac{.00097}{n},$$

equal to .0000208 and solve for n. Given the nature of the packaging, random sampling would be difficult to implement, and taking a random sample of 44 could well cost more than a cluster sample of 120. From this example, however, it is quite clear that merely assuming that the values of the variable are independent within the cluster—i.e., assuming $\rho = 0$—can lead to a significant error in the estimated precision. If we had simply calculated s^2 under this assumption, we would have been making decisions based upon a standard error of the mean of .0028 millimeters instead of the correct value of .0046 millimeters.

Problems

1. Consider the following hypothetical data:

<table>
<tr><th colspan="5">Stratum</th></tr>
<tr><th colspan="2">I</th><th>II</th><th colspan="2">III</th></tr>
<tr><td>0</td><td>7</td><td>7</td><td colspan="2">11</td></tr>
<tr><td>1</td><td>4</td><td>9</td><td colspan="2">8</td></tr>
<tr><td>3</td><td>2</td><td>6</td><td colspan="2">15</td></tr>
<tr><td>4</td><td>3</td><td>7</td><td colspan="2">12</td></tr>
<tr><td>7</td><td>2</td><td>8</td><td colspan="2"></td></tr>
<tr><td></td><td></td><td>10</td><td colspan="2"></td></tr>
</table>

Assume that these are sample observations, the *fpc* may be ignored, and the sample variances may be used as estimates of the universe strata variances.

a. Calculate $\bar{x}_1, \bar{x}_2, \bar{x}_3, \bar{x}, s_1^2, s_2^2, s_3^2,$ and s^2.
b. Assume that this is the result of proportionate sampling, and calculate $v(\bar{x}_{prop})$.
c. Assume that this is the result of an optimal allocation, and calculate the $v(\bar{x}_{min})$.

2. Consider the following results from an audit of the charge accounts of a large department store where the variable, x, is balance on account and p is the sample proportion with a past due balance.

Stratum	N_h	n_h	$\bar{x}_h$	s_h	p_h
30-Day Accounts	3,000	30	$ 38	$ 27	.50
Revolving Accounts	7,000	70	262	156	.20

a. Find $\bar{x}$, s, and p.

b. Estimate the total balance on all acounts.
c. Note that this is proportionate sampling and find $\hat{\sigma}_{\bar{x}}^2$; also find $\hat{\sigma}_p^2$.
d. Give the 95 percent confidence interval estimate on μ, the mean balance; on π, the proportion of all accounts that are past due.
e. Using the above sample statistics as point estimators of the population parameters, what would be the estimated optimal allocation for $n = 100$?

3. Assume the following to be population parameters in the three strata of nonresident tourists vacationing in the State of Nature. Out-of-state U.S. citizens are in stratum A, foreigners other than iron curtain citizens are in stratum B, and iron curtain foreigners are in stratum C. The μ_h and the σ_h are means and standard deviations of per-person daily expenditures, and the π_h are proportions of tourists who imbibe while on vacation.

Stratum	N_h	μ_h	σ_h	π_h
A	25,000	$30	$20	.3
B	15,000	60	40	.7
C	10,000	20	15	.9

a. Calculate the population parameters, μ, σ, and π.
b. For a simple random sample of 400, find the standard error of the sample mean expenditures; also find the standard error of the proportion of those who imbibe.
c. Find each of the standard errors in part b above given proportional allocation of $n = 400$ among the strata.
d. Find the optimal allocation for minimizing the variance of the mean per-person daily expenditures. What is the standard error of the mean?
e. Repeat part d for estimating the proportions of tourists who imbibe.

4. An electrical supply firm keeps inventory on 15,000 different items. Cumulative records are kept to indicate what should be on hand for each item, but there are indications that these records are not always correct (e.g., an item has been out of stock when the records have shown that there should have been several in stock). Rather than take a complete inventory, the decision was made to estimate this inventory through sampling. Four different, though not altogether unrelated, variables seemed important: (1) x = number of each specific item; (2) d = difference between this number and that shown in the records; (3) $y = xp$ = the dollar value of the item inventory where p is the item price in dollars; and

(4) $D = dp$ = the dollar value difference between the number of items in inventory and the number on record.

It was decided to stratify the inventory items by price: If $p \leq \$10.00$, classify in stratum I; if $\$10.00 < p \leq \$50.00$, stratum II; and if $p > \$50.00$, classify in stratum III.

A sample of 200 items was drawn on a proportionate basis with the following results:

Stratum	N_h	n_h	$\bar{x}_h$	s_{x_h}	$\bar{y}_h$	s_{y_h}	$\bar{d}_h$	s_{d_h}	$\bar{D}_h$	s_{D_h}
I	9,900	132	180	44	$1,120	$265	−6	14	−$31	$ 98
II	3,450	46	154	32	4,560	633	−1	12	−32	200
III	1,650	22	13	16	1,480	136	0	0	0	0
Total	15,000	200								

a. Calculate the means and standard deviations for the total sample for all 4 variables.
b. Find V_{prop} for all 4 variables.
c. Give the 95 percent confidence interval for the mean value of all 4 variables.
d. Estimate the total dollar value of the inventory; give the 95 percent confidence limits on this.
e. Estimate the total dollar difference between an inventory taken from the records and that resulting from sampling; give the 95 percent confidence limits on this difference.
f. Assume that the manager of the firm wants better precision on the inventory dollar value estimate and decides to sample another 300 items—a total of 500. Using the sample results on the variable y as population estimates, what is the optimal allocation of this 500 sample units to the 3 strata?

CHAPTER 17

STATISTICAL QUALITY CONTROL

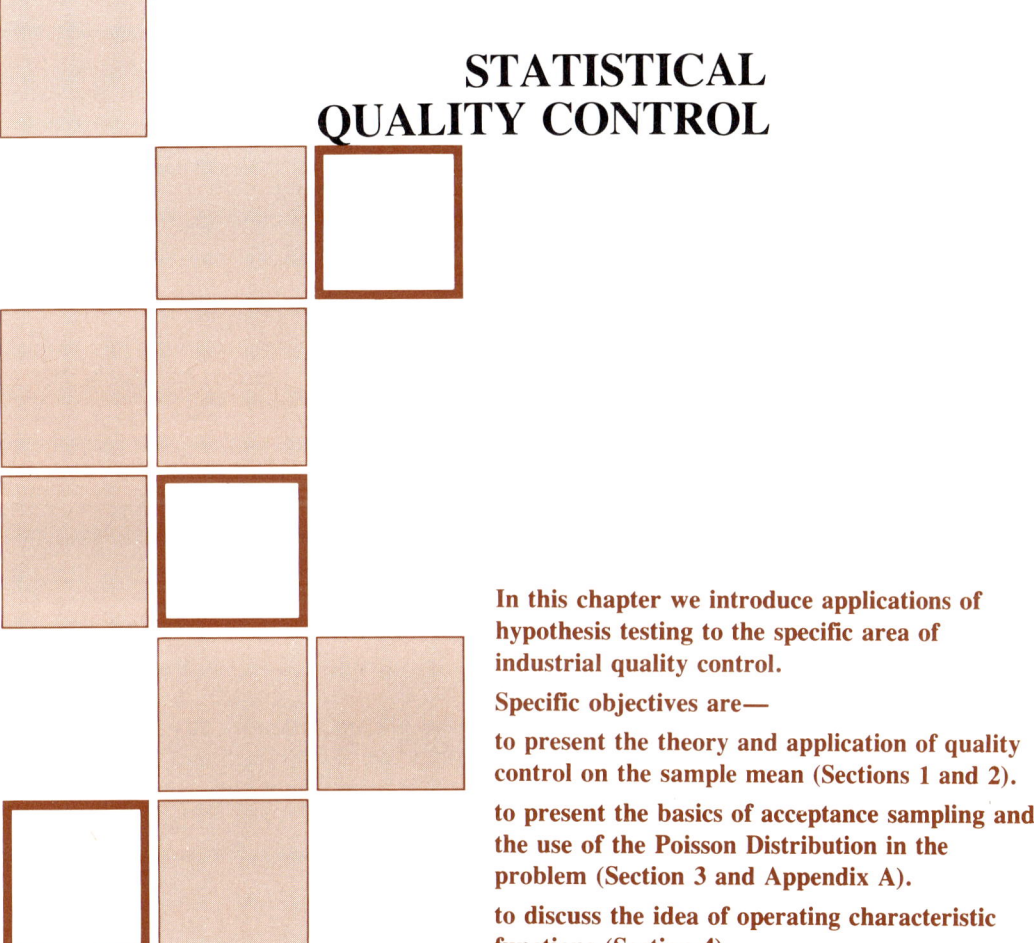

In this chapter we introduce applications of hypothesis testing to the specific area of industrial quality control.

Specific objectives are—

to present the theory and application of quality control on the sample mean (Sections 1 and 2).

to present the basics of acceptance sampling and the use of the Poisson Distribution in the problem (Section 3 and Appendix A).

to discuss the idea of operating characteristic functions (Section 4).

The management function referred to as quality assurance is currently receiving increased emphasis by top management of American corporations. One reason for the renewed interest is the consumer movement in the United States—consumers seem to be more insistent than previously concerning product quality—i.e., obtaining the level of performance from the product that the seller promised them. Another reason for the increased emphasis on quality assurance is increased competition from foreign producers. Japanese manufacturers in particular are establishing a reputation for producing top-quality products that American producers cannot ignore if they want to remain competitive in international markets. The Japanese experience is of interest here primarily because statistical methods played a key role in Japan's quest for producing high-quality products.

During the years immediately following World War II, the Japanese manufactured goods of poor or relatively low quality. In 1950 Dr. W. Edwards Deming, an American statistician, was asked to come and teach Japanese engineers statistical methods of quality control. Deming told a writer for *Quality* that, as he started teaching the first course to some 220 engineers, he became concerned that the same thing would happen in Japan as had happened in America—top management would not give proper emphasis to the application of statistical methods of quality control.[1] Arrangements were made for him to speak to 45 of the top industrial leaders in Japan. Deming told these leaders that Japan could enter the markets of the world and have the competition screaming for protection in 5 years. He told *Quality* that in 1950 he was the only person in Japan who believed this statement. These Japanese industrial leaders listened to what Deming had to say; they applied the statistical methods and techniques he taught them; they invaded the international markets; and today their ability to produce top-quality products is acknowledged throughout the world.

We do not mean to imply that the quality control function consists entirely of the use of statistical methods. An adage commonly heard among quality personnel is that quality control is 10 percent statistics and 90 percent engineering. It is important to note, however, that the statistical methods play a key role in that they indicate where the engineering effort can best be utilized. Failure to use statistical methods can result in wasting engineering effort on areas that aren't critical or that will not help solve the quality control problem.

Statistical quality control has become a highly developed and refined area of study. In this chapter we discuss some of the basic and most commonly used techniques of statistical quality control—control charts and acceptance sampling. It will be seen that these techniques are nothing more than special applications of hypothesis testing covered in Chapter 8 of this text.

---

1. "Dr. W. Edwards Deming—The Statistical Control of Quality," *Quality* 19, no. 2 (February 1980).

 ## 17.1 $\bar{x}$-Control Charts

There is inherent variability in most manufacturing processes: not all holes drilled by the drill press are exactly the same size; an injection molding machine does not always use the same amount of plastic; a bottling machine does not fill every bottle equally; etc. Because of this inherent variability, the measured value of an individual item produced by the process will deviate from a specified value even though the process is operating at an average value equal to the specified value. Similarly, a sample of 5 items produced by the process will have an average value different from the average value of all items produced by the process. For example, the average inside diameter of a sample of 5 parts produced by a stamping machine might be 2.48 inches even though the stamping machine is in fact producing parts with an average inside diameter of 2.50 inches.

Example 17.1.1

If the quality control inspector or the stamping machine operator takes a sample of 5 parts and finds that the average inside diameter of the sample of 5 parts is 2.58 inches, can it be concluded that the stamping machine is no longer operating at an average inside diameter size of 2.50 inches? (Assume the size of the inside diameter of parts stamped out by the machine follows a normal distribution.)

Clearly, there are not sufficient data provided in this example to answer the question asked. Additional information is needed with regard to the variability of the inside diameter sizes. If the stamping machine produces parts with a very large variation in inside diameter sizes, then it might be quite possible that a sample of 5 parts has an average inside diameter of 2.58 inches even though the stamping machine is producing parts with an average inside diameter of 2.50 inches. On the other hand, if the machine produces parts with only a very small variation in the sizes of the inside diameters, then it might be very unlikely that a sample of 5 parts would have an average of 2.58 inches if the machine is operating at an average of 2.50 inches. If the standard deviation of the inside diameter sizes were known, then we could make some very definite probability statements about the likelihood of the sample mean falling within a specified range of values.

The inspector or machine operator isn't going to want to stop and perform a statistical analysis of the sample results obtained out on the shop floor (assuming he or she is capable of doing it). The x-bar control chart, generally referred to by most practitioners as the $\bar{x}$-chart, is a device designed by the quality control statistician that enables the user to quickly decide whether or not the sample mean obtained is consistent with the hypothesized mean, i.e., the specified mean at which the process is supposed to be operating.

The *x*-bar control chart consists of a center line representing the desired process average or mean. A line drawn above and parallel to the mean line is referred to as the upper control limit *(UCL)*. Similarly, a line drawn below the mean line is referred to as the lower control limit *(LCL)*.

Control Limits for the $\bar{x}$-Chart

The control limits are defined as follows:

$$\text{Upper Control Limit } (UCL) = \mu + 3\sigma_{\bar{x}} = \mu + 3\frac{\sigma}{\sqrt{n}}$$

$$\text{Lower Control Limit } (LCL) = \mu - 3\sigma_{\bar{x}} = \mu - 3\frac{\sigma}{\sqrt{n}}$$

where

μ = the desired process average
σ = the process standard deviation
n = the sample size.

You will recall from Chapter 4 that the standard deviation of the sample mean, $\sigma_{\bar{x}}$, is equal to the population standard deviation divided by the square root of the sample size,

$$\frac{\sigma}{\sqrt{n}}.$$

The standard deviation, σ, of the process is determined by an engineering and statistical study performed before the control charts are set up. The most common sample size is 4 or 5 items per sample; the samples are taken periodically, every 15 minutes, every hour, or every day depending on what the quality control engineer determines is the best sampling frequency. Figure 17.1 illustrates the general format for an $\bar{x}$-chart.

Example 17.1.2

The design engineer has determined that the desired inside diameter of the parts produced by the stamping machine in Example 17.1.1 is 2.50

Figure 17.1 An $\bar{x}$-Chart

inches. A statistical engineering study performed by the quality control engineer determined that the standard deviation of the inside diameters of parts produced by the stamping machine is .029 inch. Set up the $\bar{x}$-chart for this process assuming that samples of size 5 are going to be taken periodically by the quality control inspector.

The center of the $\bar{x}$-chart will be the desired average diameter size, $\mu = 2.50$ inches. The upper and lower control limits are calculated using the definitions given above.

Upper Control Limit $(UCL) = \mu + 3\sigma_{\bar{x}} = 2.50 + 3\left(\dfrac{.029}{\sqrt{5}}\right)$

$= 2.50 + .039 = 2.539$

Lower Control Limit $(LCL) = \mu - 3\sigma_{\bar{x}} = 2.50 - 3\left(\dfrac{.029}{\sqrt{5}}\right)$

$= 2.50 - .039 = 2.461$.

The x-bar control chart for this example is illustrated in Figure 17.1.

The control limits on the $\bar{x}$-chart are set so that there is only a very small probability that a sample mean will fall outside of the control limits when the process is in fact operating at the specified average on the control chart. The probability of a sample mean falling within the so-called 3-sigma limits can be found using the standard normal table. In Chapter 5 it was shown that 99.7 percent of the items in a normally distributed population fall within ±3 standard deviations of the mean; i.e.,

$P(-3.0 < z < 3.0) = .4987 + .4987 = .9974$.

The probability is very small, $1 - .9974 = .0026$ or about .3 percent, that a sample mean will fall outside of the control chart limits if the process is in fact centered at the mean value on the control chart. (If the process output is normally distributed as assumed in this example, then the sample mean will also be normally distributed—see Chapter 4.)

If the sample means of the periodic samples fall within the control limits, then the process is said to be in control. If the sample mean falls outside of the control limits, then the process is said to be out of control, meaning that the average of the process has shifted; it is no longer operating at the specified level. The control chart is designed to protect against concluding the process is out of control when in fact it is in control. If the process is out of control, then it may be necessary to shut the whole system down. This can be expensive and, of course, it reduces the daily production rate; so the 3-sigma control chart is most often used because it protects against concluding the process should be shut down and adjusted when adjustment is in fact not needed.

Table 17.1 Sample Data for Example 17.1.3

Sample Number	Sample Values				
1	2.56	2.48	2.50	2.53	2.49
2	2.45	2.53	2.49	2.47	2.52
3	2.46	2.55	2.48	2.47	2.52
4	2.58	2.51	2.48	2.54	2.53
5	2.50	2.48	2.56	2.50	2.51
6	2.52	2.46	2.48	2.54	2.51
7	2.49	2.48	2.51	2.45	2.46
8	2.47	2.50	2.51	2.46	2.49
9	2.53	2.50	2.47	2.49	2.49
10	2.56	2.53	2.49	2.52	2.51

The averages of the 10 samples are calculated using the sample mean formula

$$\bar{x} = \frac{\Sigma x_i}{n}$$

from Chapter 2. Results are shown in Table 17.2.

Example 17.1.3 The sample values shown in Table 17.1 were obtained from the last 10 periodic samples of the parts produced by the stamping machine in Example 17.1.2. Calculate the sample means, plot them on the control chart developed in Example 17.1.2, and determine whether the process is in control.

Table 17.2 Sample Means Calculated from Data in Table 17.1

Sample Number	Sample Mean ($\bar{x}$)
1	2.512
2	2.492
3	2.496
4	2.528
5	2.510
6	2.502
7	2.478
8	2.486
9	2.496
10	2.522

Figure 17.2 Sample Means from Table 17.2 Plotted on the Control Chart from Figure 17.1

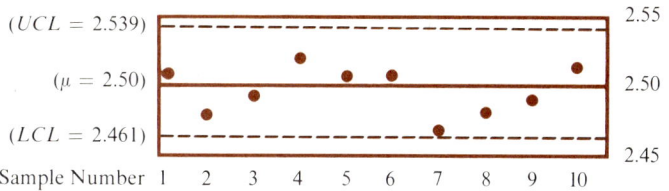

The sample means are plotted on the $\bar{x}$-chart in Figure 17.2. Note that all of the sample means fall within the control limits. Thus, we conclude the process was in control during the time period over which the samples were taken. If, when the samples are taken, the sample mean lies outside the control limits on the $\bar{x}$-chart, then, at that point in time, some action needs to be considered with regard to shutting down the process. If the process is easily adjusted with little loss in production time, then the adjustment might be made whenever a sample mean falls outside the control limits. If adjusting the process means a considerable loss of production time, then many quality control engineers will take a larger sample or a rapid series of samples to be sure that the mean of the process has in fact shifted and that the process is not being shut down unnecessarily.

17.2 $\bar{x}$-Charts and Hypothesis Testing

The use of $\bar{x}$-charts for controlling a production process is equivalent to performing a series of hypothesis tests. Each time a sample is taken, a decision is made as to whether or not the process average is equal to the average stated on the $\bar{x}$-chart. This decision is based on whether or not the sample mean falls outside the control limits on the $\bar{x}$-chart. The control limits represent a decision rule.

Example 17.2.1 Show that the application of the $\bar{x}$-chart developed as a solution to Example 17.1.2 and illustrated in Figure 17.1 is equivalent to performing a hypothesis test with the probability of Type I error, α, equal to .0026.

As indicated in Example 17.1.2, the desired inside diameter size is 2.50 inches. This would represent the hypothesized value of the population mean. The population standard deviation, σ, is assumed to be known from the engineering study, and therefore it will not be necessary to use the sample standard deviation, s, as an estimate of the population standard deviation. This permits us to use the z-statistic as a test statistic even though the sample size is small. In Chapter 6 it was stated as a theorem that the following statistic has a standard normal distribution if the population is normally distributed (as is assumed in setting up the $\bar{x}$-chart):

$$z = \frac{\bar{x} - \mu}{\frac{\sigma}{\sqrt{n}}}.$$

Using the sample mean, 2.512, calculated for the first sample of 5 items in Example 17.1.3, and the population standard deviation, $\sigma = .029$, given in Example 17.1.2, the desired hypothesis test can be performed using the methodology developed in Chapter 8.

1. $H_o: \mu = 2.50$
   $H_a: \mu \neq 2.50$.

2. $\alpha = .0026$

   $$z = \frac{\bar{x} - \mu}{\frac{\sigma}{\sqrt{n}}}.$$

3. Reject H_o if and only if $z_c < -3.0$ or $z_c > 3.0$.

4. $z_c = \dfrac{2.512 - 2.50}{\frac{.029}{\sqrt{5}}} = .93.$

   Do *not* reject H_o.

The sample result does not indicate that the process is operating at an average value different from the hypothesized value.

The decision rule value was obtained using the standard normal table, Table B. The value of α was split between the two tails as indicated in Figure 17.3. The decision rule in step 3 can be shown to be equivalent to rejecting the null hypothesis if the sample mean falls outside of the control limits of the $\bar{x}$-chart in Figure 17.1. Note that, since σ is known and μ is hypothesized, we can rewrite the test statistic in step 2 as follows:

$$z = \frac{\bar{x} - \mu}{\frac{\sigma}{\sqrt{n}}} = \frac{\bar{x} - 2.50}{\frac{.029}{\sqrt{5}}}.$$

Rejecting H_o for $z_c > 3.0$ is equivalent to rejecting H_o when

$$\frac{\bar{x} - 2.50}{\frac{.029}{\sqrt{5}}} > 3.0.$$

Figure 17.3 A Two-Tail Test

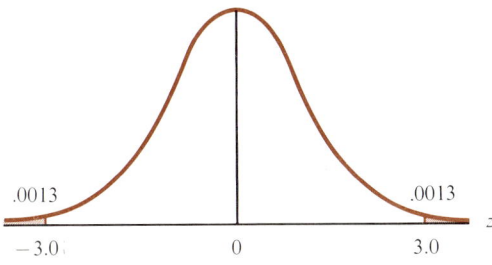

Solving algebraically for $\bar{x}$ gives us the following result:

Reject H_o if $\bar{x} > 2.50 + (3.0)\dfrac{.029}{\sqrt{5}}$

or

Reject H_o if $\bar{x} > 2.539$.

Similarly, rejecting H_o for $z_c < -3.0$ is equivalent to rejecting H_o when

$$\dfrac{\bar{x} - 2.50}{\dfrac{.029}{\sqrt{5}}} < -3.0.$$

Solving algebraically for $\bar{x}$ gives us the following result:

Reject H_o if $\bar{x} < 2.50 - (3.0)\dfrac{.029}{\sqrt{5}}$

or

Reject H_o if $\bar{x} < 2.461$.

The decision rule in step 3 can be rewritten in the following equivalent form:

Reject H_o if and only if $\bar{x} > 2.539$ or $\bar{x} < 2.461$.

Note that the upper control limit (UCL) in Figure 17.1 is 2.539 and the lower control limit (LCL) is 2.461, which are the same values as those in the restated decision rule above. Thus, the control chart limits represent a decision rule of a hypothesis test.

The answer to the question, What is the probability a sample mean will fall outside the control limits of the $\bar{x}$-chart in Figure 17.1 is .0026 if the process is operating at the average specified on the control chart. The probability that the sample mean falls within the control limits is $1 - .0026$, or .9974. A question of interest to the quality control engineer would be, What is the probability that the sample mean will fall within the control chart limits even though the process mean has shifted and the process is no longer operating at the desired average indicated on the control chart?

Example 17.2.2

If the process mean in Example 17.1.2 has suddenly shifted from the desired 2.50-inch average indicated on the control chart to an average of 2.53 inches, what is the probability that the next sample taken will not detect the shift in the process mean—i.e., what is the probability that the mean of the next sample will fall within the control limits?

If the sample mean falls within the control limits, we state that the process is in control, meaning it is operating at the desired average, and we let the process continue. The question asked in this example can be stated as,

$$P(2.461 < \bar{x} < 2.539 \mid \mu = 2.53) = ?$$

The numbers 2.461 and 2.539 are the control limits from Figure 17.1. This probability can be found using the method developed in Chapter 6 assuming a normal distribution:

$$P(a < \bar{x} < b) = P\left(\frac{a - \mu}{\frac{\sigma}{\sqrt{n}}} < z < \frac{b - \mu}{\frac{\sigma}{\sqrt{n}}}\right).$$

Using the hypothetical shifted value of the process mean, $\mu = 2.53$, and the standard deviation, $\sigma = .029$, the desired probability is found from the standard normal table:

$$P(2.461 < \bar{x} < 2.539) = P\left(\frac{2.461 - 2.53}{\frac{.029}{\sqrt{5}}} < z < \frac{2.539 - 2.53}{\frac{.029}{\sqrt{5}}}\right)$$

$$= P(-5.32 < z < .69) = .5 + .2549 = .7549.$$

There is a 75.49 percent chance that the next sample mean will fall within the control chart limits if the process is operating at an average of 2.53 inches. Figure 17.4 shows the distribution of $\bar{x}$ when the process average is 2.50 inches, the distribution of $\bar{x}$ when the process average is 2.53 inches, and the standard normal distribution used to make the probability calculation. The control limits were calculated using the desired mean, 2.50; this is indicated in Figure 17.4.

It is of interest to observe here that the .7549 probability found as a solution to the preceding example is the probability of making a Type II error if the average of the process is in fact 2.53 inches. The null hypothesis when using an $\bar{x}$-chart is the mean value indicated on the chart; in this case, it is 2.50 inches. If the process is operating at a mean value of 2.53 inches, the null hypothesis is false. But if the sample mean

Figure 17.4 Probability Distributions for Example 17.2.2

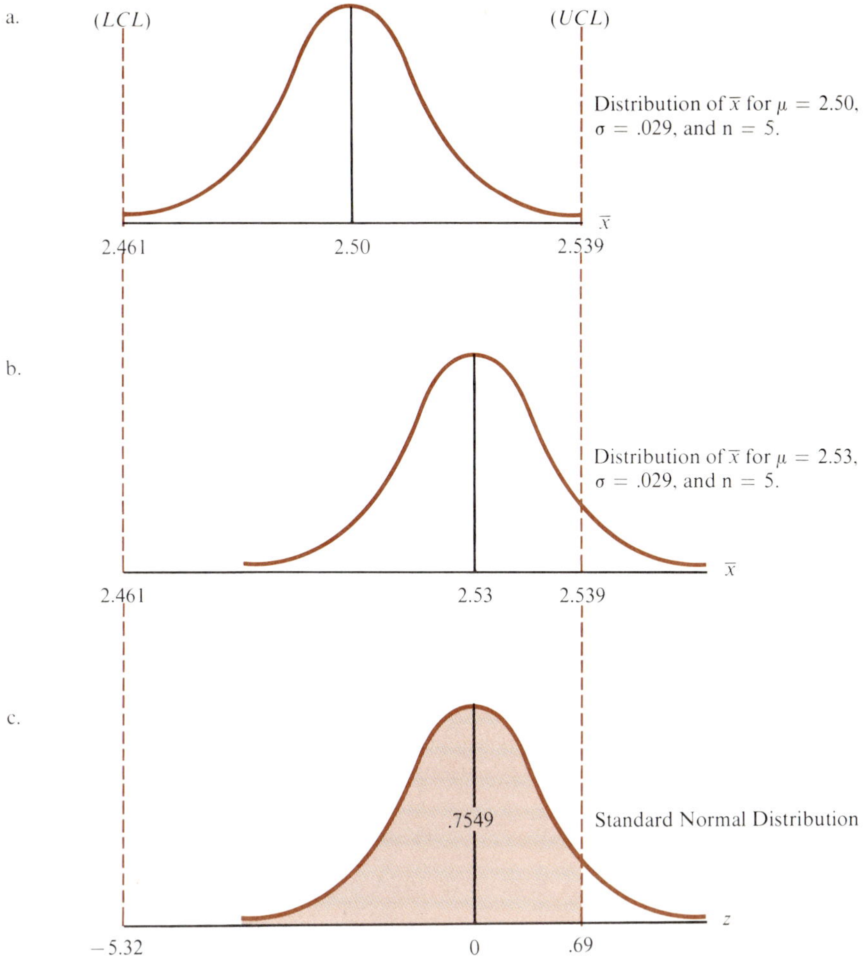

falls within the control limits, the false null hypothesis is accepted. A Type II error is committed if the null hypothesis is accepted when it is false.

The probability theory relating to the normal distribution developed in Chapter 5 can also be used to answer another question of interest to the quality control engineer.

Example 17.2.3 If the design engineer has determined that parts produced by the stamping machine in Example 17.1.2 must have an inside diameter of 2.50 ± .09 inches if they are to be useable, what percent of the parts will be out of tolerance if the process is operating at an average of 2.53

inches and with a standard deviation of .029 inches? Assume the diameter size is normally distributed.

The proportion of out-of-tolerance parts is the same as the probability that a part selected at random will be out of tolerance. This probability can be found using the technique developed in Chapter 5:

$$P(a < x < b) = P\left(\frac{a - \mu}{\sigma} < z < \frac{b - \mu}{\sigma}\right).$$

Note that in this case we are asking about the probability of an individual item, not about the mean of a sample as was done in the previous example. A part with a diameter size between 2.41 and 2.59 inches is considered good; in Example 17.1.2 the standard deviation was given as .029. If the mean of the process is 2.53 inches, the probability that a part selected at random will be good is

$$P(2.41 < x < 2.59) = P\left(\frac{2.41 - 2.53}{.029} < z < \frac{2.59 - 2.53}{.029}\right)$$
$$= P(-4.14 < z < 2.07) = .5 + .4808 = .9808.$$

This is the proportion of good parts. The proportion of bad parts is

$1 - .9808 = .0192.$

If the process is operating at an average of 2.53 inches, 1.92 percent of all parts produced will be out of tolerance with respect to the inside diameter dimensions.

17.3 Acceptance Sampling

A manufacturing firm buys items from vendors and has to decide when they arrive whether they are acceptable. The firm also manufactures some of its own parts, and a department head receiving parts produced by another company department has to determine if these parts are acceptable. In both cases, the items are generally grouped together in lots of a specified size, and the whole lot is accepted or rejected. The time and cost of inspecting each item are often prohibitive, and a decision on the acceptability of the lot is based on a sample of items randomly selected.

If the lot sizes are large, as will be assumed in this section, then the probabilities can be calculated using the binomial distribution; i.e., if the lot sizes are large, we can assume that the probabilities obtained assuming sampling with replacement will be approximately the same as the probabilities obtained allowing for the fact that sampling in this case is not done with replacement. Thus, to calculate probabilities we will use the binomial formula from Chapter 4:

$$p(x) = C_x^n \pi^x (1 - \pi)^{n-x}$$

where

n = the number of trials in the binomial experiment
π = the probability of success for each trial
x = the number of successes.

A Sampling Plan

A sampling plan is defined by two numbers:

n = the sample size
c = the acceptance number.

The acceptance number is the maximum number of defective items allowable in the sample for the lot to be accepted. If the number of defectives in the sample exceeds the acceptance number, then the whole lot is rejected.

While it would be desirable to have zero defective items in the lot of items received, it is uneconomical to do so for many processes. In such cases it is necessary to specify a maximum percentage of defectives that a lot can have and still be acceptable. This percentage or proportion is referred to as the acceptable quality level (AQL). A lot is also described in terms of the $LTPD$, the lot tolerance percent defective. This is defined as an intolerable level of percent defective items. If the proportion of items in the lot equals or exceeds the $LTPD$, then the lot is considered a "bad" lot and definitely should not be accepted. Along with these two terms, two others, *producer's risk* and *consumer's risk*, are used to evaluate acceptance sampling plans. These four terms are summarized below.

Acceptance Sampling Terms

AQL = Acceptable Quality Level = maximum percentage of defective items that a lot can have and still be considered "good."

$LTPD$ = Lot Tolerance Percent Defective = the percent of defective items that is intolerable and, therefore, for which the lot is considered "bad."

Producer's Risk = α = the probability that the sampling plan will reject a "good" lot, i.e., a lot with the percent of defective items equal to the AQL.

Consumer's Risk = β = the probability that the sampling plan will accept a lot that is "bad," i.e., a lot with the percent of defective items equal to the $LTPD$.

If the lot is "good," then the producer doesn't want it rejected; so he desires a low producer's risk. On the other hand, if the lot is "bad," the consumer doesn't want to accept it; so he desires a low consumer's risk.

Example 17.3.1

Given the following sampling plan for a large lot,

n = 5
c = 0,

find the producer's risk, α, for an *AQL* of 1 percent.

This sampling plan states that the lot should be accepted if there are no defective items in a sample of 5 items randomly selected from the lot. If the lot has 1 percent defectives, then the probability that all 5 items will be good is

P(the lot is accepted) = P(all 5 items are good)
= $P(GGGGG) = P(G)P(G)P(G)P(G)P(G) = (.99)^5 = .951$.

The probability that the lot will be rejected if it has 1 percent defectives is

$\alpha = P$(lot is rejected) = $1 - P$(lot is accepted)
$= 1 - .951 = .049$.

Thus, the producer's risk for an *AQL* of 1 percent is .049 for this sampling plan. Note that the binomial table, Table A, could have been used to find this probability. However, the table cannot be used for the next example because the largest value of n in Table A is 25.

Example 17.3.2

Given the following sampling plan for a large lot,

n = 35
c = 1,

find the producer's risk, α, for an *AQL* of 1 percent.

The *AQL* is the same as for the previous example, but the sample size is larger, and the acceptance number is 1. The lot will be accepted even if there is 1 defective item in the sample. To calculate the probabilities for this example, we note that the binomial probability function is applicable where each item sampled is a trial ($n = 35$) and the probability of a defective on each trial is the *AQL* ($\pi = .01$). Putting in the values for n and π in the binomial formula gives us

$$p(x) = C_x^{35} (.01)^x (.99)^{35-x}.$$

This formula can be used to find the probability of 0 defective items and the probability of 1 defective item in the sample:

$$P(x = 0) = p(0) = C_x^{35} (.01)^0(.99)^{35-0} = (.99)^{35} = .703$$

$$P(x = 1) = p(1) = C_1^{35} (.01)^1(.99)^{35-1} = 35(.01)(.99)^{34} = .249.$$

The lot will be accepted if either 0 defective items or 1 defective item is in the sample. These are mutually exclusive events; so the probability of acceptance is the sum of the probabilities found above:

$P(\text{accept lot}) = P(0 \text{ defects}) + P(1 \text{ defect}) = .703 + .249 = .952.$

The probability of rejecting the lot is the complement of accepting the lot:

$\alpha = 1 - P(\text{accept lot}) = 1 - .952 = .048.$

For these 2 sampling plans, the producer's risk, α, is almost the same; for $n = 5$ and $c = 0$, $\alpha = .049$ or 4.9 percent, and for $n = 35$ and $c = 1$, $\alpha = .048$ or 4.8 percent. The first sample has a much smaller sample size, which means less sampling cost per lot. Why even give the second sampling plan consideration? There is another type of risk, consumer's risk, that we should consider before passing judgment on the second sampling plan with its larger sample size.

Example 17.3.3

Given the sampling plan in Example 17.3.1,

$n = 5$
$c = 0,$

find the consumer's risk, β, if the lot tolerance percent defective (*LTPD*) is 10 percent.

The consumer's risk, β, is the probability that a "bad" lot will be accepted; a bad lot is defined as one having *LTPD* items—for this example, 10 percent defective items. If all 5 items in the sample are good, then the lot is accepted. The probability of any 1 randomly selected item being good is .9; therefore,

$\beta = P(\text{accept lot}) = P(\text{all 5 sample items are good})$
$= P(GGGGG) = P(G)P(G)P(G)P(G)P(G) = (.9)^5 = .590.$

The consumer's risk for this sampling plan, for *LTPD* equal to 10 percent, is .590.

Example 17.3.4

Given the sampling plan in Example 17.3.2,

$n = 35$
$c = 1,$

find the consumer's risk, β, if the *LTPD* is 10 percent.

The binomial probability function can be used here as was done for Example 17.1.2. For this example, we let $\pi = .10$, the *LTPD*. The binomial formula for $n = 35$ and $\pi = .10$ is

$$p(x) = C_x^{35} (.10)^x (.90)^{35-x}.$$

The probability of 0 defective items and the probability of 1 defective item in the sample can now be found:

$$P(x = 0) = p(0) = C_0^{35} (.10)^0 (.90)^{35-0} = (.90)^{35} = .025.$$

$$P(x = 1) = p(1) = C_1^{35} (.10)^1 (.90)^{35-1} = 35(.10)(.90)^{34} = .097.$$

The probability that the lot will be accepted is the sum of these 2 probabilities:

$$\beta = P(\text{lot is accepted}) = .025 + .097 = .122.$$

The consumer's risk for this sampling plan and *LTPD* equal to 10 percent is .122.

The producer's risk and the consumer's risk depend on the respective values of *AQL* and *LTPD*. These values are determined by management. They require a managerial decision. Once this decision has been made, then the producer's risk, α, and the consumer's risk, β, can be calculated for a given sampling plan.

The results of the solutions to the 4 examples in this section are summarized in Table 17.3. Both samples have almost the same producer's risk, α, for an acceptable quality level (*AQL*) of 1 percent. In this regard, one might say that both sampling plans are equally good because they give the producer equal protection. However, the second sampling plan offers greater protection to the consumer while offering the producer the same protection as the first sampling plan; the consumer's risk of 12.2 percent is considerably lower than the 59.0 percent consumer's risk of the first sampling plan. Thus, the sampling plan with the larger sample size is the preferable sampling plan. However, it does

Table 17.3 Comparison of Two Sampling Plans

Sampling Plan	Acceptable Quality Level (*AQL*)	Producer's Risk (α)	Lot Tolerance Percent Defective (*LTPD*)	Consumer's Risk (β)
$n = 5$ $c = 0$	1%	.049	10%	.590
$n = 35$ $c = 1$	1%	.048	10%	.122

cost more to take larger samples. The choice of a sampling plan depends on costs of rejecting good lots and accepting bad lots as well as the respective probabilities of these events occurring.

 ## 17.4 Operating Characteristic Curves

The operating characteristic curve provides a graphic picture of a sampling plan and is useful for comparing sampling plans. It is difficult to keep in mind four or five probabilities of acceptance for four or five different levels of percent defective items. A graph of these probabilities is easier to handle. If a sampling plan is accompanied by a graph of its operating characteristic curve, then you can read the probabilities directly from the graph and avoid lengthy probability calculations.

A typical operating characteristic curve is drawn in Figure 17.5. The horizontal axis has the percent defective in the lot, and the vertical axis shows the probability that the lot will be accepted by the sampling plan represented by the operating characteristic curve. If a lot has 6 percent defective items—i.e., the proportion of defective items is .06—then the probability that the sampling plan depicted by the operating characteristic curve in Figure 17.5 will accept the lot can be

Figure 17.5 Operating Characteristic Curve

Operating Characteristic Curves 557

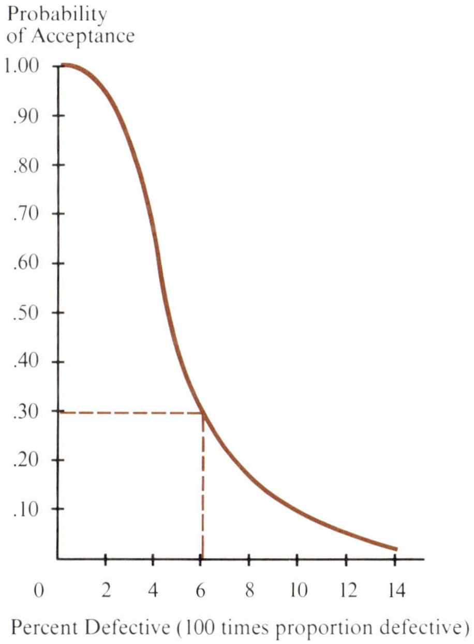

read from the graph, and it is approximately .29. If management decides that the *LTPD is* 6 percent, then the consumer's risk, β, is .29.

The operating characteristic curves for the 2 sampling plans discussed in the previous section—$n = 5$ with $c = 0$, and $n = 35$ with $c = 1$—are shown in Figure 17.6. The probabilities found in the solutions to the previous examples are also indicated on the graphs. The operating characteristic curve of the second sampling plan—$n = 35$ with $c = 1$—is steeper, and therefore the sampling plan is considered better. It discriminates better: the probability of acceptance for lots with small percentages of defective items is high, while the probability of accepting lots with large percentages of defective items is small. Generally, the larger the sample size, the steeper the operating characteristic curve.

17A Appendix: Acceptance Sampling and the Poisson Distribution

In Appendix 4D, at the end of Chapter 4, it was pointed out that the Poisson Distribution can be used to approximate the probabilities of the binomial distribution when n is large and π is small and the Poisson parameter, λ, is set equal to $n\pi$. It is common practice among quality control engineers to use the Poisson Distribution to calculate probabilities for sampling plans when the sample size is large.

Example 17A.1 Use the Poisson Distribution to find the producer's risk, α, and the consumer's risk, β, for the following sample plan:

$n = 150$
$c = 4$.

The acceptable quality level (*AQL*) for the firm is 1 percent, and the lot tolerance percent defective (*LTPD*) is 5 percent.

To determine α, we let π equal the *AQL*, 1 percent, and use the Poisson Distribution with

$$\lambda = n\pi = 150(.01) = 1.5$$

to find the probability that the number of defects will be 4 or less. This probability is found using Table F, the Poisson table. From the column for $\lambda = 1.5$ in the table we find that

$P(x \leq 4) = .9814.$

This is the probability that a lot with 1.0 percent defective items will be accepted. The producer's risk for $AQL = 1.0$ percent is the probability that the lot will be rejected,

Figure 17.6 Operating Characteristic Curves

a.

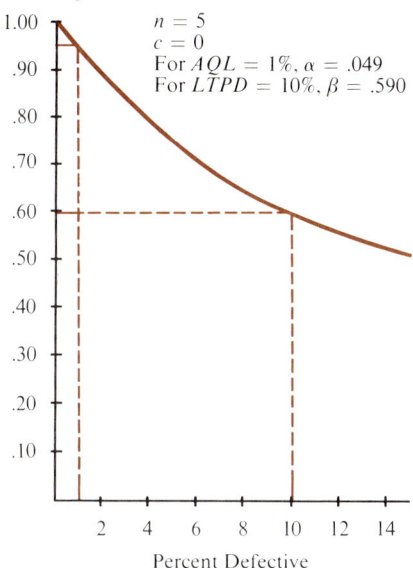

b.

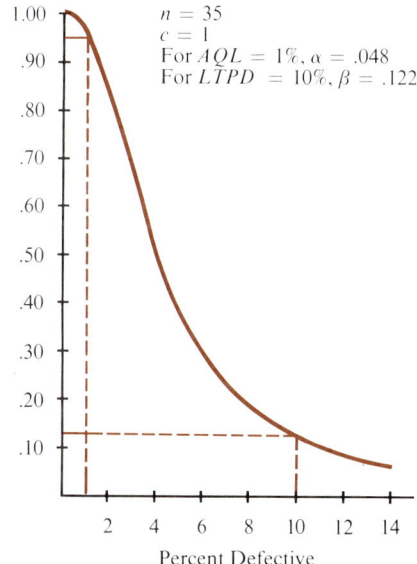

Appendix to Chapter 17

$\alpha = 1 - P(x \le 4) = 1 - .9814 = .0186.$

The consumer's risk for this example is the probability that a lot with 5 percent defective items will be accepted. To find this probability we let

$\lambda = n\pi = 150(.05) = 7.5.$

Using this value of λ, we get

$\beta = P(x \le 4) = .1321$

from the column for $\lambda = 7.5$ in Table F.

17B Appendix: Notation

The notation used in this chapter is the same as that used throughout the text and is a fairly standard notation used by most writers in applied statistics. Unfortunately, texts on statistical quality control use notation that is almost unique to that subject. This is not an insurmountable problem if you are aware of these differences when you read a statistical quality control text. We compare below some of the notation used in this text with the corresponding notation taken from one of the widely used texts in quality control, written by Grant and Leavenworth.[2]

	Notation	
Parameter or Statistic	This Text	Grant and Leavenworth
Population Mean	μ	$\bar{X}'$
Population Standard Deviation	σ	σ'
Sample Mean	$\bar{x}$	$\bar{X}$
Sample Standard Deviation	s	σ
Population Proportion	π	p'
Sample Proportion	p	p

The notation used by Grant and Leavenworth is fairly standard among quality control practitioners.

Problems

1. Steel shafts coming off a cutting machine are supposed to be 4.55 inches in length. A statistical engineering study has determined that the standard deviation of the cutting process is .096 inches. It is decided that periodic samples of 4 shafts will be taken to determine whether the cutting machine is operating properly. Deter-

---

[2]. Eugene L. Grant and Richard S. Leavenworth, *Statistical Quality Control*, 5th ed. (New York: McGraw-Hill, 1980).

mine the upper and lower control limits for an $\bar{x}$-chart assuming the process is centered at the specified average of 4.55 inches. Draw a control chart similar to that in Figure 17.1.

2. Samples of size 4 are taken from the cutting machine in Problem 1 every 45 minutes. The results of the last 8 samples are given below. Calculate the averages of the samples, plot them on the control chart made for Problem 1, and determine whether the process is in control.

Sample Number	Sample Lengths of Steel Shafts			
1	4.69	4.58	4.50	4.51
2	4.36	4.46	4.65	4.49
3	4.29	4.50	4.51	4.53
4	4.55	4.62	4.63	4.46
5	4.52	4.57	4.60	4.60
6	4.55	4.48	4.65	4.69
7	4.59	4.63	4.51	4.51
8	4.55	4.41	4.62	4.63

3. A filling machine is set to fill cans with 16.5 ounces of peaches. A statistical engineering study has determined that the standard deviation of the weight of peaches in cans filled by the machine is 1.52 ounces. Determine the upper and lower control limits of an $\bar{x}$-chart and draw the chart. Periodic samples of the contents of 5 cans will be taken to determine whether the filling machine is operating at an average fill of 16.5 ounces.

4. The sample values from the last 7 periodic samples of cans filled by the filling machine in Problem 3 are given below. Plot the sample averages on the control chart from Problem 3 and determine if the process is in control.

Sample Number	Weights of Contents of Cans in Samples				
1	16.0	16.5	16.7	17.8	17.2
2	13.9	13.1	15.3	14.7	16.9
3	13.2	14.8	14.9	15.1	14.5
4	15.8	14.7	16.3	15.6	17.2
5	16.3	14.6	14.9	16.7	14.4
6	13.4	14.3	15.1	14.6	16.4
7	15.5	14.6	13.3	11.9	15.4

5. If the setting on the cutting machine in Problem 1 suddenly shifts to 4.68 inches, what is the probability that the next sample taken will *not* detect the shift; i.e., what is the probability that the sample mean will fall within the control chart limits in Problem 1 even though the mean of the process is now 4.68 inches? (The standard

deviation of the cutting machine output is .096 inches, as given in Problem 1.)

6. Refer to Problems 1 and 5. The design engineer has specified that only steel shafts of length 4.55 ± .30 inch are usable as components in the product produced by the company. If the mean of the cutting process has in fact shifted to 4.68 inches (the standard deviation is .096 inches), what proportion of the parts will be out of tolerance?

7. Refer to Problem 6.
    a. If the process average shifts to 4.80 inches, what proportion of the parts will be out of tolerance?
    b. What is the probability that the next sample taken will detect the shift—i.e., the sample mean will fall outside the control limits determined in Problem 1?

8. Refer to Problem 3. If the average weight of fill by the filling machine changes to 15.0 ounces, what is the probability that the next sample of 5 filling weights will *not* detect the shift; i.e., what is the probability that the sample average will fall within the control limits of the $\bar{x}$-chart obtained in Problem 3?

9. Answer the question in Problem 8 assuming that the average fill of the filling machine is 14.0 ounces.

10. An appliance manufacturer receives locknuts from an outside vendor. The *AQL* (acceptable quality level) determined by management is 1 percent. If the firm uses the following sampling plan,

    $n = 25$
    $c = 0,$

    what is α, the producer's risk?

11. If it is decided by the appliance-manufacturing firm in Problem 10 that the *LTPD* (lot tolerance percent defective) is 10 percent, what is the consumer's risk, β, for the sampling plan in Problem 10?

12. If the appliance manufacturer in Problem 10 changes his sampling plan to

    $n = 25$
    $c = 1,$

    what will the producer's risk and consumer's risk be for an $AQL = 1$ percent and an $LTPD = 10$ percent as determined by the manufacturer in the previous 2 problems?

13. Given the following sampling plan,

    $n = 15$
    $c = 1,$

    find α and β if $AQL = 1$ percent and $LTPD = 10$ percent.

14. Given the following sampling plan,

    $n = 20$
    $c = 2,$

    find α and β for $AQL = 1$ percent and $LTPD = 10$ percent.

15. Given the following sampling plan,

    $n = 25$
    $c = 2,$

    find α and β for $AQL = 1$ percent and $LTPD = 10$ percent.

16. The average for a manufacturing process has been specified to be 265. The standard deviation for the process, σ, is equal to 6. The $\bar{x}$-chart control limits for a sample of size 4 would be 265 ± 9.
    a. If the process mean suddenly shifts downward by 5 units to 260, what is the probability that the shift will *not* be detected when the next sample is taken; i.e., what is the probability that the sample mean will fall within the control limits?
    b. If instead of a sample of size 4, a sample of size 16 is used for the control chart, what is the probability that the next sample taken after the shift in the process average will *not* detect the shift? (Remember, the $\bar{x}$-chart control limits are a function of sample size; so it will be necessary to recalculate the control limits to answer this question.)

17. For the sampling plans in Problems 12, 13, 14, and 15, find the points on the respective operating characteristic curves, or the probabilities of acceptance for the respective plans, if 5 percent of the items in the lot sampled are defective.

18. The operating characteristic curve for the sampling plan,

    $n = 5$
    $c = 0,$

    is plotted in Figure 17.6. The points for 1 percent and 10 percent defective items in the lot sampled also are given in Figure 17.6; i.e., for 1 percent the probability of accepting the lot is $1 - .049 = .951$, and for 10 percent it is .590. Find the points of the operating characteristic curve; i.e., find the probabilities of acceptance for the following percentages of defective items in the lot sampled: 2 percent, 4 percent, 6 percent, 8 percent, and 12 percent.

19. Answer the question in Problem 17 for a lot with 20 percent defective items.

20. The operating characteristic curve for the sampling plan,

    $n = 35$
    $c = 1,$

is plotted in Figure 17.6. The points on the curve for 1 percent and 10 percent defective items in the lot sampled are also given in Figure 17.6. Find the probabilities of acceptance—the points on the operating characteristic curve—for the following percentages of defective items in the lot sampled: 2 percent, 4 percent, 6 percent, 8 percent, and 12 percent.

21. A manufacturing process has a mean of 100 and a standard deviation equal to 5. The $\bar{x}$-chart upper control limits for a sample of size 4 is 107.5, and the lower control limit is 92.5. If the mean shifts upward by 10 percent to 110, what is the probability that the next sample will not detect the shift; i.e., what is the probability that the sample mean will fall within the control limits?

22. If the shift in the mean in Problem 21 is only 5 percent, to 105, what is the probability that the shift in the mean will not be detected by the next sample?

23. If the tolerance limits for the parts produced by the process in Problems 22 and 23 are 100 ± 16,
    a. what percent of parts will be out of tolerance if the process shifts to 110;
    b. what percent of the parts will be out of tolerance if the process only shifts to 105?

24. If the sample size is changed to 16 in Problem 21, what is the answer to that problem? (Note that control limits change with the change in sample size.)

25. An electrical manufacturer receives large lots of parts from different vendors. He has decided that an appropriate AQL for these parts is 1.5 percent. If the following sampling plan is used,

    $n = 6$
    $c = 0$,

    what is the producer's risk, α?

26. If the electrical manufacturer in Problem 25 decides that the $LTPD$ is 12 percent, what is the consumer's risk, β, for the sampling plan given in Problem 10?

27. If the electrical manufacturer in Problem 25 changes his sampling plan to

    $n = 40$
    $c = 1$,

    what will be the producer's risk and consumer's risk for $AQL = 1.5$ percent and $LTPD = 12$ percent as determined by the manufacturer?

■ The remaining problems require material covered in the appendixes.

Statistical Quality Control

28. The following acceptance sampling plan is used by a firm that has specified $AQL = .5$ percent and $LTPD = 3$ percent:

    $n = 200$
    $c = 3$.

    Use the Poisson Distribution to find
    a. The producer's risk, α;
    b. the consumer's risk, β.

29. A set of sampling plans, MIL-STD-105D, prepared by the Department of Defense,[3] is widely used in American industry. The sampling plan,

    $n = 500$
    $c = 5$,

    is given under code letter N for "normal" inspection and $AQL = .4$ percent.
    a. Find α, the producer's risk.
    b. If $LTPD = 1.5$ percent, find β, the consumer's risk.
    (Use the Poisson Distribution.)

30. Refer to the previous problem. For the same code letter, N, and the same AQL, .4 percent, the MIL-STD-105D sampling plan for "tightened" inspection is

    $n = 500$
    $c = 3$.

    Tightened inspection is designed to give more protection to the consumer. A comparison of the answers to the following questions with the solutions to parts a and b of the previous problem should verify this statement.
    a. Find α, the producer's risk.
    b. If $LTPD = 1.5$ percent, find β, the consumer's risk.
    (Use the Poisson Distribution.)

31. For code letter J, the MIL-STD-105D sampling plan is

    $n = 80$
    $c = 3$

    for $AQL = 1.5$ percent. Find the producer's risk, α.

---

3. *Military Standard—Sampling Procedures and Tables for Inspection by Attributes* (MIL-STD-105D) (Washington D.C.: U.S. Government Printing Office, 1963).

TABLES

Table A: Binomial Probabilities
Table B: Standard Normal Curve
Table C: Student-t Distribution
Table D: Chi-Square Distribution
Table E1: F-Distribution
Table E2: F-Distribution
Table F: Poisson Probabilities

Table A: Binomial Probabilities

Table Gives $P(x \leq k)$, $P(x \leq k) = \sum_{x=0}^{k} p(x) = \sum_{x=0}^{k} C_x^n \pi^x (1-\pi)^{n-x}$

$n = 5$

$\pi =$	.01	.05	.10	.20	.25	.30	.40	.50	.60	.70	.75	.80	.90	.95	.99
k															
0	.951	.774	.590	.328	.237	.168	.078	.031	.010	.002	.001	.000	.000	.000	.000
1	.999	.977	.919	.737	.633	.528	.337	.188	.087	.031	.016	.007	.000	.000	.000
2	1.000	.999	.991	.942	.896	.837	.683	.500	.317	.163	.104	.058	.009	.001	.000
3	1.000	1.000	1.000	.993	.984	.969	.913	.813	.663	.472	.367	.263	.081	.023	.001
4	1.000	1.000	1.000	1.000	.999	.998	.990	.969	.922	.832	.763	.672	.410	.226	.049

$n = 10$

$\pi =$	.01	.05	.10	.20	.25	.30	.40	.50	.60	.70	.75	.80	.90	.95	.99
k															
0	.904	.599	.349	.107	.056	.028	.006	.001	.000	.000	.000	.000	.000	.000	.000
1	.996	.914	.736	.376	.244	.149	.046	.011	.002	.000	.000	.000	.000	.000	.000
2	1.000	.988	.930	.678	.526	.383	.167	.055	.012	.002	.000	.000	.000	.000	.000
3	1.000	.999	.987	.879	.776	.650	.382	.172	.055	.011	.004	.001	.000	.000	.000
4	1.000	1.000	.998	.967	.922	.850	.633	.377	.166	.047	.020	.006	.000	.000	.000
5	1.000	1.000	1.000	.994	.980	.953	.834	.623	.367	.150	.078	.033	.002	.000	.000
6	1.000	1.000	1.000	.999	.996	.989	.945	.828	.618	.350	.224	.121	.013	.001	.000
7	1.000	1.000	1.000	1.000	1.000	.998	.988	.945	.833	.617	.474	.322	.070	.012	.000
8	1.000	1.000	1.000	1.000	1.000	1.000	.998	.989	.954	.851	.756	.624	.264	.086	.004
9	1.000	1.000	1.000	1.000	1.000	1.000	1.000	.999	.994	.972	.944	.893	.651	.401	.096

Table A: Binomial Probabilities (continued)

$n = 15$

k	$\pi = .01$	.05	.10	.20	.25	.30	.40	.50	.60	.70	.75	.80	.90	.95	.99
0	.860	.463	.206	.035	.013	.005	.000	.000	.000	.000	.000	.000	.000	.000	.000
1	.990	.829	.549	.167	.080	.035	.005	.000	.000	.000	.000	.000	.000	.000	.000
2	1.000	.964	.816	.398	.236	.127	.027	.004	.000	.000	.000	.000	.000	.000	.000
3	1.000	.995	.944	.648	.461	.297	.091	.018	.002	.000	.000	.000	.000	.000	.000
4	1.000	.999	.987	.836	.686	.515	.217	.059	.009	.001	.000	.000	.000	.000	.000
5	1.000	1.000	.998	.939	.852	.722	.403	.151	.034	.004	.001	.000	.000	.000	.000
6	1.000	1.000	1.000	.982	.943	.869	.610	.304	.095	.015	.004	.001	.000	.000	.000
7	1.000	1.000	1.000	.996	.983	.950	.787	.500	.213	.050	.017	.004	.000	.000	.000
8	1.000	1.000	1.000	.999	.996	.985	.905	.696	.390	.131	.057	.018	.000	.000	.000
9	1.000	1.000	1.000	1.000	.999	.996	.966	.849	.597	.278	.148	.061	.002	.000	.000
10	1.000	1.000	1.000	1.000	1.000	.999	.991	.941	.783	.485	.314	.164	.013	.001	.000
11	1.000	1.000	1.000	1.000	1.000	1.000	.998	.982	.909	.703	.539	.352	.056	.005	.000
12	1.000	1.000	1.000	1.000	1.000	1.000	1.000	.996	.973	.873	.764	.602	.184	.036	.000
13	1.000	1.000	1.000	1.000	1.000	1.000	1.000	1.000	.995	.965	.920	.833	.451	.171	.010
14	1.000	1.000	1.000	1.000	1.000	1.000	1.000	1.000	1.000	.995	.987	.965	.794	.537	.140

Table A: Binomial Probabilities (continued)

$n = 20$

k	$\pi = .01$	.05	.10	.20	.25	.30	.40	.50	.60	.70	.75	.80	.90	.95	.99
0	.818	.358	.122	.012	.003	.001	.000	.000	.000	.000	.000	.000	.000	.000	.000
1	.983	.736	.392	.069	.024	.008	.001	.000	.000	.000	.000	.000	.000	.000	.000
2	.999	.925	.677	.206	.091	.035	.004	.000	.000	.000	.000	.000	.000	.000	.000
3	1.000	.984	.867	.411	.225	.107	.016	.001	.000	.000	.000	.000	.000	.000	.000
4	1.000	.997	.957	.630	.415	.238	.051	.006	.000	.000	.000	.000	.000	.000	.000
5	1.000	1.000	.989	.804	.617	.416	.126	.021	.002	.000	.000	.000	.000	.000	.000
6	1.000	1.000	.998	.913	.786	.608	.250	.058	.006	.000	.000	.000	.000	.000	.000
7	1.000	1.000	1.000	.968	.898	.772	.416	.132	.021	.001	.000	.000	.000	.000	.000
8	1.000	1.000	1.000	.990	.959	.887	.596	.252	.057	.005	.001	.000	.000	.000	.000
9	1.000	1.000	1.000	.997	.986	.952	.755	.412	.128	.017	.004	.001	.000	.000	.000
10	1.000	1.000	1.000	.999	.996	.983	.872	.588	.245	.048	.014	.003	.000	.000	.000
11	1.000	1.000	1.000	1.000	.999	.995	.943	.748	.404	.113	.041	.010	.000	.000	.000
12	1.000	1.000	1.000	1.000	1.000	.999	.979	.868	.584	.228	.102	.032	.000	.000	.000
13	1.000	1.000	1.000	1.000	1.000	1.000	.994	.942	.750	.392	.214	.087	.002	.000	.000
14	1.000	1.000	1.000	1.000	1.000	1.000	.998	.979	.874	.584	.383	.196	.011	.000	.000
15	1.000	1.000	1.000	1.000	1.000	1.000	1.000	.994	.949	.762	.585	.370	.043	.003	.000
16	1.000	1.000	1.000	1.000	1.000	1.000	1.000	.999	.984	.893	.775	.589	.133	.016	.000
17	1.000	1.000	1.000	1.000	1.000	1.000	1.000	1.000	.996	.965	.909	.794	.323	.075	.001
18	1.000	1.000	1.000	1.000	1.000	1.000	1.000	1.000	.999	.992	.976	.931	.608	.264	.017
19	1.000	1.000	1.000	1.000	1.000	1.000	1.000	1.000	1.000	.999	.997	.988	.878	.642	.182

Table A: Binomial Probabilities (continued)

$n = 25$

k	$\pi = .01$	.05	.10	.20	.25	.30	.40	.50	.60	.70	.75	.80	.90	.95	.99
0	.778	.277	.072	.004	.001	.000	.000	.000	.000	.000	.000	.000	.000	.000	.000
1	.974	.642	.271	.027	.007	.002	.000	.000	.000	.000	.000	.000	.000	.000	.000
2	.998	.873	.537	.098	.032	.009	.000	.000	.000	.000	.000	.000	.000	.000	.000
3	1.000	.966	.764	.234	.096	.033	.002	.000	.000	.000	.000	.000	.000	.000	.000
4	1.000	.993	.902	.421	.214	.090	.009	.000	.000	.000	.000	.000	.000	.000	.000
5	1.000	.999	.967	.617	.378	.193	.029	.002	.000	.000	.000	.000	.000	.000	.000
6	1.000	1.000	.991	.780	.561	.341	.074	.007	.000	.000	.000	.000	.000	.000	.000
7	1.000	1.000	.998	.891	.727	.512	.154	.022	.001	.000	.000	.000	.000	.000	.000
8	1.000	1.000	1.000	.953	.851	.677	.274	.054	.004	.000	.000	.000	.000	.000	.000
9	1.000	1.000	1.000	.983	.929	.811	.425	.115	.013	.000	.000	.000	.000	.000	.000
10	1.000	1.000	1.000	.994	.970	.902	.586	.212	.034	.002	.000	.000	.000	.000	.000
11	1.000	1.000	1.000	.998	.989	.956	.732	.345	.078	.006	.001	.000	.000	.000	.000
12	1.000	1.000	1.000	1.000	.997	.983	.846	.500	.154	.017	.003	.000	.000	.000	.000
13	1.000	1.000	1.000	1.000	.999	.994	.922	.655	.268	.044	.011	.002	.000	.000	.000
14	1.000	1.000	1.000	1.000	1.000	.998	.966	.788	.414	.098	.030	.006	.000	.000	.000
15	1.000	1.000	1.000	1.000	1.000	1.000	.987	.885	.575	.189	.071	.017	.000	.000	.000
16	1.000	1.000	1.000	1.000	1.000	1.000	.996	.946	.726	.323	.149	.047	.000	.000	.000
17	1.000	1.000	1.000	1.000	1.000	1.000	.999	.978	.846	.488	.273	.109	.002	.000	.000
18	1.000	1.000	1.000	1.000	1.000	1.000	1.000	.993	.926	.659	.439	.220	.009	.000	.000
19	1.000	1.000	1.000	1.000	1.000	1.000	1.000	.998	.971	.807	.622	.383	.033	.001	.000
20	1.000	1.000	1.000	1.000	1.000	1.000	1.000	1.000	.991	.910	.786	.579	.098	.007	.000
21	1.000	1.000	1.000	1.000	1.000	1.000	1.000	1.000	.998	.967	.904	.766	.236	.034	.000
22	1.000	1.000	1.000	1.000	1.000	1.000	1.000	1.000	1.000	.991	.968	.902	.463	.127	.002
23	1.000	1.000	1.000	1.000	1.000	1.000	1.000	1.000	1.000	.998	.993	.973	.729	.358	.026
24	1.000	1.000	1.000	1.000	1.000	1.000	1.000	1.000	1.000	1.000	.999	.996	.928	.723	.222

Table B: Standard Normal Curve

Value in Table is $P(0 \leq z \leq z_0)$

z_0	.00	.01	.02	.03	.04	.05	.06	.07	.08	.09
0.0	.0000	.0040	.0080	.0120	.0160	.0199	.0239	.0279	.0319	.0359
0.1	.0398	.0438	.0478	.0517	.0557	.0596	.0636	.0675	.0714	.0753
0.2	.0793	.0832	.0871	.0910	.0948	.0987	.1026	.1064	.1103	.1141
0.3	.1179	.1217	.1255	.1293	.1331	.1368	.1406	.1443	.1480	.1517
0.4	.1554	.1591	.1628	.1664	.1700	.1736	.1772	.1808	.1844	.1879
0.5	.1915	.1950	.1985	.2019	.2054	.2088	.2123	.2157	.2190	.2224
0.6	.2257	.2291	.2324	.2357	.2389	.2422	.2454	.2486	.2517	.2549
0.7	.2580	.2611	.2642	.2673	.2704	.2734	.2764	.2794	.2823	.2852
0.8	.2881	.2910	.2939	.2967	.2995	.3023	.3051	.3078	.3106	.3133
0.9	.3159	.3186	.3212	.3238	.3264	.3289	.3315	.3340	.3365	.3389
1.0	.3413	.3438	.3461	.3485	.3508	.3531	.3554	.3577	.3599	.3621
1.1	.3643	.3665	.3686	.3708	.3729	.3749	.3770	.3790	.3810	.3830
1.2	.3849	.3869	.3888	.3907	.3925	.3944	.3962	.3980	.3997	.4015
1.3	.4032	.4049	.4066	.4082	.4099	.4115	.4131	.4147	.4162	.4177
1.4	.4192	.4207	.4222	.4236	.4251	.4265	.4279	.4292	.4306	.4319
1.5	.4332	.4345	.4357	.4370	.4382	.4394	.4406	.4418	.4429	.4441
1.6	.4452	.4463	.4474	.4484	.4495	.4505	.4515	.4525	.4535	.4545
1.7	.4554	.4564	.4573	.4582	.4591	.4599	.4608	.4616	.4625	.4633
1.8	.4641	.4649	.4656	.4664	.4671	.4678	.4686	.4693	.4699	.4706
1.9	.4713	.4719	.4726	.4732	.4738	.4744	.4750	.4756	.4761	.4767
2.0	.4772	.4778	.4783	.4788	.4793	.4798	.4803	.4808	.4812	.4817
2.1	.4821	.4826	.4830	.4834	.4838	.4842	.4846	.4850	.4854	.4857
2.2	.4861	.4864	.4868	.4871	.4875	.4878	.4881	.4884	.4887	.4890
2.3	.4893	.4896	.4898	.4901	.4904	.4906	.4909	.4911	.4913	.4916
2.4	.4918	.4920	.4922	.4925	.4927	.4929	.4931	.4932	.4934	.4936
2.5	.4938	.4940	.4941	.4943	.4945	.4946	.4948	.4949	.4951	.4952
2.6	.4953	.4955	.4956	.4957	.4959	.4960	.4961	.4962	.4963	.4964
2.7	.4965	.4966	.4967	.4968	.4969	.4970	.4971	.4972	.4973	.4974
2.8	.4974	.4975	.4976	.4977	.4977	.4978	.4979	.4979	.4980	.4981
2.9	.4981	.4982	.4982	.4983	.4984	.4984	.4985	.4985	.4986	.4986
3.0	.4987	.4987	.4987	.4988	.4988	.4989	.4989	.4989	.4990	.4990

From ELEMENTS OF BUSINESS STATISTICS by R. C. Gulezian. Copyright © 1979 by W. B. Saunders Company. Reprinted by permission of Holt, Rinehart and Winston.

Table C: Student-t Distribution

Table gives t_0 such that $P(t \geq t_0) = \alpha$

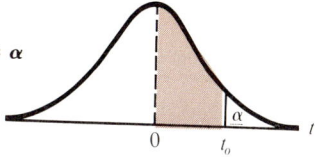

df	α = .10	.05	.025	.01	.005
1	3.078	6.314	12.706	31.821	63.657
2	1.886	2.920	4.303	6.965	9.925
3	1.638	2.353	3.182	4.541	5.841
4	1.533	2.132	2.776	3.747	4.604
5	1.476	2.015	2.571	3.365	4.032
6	1.440	1.943	2.447	3.143	3.707
7	1.415	1.895	2.365	2.998	3.499
8	1.397	1.860	2.306	2.896	3.355
9	1.383	1.833	2.262	2.821	3.250
10	1.372	1.812	2.228	2.764	3.169
11	1.363	1.796	2.201	2.718	3.106
12	1.356	1.782	2.179	2.681	3.055
13	1.350	1.771	2.160	2.650	3.012
14	1.345	1.761	2.145	2.624	2.977
15	1.341	1.753	2.131	2.602	2.947
16	1.337	1.746	2.120	2.583	2.921
17	1.333	1.740	2.110	2.567	2.898
18	1.330	1.734	2.101	2.552	2.878
19	1.328	1.729	2.093	2.539	2.861
20	1.325	1.725	2.086	2.528	2.845
21	1.323	1.721	2.080	2.518	2.831
22	1.321	1.717	2.074	2.508	2.819
23	1.319	1.714	2.069	2.500	2.807
24	1.318	1.711	2.064	2.492	2.797
25	1.316	1.708	2.060	2.485	2.787
26	1.315	1.706	2.056	2.479	2.779
27	1.314	1.703	2.052	2.473	2.771
28	1.313	1.701	2.048	2.467	2.763
29	1.311	1.699	2.045	2.462	2.756
∞	1.282	1.645	1.960	2.326	2.576

From "Table of Percentage Points of the *t*-Distribution." Computed by Maxine Merrington, *Biometrika*, Vol. 32 (1941), p. 300. Reproduced by permission of Professor E. J. Snell.

Table D: Chi-Square Distribution

Table gives k such that $P(\chi^2 \geq k) = \alpha$

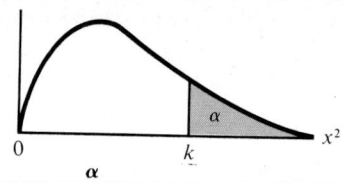

df	0.995	0.990	0.975	0.950	0.900
1	0.0000393	0.0001571	0.0009821	0.0039321	0.0157908
2	0.0100251	0.0201007	0.0506356	0.102587	0.210720
3	0.0717212	0.114832	0.215795	0.351846	0.584375
4	0.206990	0.297110	0.484419	0.710721	1.063623
5	0.411740	0.554300	0.831211	1.145476	1.61031
6	0.675727	0.872085	1.237347	1.63539	2.20413
7	0.989265	1.239043	1.68987	2.16735	2.83311
8	1.344419	1.646482	2.17973	2.73264	3.48954
9	1.734926	2.087912	2.70039	3.32511	4.16816
10	2.15585	2.55821	3.24697	3.94030	4.86518
11	2.60321	3.05347	3.81575	4.57481	5.57779
12	3.07382	3.57056	4.40379	5.22603	6.30380
13	3.56503	4.10691	5.00874	5.89186	7.04150
14	4.07468	4.66043	5.62872	6.57063	7.78953
15	4.60094	5.22935	6.26214	7.26094	8.54675
16	5.14224	5.81221	6.90766	7.96164	9.31223
17	5.69724	6.40776	7.56418	8.67176	10.0852
18	6.26481	7.01491	8.23075	9.39046	10.8649
19	6.84398	7.63273	8.90655	10.1170	11.6509
20	7.43386	8.26040	9.59083	10.8508	12.4426
21	8.03366	8.89720	10.28293	11.5913	13.2396
22	8.64272	9.54249	10.9823	12.3380	14.0415
23	9.26042	10.19567	11.6885	13.0905	14.8479
24	9.88623	10.8564	12.4011	13.8484	15.6587
25	10.5197	11.5240	13.1197	14.6114	16.4734
26	11.1603	12.1981	13.8439	15.3791	17.2919
27	11.8076	12.8786	14.5733	16.1513	18.1138
28	12.4613	13.5648	15.3079	16.9279	18.9392
29	13.1211	14.2565	16.0471	17.7083	19.7677
30	13.7867	14.9535	16.7908	18.4926	20.5992
40	20.7065	22.1643	24.4331	26.5093	29.0505
50	27.9907	29.7067	32.3574	34.7642	37.6886
60	35.5346	37.4848	40.4817	43.1879	46.4589
70	43.2752	45.4418	48.7576	51.7393	55.3290
80	51.1720	53.5400	57.1532	60.3915	64.2778
90	59.1963	61.7541	65.6466	69.1260	73.2912
100	67.3276	70.0648	74.2219	77.9295	82.3581

Table D: Chi-Square Distribution (continued)

		α			
0.100	0.050	0.025	0.010	0.005	df
2.70554	3.84146	5.02389	6.63490	7.87944	1
4.60517	5.99147	7.37776	9.21034	10.5966	2
6.25139	7.81473	9.34840	11.3449	12.8381	3
7.77944	9.48773	11.1433	13.2767	14.8602	4
9.23635	11.0705	12.8325	15.0863	16.7496	5
10.6446	12.5916	14.4494	16.8119	18.5476	6
12.0170	14.0671	16.0128	18.4753	20.2777	7
13.3616	15.5073	17.5346	20.0902	21.9550	8
14.6837	16.9190	19.0228	21.6660	23.5893	9
15.9871	18.3070	20.4831	23.2093	25.1882	10
17.2750	19.6751	21.9200	24.7250	26.7569	11
18.5494	21.0261	23.3367	26.2170	28.2995	12
19.8119	22.3621	24.7356	27.6883	29.8194	13
21.0642	23.6848	26.1190	29.1413	31.3193	14
22.3072	24.9958	27.4884	30.5779	32.8013	15
23.5418	26.2962	28.8454	31.9999	34.2672	16
24.7690	27.5871	30.1910	33.4087	35.7185	17
25.9894	28.8693	31.5264	34.8053	37.1564	18
27.2036	30.1435	32.8523	36.1908	38.5822	19
28.4120	31.4104	34.1696	37.5662	39.9968	20
29.6151	32.6705	35.4789	38.9321	41.4010	21
30.8133	33.9244	36.7807	40.2894	42.7956	22
32.0069	35.1725	38.0757	41.6384	44.1813	23
33.1963	36.4151	39.3641	42.9798	45.5585	24
34.3816	37.6525	40.6465	44.3141	46.9278	25
35.5631	38.8852	41.9232	45.6417	48.2899	26
36.7412	40.1133	43.1944	46.9630	49.6449	27
37.9159	41.3372	44.4607	48.2782	50.9933	28
39.0875	42.5569	45.7222	49.5879	52.3356	29
40.2560	43.7729	46.9792	50.8922	53.6720	30
51.8050	55.7585	59.3417	63.6907	66.7659	40
63.1671	67.5048	71.4202	76.1539	79.4900	50
74.3970	79.0819	83.2976	88.3794	91.9517	60
85.5271	90.5312	95.0231	100.425	104.215	70
96.5782	101.879	106.629	112.329	116.321	80
107.565	113.145	118.136	124.116	128.299	90
118.498	124.342	129.561	135.807	140.169	100

"Tables of the Percentage Points of the χ^2-Distribution." *Biometrika*, Vol. 32 (1941), pp. 188–189, by Catherine M. Thompson. Reproduced by permission of Professor E. J. Snell.

Table E-1: F-Distribution ($\alpha = .05$)

Table gives k such that $P(F \geq k) = .05$.

Degrees of Freedom

df_2 \ df_1	1	2	3	4	5	6	7	8	9
1	161.4	199.5	215.7	224.6	230.2	234.0	236.8	238.9	240.5
2	18.51	19.00	19.16	19.25	19.30	19.33	19.35	19.37	19.38
3	10.13	9.55	9.28	9.12	9.01	8.94	8.89	8.85	8.81
4	7.71	6.94	6.59	6.39	6.26	6.16	6.09	6.04	6.00
5	6.61	5.79	5.41	5.19	5.05	4.95	4.88	4.82	4.77
6	5.99	5.14	4.76	4.53	4.39	4.28	4.21	4.15	4.10
7	5.59	4.74	4.35	4.12	3.97	3.87	3.79	3.73	3.68
8	5.32	4.46	4.07	3.84	3.69	3.58	3.50	3.44	3.39
9	5.12	4.26	3.86	3.63	3.48	3.37	3.29	3.23	3.18
10	4.96	4.10	3.71	3.48	3.33	3.22	3.14	3.07	3.02
11	4.84	3.98	3.59	3.36	3.20	3.09	3.01	2.95	2.90
12	4.75	3.89	3.49	3.26	3.11	3.00	2.91	2.85	2.80
13	4.67	3.81	3.41	3.18	3.03	2.92	2.83	2.77	2.71
14	4.60	3.74	3.34	3.11	2.96	2.85	2.76	2.70	2.65
15	4.54	3.68	3.29	3.06	2.90	2.79	2.71	2.64	2.59
16	4.49	3.63	3.24	3.01	2.85	2.74	2.66	2.59	2.54
17	4.45	3.59	3.20	2.96	2.81	2.70	2.61	2.55	2.49
18	4.41	3.55	3.16	2.93	2.77	2.66	2.58	2.51	2.46
19	4.38	3.52	3.13	2.90	2.74	2.63	2.54	2.48	2.42
20	4.35	3.49	3.10	2.87	2.71	2.60	2.51	2.45	2.39
21	4.32	3.47	3.07	2.84	2.68	2.57	2.49	2.42	2.37
22	4.30	3.44	3.05	2.82	2.66	2.55	2.46	2.40	2.34
23	4.28	3.42	3.03	2.80	2.64	2.53	2.44	2.37	2.32
24	4.26	3.40	3.01	2.78	2.62	2.51	2.42	2.36	2.30
25	4.24	3.39	2.99	2.76	2.60	2.49	2.40	2.34	2.28
26	4.23	3.37	2.98	2.74	2.59	2.47	2.39	2.32	2.27
27	4.21	3.35	2.96	2.73	2.57	2.46	2.37	2.31	2.25
28	4.20	3.34	2.95	2.71	2.56	2.45	2.36	2.29	2.24
29	4.18	3.33	2.93	2.70	2.55	2.43	2.35	2.28	2.22
30	4.17	3.32	2.92	2.69	2.53	2.42	2.33	2.27	2.21
40	4.08	3.23	2.84	2.61	2.45	2.34	2.25	2.18	2.12
60	4.00	3.15	2.76	2.53	2.37	2.25	2.17	2.10	2.04
120	3.92	3.07	2.68	2.45	2.29	2.17	2.09	2.02	1.96
∞	3.84	3.00	2.60	2.37	2.21	2.10	2.01	1.94	1.88

From ELEMENTS OF BUSINESS STATISTICS by R. C. Gulezian. Copyright © 1979 by W. B. Saunders Company. Reprinted by permission of Holt, Rinehart and Winston.

Table E-1 F-Distribution (continued) ($\alpha = .05$)

10	12	15	20	24	30	40	60	120	∞	df_1 / df_2
241.9	243.9	245.9	248.0	249.1	250.1	251.1	252.2	253.3	254.3	1
19.40	19.41	19.43	19.45	19.45	19.46	19.47	19.48	19.49	19.50	2
8.79	8.74	8.70	8.66	8.64	8.62	8.59	8.57	8.55	8.53	3
5.96	5.91	5.86	5.80	5.77	5.75	5.72	5.69	5.66	5.63	4
4.74	4.68	4.62	4.56	4.53	4.50	4.46	4.43	4.40	4.36	5
4.06	4.00	3.94	3.87	3.84	3.81	3.77	3.74	3.70	3.67	6
3.64	3.57	3.51	3.44	3.41	3.38	3.34	3.30	3.27	3.23	7
3.35	3.28	3.22	3.15	3.12	3.08	3.04	3.01	2.97	2.93	8
3.14	3.07	3.01	2.94	2.90	2.86	2.83	2.79	2.75	2.71	9
2.98	2.91	2.85	2.77	2.74	2.70	2.66	2.62	2.58	2.54	10
2.85	2.79	2.72	2.65	2.61	2.57	2.53	2.49	2.45	2.40	11
2.75	2.69	2.62	2.54	2.51	2.47	2.43	2.38	2.34	2.30	12
2.67	2.60	2.53	2.46	2.42	2.38	2.34	2.30	2.25	2.21	13
2.60	2.53	2.46	2.39	2.35	2.31	2.27	2.22	2.18	2.13	14
2.54	2.48	2.40	2.33	2.29	2.25	2.20	2.16	2.11	2.07	15
2.49	2.42	2.35	2.28	2.24	2.19	2.15	2.11	2.06	2.01	16
2.45	2.38	2.31	2.23	2.19	2.15	2.10	2.06	2.01	1.96	17
2.41	2.34	2.27	2.19	2.15	2.11	2.06	2.02	1.97	1.92	18
2.38	2.31	2.23	2.16	2.11	2.07	2.03	1.98	1.93	1.88	19
2.35	2.28	2.20	2.12	2.08	2.04	1.99	1.95	1.90	1.84	20
2.32	2.25	2.18	2.10	2.05	2.01	1.96	1.92	1.87	1.81	21
2.30	2.23	2.15	2.07	2.03	1.98	1.94	1.89	1.84	1.78	22
2.27	2.20	2.13	2.05	2.01	1.96	1.91	1.86	1.81	1.76	23
2.25	2.18	2.11	2.03	1.98	1.94	1.89	1.84	1.79	1.73	24
2.24	2.16	2.09	2.01	1.96	1.92	1.87	1.82	1.77	1.71	25
2.22	2.15	2.07	1.99	1.95	1.90	1.85	1.80	1.75	1.69	26
2.20	2.13	2.06	1.97	1.93	1.88	1.84	1.79	1.73	1.67	27
2.19	2.12	2.04	1.96	1.91	1.87	1.82	1.77	1.71	1.65	28
2.18	2.10	2.03	1.94	1.90	1.85	1.81	1.75	1.70	1.64	29
2.16	2.09	2.01	1.93	1.89	1.84	1.79	1.74	1.68	1.62	30
2.08	2.00	1.92	1.84	1.79	1.74	1.69	1.64	1.58	1.51	40
1.99	1.92	1.84	1.75	1.70	1.65	1.59	1.53	1.47	1.39	60
1.91	1.83	1.75	1.66	1.61	1.55	1.50	1.43	1.35	1.25	120
1.83	1.75	1.67	1.57	1.52	1.46	1.39	1.32	1.22	1.00	∞

Table E-2: F-Distribution ($\alpha = .01$)

Table gives k such that $P(F \geq k) = .01$

Degrees of Freedom

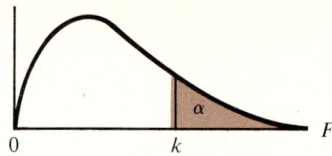

df_2 \ df_1	1	2	3	4	5	6	7	8	9
1	4052	4999.5	5403	5625	5764	5859	5928	5982	6022
2	98.50	99.00	99.17	99.25	99.30	99.33	99.36	99.37	99.39
3	34.12	30.82	29.46	28.71	28.24	27.91	27.67	27.49	27.35
4	21.20	18.00	16.69	15.98	15.52	15.21	14.98	14.80	14.66
5	16.26	13.27	12.06	11.39	10.97	10.67	10.46	10.29	10.16
6	13.75	10.92	9.78	9.15	8.75	8.47	8.26	8.10	7.98
7	12.25	9.55	8.45	7.85	7.46	7.19	6.99	6.84	6.72
8	11.26	8.65	7.59	7.01	6.63	6.37	6.18	6.03	5.91
9	10.56	8.02	6.99	6.42	6.06	5.80	5.61	5.47	5.35
10	10.04	7.56	6.55	5.99	5.64	5.39	5.20	5.06	4.94
11	9.65	7.21	6.22	5.67	5.32	5.07	4.89	4.74	4.63
12	9.33	6.93	5.95	5.41	5.06	4.82	4.64	4.50	4.39
13	9.07	6.70	5.74	5.21	4.86	4.62	4.44	4.30	4.19
14	8.86	6.51	5.56	5.04	4.69	4.46	4.28	4.14	4.03
15	8.68	6.36	5.42	4.89	4.56	4.32	4.14	4.00	3.89
16	8.53	6.23	5.29	4.77	4.44	4.20	4.03	3.89	3.78
17	8.40	6.11	5.18	4.67	4.34	4.10	3.93	3.79	3.68
18	8.29	6.01	5.09	4.58	4.25	4.01	3.84	3.71	3.60
19	8.18	5.93	5.01	4.50	4.17	3.94	3.77	3.63	3.52
20	8.10	5.85	4.94	4.43	4.10	3.87	3.70	3.56	3.46
21	8.02	5.78	4.87	4.37	4.04	3.81	3.64	3.51	3.40
22	7.95	5.72	4.82	4.31	3.99	3.76	3.59	3.45	3.35
23	7.88	5.66	4.76	4.26	3.94	3.71	3.54	3.41	3.30
24	7.82	5.61	4.72	4.22	3.90	3.67	3.50	3.36	3.26
25	7.77	5.57	4.68	4.18	3.85	3.63	3.46	3.32	3.22
26	7.72	5.53	4.64	4.14	3.82	3.59	3.42	3.29	3.18
27	7.68	5.49	4.60	4.11	3.78	3.56	3.39	3.26	3.15
28	7.64	5.45	4.57	4.07	3.75	3.53	3.36	3.23	3.12
29	7.60	5.42	4.54	4.04	3.73	3.50	3.33	3.20	3.09
30	7.56	5.39	4.51	4.02	3.70	3.47	3.30	3.17	3.07
40	7.31	5.18	4.31	3.83	3.51	3.29	3.12	2.99	2.89
60	7.08	4.98	4.13	3.65	3.34	3.12	2.95	2.82	2.72
120	6.85	4.79	3.95	3.48	3.17	2.96	2.79	2.66	2.56
∞	6.63	4.61	3.78	3.32	3.02	2.80	2.64	2.51	2.41

From ELEMENTS OF BUSINESS STATISTICS by R. C. Gulezian. Copyright © 1979 by W. B. Saunders Company. Reprinted by permission of Holt, Rinehart and Winston.

Table E-2: *F*-Distribution (continued) ($\alpha = .01$)

10	12	15	20	24	30	40	60	120	∞	df_1 / df_2
6056	6106	6157	6209	6235	6261	6287	6313	6339	6366	1
99.40	99.42	99.43	99.45	99.46	99.47	99.47	99.48	99.49	99.50	2
27.23	27.05	26.87	26.69	26.60	26.50	26.41	26.32	26.22	26.13	3
14.55	14.37	14.20	14.02	13.93	13.84	13.75	13.65	13.56	13.46	4
10.05	9.89	9.72	9.55	9.47	9.38	9.29	9.20	9.11	9.02	5
7.87	7.72	7.56	7.40	7.31	7.23	7.14	7.06	6.97	6.88	6
6.62	6.47	6.31	6.16	6.07	5.99	5.91	5.82	5.74	5.65	7
5.81	5.67	5.52	5.36	5.28	5.20	5.12	5.03	4.95	4.86	8
5.26	5.11	4.96	4.81	4.73	4.65	4.57	4.48	4.40	4.31	9
4.85	4.71	4.56	4.41	4.33	4.25	4.17	4.08	4.00	3.91	10
4.54	4.40	4.25	4.10	4.02	3.94	3.86	3.78	3.69	3.60	11
4.30	4.16	4.01	3.86	3.78	3.70	3.62	3.54	3.45	3.36	12
4.10	3.96	3.82	3.66	3.59	3.51	3.43	3.34	3.25	3.17	13
3.94	3.80	3.66	3.51	3.43	3.35	3.27	3.18	3.09	3.00	14
3.80	3.67	3.52	3.37	3.29	3.21	3.13	3.05	2.96	2.87	15
3.69	3.55	3.41	3.26	3.18	3.10	3.02	2.93	2.84	2.75	16
3.59	3.46	3.31	3.16	3.08	3.00	2.92	2.83	2.75	2.65	17
3.51	3.37	3.23	3.08	3.00	2.92	2.84	2.75	2.66	2.57	18
3.43	3.30	3.15	3.00	2.92	2.84	2.76	2.67	2.58	2.49	19
3.37	3.23	3.09	2.94	2.86	2.78	2.69	2.61	2.52	2.42	20
3.31	3.17	3.03	2.88	2.80	2.72	2.64	2.55	2.46	2.36	21
3.26	3.12	2.98	2.83	2.75	2.67	2.58	2.50	2.40	2.31	22
3.21	3.07	2.93	2.78	2.70	2.62	2.54	2.45	2.35	2.26	23
3.17	3.03	2.89	2.74	2.66	2.58	2.49	2.40	2.31	2.21	24
3.13	2.99	2.85	2.70	2.62	2.54	2.45	2.36	2.27	2.17	25
3.09	2.96	2.81	2.66	2.58	2.50	2.42	2.33	2.23	2.13	26
3.06	2.93	2.78	2.63	2.55	2.47	2.38	2.29	2.20	2.10	27
3.03	2.90	2.75	2.60	2.52	2.44	2.35	2.26	2.17	2.06	28
3.00	2.87	2.73	2.57	2.49	2.41	2.33	2.23	2.14	2.03	29
2.98	2.84	2.70	2.55	2.47	2.39	2.30	2.21	2.11	2.01	30
2.80	2.66	2.52	2.37	2.29	2.20	2.11	2.02	1.92	1.80	40
2.63	2.50	2.35	2.20	2.12	2.03	1.94	1.84	1.73	1.60	60
2.47	2.34	2.19	2.03	1.95	1.86	1.76	1.66	1.53	1.38	120
2.32	2.18	2.04	1.88	1.79	1.70	1.59	1.47	1.32	1.00	∞

Table F: Poisson Probabilities

Table gives $P(x \leq k)$,

$$P(x \leq k) = \sum_{x=0}^{k} p(x) = \sum_{x=0}^{k} \frac{\lambda^x e^{-\lambda}}{x!}$$

	$\lambda = .1$	.2	.3	.4	.5	.6	.7	.8	.9	1.0
k										
0	.9048	.8187	.7408	.6703	.6065	.5488	.4966	.4493	.4066	.3679
1	.9953	.9825	.9631	.9384	.9098	.8781	.8442	.8088	.7725	.7358
2	.9998	.9989	.9964	.9921	.9856	.9769	.9659	.9526	.9371	.9197
3	1.0000	.9999	.9997	.9992	.9982	.9966	.9942	.9909	.9865	.9810
4	1.0000	1.0000	1.0000	.9999	.9998	.9996	.9992	.9986	.9977	.9963
5	1.0000	1.0000	1.0000	1.0000	1.0000	1.0000	.9999	.9998	.9997	.9994
6	1.0000	1.0000	1.0000	1.0000	1.0000	1.0000	1.0000	1.0000	1.0000	.9999
7	1.0000	1.0000	1.0000	1.0000	1.0000	1.0000	1.0000	1.0000	1.0000	1.0000

	$\lambda = 1.1$	1.2	1.3	1.4	1.5	1.6	1.7	1.8	1.9	2.0
k										
0	.3329	.3012	.2725	.2466	.2231	.2019	.1827	.1653	.1496	.1353
1	.6990	.6626	.6268	.5918	.5578	.5249	.4932	.4628	.4337	.4060
2	.9004	.8795	.8571	.8335	.8088	.7834	.7572	.7306	.7037	.6767
3	.9743	.9662	.9569	.9463	.9344	.9212	.9068	.8913	.8747	.8571
4	.9946	.9923	.9893	.9857	.9814	.9763	.9704	.9636	.9559	.9473
5	.9990	.9985	.9978	.9968	.9955	.9940	.9920	.9896	.9868	.9834
6	.9999	.9997	.9996	.9994	.9991	.9987	.9981	.9974	.9966	.9955
7	1.0000	1.0000	.9999	.9999	.9998	.9997	.9996	.9994	.9992	.9989
8	1.0000	1.0000	1.0000	1.0000	1.0000	1.0000	.9999	.9999	.9998	.9998
9	1.0000	1.0000	1.0000	1.0000	1.0000	1.0000	1.0000	1.0000	1.0000	1.0000

Table F: Poisson Probabilities (continued)

$\lambda =$	2.1	2.2	2.3	2.4	2.5	2.6	2.7	2.8	2.9	3.0
k										
0	.1225	.1108	.1003	.0907	.0821	.0743	.0672	.0608	.0550	.0498
1	.3796	.3546	.3309	.3084	.2873	.2674	.2487	.2311	.2146	.1991
2	.6496	.6227	.5960	.5697	.5438	.5184	.4936	.4695	.4460	.4232
3	.8386	.8194	.7993	.7787	.7576	.7360	.7141	.6919	.6696	.6472
4	.9379	.9275	.9162	.9041	.8912	.8774	.8629	.8477	.8318	.8153
5	.9796	.9751	.9700	.9643	.9580	.9510	.9433	.9349	.9258	.9161
6	.9941	.9925	.9906	.9884	.9858	.9828	.9794	.9756	.9713	.9665
7	.9985	.9980	.9974	.9967	.9958	.9947	.9934	.9919	.9901	.9881
8	.9997	.9995	.9994	.9991	.9989	.9985	.9981	.9976	.9969	.9962
9	.9999	.9999	.9999	.9998	.9997	.9996	.9995	.9993	.9991	.9989
10	1.0000	1.0000	1.0000	1.0000	.9999	.9999	.9999	.9998	.9998	.9997
11	1.0000	1.0000	1.0000	1.0000	1.0000	1.0000	1.0000	1.0000	.9999	.9999
12	1.0000	1.0000	1.0000	1.0000	1.0000	1.0000	1.0000	1.0000	1.0000	1.0000

$\lambda =$	3.1	3.2	3.3	3.4	3.5	3.6	3.7	3.8	3.9	4.0
k										
0	.0450	.0408	.0369	.0334	.0302	.0273	.0247	.0224	.0202	.0183
1	.1847	.1712	.1586	.1468	.1359	.1257	.1162	.1074	.0992	.0916
2	.4012	.3799	.3594	.3397	.3208	.3027	.2854	.2689	.2531	.2381
3	.6248	.6025	.5803	.5584	.5366	.5152	.4942	.4735	.4532	.4335
4	.7982	.7806	.7626	.7442	.7254	.7064	.6872	.6678	.6484	.6288
5	.9057	.8946	.8829	.8705	.8576	.8441	.8301	.8156	.8006	.7851
6	.9612	.9554	.9490	.9421	.9347	.9267	.9182	.9091	.8995	.8893
7	.9858	.9832	.9802	.9769	.9733	.9692	.9648	.9599	.9546	.9489
8	.9953	.9943	.9931	.9917	.9901	.9883	.9863	.9840	.9815	.9786
9	.9986	.9982	.9978	.9973	.9967	.9960	.9952	.9942	.9931	.9919
10	.9996	.9995	.9994	.9992	.9990	.9987	.9984	.9981	.9977	.9972
11	.9999	.9999	.9998	.9998	.9997	.9996	.9995	.9994	.9993	.9991
12	1.0000	1.0000	1.0000	.9999	.9999	.9999	.9999	.9998	.9998	.9997
13	1.0000	1.0000	1.0000	1.0000	1.0000	1.0000	1.0000	1.0000	.9999	.9999
14	1.0000	1.0000	1.0000	1.0000	1.0000	1.0000	1.0000	1.0000	1.0000	1.0000

Table F: Poisson Probabilities (continued)

k	λ = 4.1	4.2	4.3	4.4	4.5	4.6	4.7	4.8	4.9	5.0
0	.0166	.0150	.0136	.0123	.0111	.0101	.0091	.0082	.0074	.0067
1	.0845	.0780	.0719	.0663	.0611	.0563	.0518	.0477	.0439	.0404
2	.2238	.2102	.1974	.1851	.1736	.1626	.1523	.1425	.1333	.1247
3	.4142	.3954	.3772	.3594	.3423	.3257	.3097	.2942	.2793	.2650
4	.6093	.5898	.5704	.5512	.5321	.5132	.4946	.4763	.4582	.4405
5	.7693	.7531	.7367	.7199	.7029	.6858	.6684	.6510	.6335	.6160
6	.8786	.8675	.8558	.8436	.8311	.8180	.8046	.7908	.7767	.7622
7	.9427	.9361	.9290	.9214	.9134	.9049	.8960	.8867	.8769	.8666
8	.9755	.9721	.9683	.9642	.9597	.9549	.9497	.9442	.9382	.9319
9	.9905	.9889	.9871	.9851	.9829	.9805	.9778	.9749	.9717	.9682
10	.9966	.9959	.9952	.9943	.9933	.9922	.9910	.9896	.9880	.9863
11	.9989	.9986	.9983	.9980	.9976	.9971	.9966	.9960	.9953	.9945
12	.9997	.9996	.9995	.9993	.9992	.9990	.9988	.9986	.9983	.9980
13	.9999	.9999	.9998	.9998	.9997	.9997	.9996	.9995	.9994	.9993
14	1.0000	1.0000	1.0000	.9999	.9999	.9999	.9999	.9999	.9998	.9998
15	1.0000	1.0000	1.0000	1.0000	1.0000	1.0000	1.0000	1.0000	.9999	.9999
16	1.0000	1.0000	1.0000	1.0000	1.0000	1.0000	1.0000	1.0000	1.0000	1.0000

Table F: Poisson Probabilities (continued)

k	λ = 5.1	5.2	5.3	5.4	5.5	5.6	5.7	5.8	5.9	6.0
0	.0061	.0055	.0050	.0045	.0041	.0037	.0033	.0030	.0027	.0025
1	.0372	.0342	.0314	.0289	.0266	.0244	.0224	.0206	.0189	.0174
2	.1165	.1088	.1016	.0948	.0884	.0824	.0768	.0715	.0666	.0620
3	.2513	.2381	.2254	.2133	.2017	.1906	.1800	.1700	.1604	.1512
4	.4231	.4061	.3895	.3733	.3575	.3422	.3272	.3127	.2987	.2851
5	.5984	.5809	.5635	.5461	.5289	.5119	.4950	.4783	.4619	.4457
6	.7474	.7324	.7171	.7017	.6860	.6703	.6544	.6384	.6224	.6063
7	.8560	.8449	.8335	.8217	.8095	.7970	.7841	.7710	.7576	.7440
8	.9252	.9181	.9106	.9027	.8944	.8857	.8766	.8672	.8574	.8472
9	.9644	.9603	.9559	.9512	.9462	.9409	.9352	.9292	.9228	.9161
10	.9844	.9823	.9800	.9775	.9747	.9718	.9686	.9651	.9614	.9574
11	.9937	.9927	.9916	.9904	.9890	.9875	.9859	.9841	.9821	.9799
12	.9976	.9972	.9967	.9962	.9955	.9949	.9941	.9932	.9922	.9912
13	.9992	.9990	.9988	.9986	.9983	.9980	.9977	.9973	.9969	.9964
14	.9997	.9997	.9996	.9995	.9994	.9993	.9991	.9990	.9988	.9986
15	.9999	.9999	.9999	.9998	.9998	.9998	.9997	.9996	.9996	.9995
16	1.0000	1.0000	1.0000	.9999	.9999	.9999	.9999	.9999	.9999	.9998
17	1.0000	1.0000	1.0000	1.0000	1.0000	1.0000	1.0000	1.0000	1.0000	.9999
18	1.0000	1.0000	1.0000	1.0000	1.0000	1.0000	1.0000	1.0000	1.0000	1.0000

Table F: Poisson Probabilities (continued)

k	λ = 6.1	6.2	6.3	6.4	6.5	6.6	6.7	6.8	6.9	7.0
0	.0022	.0020	.0018	.0017	.0015	.0014	.0012	.0011	.0010	.0009
1	.0159	.0146	.0134	.0123	.0113	.0103	.0095	.0087	.0080	.0073
2	.0577	.0536	.0498	.0463	.0430	.0400	.0371	.0344	.0320	.0296
3	.1425	.1342	.1264	.1189	.1118	.1052	.0988	.0928	.0871	.0818
4	.2719	.2592	.2469	.2351	.2237	.2127	.2022	.1920	.1823	.1730
5	.4298	.4141	.3988	.3837	.3690	.3547	.3406	.3270	.3137	.3007
6	.5902	.5742	.5582	.5423	.5265	.5108	.4953	.4799	.4647	.4497
7	.7301	.7160	.7017	.6873	.6728	.6581	.6433	.6285	.6136	.5987
8	.8367	.8259	.8148	.8033	.7916	.7796	.7673	.7548	.7420	.7291
9	.9090	.9016	.8939	.8858	.8774	.8686	.8596	.8502	.8405	.8305
10	.9531	.9486	.9437	.9386	.9332	.9274	.9214	.9151	.9084	.9015
11	.9776	.9750	.9723	.9693	.9661	.9627	.9591	.9552	.9510	.9467
12	.9900	.9887	.9873	.9857	.9840	.9821	.9801	.9779	.9755	.9730
13	.9958	.9952	.9945	.9937	.9929	.9920	.9909	.9898	.9885	.9872
14	.9984	.9981	.9978	.9974	.9970	.9966	.9961	.9956	.9950	.9943
15	.9994	.9993	.9992	.9990	.9988	.9986	.9984	.9982	.9979	.9976
16	.9998	.9997	.9997	.9996	.9996	.9995	.9994	.9993	.9992	.9990
17	.9999	.9999	.9999	.9999	.9998	.9998	.9998	.9997	.9997	.9996
18	1.0000	1.0000	1.0000	1.0000	.9999	.9999	.9999	.9999	.9999	.9999
19	1.0000	1.0000	1.0000	1.0000	1.0000	1.0000	1.0000	1.0000	1.0000	1.0000

Table F: Poisson Probabilities (continued)

k	λ = 7.1	7.2	7.3	7.4	7.5	7.6	7.7	7.8	7.9	8.0
0	.0008	.0007	.0007	.0006	.0006	.0005	.0005	.0004	.0004	.0003
1	.0067	.0061	.0056	.0051	.0047	.0043	.0039	.0036	.0033	.0030
2	.0275	.0255	.0236	.0219	.0203	.0188	.0174	.0161	.0149	.0138
3	.0767	.0719	.0674	.0632	.0591	.0554	.0518	.0485	.0453	.0424
4	.1641	.1555	.1473	.1395	.1321	.1249	.1181	.1117	.1055	.0996
5	.2881	.2759	.2640	.2526	.2414	.2307	.2203	.2103	.2006	.1912
6	.4349	.4204	.4060	.3920	.3782	.3646	.3514	.3384	.3257	.3134
7	.5838	.5689	.5541	.5393	.5246	.5100	.4956	.4812	.4670	.4530
8	.7160	.7027	.6892	.6757	.6620	.6482	.6343	.6204	.6065	.5925
9	.8202	.8096	.7988	.7877	.7764	.7649	.7531	.7411	.7290	.7166
10	.8942	.8867	.8788	.8707	.8622	.8535	.8445	.8352	.8257	.8159
11	.9420	.9371	.9319	.9265	.9208	.9148	.9085	.9020	.8952	.8881
12	.9703	.9673	.9642	.9609	.9573	.9536	.9496	.9454	.9409	.9362
13	.9857	.9841	.9824	.9805	.9784	.9762	.9739	.9714	.9687	.9658
14	.9935	.9927	.9918	.9908	.9897	.9886	.9873	.9859	.9844	.9827
15	.9972	.9969	.9964	.9959	.9954	.9948	.9941	.9934	.9926	.9918
16	.9989	.9987	.9985	.9983	.9980	.9978	.9974	.9971	.9967	.9963
17	.9996	.9995	.9994	.9993	.9992	.9991	.9989	.9988	.9986	.9984
18	.9998	.9998	.9998	.9997	.9997	.9996	.9996	.9995	.9994	.9993
19	.9999	.9999	.9999	.9999	.9999	.9999	.9998	.9998	.9998	.9997
20	1.0000	1.0000	1.0000	1.0000	1.0000	.9999	.9999	.9999	.9999	.9999
21	1.0000	1.0000	1.0000	1.0000	1.0000	1.0000	1.0000	1.0000	1.0000	1.0000

Table F: Poisson Probabilities (continued)

k	λ = 8.1	8.2	8.3	8.4	8.5	8.6	8.7	8.8	8.9	9.0
0	.0003	.0003	.0002	.0002	.0002	.0002	.0002	.0002	.0001	.0001
1	.0028	.0025	.0023	.0021	.0019	.0018	.0016	.0015	.0014	.0012
2	.0127	.0118	.0109	.0100	.0093	.0086	.0079	.0073	.0068	.0062
3	.0396	.0370	.0346	.0323	.0301	.0281	.0262	.0244	.0228	.0212
4	.0940	.0887	.0837	.0789	.0744	.0701	.0660	.0621	.0584	.0550
5	.1822	.1736	.1653	.1573	.1496	.1422	.1352	.1284	.1219	.1157
6	.3013	.2896	.2781	.2670	.2562	.2457	.2355	.2256	.2160	.2068
7	.4391	.4254	.4119	.3987	.3856	.3728	.3602	.3478	.3357	.3239
8	.5786	.5647	.5507	.5369	.5231	.5094	.4958	.4823	.4689	.4557
9	.7041	.6915	.6788	.6659	.6530	.6400	.6269	.6137	.6006	.5874
10	.8058	.7955	.7850	.7743	.7634	.7522	.7409	.7294	.7178	.7060
11	.8807	.8731	.8652	.8571	.8487	.8400	.8311	.8220	.8126	.8030
12	.9313	.9261	.9207	.9150	.9091	.9029	.8965	.8898	.8829	.8758
13	.9628	.9595	.9561	.9524	.9486	.9445	.9403	.9358	.9311	.9261
14	.9810	.9791	.9771	.9749	.9726	.9701	.9675	.9647	.9617	.9585
15	.9908	.9898	.9887	.9875	.9862	.9848	.9832	.9816	.9798	.9780
16	.9958	.9953	.9947	.9941	.9934	.9926	.9918	.9909	.9899	.9889
17	.9982	.9979	.9977	.9973	.9970	.9966	.9962	.9957	.9952	.9947
18	.9992	.9991	.9990	.9989	.9987	.9985	.9983	.9981	.9978	.9976
19	.9997	.9997	.9996	.9995	.9995	.9994	.9993	.9992	.9991	.9989
20	.9999	.9999	.9998	.9998	.9998	.9998	.9997	.9997	.9996	.9996
21	1.0000	1.0000	.9999	.9999	.9999	.9999	.9999	.9999	.9998	.9998
22	1.0000	1.0000	1.0000	1.0000	1.0000	1.0000	1.0000	1.0000	.9999	.9999
23	1.0000	1.0000	1.0000	1.0000	1.0000	1.0000	1.0000	1.0000	1.0000	1.0000

Table F: Poisson Probabilities (continued)

k	λ = 9.1	9.2	9.3	9.4	9.5	9.6	9.7	9.8	9.9	10.0
0	.0001	.0001	.0001	.0001	.0001	.0001	.0001	.0001	.0001	.0000
1	.0011	.0010	.0009	.0009	.0008	.0007	.0007	.0006	.0005	.0005
2	.0058	.0053	.0049	.0045	.0042	.0038	.0035	.0033	.0030	.0028
3	.0198	.0184	.0172	.0160	.0149	.0138	.0129	.0120	.0111	.0103
4	.0517	.0486	.0456	.0429	.0403	.0378	.0355	.0333	.0312	.0293
5	.1098	.1041	.0986	.0935	.0885	.0838	.0793	.0750	.0710	.0671
6	.1978	.1892	.1808	.1727	.1649	.1574	.1502	.1433	.1366	.1301
7	.3123	.3010	.2900	.2792	.2687	.2584	.2485	.2388	.2294	.2202
8	.4426	.4296	.4168	.4042	.3918	.3796	.3676	.3558	.3442	.3328
9	.5742	.5611	.5479	.5349	.5218	.5089	.4960	.4832	.4705	.4579
10	.6941	.6820	.6699	.6576	.6453	.6329	.6205	.6080	.5955	.5830
11	.7932	.7832	.7730	.7626	.7520	.7412	.7303	.7193	.7081	.6968
12	.8684	.8607	.8529	.8448	.8364	.8279	.8191	.8101	.8009	.7916
13	.9210	.9156	.9100	.9042	.8981	.8919	.8853	.8786	.8716	.8645
14	.9552	.9517	.9480	.9441	.9400	.9357	.9312	.9265	.9216	.9165
15	.9760	.9738	.9715	.9691	.9665	.9638	.9609	.9579	.9546	.9513
16	.9878	.9865	.9852	.9838	.9823	.9806	.9789	.9770	.9751	.9730
17	.9941	.9934	.9927	.9919	.9911	.9902	.9892	.9881	.9870	.9857
18	.9973	.9969	.9966	.9962	.9957	.9952	.9947	.9941	.9935	.9928
19	.9988	.9986	.9985	.9983	.9980	.9978	.9975	.9972	.9969	.9965
20	.9995	.9994	.9993	.9992	.9991	.9990	.9989	.9987	.9986	.9984
21	.9998	.9998	.9997	.9997	.9996	.9996	.9995	.9995	.9994	.9993
22	.9999	.9999	.9999	.9999	.9999	.9998	.9998	.9998	.9997	.9997
23	1.0000	1.0000	1.0000	1.0000	.9999	.9999	.9999	.9999	.9999	.9999
24	1.0000	1.0000	1.0000	1.0000	1.0000	1.0000	1.0000	1.0000	1.0000	1.0000

Table F: Poisson Probabilities (continued)

k	λ = 10.1	10.2	10.3	10.4	10.5	10.6	10.7	10.8	10.9	11.0
0	.0000	.0000	.0000	.0000	.0000	.0000	.0000	.0000	.0000	.0000
1	.0005	.0004	.0004	.0003	.0003	.0003	.0003	.0002	.0002	.0002
2	.0026	.0023	.0022	.0020	.0018	.0017	.0016	.0014	.0013	.0012
3	.0096	.0089	.0083	.0077	.0071	.0066	.0062	.0057	.0053	.0049
4	.0274	.0257	.0241	.0225	.0211	.0197	.0185	.0173	.0162	.0151
5	.0634	.0599	.0566	.0534	.0504	.0475	.0448	.0423	.0398	.0375
6	.1240	.1180	.1123	.1069	.1016	.0966	.0918	.0872	.0828	.0786
7	.2113	.2027	.1944	.1863	.1785	.1710	.1636	.1566	.1498	.1432
8	.3217	.3108	.3001	.2896	.2794	.2694	.2597	.2502	.2410	.2320
9	.4455	.4332	.4210	.4090	.3971	.3854	.3739	.3626	.3515	.3405
10	.5705	.5580	.5456	.5331	.5207	.5084	.4961	.4840	.4719	.4599
11	.6853	.6738	.6622	.6505	.6387	.6269	.6150	.6031	.5912	.5793
12	.7820	.7722	.7623	.7522	.7420	.7316	.7210	.7104	.6996	.6887
13	.8571	.8494	.8416	.8336	.8253	.8169	.8083	.7995	.7905	.7813
14	.9112	.9057	.9000	.8940	.8879	.8815	.8750	.8682	.8612	.8540
15	.9477	.9440	.9400	.9359	.9317	.9272	.9225	.9177	.9126	.9074
16	.9707	.9684	.9658	.9632	.9604	.9574	.9543	.9511	.9477	.9441
17	.9844	.9830	.9815	.9799	.9781	.9763	.9744	.9723	.9701	.9678
18	.9921	.9913	.9904	.9895	.9885	.9874	.9863	.9850	.9837	.9823
19	.9962	.9957	.9953	.9948	.9942	.9936	.9930	.9923	.9915	.9907
20	.9982	.9980	.9978	.9975	.9972	.9969	.9966	.9962	.9958	.9953
21	.9992	.9991	.9990	.9989	.9987	.9986	.9984	.9982	.9980	.9977
22	.9997	.9996	.9996	.9995	.9994	.9994	.9993	.9992	.9991	.9990
23	.9999	.9998	.9998	.9998	.9998	.9997	.9997	.9996	.9996	.9995
24	.9999	.9999	.9999	.9999	.9999	.9999	.9999	.9998	.9998	.9998
25	1.0000	1.0000	1.0000	1.0000	1.0000	1.0000	.9999	.9999	.9999	.9999
26	1.0000	1.0000	1.0000	1.0000	1.0000	1.0000	1.0000	1.0000	1.0000	1.0000

Table F: Poisson Probabilities (continued)

k	λ = 11.1	11.2	11.3	11.4	11.5	11.6	11.7	11.8	11.9	12.0
0	.0000	.0000	.0000	.0000	.0000	.0000	.0000	.0000	.0000	.0000
1	.0002	.0002	.0002	.0001	.0001	.0001	.0001	.0001	.0001	.0001
2	.0011	.0010	.0009	.0009	.0008	.0007	.0007	.0006	.0006	.0005
3	.0046	.0042	.0039	.0036	.0034	.0031	.0029	.0027	.0025	.0023
4	.0141	.0132	.0123	.0115	.0107	.0100	.0094	.0087	.0081	.0076
5	.0353	.0333	.0313	.0295	.0277	.0261	.0245	.0230	.0217	.0203
6	.0746	.0708	.0671	.0636	.0603	.0571	.0541	.0512	.0484	.0458
7	.1369	.1307	.1249	.1192	.1137	.1085	.1035	.0986	.0940	.0895
8	.2232	.2147	.2064	.1984	.1906	.1830	.1757	.1686	.1617	.1550
9	.3298	.3192	.3089	.2987	.2888	.2791	.2696	.2603	.2512	.2424
10	.4480	.4362	.4246	.4131	.4017	.3905	.3794	.3685	.3578	.3472
11	.5673	.5554	.5435	.5316	.5198	.5080	.4963	.4847	.4731	.4616
12	.6777	.6666	.6555	.6442	.6329	.6216	.6102	.5988	.5874	.5760
13	.7719	.7624	.7528	.7430	.7330	.7230	.7128	.7025	.6920	.6815
14	.8467	.8391	.8313	.8234	.8153	.8069	.7985	.7898	.7810	.7720
15	.9020	.8963	.8905	.8845	.8783	.8719	.8653	.8585	.8516	.8444
16	.9403	.9364	.9323	.9280	.9236	.9190	.9142	.9092	.9040	.8987
17	.9654	.9628	.9601	.9572	.9542	.9511	.9478	.9444	.9408	.9370
18	.9808	.9792	.9775	.9757	.9738	.9718	.9697	.9674	.9651	.9626
19	.9898	.9889	.9879	.9868	.9857	.9845	.9832	.9818	.9803	.9787
20	.9948	.9943	.9938	.9932	.9925	.9918	.9910	.9902	.9893	.9884
21	.9975	.9972	.9969	.9966	.9962	.9958	.9954	.9950	.9945	.9939
22	.9988	.9987	.9985	.9984	.9982	.9980	.9978	.9975	.9972	.9970
23	.9995	.9994	.9993	.9992	.9992	.9991	.9989	.9988	.9987	.9985
24	.9998	.9997	.9997	.9997	.9996	.9996	.9995	.9995	.9994	.9993
25	.9999	.9999	.9999	.9999	.9998	.9998	.9998	.9998	.9997	.9997
26	1.0000	1.0000	.9999	.9999	.9999	.9999	.9999	.9999	.9999	.9999
27	1.0000	1.0000	1.0000	1.0000	1.0000	1.0000	1.0000	1.0000	1.0000	.9999
28	1.0000	1.0000	1.0000	1.0000	1.0000	1.0000	1.0000	1.0000	1.0000	1.0000

Table F: Poisson Probabilities (continued)

$\lambda =$	12.1	12.2	12.3	12.4	12.5	12.6	12.7	12.8	12.9	13.0
k										
0	.0000	.0000	.0000	.0000	.0000	.0000	.0000	.0000	.0000	.0000
1	.0001	.0001	.0001	.0001	.0001	.0000	.0000	.0000	.0000	.0000
2	.0005	.0004	.0004	.0004	.0003	.0003	.0003	.0003	.0002	.0002
3	.0021	.0020	.0018	.0017	.0016	.0014	.0013	.0012	.0011	.0011
4	.0071	.0066	.0062	.0057	.0053	.0050	.0046	.0043	.0040	.0037
5	.0191	.0179	.0168	.0158	.0148	.0139	.0130	.0122	.0115	.0107
6	.0433	.0410	.0387	.0366	.0346	.0326	.0308	.0291	.0274	.0259
7	.0852	.0811	.0772	.0734	.0698	.0664	.0631	.0599	.0569	.0540
8	.1486	.1424	.1363	.1305	.1249	.1195	.1143	.1093	.1044	.0998
9	.2338	.2254	.2172	.2092	.2014	.1939	.1866	.1794	.1725	.1658
10	.3368	.3266	.3166	.3067	.2971	.2876	.2783	.2693	.2604	.2517
11	.4502	.4389	.4278	.4167	.4058	.3950	.3843	.3738	.3634	.3532
12	.5645	.5531	.5417	.5303	.5190	.5077	.4964	.4853	.4741	.4631
13	.6709	.6603	.6495	.6387	.6278	.6169	.6060	.5950	.5840	.5730
14	.7629	.7536	.7442	.7347	.7250	.7153	.7054	.6954	.6853	.6751
15	.8371	.8296	.8219	.8140	.8060	.7978	.7895	.7810	.7724	.7636
16	.8932	.8875	.8816	.8755	.8693	.8629	.8563	.8495	.8426	.8355
17	.9331	.9290	.9248	.9204	.9158	.9111	.9062	.9011	.8959	.8905
18	.9600	.9572	.9543	.9513	.9481	.9448	.9414	.9378	.9341	.9302
19	.9771	.9753	.9734	.9715	.9694	.9672	.9649	.9625	.9600	.9573
20	.9874	.9863	.9852	.9840	.9827	.9813	.9799	.9783	.9767	.9750
21	.9934	.9927	.9921	.9914	.9906	.9898	.9889	.9880	.9870	.9859
22	.9966	.9963	.9959	.9955	.9951	.9946	.9941	.9936	.9930	.9924
23	.9984	.9982	.9980	.9978	.9975	.9973	.9970	.9967	.9964	.9960
24	.9992	.9991	.9990	.9989	.9988	.9987	.9985	.9984	.9982	.9980
25	.9997	.9996	.9996	.9995	.9994	.9994	.9993	.9992	.9991	.9990
26	.9998	.9998	.9998	.9998	.9997	.9997	.9997	.9996	.9996	.9995
27	.9999	.9999	.9999	.9999	.9999	.9999	.9999	.9998	.9998	.9998
28	1.0000	1.0000	1.0000	1.0000	1.0000	.9999	.9999	.9999	.9999	.9999
29	1.0000	1.0000	1.0000	1.0000	1.0000	1.0000	1.0000	1.0000	1.0000	1.0000

Answers to Even-Numbered Problems

Chapter 2

2. $\bar{x} = 3$; $md = 3$; $Rg = 17$; $s^2 = 40.67$; $s = 6.38$.

4. a. (i) $2305 (ii) $535 (iii) $885;
   b. (i) $192.08 (ii) $44.58 (iii) $73.75;
   c. (i) $133.52 (ii) $43.63 (iii) $85.16.

6. 81.1 percent.

10. Adjusted mode = 114.3
    $Q_1 = 113.6$
    $Q_2 = 114.8$
    $Q_3 = 118.2$.

12. a. Using seven intervals of six units each, lower endpoint of lowest interval is 18.5; $\bar{x} \cong 39.95$, $s \cong 12.66$, $md \cong 39.93$;
    b. Ungrouped data: $\bar{x} = 39.15$; $md = 39.00$; $s = 12.68$.

14. a. $\bar{x} = 2.61$; b. $s = 1.54$; c. $md = 2$.

18. a. The mean equals the mode for any symmetric distribution; it is also possible to construct nonsymmetric distributions with the mean and mode equal;
    b. In any distribution with a positive skew, the mean is larger than the median.

20. $\bar{x} = 37.5$.

22. 833, estimated duck population.

24. $\bar{x} = 14$.

26. a. $\Sigma x = 42$; b. $s = 2.11$.

28. a. $\Sigma x = 102$; b. $s^2 = 9.20$, $s = 3.03$.

30. a. $\bar{x} = 1,852.5$ hours; $s = 139.9$ hours;
    b. $Md = 1,830.0$ hours; $Rg = 750$ hours; $D_9 = 2.057.5$ hours.

32. a. 1; b. −3; c. 5; d. 0; e. 40.

34. a. 25; b. 155; c. 18; d. −12.

36. 8.9 percent.

38. $HM = 45$ miles per hour.

Chapter 3

2. a. Not independent; b. Not mutually exclusive.

4. a. .5; b. .4; c. .67; d. .7.

6. .76.

8. a. .0015; b. .0785.

10. a. .49; b. .42.

12. .955.

14. a. .343; b. .189; c. .784.

16. .92.

18. .6561.

20. .97.

22. .75.

24. .99998.

26. .015.

28. a. .452; b. .788; c. .024.

30. .083.

34. a. .2 and .68; b. .25.

36. .0285, or 2.85 percent.

38. .0116.

40. .632.

42. .429.

Chapter 4

2. a. 14; b. 6 and .192; c. .001; d. .072; e. .952.

4. a. .115; b. .238; c. .048.

6. a. .227; b. .072; c. .967; d. .830.

8. 648.75.

10. a. $p(x)$ = .02 for x = 50,000
    = .12 for x = 35,000
    = .50 for x = 20,000
    = .25 for x = 10,000
    = .08 for x = 0
    = .03 for x = $-10{,}000$
    = 0 otherwise.

   b. 17,400; c. 122,224,000.

12. a. .078; b. .317; c. .010; d. .912.

14. b. 10.65; c. 11.03.

16. a. .206; b. .370; c. .219.

18. a. .358; b. .271.

20. a. −3; b. 144.

22. a. .1157; b. .8843.

24. a. The function never assumes a negative value and the total area under the curve is equal to one.

   b. .33; c. 1.

26. .856.

28. .1559.

30. a. .0498, .2240 and .8153; b. .0839.

32. a. .3012; b. .0077; c. .8795; d. .3614.

34. a. 0; b. 0; c. .017; d. .2; e. 30 and 17.32.

Chapter 5

2. a. 1.55; b. −1.04; c. −1.18; d. 1.41; e. 1.96; f. 1.55

4. a. 170.72; b. 182.56; c. 144.16.

6. a. .1056; b. 8.252.

8. a. .0301; b. .7565.

10. a. .7054; b. .9162; c. .8154; d. 2.26 and 4.30.

12. .8490.

14. Supplier A; 94.29 percent.

16. .9119.

18. a. 99.84 percent; b. 80.38 percent; c. 24.8.

20. 53.46 percent.

22. a. .0423; b. .2877.

24. a. .6915; b. .0579.

Chapter 6

2. a. .9446;
   b. Approximately 0;
   c. .1446;
   d. .3642;
   e. .0068.

4. a. $E(p) = \frac{1}{6}$; $\sigma_p = .011$;
   b. .7602;
   c. 192;

6. a. $118;

b. $2;
c. .8413;
d. .5675 (assume x is normally distributed);
e. $E(\bar{x}) = \$118$; $\sigma_{\bar{x}} = \$1.44$; $P(\bar{x} \leq 120) = .9177$.

8. .4168.

10.

		n		
		25	100	400
π	.1(.9)	.06	.03	.015
	.2(.8)	.08	.04	.020
	.5	.10	.05	.025

12. .9599. (Consumption cannot be normally distributed because 0 is less than 1 σ below the mean.)

Chapter 7

2. $1.2 \pm .064$.
4. a. .818; b. $.818 \pm .043$.
6. $.842 \pm .044$.
8. $7.416 \pm .743$.
10. $.54 \pm .024$.
12. 79.
14. a. 757; b. 484.
16. 107.
18. $.133 \pm .095$.
20. 1,267.
22. 196.
24. a. $4.35 \pm .29$; b. 167.81 to 191.79.
26. a. 16280 ± 307; b. 228,094,000 to 236,862,000.
28. 18.67 ± 3.14.
30. a. 1695; b. 629.
32. 424.

Chapter 8

2. Reject H_0 if $z_c < -2.33$; $z_c = -3.13$; reject H_0.
4. Reject H_0 if $z_c < -1.88$ or $z_c > 1.88$; $z_c = -.74$; do not reject H_0.
6. Reject H_0 if $z_c > 1.55$; $z_c = 2.83$; reject H_0.

8. Reject H_0 if $t_c > 2.602$; $t_c = 5.33$; reject H_0.
10. Reject H_0 if $z_c < -2.58$ or $z_c > 2.58$; $z_c = -6.67$; reject H_0.
12. Reject H_0 if $z_c > 2.05$; $z_c = 4.32$; reject H_0.
14. Reject H_0 if $z_c < -2.58$ or $z_c > 2.58$; $z_c = .95$; do not reject H_0.
16. Reject H_0 if $t_c < -1.717$; $t_c = -3.94$; reject H_0.
18. Reject H_0 if $z_c < -1.96$ or $z_c > 1.96$; $z_c = 2.0$; reject H_0.
20. Reject H_0 if $z_c > 1.75$; $z_c = 1.80$; reject H_0.
22. Reject H_0 if $z_c < -1.88$ or $z_c > 1.88$; $z_c = 2.10$; reject H_0.
24. Reject H_0 if $z_c < -2.05$; $z_c = -2.61$; reject H_0.
26. Reject H_0 if $z_c < -1.65$; $z_c = -6.06$; reject H_0.
28. Reject H_0 if $t_c > 3.707$ or $t_c < -3.707$; $t_c = .43$; do not reject H_0.
30. Reject H_0 if $z_c < -1.88$; $z_c = -2.92$; reject H_0.
32. Reject H_0 if $t_c < -1.533$; $t_c = -2.19$; reject H_0.
34. Reject H_0 if $z_c > 1.41$; z_c is obviously negative (i.e., $\bar{x} < 30$); therefore, do not reject H_0.
36. a. .5199; b. .6293.
38. a. .9003; b. .9656; c. .2877.

Chapter 9

2. Reject H_0 if $z_c > 1.75$; $z_c = 8.42$; reject H_0.
4. Reject H_0 if $t_c < -1.734$ or $t_c > 1.734$; $t_c = -1.13$; do not reject H_0.
6. Reject H_0 if $t_c < -2.779$ or $t_c > 2.779$; $t_c = 3.42$; reject H_0.
8. Reject H_0 if $z_c > 1.65$; $z_c = 2.97$; reject H_0.
10. 41.67 ± 5.59.
12. 11.89 ± 4.30.
14. Reject H_0 if $z_c < -1.96$ or $z_c > 1.96$; $z_c = 1.77$; do not reject H_0.
16. Reject H_0 if $z_c < -1.88$; $z_c = -4.88$; reject H_0.
18. -11.5 ± 2.97.
20. Reject H_0 if $z_c < -1.65$ or $z_c > 1.65$; $z_c = 1.79$; reject H_0.
22. 34.36 ± 6.63.
24. Reject H_0 if $z_c > 2.05$; $z_c = 1.85$; do not reject H_0.
26. Reject H_0 if $z_c > 1.28$; $z_c = .93$; do not reject H_0.

Chapter 10

2. a. $\hat{y} = 20.73 - 1.404x$;
   b. $r = .775$.

4. a. $\hat{y} = 6.90 - 2.00x$;
   b. $r = -.667$; $s_e = 1.70$, $s_b = 0.17$;
   c. 95 percent confidence on B, $-2.99 \pm .33$;
   d. $\hat{\mu} = 4.90$; 95 percent confidence on μ, $4.90 \pm .33$.

6. a. (i) $\bar{x} = \$11,200$; (ii) $r = .80$; (iii) $s_e = \$2,370$; (iv) $s_b = .057$;
   b. (i) For $x < \$7,600$; (ii) 64 percent; (iii) Test $z = 13.18$, reject H_0: $B = 0$; (iv) Test $z = -1.76$; $H_0 B \geq .85$ is borderline since rejection depends on choice of α; (v) $\hat{y} = \$16,800$; (vi) 95 percent confidence on y; $\$16,800 \pm \$5,500$; (vii) $\hat{\mu} = \$13,800$; (viii) 95 percent confidence on μ, $\$13,800 \pm \$500$.

8. a. $\hat{y} = 1.84 + .40x$;
   c. $\hat{y} = 3.84$.

10. a. $\hat{y} = 14 - 3x$;
    b. $\hat{x} = \frac{14}{3} - \frac{1}{3}y$;
    c. $r = -1$;
    d. $s_e = 0$.

12. a. $\hat{y} = -165 + 4.80x$;
    b. $r = .60$;
    c. $s_e = 16$;
    d. $s_b = .37$.
       (i) Test $z = 12.96$, reject H_0: $B = 0$;
       (ii) Test $z = -5.95$, reject H_0: $B = 7.0$.
    e. $\hat{\mu} = 180.6$ pounds;
       95 percent confidence on $\mu = 180.6 \pm 2.3$;
       95 percent confidence on $y = 180.6 \pm 31.4$.

14. a. $b = -120$: with each $1 increase in price, estimated sales decrease by $120,000 and vice versa;
    $a = 2,800$: realistically, it should not be interpreted; literally, if price were zero, sales would be estimated at $2,800,000.
    b. $r = -.857$;
    c. $s_e = 7.65$;
    d. $\$2,380,000 \pm \$13,800$.

16. b. $\hat{y} = \$1,810$;
    c. $\hat{y} = \$2,440$;

d. $b = -900$: for each $1 increase in price, estimated sales decrease by $900 and vice versa; $a = 4,060$: if a selling price of $0 were reasonable or possible, estimated sales would be $4,060.

18. a. $\hat{y} = 8.93 + .0242x$; other results: $\bar{y} = 57.02$; $s_y = 19.76$, $\bar{x} = 1,986$, $s_x = 751.7$, and $F = 78.3$;

    b. $s_e = 7.967$, $r = .9210$, and $s_b = .00274$;
    reject H_0: $B = 0$, test $t = 8.83$;
    conf. $(.0194 \leq B \leq .0290) = .90$;

    c. $\hat{y} = 57.33$ or $57,300;
    conf. ($42,900 \leq \hat{y} \leq $71,800) = .90$;
    $\hat{\mu} = 57.33$ or $57,300;
    conf ($53,800 \leq \hat{\mu} \leq $60,800) = .90$;

    d. $\hat{y} = 13.69 + 11.36x$; other results $\bar{x} = 3.81$, $s_x = 1.33$, $r = .7634$, $s_e = 13.21$, $F = 19.56$, $s_b = 2.57$;

    e. $\hat{y} = 16.82 + .0146x_1 + 5.05x_2 - .390x_3$;
    standard deviations of the slope coefficents are .0044, 2.160 and .192, respectively, for b_1, b_2, and b_3;
    other results: $\bar{x}_3 = 20.75$, $s_3 = 12.05$, $R = .950$, $s_e = 6.893$, and $F = 37.10$.

    Correlation Matrix

	y	x_1	x_2	x_3
y	1.000	.921	.763	-.554
x_1		1.000	.720	-.508
x_2			1.000	-.099
x_3				1.000

20. a. $y = -13 + 5x$;
    b. $y = -3 + 2x$.

22. a. $y = -3 + \frac{2}{3}x$;
    b. Slope $= \frac{3}{5}$; y - intercept $= -4$;
    c. $y = -\frac{11}{5} + \frac{3}{5}x$.

Chapter 11

2. a. $a = 150$: the 1964 GNP trend value is 150 billion buckniks.
   $b = 7$: the average growth in annual GNP per five-year period $= 7$ billion buckniks;
   b. 172.4 billion buckniks;
   c. $\hat{y} = 158.4 + 7x$;
   d. 1,857.

4. a. Yes; sum of the 12 seasonal indexes is 1,200;

b. February, 1980 → $x = 50$;
   May, 1982 → $x = 77$;
   December, 1981 → $x = 72$;

c. $\hat{y} = \$100,000$ for February, 1980;
   $\hat{y} = \$146,855$ for May, 1982;
   $\hat{y} = \$190,800$ for December, 1981;
   First six-month forecast, 1980 = \$750,955.

6. a. Deflated series, 1969 through 1977, respectively: 3.15, 3.53, 3.59, 3.34, 3.35, 3.01, 2.98, 2.88, 3.01;

   b. $\hat{y} = 4.47 + 0.196x$ where x is in years and 1973 is the origin;

   c. $\hat{y} = 3.20 - .067x$ where x is in years and 1973 is the origin.
   Interpretation of coefficient in parts b and c:
   $a = 4.47$ in the first equation is the 1973 trend sales in billions of 1973 dollars, but when the series is deflated, $a = 3.20$, which is the 1973 trend in billions of 1967 dollars;
   $b = 0.196$ indicates an average annual increase of \$196 million (in current dollars), but when the series is deflated, $b = -.067$, indicating an average annual decrease in real dollar sales of \$67 million.

8. a. $\hat{y} = (750.6)(1.067)^x$;

   b. $a = 750.6$: the trend value of sales in 1956 was \$750,600;
   $b = 1.067$: over the data period, increase in annual sales averaged 6.7 percent per year;

   c. By 1958, trend sales exceed \$1 million;

   d. \$4,306,000.

10. a. $\hat{y} = 11.45 + .326x$ where $\hat{y}$ is estimated unemployment in thousands and x is time in quarters with Q_2, 1976, origin. Other results: $\bar{y} = 12.26$, $s_y = 3.94$, $\bar{x} = 2.50$, $s_x = 7.07$, $r = .584$, and $F = 11.37$. The trend value unemployment in the second quarter, 1976, is 11,450, and the average quarterly increase in unemployment is 326 persons over the six years.

    b. $\hat{y} = 11.58 + 1.383x$ where $\hat{y}$ is estimated unemployment in thousands and x is in years with 1976 origin.
    The 1976 trend unemployment of 11,580 for the annual line is, for all practical purposes, the same as that derived from the quarterly data, considering that the annual trend line origin is half a quarter later than the quarterly trend origin. The quarterly increment, 326, multiplied by 4 is 1,304; this is 79 persons fewer than the increment derived from the annual data. The lower slope in the quarterly line is probably a result of high unemployment early in the year and lower unemployment later in the year.

c.

	Moving Average	Quarterly Index Series
1974 Q_3	8.96	.736
Q_4	9.61	.916
1975 Q_1	10.16	1.319
Q_2	10.22	1.115
Q_3	10.31	.727
Q_4	10.36	.811
1976 Q_1	10.30	1.408
Q_2	10.40	1.029
Q_3	10.76	.715
Q_4	11.28	.798
1977 Q_1	11.72	1.433
Q_2	12.30	1.016
Q_3	13.19	.720
Q_4	14.19	.832
1978 Q_1	14.99	1.408
Q_2	15.45	1.048
Q_3	15.60	.782
Q_4	15.48	.827
1979 Q_1	15.50	1.374
Q_2	15.80	.949

Set of Seasonal Indexes (Adjusted Medians)

Q_1	1.411
Q_2	1.031
Q_3	.729
Q_4	.829
Total	4.000

d. **Quarterly Unemployment Forecast for 1980 and 1981**

	x	Trend $\hat{y}$	Seasonal Index	Forecast
1980 Q_1	15	16.34	1.411	23,100
Q_2	16	16.67	1.031	17,200
Q_3	17	16.99	.729	12,400
Q_4	18	17.32	.829	14,400
1981 Q_1	19	17.64	1.411	24,900
Q_2	20	17.97	1.031	18,500
Q_3	21	18.30	.729	13,300
Q_4	22	18.62	.829	15,400

e. $\hat{y} = 11.39 + .319x + .00126x^2$;
$R = .584$, $s_e = 3.35$, and $F = 5.43$.
(Note that the inclusion of the second-degree term does not improve the fit in this problem.)

f. Multivariable equation y is unemployment in thousands; x is in quarters with Q_2, 1976, origin; d_1, d_2, and d_3 are (1, 0) according as Q_1, Q_3, and Q_4, respectively;
$\hat{y} = 11.508 + .3875x + 4.571 d_1 - 3.154 d_2 - 2.292 d_3$ with $R = .965$, $s_e = 1.142$, and $F = 63.84$.
(All coefficients are significant at the .01 level.)

	x	Forecast, 1000 $\hat{y}$
1980 Q_1	15	21,900
Q_2	16	17,700
Q_3	17	14,900
Q_4	18	16,200
1981 Q_1	19	23,400
Q_2	20	19,300
Q_3	21	16,500
Q_4	22	17,700

12. a.

Year	Quarter	Percentage Unemployment	Moving Average	Quarterly Index Series
1974	1	5.00		
	2	3.24		
	3	2.66	3.67	.725
	4	3.53	3.88	.908
1975	1	5.48	4.08	1.344
	2	4.50	4.08	1.102
	3	2.93	4.11	.712
	4	3.31	4.14	.799
1976	1	5.94	4.12	1.441
	2	4.27	4.15	1.028
	3	3.01	4.24	.710
	4	3.47	4.36	.797
1977	1	6.49	4.47	1.452
	2	4.64	4.66	.996
	3	3.56	4.93	.722
	4	4.40	5.22	.843
1978	1	7.73	5.45	1.418
	2	5.73	5.57	1.029
	3	4.32	5.58	.774
	4	4.57	5.50	.830
1979	1	7.65	5.48	1.397
	2	5.22	5.53	.947
	3	4.61		
	4	4.68		

b.

Set of Seasonal Indexes (Adjusted Averages):

Q_1	1.412
Q_2	1.023
Q_3	.729
Q_4	.836
	4.000

c. Reported to public: 5.9 percent, 6.4 percent, 5.6 percent, and 7.4 percent.

Chapter 12

2.

	Union Series 1945 Base	
	Wage Rates	Steel Prices
1939	83	67
1945	100	100
1950	136	115
1955	140	125

Divide union wage rate series by 1.2 and union steel price series by 1.5. We see the union series is identical with Semco's. From 1939 to 1945, growth in steel prices exceeded that in wage rates, but the trend reversed from 1945 to 1955. The question can be settled by deciding at what point in time there was equity between wage rates and steel prices.

4. Laspeyres price index = 82.6,
   Paasche price index = 78.1,
   Fisher price index = 80.3;
   Laspeyres quantity index = 74.2,
   Paasche quantity index = 70.1,
   Fisher quantity index = 72.1;
   index of total expenditures = .579,
   (Laspeyres price) (Paasche quantity) = .579,
   (Paasche price) (Laspeyres quantity) = .579.

6. a. Price relatives:
   milk, 195.0
   bread, 152.0
   Swiss cheese, 208.3;
   b. Laspeyres price index = 172.4;
   c. Paasche price index = 168.4;
   d. Laspeyres consumption index = 103.6.

8. a.

Year	Quarter	CPI	MA	Deflated Series
1967	1	.957	1,851	1,875
	2	.994	1,860	1,871
	3	1.005	1,374	1,865
	4	1.013	1,894	1,870
1968	1	1.023	1,916	1,873
	2	1.034	1,957	1,893
	3	1.048	1,996	1,905
	4	1.061	2,014	1,898
1969	1	1.071	2,034	1,899
	2	1.090	2,079	1,907
	3	1.107	2,130	1,924
	4	1.122	2,180	1,943
1970	1	1.139	2,230	1,958
	2	1.157	2,266	1,959
	3	1.169	2,303	1,970
	4	1.185	2,344	1,978

10. a.

Production Indexes, 1971 Base

	1973	1977
Laspeyres	134.0	197.5
Paasche	131.2	173.9

b.

Price Indexes, 1971 Base		
	1973	1977
Laspeyres	126.3	161.4
Paasche	123.6	142.1

c. Deflating prices change all prices for a given year by a constant. Two propositions follow:
(i) The production indexes do not change.
(ii) Price indexes are changed by the deflating factor.

Chapter 13

2. a. Conf: $(19.78 \leq \sigma^2 \leq 72.93) = .95$;
   b. Conf: $(514 \leq \sigma^2 \leq 2,952) = .95$;
   c. Conf: $(46.8 \leq \sigma^2 \leq 722) = .95$.

4. a. $\bar{x} = \$1,119$; $s^2 = 154,913$; $s = \$393.60$;
   b. Conf. $(\$869 \leq \mu \leq \$1369) = .95$;
   c. Conf. $(\$279 \leq \sigma \leq \$668) = .95$;
   d. It was necessary to make the normalcy assumption on the variable, x.

6. $H_0: \mu_1 = \mu_2 = \mu_3$; H_a: at least one pair of means are not equal. ANOVA test $F = 1.20$, $F_c = 3.68$; therefore, we do not reject H_0. Mean family incomes in all three cities could be equal.

8. $H_0: \mu_1 = \mu_2 = \mu_3$; H_a: at least one pair of means are not equal. ANOVA test $F = 1.88$.
   With $\alpha = .05$, $df_1 = 2$, and $df_2 = 10$, $F_c = 4.10$; therefore, we do not reject H_0. Mean mileage could be the same on all three brands.

Chapter 14

2. Reject H_0 if $\chi_c^2 > 5.991$; $\chi_c^2 = 12.51$; reject H_0.
4. Reject H_0 if $\chi_c^2 > 11.345$; $\chi_c^2 = 1.89$; do not reject H_0.
6. Reject H_0 if $\chi_c^2 > 16.812$; $\chi_c^2 = 24.45$; reject H_0.
8. Reject H_0 if $z_c < -2.58$ or $z_c > 2.58$; $z_c = 1.12$; do not reject H_0.
10. Reject H_0 if $\chi_c^2 > 9.488$; $\chi_c^2 = 12.03$; reject H_0.
12. Reject H_0 if $\chi_c^2 > 16.919$; $\chi_c^2 = 30.10$; reject H_0.
14. Reject H_0 if $z_c < -1.65$ or $z_c > 1.65$; $z_c = 2.54$; reject H_0.
16. Reject H_0 if $\chi_c^2 > 5.991$; $\chi_c^2 = 5.51$; do not direct H_0.
18. Reject H_0 if $\chi_c^2 > 9.210$; $\chi_c^2 = 13.87$; reject H_0.
20. Reject H_0 if $\chi_c^2 > 7.815$; $\chi_c^2 = 8.97$; reject H_0.
22. Reject H_0 if $\chi_c^2 > 6.251$; $\chi_c^2 = 4.01$; do not reject H_0.

24. Reject H_0 if $t_c < -2.861$ or $t_c > 2.861$; $t_c = 4.57$; reject H_0.

26. Rejct H_0 of $\chi_c^2 > 13.2767$; $\chi_c^2 = 27.49$; reject H_0.

28. a. .460; b. .643; c. .229.

30. $-.934$.

32. -1.

34. Reject H_0 if $\chi_c^2 > 18.475$; $\chi_c^2 = 3.02$; do not reject H_0.

36. Reject H_0 if $\chi_c^2 > 15.507$; $\chi_c^2 = 56.4$; reject H_0.

Chapter 15

2. a. $Q^* = 9$;
   b. $EMV = \$51.78$;
   c. $EVPI = \$7.23$;
   d. Expected payoff with perfect information is $59.01.

4. b. Order 30 tons in September with minimum expected cost of $548;
   c. $EVPI = \$38.00$;
   d.

   Cost Table

	September Coal Order in Tons		
	20	30	50
Mild Winter	300	370	510
Normal	490	450	590
Very Cold	870	830	750
Expected Cost	$566	$548	$622

   Opportunity Loss Table

	September Coal Order in Tons		
	20	30	50
Mild Winter	0	70	210
Normal	40	0	140
Cold	120	80	0
EOL	$56	$38	$112

   e. Expected coal requirements = 34 tons,
   Expected cost @ \$19 per ton = \$646,
   Expected cost of irrationality = \$646 − 548 = \$98.

6. a.

   Cost Table

	0	10	11	12	13
$x \leq 10$		4,200	4,500	4,800	5,100
$x = 9$		5,200	5,500	5,800	6,100
$x = 8$		6,200	6,500	6,800	7,100

   b. Expected cost for Q of 10, 11, 12, and 13, respectively: \$4,700, \$4,620, \$4,850, \$5,100;

c. Set up for 11 units on first run with minimum expected cost of $4,620.

8. y = profit, L = opportunity loss, Q = order quantity, and d = demand.
   a. If $d \geq Q$, then $y = 5.50Q$, and $L = 5.50(d - Q)$;
      if $d < Q$, then $y = 8.50d - 3.00Q$ and $L = 3.00(Q - d)$;
   b. $E(d) = 9.3$;
   c. For $Q = 9$, $EOL = \$3.35$; for $Q = 10$, $EOL = \$2.95$;
   d. $Q^* = 10$; cost of uncertainty = $EVPI = \$2.95$;
   e. For $Q = 10$, $EMV = \$48.20$.

10. a. For A_1, $U_1 = .610$; for A_2, $U_2 = .626$;
       $CE_1 = 1.833$, or $\$1,833$;
       $CE_2 = 2.1$, or $\$2,100$;
    b. Select action 2;
    c. $EMV_1 = \$3,000$; $EMV_2 = \$2,100$; select action 1;

12. y = payoff in thousands of dollars, u = utility, and p = probability of winning $20,000.
    a. $p = .8$;
    b. $E(y) = 16$, or $\$16,000$.
       Because $CE = 15$, Mrs. Smith is risk averse.

Chapter 16

2. a. $\bar{x} = \$194.80$; $p = .29$; $s = 166.7$.
   b. $\$1,948,000$.
   c. $\hat{V}(\bar{x}) = \$172.54$; $\hat{\sigma}_{\bar{x}} = \$13.13$; $\hat{V}(p) = .00187$; $\hat{\sigma}_p = .043$.
   d. For μ, $\$194.80 \pm \$25.70$;
      for p, $.290 \pm .084$;
   e. $n_1 = 7$; $n_2 = 93$.

4. a. $\bar{x} = 168.5$; $s_x^2 = 4,337.0$; $s_x = 65.86$;
      $\bar{y} = \$1,950.80$; $s_y^2 = 2,186,297$; $s_y = \$1,478.60$;
      $\bar{d} = -4.19$; $s_d^2 = 168.90$; $s_d = 13.00$;
      $\bar{D} = -\$27.82$; $s_D^2 = 15,634.4$; $s_D = \$125.04$.
   b. $\hat{V}(\bar{x}) = 7.71$; $\hat{\sigma}_{\bar{x}} = 2.78$;
      $\hat{V}(\bar{y}) = 7,202.7$; $\hat{\sigma}_{\bar{y}} = \$84.90$;
      $\hat{V}(\bar{d}) = .8124$; $\hat{\sigma}_{\bar{d}} = .901$;
      $V(\bar{D}) = 77.69$; $\hat{\sigma}_{\bar{D}} = \$8.81$.
   c. For μ_x, 168.5 ± 5.45;
      for μ_y, $\$1,950.80 \pm \$166.40$;
      for μ_d, -4.19 ± 1.78;
      for μ_D, $-(\$27.82 \pm \$17.27)$;

d. Let T = total dollars; $\hat{T} = \$29{,}262{,}000$;
95 percent confidence interval is $\$29{,}262{,}000 \pm \$2{,}496{,}000$;

e. Let R be total dollar difference;
$\hat{R} = \$-417{,}300$;
95 percent confidence interval is $-(\$417{,}300 \pm \$259{,}050)$;

f. $n_1 = 261$; $n_2 = 217$; $n_3 = 22$.

Chapter 17

2. Sample means: 4.57, 4.49, 4.46, 4.57, 4.57, 4.59, 4.56, 4.55. Process is in control.

4. Sample means: 16.8, 14.8, 14.5, 15.9, 15.4, 14.8, 14.1. Process is not in control.

6. .0384.

8. .7704.

10. .222.

12. .026 and .271.

14. .001 and .677.

16. a. .9082; b. .3707.

18. (.02, .904), (.04, .815), (.06, .734), (.08, .659), (.12, .528);

20. (.02, .845), (.04, .589), (.06, .371), (.08, .218), (.12, .066);

22. .8413;

24. 0 (very small);

26. .464;

28. a. .019; b. .1512.

30. a. .1429; b. .0591.

Index

Acceptance Number, 553
Acceptance Sampling, 552
Accuracy, 7
Alternate Hypothesis, 235
Analysis of Variance, 411
    One-Way, 416
    Regression, 330, 427
    Two-Way, 423
    Unequal Sample Sizes, 421
Analytical Geometry of Straight Line, 333
AQL, 553
Arithmetic Mean, 11, 12, 38, 49
Array, 14
Attribute, 31, 172
Attribute Sampling, 118
Average (See Mean, *also* Arithmetic Mean)

Bar Chart, 23, 24, 31, 32
Basic Probability Formulas, 82
Bayes' Theorem, 91, 98
Bias, 5
Binomial Distribution, 109
Binomial Experiment, 117
Binomial Probability Function, 121
    Mean, 122
    Table, 568
    Variance, 122
Binomial Theorem, 138

Central Limit Theorem, 182
Certainty, Decisions under, 492
Chebyshev's Inequality, 196
Chi-Square Distribution, 435
    Table, 574
Class Intervals, 27
Classical Definition of Probability, 66
Cluster Sampling, 533
Coding, 48, 49
Coefficient of Correlation
    Simple Linear, 317, 320
    Multiple, 331
Coefficient of Determination, 317, 331
Combination Formula, 94, 120
Computer Regression Analysis, 323
Conditional Probability, 78
Confidence Interval Estimation
    Difference of Means, 278, 281
    Difference of Proportions, 279
    Population Mean, 203
    Population Proportion, 212
    Population Variance, 413
    Regression, 312
    Total, 206

Consumer Price Index, 392, 394, 404, 406
Consumer's Risk, 553
Contingency Table Tests, 440
Continuous Random Variables, 126
Control Charts, 543
Control Limits, 544
Correlation, 317, 320, 331
Cost of Uncertainty, 492
Cumulative Frequency, 41
Cycle, 357, 361, 365

Data
    Grouped (*See* Frequency Distribution), 38
Deciles, 21
Decision Making, 481
Decision Rule, 238
Decision Trees, 494
Degrees of Freedom, 207
    Student-t Statistic, 185, 207
    Chi-Square Statistic, 413, 435
    F-Statistic, 415
Deming, W. Edwards, 542
Discrete Random Variable, 110
Dispersion, Measure of, 16
Distribution-Free Tests, 433
Dummy Variable, 380

Errors, Types of
    Response, 5
    Non Response, 5
    Statistical, 6
Estimation (*see* Confidence Interval Estimation)
Event, 63
Expected Value
    Definition, 112
    Monetary, 483, 485
    Of Perfect Information (EVPI), 488
    Opportunity Loss, 489
Expected Values for Contingency Tables, 442
Experiment, 63

F-Distribution, 330, 415, 419, 425
    Table, 576
Finite Population Correction Factor (fpc), 190, 221, 519, 529
Fisher's Ideal Index, 400, 402, 405
Forecasting, 356, 381
Frequency Distribution
    Artificial, 27, 33, 41, 49
    Natural, 25, 39

Frequency Polygon, 28, 29
Friedman, M., 502

Galton, Sir Francis, 3
General Chi-Square Test Statistic, 435
Geometric Mean, 58, 59
Goodness-of-Fit Test
   Multinomial Population, 436
   Normal Distribution, 445
   Poisson Distribution, 463
Gosset, William S., 185
Grouped Data (see Frequency
   Distribution)

Harmonic Mean, 59
Histogram, 28, 33
Homoscedasticity, 308
Hypergeometric Probability Function, 134
Hypothesis Testing, 230
   Analysis of Variance, 416
   Goodness-of-Fit Test, 436, 445, 463
   Independence, 440
   Kruskall-Wallis Test, 457
   Multiple Regression, 331
   Paired Difference Test, 283
   Population Mean, 234, 243
   Population Proportion, 246
   Population Variance, 414
   Randomness, 448
   Rank Correlation Coefficient, 461
   Simple Linear Regression, 312
   Two Population Means, 262, 268, 283
   Two Population Proportions, 272
   Two Population Variances, 414
   Wilcoxon Test, 452
   $\bar{x}$-Charts, 547

Implicit Price Deflator, 393
Independent Events, 78
Index of Industrial Production, 394
Index Numbers, 391
   Fisher's Ideal, 400, 402, 405
   Laspeyres, 398, 402
   Paasche, 399, 402
   Seasonal, 367
Interaction, 423
Inventory Problem, 485
Irregular Component of Time Series, 358, 361

Kruskal-Wallis Test, 457

Laspeyres Index, 398, 402
Least Squares
   Criterion, 303

Least Squares—(Continued)
   Method, 339
   Equations, 303, 328
Line Chart, 25
Linear Correlation, 317, 320, 329
Linear Regression, 301
LTPD, 553

Maximax Criterion, 493
Mean (Average)
   Grouped Data, 38
   Population, 112, 133
   Sample, 11
   Standard Deviation of, 174
   Theorems Related, 173, 174, 175, 183, 185
   Weighted, 12
Mean Absolute Deviation, 20
Mean of the Binomial Random
   Variable, 122
Measurement, 7
Measure of Centrality, 10
Measure of Dispersion, 16
Median, 14, 41
Military-Standard-105D, 565
Minimax Criterion, 493
Mode, 15, 43
Moving Average, 361
Multiple Regression, 327
Mutually Exclusive Events, 74

Non-Linear Regression, 341
Non-Linear Trend, 374
Non-Parametric Statistics, 433
   Contingency Tables, 440
   General Chi-Square Test Statistic, 435
   Kruskal-Wallis Test, 457
   Multinomial Goodness-of-Fit Test, 436
   Normal Goodness-of-Fit Test, 445
   Poisson Goodness-of-Fit Test, 463
   Rank Correlation Coefficient, 460
   Runs Tests for Randomness, 448
   Wilcoxon Rank Sum Test, 452
Normal Approximation to the Binomial
   Distribution, 162
Normal Distribution, 149
   Distribution of Sample Mean, 175, 177, 180
   Distribution of Sample Proportion, 189
   Linear Combination, 195
   Table, 572
Normal Equations, 303
Normal Probability Density Function, 152
Null Hypothesis, 235

Occam, William of (Occam's Razor), 301
Ogive, 43

One-Tail Hypothesis Test, 240
One-Way Analysis Variance, 416
Operating Characteristic Curves, 557
Opportunity Loss, 489
Opportunity Loss Table, 490
Outcome, 63

Paasches Index, 399, 402
Paired Difference Test, 283
Parabola, 375
Parameter, 10
Payoff Table, 486
Percentiles, 21, 42
Perfect Information, 488
Pie Chart, 23, 31
Point Estimate, 201
Poisson Probability Function, 137
   Table, 580
Population
   Definition, 3
   Mean, 112, 133
   Parameter, 4
   Variance, 113, 134
Population Tree, 34
Price Relative, 397
Probability, 61
Probability Axioms, 77
Probability Density Function, 126
Probability Function, 110
Probability Space, 63
Probability Trees, 95
Process Control, 545
Producer's Price Index, 393
Producer's Risk, 553
Proportion, 211, 246
   Theorems Related to, 187, 188, 189

Quality Control, 541
Quantity Index, 401
Quartiles, 21

Random Sample, 4, 172
Range, 16
Rank Sum Test, 453
Regression Analysis
   Computer, Use of, 323, 327
   Linear, 301
   Multiple, 327
Relative Frequency, 30
Relative Frequency Definition of
   Probability, 71
Risk, 482
Rounding, 7
Runs Tests, 448
   Sign Test for Randomness, 449
   Runs Above and Below Median Test, 451

Sample
   Definition, 4
   Mean, 11
   Proportion, 211
   Statistic, 4
   Variance, 17
Sample Size Determination
   For Estimating the Mean, 215
   For Estimating the Proportion, 217
   For Finite Populations, 223
   For Stratified Sampling, 526
Sampling
   Attribute, 172
   Variable, 172
   Theorems, 173, 174, 175, 183, 185, 187, 189
Sampling Designs
   Cluster, 533
   Simple Random, 529
   Stratified, 516
   Systematic, 532
Savage, I., 502
Scattergram, 302, 306
Seasonal Component of Time Series, 357, 365
Seasonal Indices, 367
Significance Level, 235
Skewness, 37
Slope, 301, 333
Spearman's Rank Correlation
   Coefficient, 460
Standard Deviation, 113
   Alternate Formula, 18
   Definition, 17
   Grouped Data, 39, 49
   Of the Sample Mean, 174
   Of the Sample Proportion, 188
Standard Error of
   The Mean, 174, 190, 519
   The Proportion, 188, 190, 523
   Regression, 310, 320
Standard Normal Distribution, 152
   Table, 572
Standard Normal Random Variable, 153
Statistic, 10
Stratified Sampling, 516
Students'-t Distribution, 184
   Table, 573
Subjective Definition of Probability, 72
Subjective Probability, 482
Summation Symbol, 45
Systematic Sampling, 532

t-Distribution (*see* Students' t)
Tests of Independence, 440
Test Statistic, 235
Time Series, 355
   Charting, 24, 35
Trend Analysis, 357

Two Population Means (*see* Confidence Interval Estimation and Hypothesis Testing)
Two Population Proportions (*see* Confidence Interval Estimation and Hypothesis Testing)
Two-Tail Hypothesis Test, 239
Type I Error, 233
Type II Error, 233, 249, 550

Unbiased Estimator, 201
Uncertainty, 482
Utility, 483, 500

Variance
  Analysis of, 411
  Definition, 17, 113
Variance of the Binomial Random Variable, 122
Venn Diagrams, 75

Weighted Average, 12
Wilcoxon Test, 452

$\bar{x}$-Charts, 543